Introduction to Random Signals, Estimation Theory, and Kalman Filtering

M. Sami Fadali

Introduction to Random Signals, Estimation Theory, and Kalman Filtering

Springer

M. Sami Fadali
Department of Electrical and Biomedical
Engineering
University of Nevada, Reno
Reno, NV, USA

ISBN 978-981-99-8062-8 ISBN 978-981-99-8063-5 (eBook)
https://doi.org/10.1007/978-981-99-8063-5

This Springer imprint is published by the registered company Springer Nature Singapore Pte Ltd.
The registered company address is: 152 Beach Road, #21-01/04 Gateway East, Singapore 189721, Singapore

Paper in this product is recyclable.

Preface

In 1998, our faculty decided that all our graduate students should be familiar with random signals and estimation that we should require them to take a course on this subject. I was given the task of developing the course. It seemed that this would be an easy task because there were many available textbooks on the subject. However, upon close examination of the available textbooks, I discovered that none of them would be suitable for the course. There were chapters in several texts that were appropriate for the course, but no single text contained all the material that I wished to cover, and more critically, the material was at a level that is accessible to a student with an undergraduate engineering degree. There were excellent textbooks on Kalman filtering, excellent books on random signals, excellent books on estimation, but many assumed a level of mathematical maturity and a deep knowledge of system theory that was beyond what can be reasonably expected from the average engineer. An additional problem was that many of the textbooks did not make use of the computer packages that have become an indispensable part of the toolkit in this field. In particular, many books were written before the development of MATLAB© and its myriad of toolboxes. Moreover, many of the available textbooks omitted coverage of some basic statistical properties of estimators that I believe to be an essential foundation for understanding estimators.

For years, I was content with complementing the textbook I used to teach the class with material from several other sources. This was an adequate but not completely satisfactory solution because different sources use different styles and notation, which can be confusing for some students. Then came the COVID pandemic. This was a period where we were confined to home for several months. Because I had developed a battery of slides for my class presentations, I proceeded to add text and create chapters of notes for my class. By the time the COVID confinement was over, I had completed a sufficient number of chapters to use my notes for teaching. Gradually adding chapters and appendices, I accumulated enough material to contact Springer Nature who graciously accepted to publish the textbook.

Based on the above history, for whom is the textbook intended? The book is intended for first-year graduate engineering students or practicing engineers who have completed a course on probability and statistics and have some knowledge of state equations and basic system theory. The prerequisite topics are part of most undergraduate engineering programs and can be quickly reviewed using the material provided in appendices. The book is suitable for a course on random signals and state estimation for first-year graduate engineering students. Depending on the background of the audience, most of the material can be covered in one or two semesters.

Upon completing a course using this book, the student should be ready for a variety of more specialized courses including courses on signal detection, random signals, or Kalman filtering. There is no shortage of textbooks for courses on these topics. The use of MATLAB in this first course will allow students to use it effectively in the more specialized courses.

In writing this book, I have relied on feedback from generations of students who took my course. I thank them for their patience and hard work, without which I would not have been able to complete the book.

Reno, USA M. Sami Fadali

Contents

About the Author

M. Sami Fadali earned a B.S. in Electrical Engineering from Cairo University in 1974, an M.S. from the Control Systems Center, UMIST, England, in 1977, and a Ph.D. from the University of Wyoming in 1980. He was an Assistant Professor of Electrical Engineering at the University of King Abdul Aziz in Jeddah, Saudi Arabia from 1981–1983. From 1983 to 1985, he was a Postdoctoral Fellow at Colorado State University. In 1985, he joined the Electrical Engineering Department at the University of Nevada, Reno, where he is currently Professor of Electrical Engineering. In 1994, he was a Visiting Professor at Oakland University and GM Research and Development Laboratories. He spent the summer of 2000 as a Senior Engineer at TRW, San Bernardino. His research interests are in fuzzy logic stability and control, state estimation and fault detection, and applications to power systems, renewable energy, and physiological systems.

Acronyms

BLUE	Best linear unbiased estimator
BLWN	Band-limited white noise
CDF	Cumulative distribution function
EKF	Extended Kalman filter
i.i.d.	Independent identically distributed
MLE	Maximum-likelihood estimator
pdf	Probability distribution function
pmf	Probability mass function
PSD	Power spectral density
UKF	Unscented Kalman filter
WLOG	Without loss of generality

List of Figures

List of Tables

Chapter 1
Review of Probability Theory

1.1 Interpretations of Probability

The probability of a random event provides a measure of its likelihood. This measure has different interpretations. The oldest interpretation uses the assumption that all outcomes of a random event are equally likely. This works well for the probability of the outcome of games of chance where the number of outcomes is relatively small and there is no reason to prefer one outcome over others, for example, when throwing a die where there are six possible outcomes with no reason to favor any one of them over others. With this interpretation, probability is defined as

$$P_A = \frac{N_A}{N} \in [0, 1], \tag{1.1}$$

where N_A = number of ways event A can occur, and N = total number of possible outcomes.

Clearly, this approach has limited applicability because (i) it cannot handle outcomes that are not equally likely, (ii) it is not practical in cases where there is a large (finite) number of outputs, and (iii) it cannot handle an (uncountably) infinite number of outcomes.

Another interpretation of probability is that it is the relative frequency with which the outcome of a random event occurs. For this interpretation, the probability is given by (1.1) with the numbers N and N_A obtained by repeating a random experiment a very large number of times until their ratio approaches a constant value. This approach also has problems because many experiments can only be repeated a small number of times. In addition, the assumption that the relative frequency approaches a constant limit is not true in practice. Instead, the ratio hovers around a constant value.

A completely different interpretation of probability is to view it as a subjective belief in the likelihood of an event. Proponents of this Bayesian or subjective approach criticize approaches that base probability on the results of repeating

M. S. Fadali, *Introduction to Random Signals, Estimation Theory, and Kalman Filtering*, https://doi.org/10.1007/978-981-99-8063-5_1

a random experiment, called frequentists, because the results will depend on the number of repetitions.

Regardless of the interpretation of probability, the theory of probability can be developed using an axiomatic approach.

1.2 Axiomatic Definition of Probability

Accept a set of axioms based on experience:

- Derive a complete theory based on axioms.
- **Random Experiment**: Experiment with nondeterministic outputs.
- **Probability Space**:
  1. Sample space S.
  2. Event space E.
  3. Probability measure $P(E)$.

The Sample Space S

The sample space is the set of outcomes of random experiment $s_i \in S$. The outcomes s_i are known as elementary events. They can be finite, countably infinite, or infinite but are always assumed to be disjoint.

Example 1.1 Characterizer the sample space for the following random experiments:

(i) Randomly choose a hexadecimal number.
(ii) Receive a two-bit binary random message.

Solution

(i) There are 16 possible hexadecimal numbers. The sample space S has the 16 elements

$$\{0, 1, 2, \ldots, 9, A, B, C, D, E, F\}.$$

(ii) In this case, the message is made of ones and zeros but the order matters. The number of outcomes is the permutation of ones and zeros with repetition. In general, the number of possible messages of length n where every element has one of k possible values is

$$k \times k \times \cdots \times k,$$

that is k^n. In this case, $k = 2$ and $n = 2$, and there are $2^2 = 4$ possible outcomes and the sample space S is:

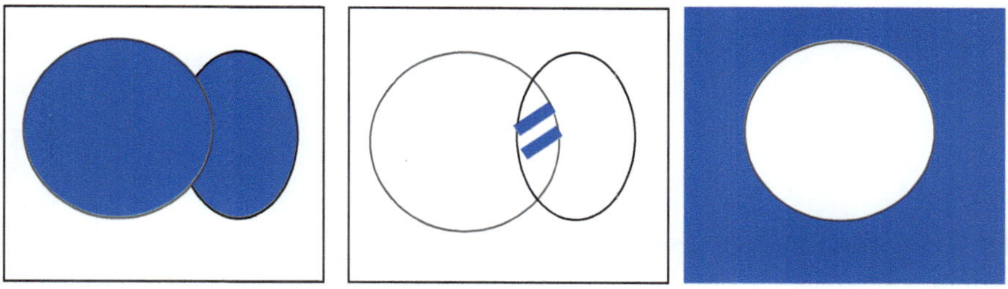

Fig. 1.1 Venn diagrams for set operations in probability theory

$$\{(1, 1), (1, 0), (0, 1), (0, 0)\}.$$

Note that in this case the experiment has a vector outcome. ■

The Event Space E

An event is a set of one or more outcomes of a random experiment. The event space is the set of all events. The event space is closed under union, intersection, complement (see Fig. 1.1 for a depiction of the set operations). In other words, the union or intersection of any two events, and the complement of any event is an event. Hence, the set S of all elementary events and its complement, the empty set $\emptyset$, are members of the event space.

If the event space is discrete, then it is the power set of S that is the set of all its subsets. For events on the real line $\mathcal{R}$ to form a space that is closed under countably infinite union, intersection, or complement, the events are defined as sets of open, closed, or semi-open subsets of $\mathcal{R}$. complement, the events are defined as sets of open, closed, or semi-open subsets of $\mathcal{R}$. The latter is more complicated, all the properties we deal with can be discussed in terms of the discrete case but will apply to the continuous case (Fig. 1.1).

Probability Measure

The probability measure assigns a number in [0, 1] to each event in the event space and is denoted by the mapping:

$$P(\cdot) : E \to [0, 1] \subset \mathcal{R}. \tag{1.2}$$

It satisfies the following axioms of probability.

Axioms of Probability

$$P(A) \geq 0, \forall A \in E \tag{1.3}$$

$$P(S) = 1 \tag{1.4}$$

$$P\left(\bigcup_i A_i\right) = \sum_i P(A_i), \tag{1.5}$$

for A_i disjoint (mutually exclusive) and the number of sets A_i finite or countably infinite

Axiom (iii) gives the probability of obtaining one or more of a set of events. The joint probability $P(A \cap B)$ is the probability that events A and B occur simultaneously. The axiom of infinite additivity extends axiom (iii) to countably infinite unions.

Example 1.2 Define the sample space and the event space for the number of ones in a 2-bit random message where all messages are equally probable. What is the probability of getting 1 or 2 ones?

Solution

From Example 1.2 we know that there are $2^2 = 4$ possible outcomes for the experiment with anywhere between 0 and 2 ones. The sample space is $S = \{0, 1, 2\}$, with 3 elements, and the event space is the power set of S with 2^3 elements

$$\{S, \emptyset, \{0\}, \{1\}, \{2\}, \{0, 1\}, \{0, 2\}, \{1, 2\}\}.$$

The probability of getting 1 or 2 ones is the sum of the two probabilities since the two are mutually exclusive:

$$P(1 \cup 2) = P(1) + P(2) = \frac{2}{4} + \frac{1}{4} = \frac{3}{4}.$$

Alternatively, the probability can be calculated using.

$$P(1 \cup 2) = 1 - P(0) = 1 - \frac{1}{4} = \frac{3}{4}.$$

■

From the Venn diagram of Fig. 1.1, we can deduce the identity

$$A \cup B = A \cup \left(B \cap A^c\right), \tag{1.6}$$

where A^c is the complement of the set A. Hence, it is intuitively obvious that if events are not mutually exclusive, to avoid counting joint occurrence twice, the probability of event A or event B is given by

$$P(A \cup B) = P(A) + P(B) - P(A \cap B). \tag{1.7}$$

Example 1.3 What is the probability of obtaining an odd number of ones or a number that is less than 4 in two independent 3-bit random messages if all outcomes are equally probable?

Solution

For each two-bit message, there can be {0, 1, 2, or 3} ones. There are 2^4 equally probable outcomes:

$$S = \left\{ \begin{array}{l} (0,0),(0,1),(0,2),(0,3),(1,0),(1,1),(1,2),(1,3), \\ (2,0),(2,1),(2,2),(2,3),(3,0),(3,1),(3,2),(3,3) \end{array} \right\}.$$

The corresponding sums are {0, 1, 2, 3, 1, 2, 3, 4, 2, 3, 4, 5, 3, 4, 5, 6}, that is the number of {0, 1, 2, 3, 4, 5, 6} ones is, respectively, {1, 2, 3, 4, 3, 2, 1}. Let the number of ones in the messages be N. Then we have the following probabilities:

$$P(N < 4) = \frac{10}{16}, \quad P(N \text{ odd}) = \frac{8}{16}, \quad P(N < 4 \cap N \text{ odd}) = \frac{6}{16}$$

The probability of an even number or a number less than 5 is

$$\begin{aligned} P(N < 4 \cup N \text{ odd}) &= P(N < 4) + P(N \text{ odd}) - P(N < 4 \cap N \text{ odd}) \\ &= \frac{10}{16} + \frac{8}{16} - \frac{6}{16} = \frac{12}{16} \end{aligned}$$

Alternatively, we have

$$\begin{aligned} P(N < 4 \cup N \text{ odd}) &= 1 - P(N \geq 4 \cup N \text{ even}) \\ &= 1 - P(N = 4) - P(N = 6) \\ &= 1 - \frac{3}{16} - \frac{1}{16} = \frac{12}{16} \end{aligned}$$

■

1.3 Marginal Probability

Consider two sets of events $A_i, i = 1, 2, \ldots, m, B_j, j = 1, 2, \ldots, n$ with joint probability

$$P(A_i \cap B_j), i = 1, 2, \ldots, m, j = 1, 2, \ldots, n \tag{1.8}$$

The probabilities are summarized in Table 1.1. The marginal probabilities are the probabilities written in the margin outside the m by n probability array of the table. The marginal probabilities are

$$P(A_i) = \sum_{j=1}^{n} P\left(A_i \cap B_j\right), i = 1, 2, \ldots, m \tag{1.9}$$

$$P\left(B_j\right) = \sum_{i=1}^{m} P\left(A_i \cap B_j\right), j = 1, 2, \ldots, n. \tag{1.10}$$

From the probability axiom (ii) of (1.4), we have (Table 1.1)

$$\sum_{i=1}^{m} P(A_i) = \sum_{j=1}^{n} P\left(B_j\right) = 1. \tag{1.11}$$

Example 1.4 A two-bit message is sent through a communication channel with equal probabilities of 1 and zero. Calculate the entries of Table 1.1 for the message.

Solution

Let B_i denote the ith bit, $i = 1, 2$, with values in $\{0, 1\}$ with equal probability. Since the two outcomes are equally probably, they both have probability equal to 1/2. The random contents of the message have four equally probable outcomes $(B_1, B_2) \in \{(1, 1), (1, 0), (0, 1), (0, 0)\}$ (Table 1.2).

Table 1.1 Joint and marginal probabilities

Event	B_1	…	B_n	Marginal probabilities
A_1	$P(A_1 \cap B_1)$	…	$P(A_1 \cap B_1)$	$P(A_1)$
⋮	⋮	⋮	⋮	⋮
A_m	$P(A_m \cap B_1)$	…	$P(A_m \cap B_n)$	$P(A_m)$
Marginal probabilities	$P(B_1)$	…	$P(B_n)$	Sum = 1

Table 1.2 Joint and marginal probabilities for the two-bit message

Event	$B_2 = 1$	$B_2 = 0$	Marginal probabilities
$B_1 = 1$	1/4	1/4	1/2
$B_1 = 0$	1/4	1/4	1/2
Marginal probabilities	1/2	1/2	1

1.4 Conditional Probability

Conditional probability is the probability of event A given that event B *has occurred* (i.e., B is the certain event). The conditional probability is given by

$$P(A|B) = \frac{P(A \cap B)}{P(B)} \tag{1.12}$$

Clearly, conditional probability is not defined for B impossible ($P(B) = 0$). The definition agrees with the relative frequency notion:

$$P(A|B) = \frac{N_{\text{outcomes with both } A \text{ and } B}/N_{\text{total}}}{N_{\text{outcomes with } B}/N_{\text{total}}} \tag{1.13}$$

If B occurs, A may or may not occur. As a result, we have that, in general

$$P(A|B) + P\left(A|B^c\right) \neq P(A) \tag{1.14}$$

The proof is left as an exercise for the reader.

Example 1.5 A manufacturer who buys a component from two vendors compiled data on the percentage of defective components. On average, 40% of the components came from supplier 1 and 60% from supplier 2. On average, the components from supplier 1 were 5% defective while those from supplier 2 were 2% defective.

(a) What is the probability of getting a defective component?
(b) If we know that a component is from supplier 1, what is the probability of a defective component?

Solution

S_i = supplier i, i= 1, 2, D = defective component

(a) The probability of a defective component is the sum

$$P(S_1 \cap D) + P(S_2 \cap D) = \frac{5}{100} + \frac{2}{100} = \frac{7}{100}.$$

(b) If we know that a component is from supplier 1, the probability of a defective component is

$$P(D|S_1) = P(D \cap S_1)/P(S_1) = \frac{\frac{5}{100}}{\frac{40}{100}} = 1/8.$$

1.5 Independence

Two events are said to be mutually independent if the occurrence of one event does not affect the likelihood of the other. Thus, the conditional probability becomes

$$P(A|B) = \frac{P(A \cap B)}{P(B)} = P(A). \tag{1.15}$$

The joint probability of mutually independent events is the product

$$P(A \cap B) = P(A).P(B). \tag{1.16}$$

Clearly, if A is independent of B then B is independent of A

$$P(B|A) = \frac{P(A \cap B)}{P(A)} = P(B). \tag{1.17}$$

Example 1.6 Give two examples of independent events and an example of dependent events.

Solution

The following are three examples of independent events:

- The outcome of rolling a die twice.
- The random failure of two integrated circuits on two different circuit boards.

An example of dependent events is the random failure of one of the columns supporting a structure with randomly varying weight W and the failure of the entire structure. If a column fails, then the probability of the failure of the structure increases. For example, if a structure supported by four columns fails when the supported weight W exceeds the sum of the bearing capacities C_i, $i = 1, 2, 3, 4$, of the four columns, then the probability of its failure with four intact columns is

$$P(\text{failure}) = P(W > C_1 + C_2 + C_3 + C_4).$$

After the failure of, say the first column, the probability of failure becomes

$$P(\text{failure}) = P(W > C_2 + C_3 + C_4).$$

Example 1.7 If we make two letter selections from the English alphabet with equally probably letters, what is the probability of a vowel for the second letter?

(a) If a letter can be selected more than once.
(b) If each letter can only be selected once and the first letter is a vowel.
(c) If each letter can only be selected once and the first letter is a consonant.

Solution

(a) The English alphabet has 26 letters including 5 vowels. The probability of a vowel both for the first and second draw is

$$\Pr(\text{Vowel}) = 5/26.$$

(b) If the first letter selected is a vowel and it cannot be selected again, then only 3 vowels are available for selection in the second draw and the probability of a vowel is

$$\Pr(\text{Vowel 2}|\text{Vowel 1}) = \frac{4}{25} \neq \frac{5}{26}.$$

(c) If the first letter selected is a consonant and it cannot be selected again, then 5 vowels are available for selection in the second draw and the probability of a vowel is

$$\Pr(\text{Vowel 2}|\text{Consonant 1}) = \frac{5}{25} \neq \frac{5}{26}.$$

■

1.6 Bayes' Theorem

Bayes' theorem allows us to express the conditional probability of an event A_i given an event B, in terms of the conditional probability of B given A_i. For mutually exclusive events $A_i, i = 1, 2, \ldots, m$, the conditional probability satisfies

$$P(A_i|B) = \frac{P(B|A_i)P(A_i)}{\sum_{j=1}^{m} P(B|A_j)P(A_j)}. \tag{1.18}$$

The result is proved by substituting the following two identities in the definition of the conditional probability:

i. $P(A_i \cap B) = P(B|A_i)P(A_i)$
ii. $P(B) = \sum_{j=1}^{m} P(B \cap A_j) = \sum_{j=1}^{m} P(B|A_j)P(A_j)$.

Example 1.8 Consider a binary channel with random input $u \in \{0, 1\}$ and random output $y \in \{0, 1\}$. The channel is characterized by the probabilities (Fig. 1.2).

$$P(y = 1|u = 0) = q, \quad P(y = 0|u = 1) = p, \quad P(u = 1) = p_1.$$

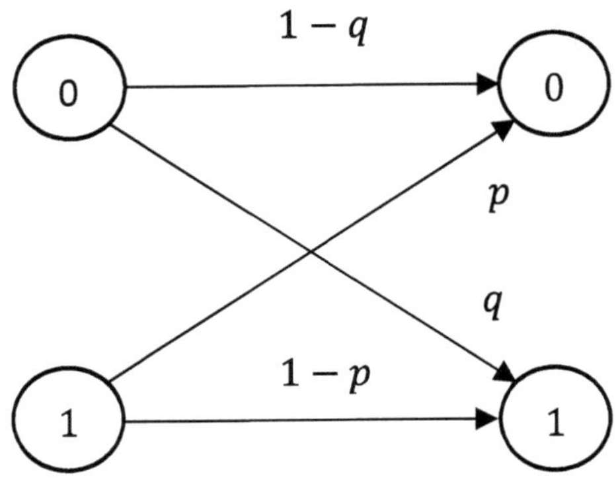

Fig. 1.2 Binary channel

(a) What is the probability of receiving a 1?
(b) What is the probability of receiving a 0?
(c) Use Bayes theorem to determine the probability of transmitting a 1 if a zero is received.

Solution

(a) The probability of receiving a 1 is

$$\begin{aligned} P(y=1) &= P(y=1|u=0)P(u=0) + P(y=1|u=1)P(u=1) \\ &= q(1-p_1) + (1-p)p_1 = q + p_1(1-p-q). \end{aligned}$$

(b) The probability of receiving a 0 is

$$\begin{aligned} P(y=0) &= P(y=0|u=0)P(u=0) + P(y=0|u=1)P(u=1) \\ &= (1-q)(1-p_1) + pp_1 \\ &= 1-q-p_1(1-p-q. \end{aligned}$$

Clearly, we can also use $P(y=0) = 1 - P(y=1)$.

(c) Bayes theorem for this example gives

$$P(u=i|y=j) = \frac{P(y=j|u=i)P(u=i)}{\sum_{l=1}^{2} P(y=l|u=i)P(u=i)}, \quad i, j \in \{1, 2\}.$$

Substituting for the probabilities gives

$$P(u=1|y=0) = \frac{P(y=1|u=0)P(u=0)}{\sum_{l=1}^{2} P(y=l|u=0)P(u=0)}$$

$$= \frac{q(1-p_1)}{(1-q)(1-p_1)+pp_1}$$
$$= \frac{q(1-p_1)}{1-q-p_1(1-p-q)}.$$

Problems

1.1 The number of goals that a soccer player will score in a single game is a random variable X with Probability

$$P(0) = 0.4,\ P(1) = 0.3,\ P(2) = 0.16,\ P(3) = 0.08,\ P(4) = 0.05,\ P(5) = 0.01$$

(a) Why must the probability of scoring more goals be negligible (clearly number of goals must be positive)?
(b) Find the mean, variance, and standard deviation of the number of goals scored.

1.2 A basketball player scores with a three-point field goal randomly with probability $p \in [0, 1]$.

(a) Find the mean and variance of the player's score with one three-point field goal attempt.
(b) Find the probability of scoring in n out of N shots.
(c) If the probability of scoring with a two-point field goal is $q \in [0, 1]$, $2q < 3p < 3q$, find the ratio of probabilities of scoring n out of 10 three-pointers over m out of 10 two pointers.

If a player achieves $q = 0.8$, $p = 0.6$, find the ratio of probabilities in (c) for $n = 6$, $m = 8$, $N = 10$. Explain why NBA players may prefer to take 3-pointers over 2-pointers when they have the opportunity.

1.3 Prove the inequality

$$P(A_1 \cap A_2) \geq P(A_1) + P(A_2) - 1$$

Comment on the inequality in the case of events A_1 and A_2 of low probability.

1.4 Starting with the result of Problem 3, generalize the result by induction to

$$P\left(\bigcap_{i=1}^{n} A_i\right) \geq \sum_{i=1}^{n} P(A_i) - (n-1)$$

1.5 Prove the inequality

$$P\left(\bigcup_{i=1}^{n} A_i\right) \leq \sum_{i=1}^{n} P(A_i)$$

1.6 Prove that the probability of the complement A^c is $P(A^c) = 1 - P(A)$, then show that the probability of the empty set $\emptyset$ is zero.

1.7 Prove that if A is a subset of B then $P(A) \leq P(B)$.

1.8 Prove the following identities:

(a) $P(A|B, C) = P(B|A, C)P(A|C)/P(B|C)$

(b) $P(A, B|C) = P(B|A, C)P(A|C)$

1.9 Prove that if A and B are mutually independent then A and B^c are mutually independent and A^c and B are mutually independent.

1.10 Prove that, in general,

$$P(A|B) + P\left(A|B^c\right) \neq P(A)$$

What is the form of the LHS if

(a) A A and B are mutually independent?

(b) A and B are mutually exclusive?

1.11 The Information Associated with the Occurrence of an Event A is Defined as

$$I(A) = -\log P(A)$$

(a) Show that information is always nonnegative.

(b) Show that the information associated with the simultaneous occurrence of two independent event is the sum of their individual information.

1.12 Mutual information is a measure of the information provided about a random event A by the occurrence of a random event B. It can be used to assess the information provided by a received message about the transmitted message in a communication channel. Mutual information is defined as

$$I(A; B) = \log \frac{P(A|B)}{P(B)}$$

Show that

(a) The mutual information is symmetric in that $I(A; B) = I(B; A)$.

(b) The mutual information can be written as $I(A; B) = I(A) + I(B) - I(A \cap B)$, where $I(A) = -\log P(A)$.

(c) If A and B are independent, then the mutual information is zero.

(d) If the occurrence of B implies the occurrence of A, that is $P(A|B) = 1$, then $I(A; B) = -\log P(A)$.

1.13 Three missiles are to be independently fired at a target with equal probability of success. An observer identifies the missile that hit the target with equal probability. Denote the event of a missile hitting the target by M_i and the event of the observer identifying a missile by O_i, $i = 1, 2, 3$, where i denotes the number of the missile.

(a) Form a table for the joint and marginal probabilities M_i, O_i, $i = 1, 2, 3$.
(b) Determine the probability that the first missile hit the target if the observer determined that the second missile hit the target.
(c) Determine the probability that the first missile hit the target if the observer determined that the second missile did not hit the target.

1.14 A communication channel delivers a 3-bit message with independent bits, each bit having a value of 1 or 0 with equal probability. Determine the following probabilities.

(a) The probability of an all-one or all-zero message.
(b) The probability of a message that has at least one zero.
(c) The probability of a message that has two zeros.
(d) The probability of a message that has two zeros given that the first bit is one using two different approaches.

1.15 Consider a Binary Symmetric Channel with Random Input $u \in \{0, 1\}$ and Random Output $y \in \{0, 1\}$. The channel is characterized by the probabilities (see Fig. P.1.1)

$$P(y = 1|u = 0) = 0.1,\ P(y = 0|u = 1) = 0.1,\ P(y = 1) = 0.5$$

(a) What is the probability of receiving a 1?
(b) What is the probability of receiving a 0?
(c) Use Bayes theorem to complete a table of the probabilities $P(u = i|y = j), i, j \in \{1, 2\}$.

Fig. P.1.1 Binary channel

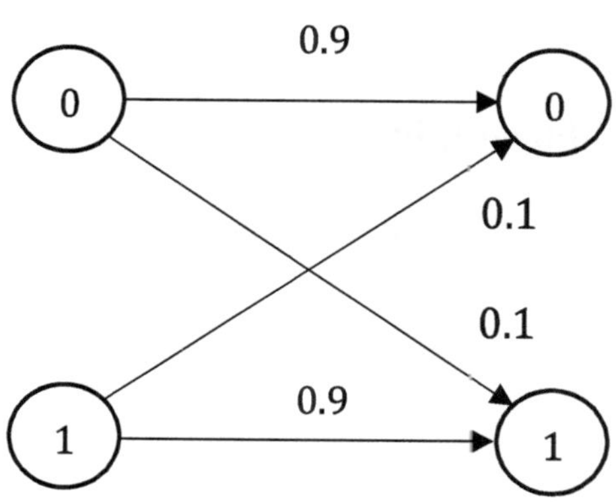

Fig. P.1.2 Electric circuit

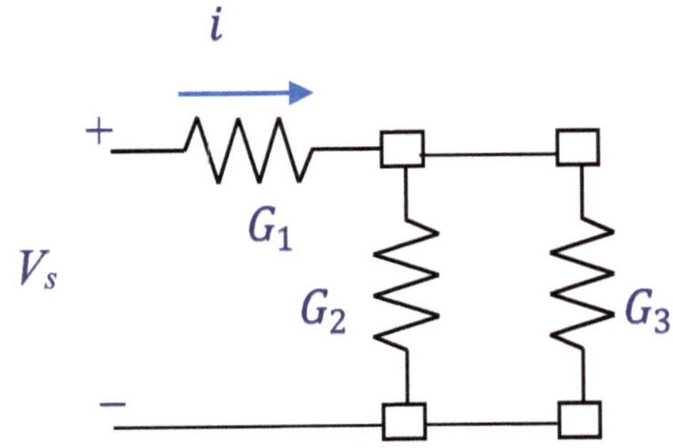

1.16 Show that if mutually independent elements $R_i, i = 1, 2, \ldots, n$, whose nonzero probabilities of failure are $p_i, i = 1, 2, \ldots, n$, respectively, then the probability that the system will fail is higher for the system with the elements in series than it is for the system with the elements in parallel.

1.17 Find the probability of zero current *i* in the circuit of Fig. P.1.2 if the probability of failure of the resistors is $p_i, i = 1, 2, 3$, respectively, and their failure is mutually independent.

1.18 For the electric circuit of Fig. P.1.2, find the probability of a voltage across the parallel resistors greater than $V_s/3$ if all the resistances are equal.

1.19 A test for a virus is required to detect a viral infection with probability $P_{I|T}$. Data is collected to estimate the probability $P_{I|T}$ of the test detecting infection when administered to infected patients and the probability P_I of infection among a city's population.

a. Write a general expression for the probability $P_{I|T}$ as a function of $P_{T|I}$ and P_I.
b. Calculation and plot $P_{I|T}$ versus $P_{T|I}$ for the probability of infection

$$P_I = 5, 10, 15, \ldots, 25, 30\%$$

and a probability of a positive test for a healthy patient equal to 0.05.
c. Comment on the results and discuss the effect of the rate of infection on the required efficacy for the test to achieve a specified probability P_v.

Bibliography

1. Casella, G., & Berger, R. (1990). *Statistical inference*. Duxbury Press.
2. DeGroot, M. H., & Schervish, M. J. (2012). *Probability and statistics*. Addison-Wesley.

Chapter 2
Random Variables

2.1 Mathematical Characterization of a Random Variable

A random variable is a function mapping the elements of the sample space S to the real line $\mathcal{R}$ as depicted in Fig. 2.1. The real values associated with elementary events of the sample space form an equivalent event. The probability of a real value is the sum of the probabilities of the elementary events associated with it. A random variable is discrete if it has a finite or countably infinite set of values; otherwise, it is continuous. The equations governing continuous random variables can be used for discrete random variables by using Dirac delta functions but the equations for the discrete variables are generally simpler.

Example 2.1 Die Characterize the random variable associated with throwing a fair die.

Solution

If we throw a fair die then the sample space consists of getting a face with 1–6 dots. The associated discrete random variable maps i dots to the number $i, i = 1, \ldots, 6$. ■

Example 2.2 Explain how measurement of a physical quantity can be viewed as a random variable.

Solution

When measuring any physical quantity with additive random noise, we associate a real number with the physical quantity, and this defines a continuous random variable. ■

To assign probabilities to the values of a random variable, we use the following definitions.

M. S. Fadali, *Introduction to Random Signals, Estimation Theory, and Kalman Filtering*, https://doi.org/10.1007/978-981-99-8063-5_2

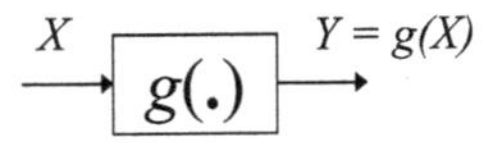

Fig. 2.1 Random variable ***X*** maps the sample space ***S*** to the real line

Definition 2.1 Probability Mass Function A probability mass function assigns a probability to each value x_i of a discrete random variable X in the event space E, i.e., to each equivalent event.

$$p_X(.) : E \rightarrow \mathcal{R}. \tag{2.1}$$

To satisfy the probability axioms, the probability mass function must be nonnegative

$$p_X(x_i) \geq 0, \forall x_i \in \Omega, \tag{2.2}$$

and its probability weights must add to unity since they represent the probability of the certain event

$$\sum_{x_i \in \Omega} p_X(x_i) = 1. \tag{2.3}$$

Example 2.3 Poisson Distribution The number of data packets in a communication link is modeled as a random variable governed by the Poisson distribution:

$$P_N(n) = \frac{\lambda^n \mathrm{e}^{-\lambda}}{n!}, n = 0, 1, 2, \ldots, \tag{2.4}$$

where λ is the average number of arriving packets. Show that the Poisson pmf satisfies (2.3).

Solution

$$\sum_{n=0}^{\infty} p_N(n_i) = \sum_{n=0}^{\infty} \frac{\lambda^n \mathrm{e}^{-\lambda}}{n!} = \mathrm{e}^{-\lambda} \sum_{n=0}^{\infty} \frac{\lambda^n}{n!} = \mathrm{e}^{-\lambda}\mathrm{e}^{\lambda} = 1.$$

■

For the probability that a random variable does not exceed a specific value, we need the following definition.

Definition 2.2 Cumulative Distribution Function The cumulative distribution function (CDF) of a random variable X is a function defined $\forall x \in \mathcal{R}$

$$P_X(x) = P(X \leq x), -\infty < x < \infty, \tag{2.5}$$

that has the following properties:

1. $P_X(x) \to 0$ as $x \to -\infty$.
2. $P_X(x) \to 1$ as $x \to \infty$.
3. $P_X(x)$ is a nondecreasing function of x, $x_1 < x_2 \Rightarrow P_X(x_1) \le P_X(x_2)$.

Continuous random variables can have infinitely many values and, because these are values on the real line, they are not countably infinite. Because an event is an interval on the real line, the probability of single value is infinitesimal and probability must be characterized by a density function.

Definition 2.3 Probability Density Function The probability density function (pdf) of a continuous random variable X is a nonnegative function $p(.)$ defined on the real line such that:

1. $\int_{-\infty}^{\infty} p_X(x)\mathrm{d}x = 1$.
2. $p_X(x) \ge 0, \forall x$.
3. $p_X(x) = \frac{dP_X(x)}{\mathrm{d}x}$.

The first two properties follow from the axioms of probability while the third property relates the pdf to the CDF.

For every real interval $(a, b]$, the probability of an X value in the interval is

$$P(a < X \le b) = \int_a^b p_X(x)\mathrm{d}x. \tag{2.6}$$

The integral formula yields an expression for the CDF (Fig. 2.2)

$$P_X(x) = \int_{-\infty}^{x} p_X(y)\mathrm{d}y. \tag{2.7}$$

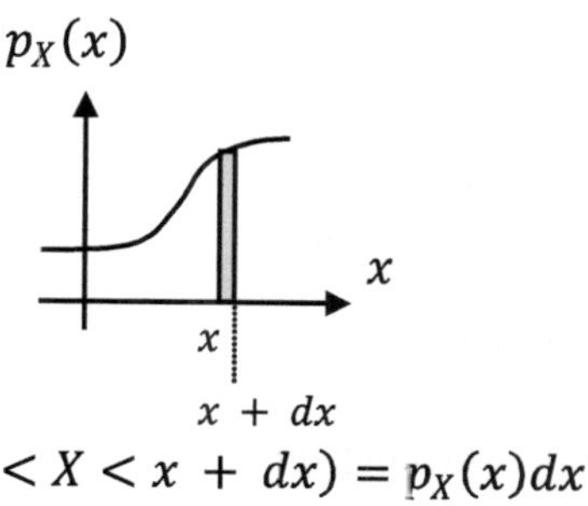

Fig. 2.2 Properties of the pdf

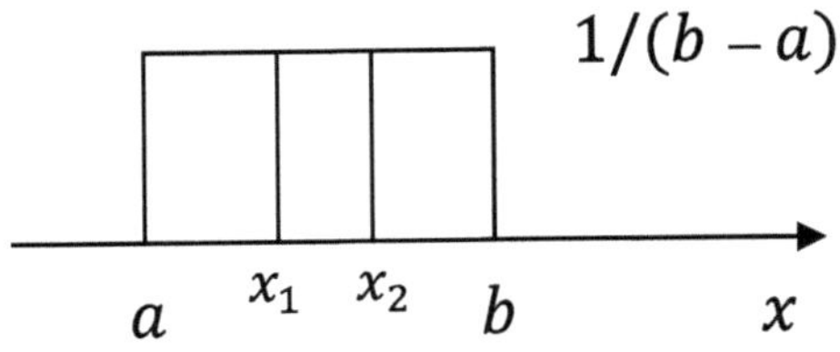

Fig. 2.3 Uniform distribution $U[a, b]$

Example 2.4 Uniform Distribution A uniform distribution indicates equally probably outcome over a finite range as shown in Fig. 2.3. The pdf is given by

$$p_X(x) = \begin{cases} \frac{1}{b-a}, & a \le x \le b \\ 0, & \text{elsewhere} \end{cases}. \tag{2.8}$$

From the axioms of probability, the area under the rectangle must be unity. Hence, the height for a pdf with support $[a, b]$ must be $1/(b-a)$. The probability of a value in any subinterval of $[a, b]$ is proportional to the length of the subinterval, and for an interval $[x_1, x_2]$, the probability is

$$P(x_1 < X \le x_2) = \frac{x_2 - x_1}{b - a}.$$

Using (2.7), we obtain the CDF

$$P(0 < X \le x) = \begin{cases} 0, & x \le a \\ \frac{x-a}{b-a}, & a < x \le b \\ 1 & x > b \end{cases}. \tag{2.9}$$

MATLAB generates uniformly distributed data from $U[0, 1]$ with the command.

> > **rand(m, n)** % a = 0, b = 1, m by n.

Example 2.5 Quantization Quantization maps an analog signal to a discrete set of values for representation using a finite word length. The number of quantization levels depends on the word length. The effect of quantization is represented by uniformly distributed quantization noise. Quantization is either by (i) rounding up or down to the closest level, or (ii) by truncation to the lower level. If the quantization level is q, obtain the noise pdf and the probability of noise in the range $[-q/4, q/4]$ for (a) the rounding quantizer, (b) the truncation quantizer. Which quantizer results in a smaller error on average?

Solution

(a) Rounding quantizer

The error is in the range $[-q/2, q/2]$ and the noise has the pdf $U[-q/2, q/2]$.

$$P\left(-\frac{q}{4} < X \leq \frac{q}{4}\right) = \left(\frac{q}{4} + \frac{q}{4}\right) = \frac{q}{2}$$

(b) Truncation quantizer

The error is in the range $[0, q]$ and the noise has the pdf $U[-q/2, q/2]$.

$$P\left(-\frac{q}{4} < X \leq \frac{q}{4}\right) = \left(\frac{q}{4} - 0\right) = \frac{q}{4}.$$

Outside the range, we have

$$P\left(|X| > \frac{q}{4}\right) = 1 - \left(-\frac{q}{4} < X \leq \frac{q}{4}\right).$$

The results show that the rounding quantizer gives a smaller error on average because

$$P\left(|X| > \frac{q}{4}\right) = \begin{cases} q/2, & \text{rounding quantizer} \\ 3q/4, & \text{truncation quantizer} \end{cases}.$$

■

Using Dirac deltas, we can write any pmf as a pdf. Delta functions are also used for mixed continuous and discrete distributions.

Example 2.6 Poisson pdf Write an expression to represent the Poisson pmf of Example 2.3 as a pdf.

Solution

$$p_N(n) = \sum_{i=0}^{\infty} \frac{\lambda^i e^{-\lambda}}{i!} \delta(n - i).$$

■

Example 2.7 A half-wave rectifier has a random input with pdf $\boldsymbol{U}[-\mathbf{1}, \mathbf{1}]$. Write an expression for the pdf of the output.

Solution

From symmetry, the probability of a negative input is 0.5, and all negative inputs result in a zero output. This is represented by an impulse with strength 0.5. Positive inputs give outputs of equal magnitude to the input. The distribution of the output of the half-wave rectifier is

$$p_V(v) = 0.5\delta(v) + 0.5[1(v) - 1(v - 1)],$$

where 1(.) denotes the step function (Fig. 2.4).

■

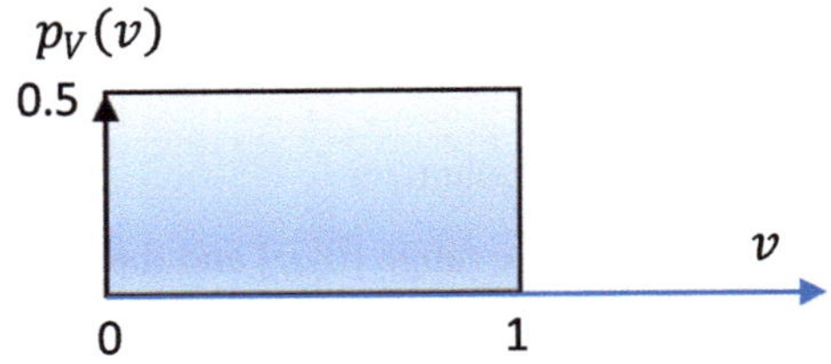

Fig. 2.4 Probability density for the output of a half-wave rectifier

2.2 Expectation of a Random Variable

The expected value or mean of X is its average value over the entire population, that is, the **ensemble** average. For discrete random variable, the expectation is

$$E(X) = \sum_i x_i P(x_i). \tag{2.10}$$

The definition is justified by the relative frequency view of probability $P(x_i) = N_i/N$ where each value is weighted with its relative frequency. For a continuous random variable, the expectation is

$$E(X) = \int_{-\infty}^{\infty} x f_X(x) \mathrm{d}x. \tag{2.11}$$

The following are useful properties of expectation:

$$E\{c\} = c, c \text{ constant} \tag{2.12}$$

$$E\{cX\} = cE\{X\}, c \text{ constant} \tag{2.13}$$

$$E\{X + Y\} = E\{X\} + E\{Y\} \tag{2.14}$$

$$X \geq 0 \Rightarrow E\{X\} \geq 0 \tag{2.15}$$

$$X \geq Y \Rightarrow E\{X\} \geq E\{Y\} \tag{2.16}$$

$$|E\{X\}| \leq E\{|X|\}. \tag{2.17}$$

The proofs are left as an exercise. The following theorem provides an important inequality.

Theorem 2.1 Jensen's Inequality *For a random variable X and a function $g(\cdot)$*

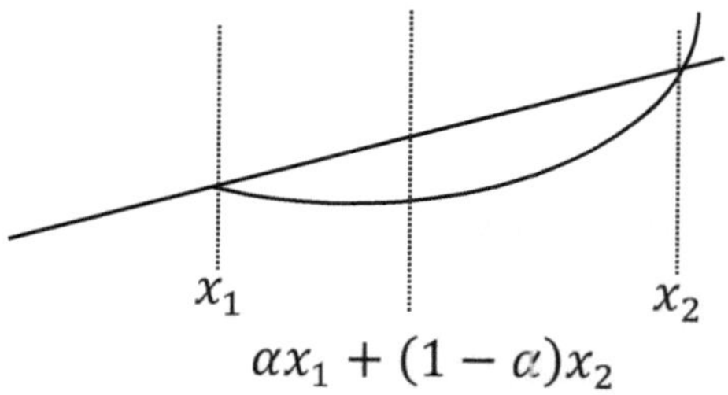

Fig. 2.5 Strictly convex function

$$g(E\{X\}) \leq E\{g(X)\}, g(\cdot)\text{convex} \tag{2.18}$$

$$g(E\{X\}) < E\{g(X)\}, g(\cdot)\text{strictly convex.} \tag{2.19}$$

A convex function is one that satisfies the inequality

$$f(\alpha x_1 + (1 - \alpha)x_2) \leq \alpha f(x_1) + (1 - \alpha) f(x_2), \forall \alpha \in (0, 1). \tag{2.20}$$

A function is strictly convex, as depicted in Fig. 2.5, *if the inequality is replaced by a strict inequality (<). See Hogg et al. Math Statistics, p. 70 for a proof of Jensen's inequality.*

Example 2.8 Show that $E\{X\}^2 < E\{X^2\}$.

Solution

The function X^2 is strictly convex. The result follows by direct application of Jensen's inequality. This confirms that the known expression for the variance

$$\sigma^2 = E\{X^2\} - E\{X\}^2$$

always results in a positive number. ■

2.3 Moments

The kth moment of a random variable is the expectation of its kth power. Thus, the first moment is the mean, and the second moment is the mean square. For a discrete random variable, the kth moment is

$$E(X^k) = \sum_i x_i^k P(x_i). \tag{2.21}$$

For a continuous random variable, the kth moment is

$$E\left(X^k\right) = \int_{-\infty}^{\infty} x^k f_X(x)\mathrm{d}x. \tag{2.22}$$

The variance of a random variable is the second moment about the mean

$$\mathrm{Var}(X) = \sigma_X^2 = E\left\{[X - E(X)]^2\right\}. \tag{2.23}$$

The standard deviation is the square root of variance. The variance of a discrete random variable is

$$\mathrm{Var}(X) = \sum_i [x_i - E(X)]^2 P(x_i), \tag{2.24}$$

while that of a continuous random variable is

$$\mathrm{Var}(X) = \int_{-\infty}^{\infty} [x - E(X)]^2 f_X(x)\mathrm{d}x. \tag{2.25}$$

The following is an important property of the variance that is derived later

$$\mathrm{Var}(X) = E\left\{X^2\right\} - E\{X\}^2. \tag{2.26}$$

Definition 2.4 Uncorrelated Random Variables Random variables are uncorrelated if their expectation satisfies

$$E\{XY\} = E\{X\}E\{Y\}. \tag{2.27}$$

Theorem 2.2 *The variance* σ_{sum}^2 *of the sum of uncorrelated random variables* X_i, $i = 1, \ldots, n$, *with means* m_i *and variance* σ_i^2, *respectively, is the sum of their variances.*

$$\sigma_{\text{sum}}^2 = \sum_{i=1}^{n} \sigma_i^2,\ X_i \text{uncorrelated}. \tag{2.28}$$

Proof We recall that we can write the variance in terms of the first and second moments

$$\mathrm{Var}(X) = E\left\{X^2\right\} - E\{X\}^2$$

and apply this identity to the variance of the sums

$$\sigma_{\text{sum}}^2 = \mathrm{Var}\left(\sum_{i=1}^{n} X_i\right) = E\left\{\left(\sum_{i=1}^{n} X_i\right)^2\right\} - \left(E\left\{\sum_{i=1}^{n} X_i\right\}\right)^2.$$

Expanding the second moment and evaluating the second expectation gives

$$\sigma_{\text{sum}}^2 = E\left\{\sum_{i=1}^{n} X_i^2 + \sum_{i=1}^{n} \sum_{\substack{j=1 \\ j \neq i}}^{n} X_i X_j\right\} - \left(\sum_{i=1}^{n} m_i\right)^2.$$

Since X_i and X_j are uncorrelated, evaluating the expectation of the first term and expanding the second term gives

$$\begin{aligned}\sigma_{\text{sum}}^2 &= \sum_{i=1}^{n} E\{X_i^2\} + \sum_{i=1}^{n} \sum_{\substack{j=1 \\ j \neq i}}^{n} m_i m_j - \sum_{i=1}^{n} m_i^2 - \sum_{i=1}^{n} \sum_{\substack{j=1 \\ j \neq i}}^{n} m_i m_j \\ &= \sum_{i=1}^{n} \left[E\{X_i^2\} - m_i^2\right].\end{aligned}$$

Using the variance identity proves the result

$$\sigma_{sum}^2 = \sum_{i=1}^{n} \sigma_i^2,\ X_i \text{uncorrelated}.$$

■

Example 2.9 Quantizer For each of the quantizers of Example 2.5, write an expression for the pdf of the quantization noise and use it to obtain its mean and variance.

Solution

(a) Rounding quantizer.

Assuming equally likely noise values, the pdf is

$$p_X(x) = \begin{cases} 1/q, & -q/2 < x < q/2 \\ 0, & \text{elsewhere} \end{cases}.$$

Using the pdf, we calculate the mean

$$E(X) = \int_{-q/2}^{q/2} \frac{x}{A} \mathrm{d}x = \left.\frac{x^2}{2A}\right]_{-q/2}^{q/2} = 0.$$

For zero-mean, the variance is equal to the mean square:

$$E\left(X^2\right) = \int_{-q/2}^{q/2} \frac{x^2}{A} \mathrm{d}x = \left.\frac{x^3}{3A}\right]_{-q/2}^{q/2} = \frac{q^2}{12}.$$

(b) Truncation quantizer

Assuming equally likely noise values, the pdf is

$$p_X(x) = \begin{cases} 1/q, & 0 < x < q \\ 0, & \text{elsewhere} \end{cases}.$$

Using the pdf, we calculate the mean

$$E(X) = \int_{0}^{q} \frac{x}{A} \mathrm{d}x = \left.\frac{x^2}{2A}\right]_{0}^{q} = \frac{q}{2}$$

and the mean square

$$E\left(X^2\right) = \int_{0}^{q} \frac{x^2}{A} \mathrm{d}x = \left.\frac{x^3}{3A}\right]_{0}^{q} = \frac{q^2}{3}$$

The variance is

$$\sigma_X^2 = E\left\{X^2\right\} - E\{X\}^2 = \frac{q^2}{3} - \left(\frac{q}{2}\right)^2 = \frac{q^2}{12}.$$

■

2.3.1 *Moment Generating Function and Characteristic Function*

Moment generating functions provide a convenient means of calculating moments. The moment generating function is defined as

$$M_X(s) = E\left\{\mathrm{e}^{-sx}\right\} = \int_{-\infty}^{\infty} \mathrm{e}^{-sx} p_X(x) \mathrm{d}x. \tag{2.29}$$

We observe that the moment generating function is the two-sided Laplace transform of the pdf

$$M_X(s) = \mathscr{L}_2\{p_X(x)\}. \tag{2.30}$$

Some authors prefer to use a different definition with the exponential e^{-sx} replaced by e^{sx}. With either definition, we find the moment generating function for a given pdf by using the two-sided Laplace transform table.

To find moments from the moment generating function, we use the series expansion

$$M_X(s) = \int_{-\infty}^{\infty} e^{-sx} p_X(x)\mathrm{d}x = \int_{-\infty}^{\infty} \left[1 - sx + \frac{(sx)^2}{2!} - \cdots\right] p_X(x)\mathrm{d}x$$
$$= \sum_{k=0}^{\infty} \frac{(-s)^k}{k!} \int_{-\infty}^{\infty} x^k p_X(x)\mathrm{d}x.$$

The expansion shows that the kth moment is given by

$$E\left(X^k\right) = (-1)^k \left.\frac{\mathrm{d}^k \Psi_X(s)}{\mathrm{d}s^k}\right]_{s\to 0}. \tag{2.31}$$

The RHS is evaluated by first differentiating to remove the lower-order terms, then substituting $s \to 0$ to remove the higher-order terms leaving the kth moment.

An alternative to the moment generating function is the characteristic function

$$\Psi_X(\omega) = E\{\exp(j\omega X)\} = \int_{-\infty}^{\infty} \exp(j\omega x) f_X(x)\mathrm{d}x = \mathcal{F}\{f_X(x)\}]_{\omega\to-\omega}. \tag{2.32}$$

The characteristic function for a given pdf is obtained using the Fourier transforms table.

To find moment from the characteristic function, we use the series expansion

$$\Psi_X(\omega) = \sum_{k=0}^{\infty} \frac{(j\omega)^k}{k!} \int_{-\infty}^{\infty} x^k f_X(x)dx. \tag{2.33}$$

The expansion shows that the kth moments is given by

$$E\left(X^k\right) = \frac{1}{j^k} \times \left.\frac{d^k \Psi_X(\omega)}{d\omega^k}\right]_{\omega\to 0}. \tag{2.34}$$

2.4 Normal or Gaussian Density

The most commonly used pdf in applications is the normal or Gaussian pdf

$$p_X(x) = \frac{1}{\sqrt{2\pi\sigma_x^2}} \exp\left\{-\frac{(x - m_x)^2}{2\sigma_x^2}\right\}. \tag{2.35}$$

The function is symmetric about the mean m_x.where it has the peak value $1/\sqrt{2\pi\sigma_x^2} = 0.3989/\sigma_x$. The standard deviation determines the shape of the pdf plot with flatter curves for a large standard deviation and sharper peaks for a small one. For the normal distribution, the mean is equal to the mode (most likely value) and to the median (the value with equal probability of falling above or below it).

A normal distribution with zero-mean and unit variance is the standard normal distribution,

$$\Phi(x) = \frac{1}{\sqrt{2\pi}} \int_{-\infty}^{x} \mathrm{e}^{-t^2/2} \mathrm{d}t. \tag{2.36}$$

Why is the Gaussian distribution so important?

- It fits many physical phenomena.
- It is completely described by mean and variance.
- For Gaussian random variables, independent⇔ uncorrelated.
- Central limit: distribution of sum of any independent vars. approaches normal as n goes to infinity (Fig. 2.6).

2.4.1 Right Tail Probability

The probability of exceeding a given value above the mean is known as the right tail probability or complementary cumulative distribution:

$$\begin{aligned} Q(x) &= \frac{1}{\sqrt{2\pi}} \int_{x}^{\infty} e^{-t^2/2} \mathrm{d}t \\ &= 1 - \Phi(x). \end{aligned} \tag{2.37}$$

MATLAB allows us to plot the normal distribution and show the right tail probability with the command.

> > **p = normspec(specs,mu,sigma,region)**

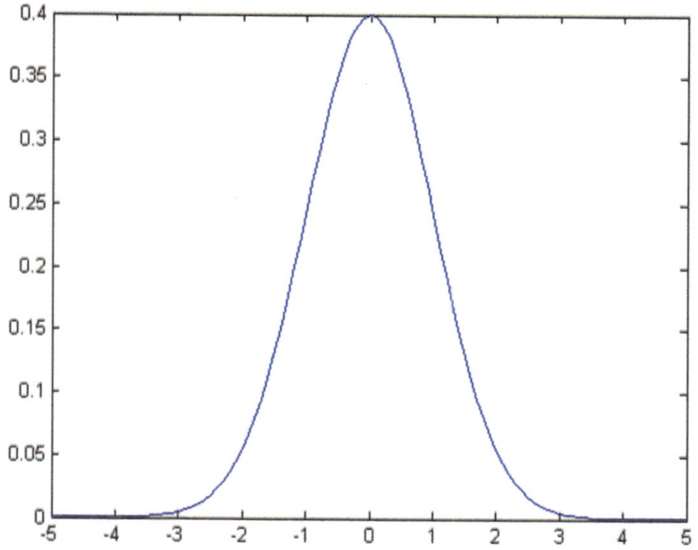

Fig. 2.6 Standard **normal pdf**

where specs gives the upper and lower limits, (mu,sigma) are the (mean, standard deviation) of the distribution, and region specifies the shading region, which is either "inside" or "outside." For a standard normal distribution and a range $(-\infty, 1)$, the following command gives the plot of Fig. 2.7 and the value of the right tail probability:

> >p = normspec([-Inf,1],0,1,'outside')

To shade the range below 1 and obtain the CDF value, as in Fig. 2.8, we use the command:

Fig. 2.7 Right tail probability

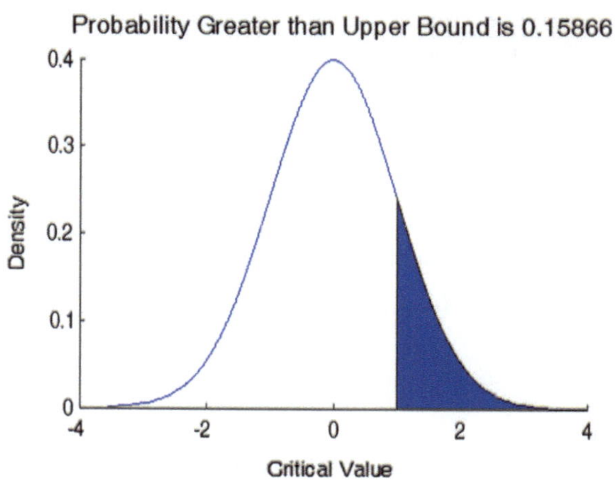

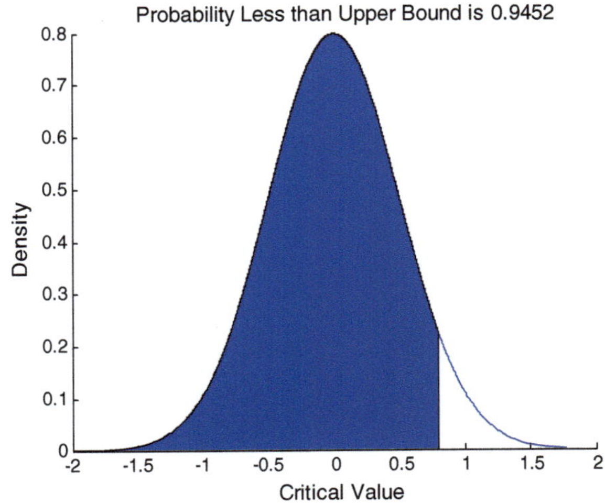

Fig. 2.8 Cumulative distribution

> > p = normspec([-Inf,1],0,1,'inside');

The right tail probability is important in many applications such as the calculation of the probability of false alarm, the probability of detection, etc.

Because the right tail probability Q is monotonically decreasing, it is one–one and is therefore invertible. If the right tail probability is equal to P_{FA}, then the inverse is

$$Q(\gamma) = \frac{1}{\sqrt{2\pi}} \int_{\gamma}^{\infty} e^{-t^2/2} dt = P_{FA} \tag{2.38}$$

$$\gamma = Q^{-1}(P_{FA}). \tag{2.39}$$

The error function provides a method of calculating the area under a normal curve:

$$\text{erf}(x) = \frac{2}{\sqrt{\pi}} \int_{0}^{x} e^{-t^2} dt. \tag{2.40}$$

The complementary error function is

$$\text{erfc}(x) = 1 - \text{erf}(x) = \frac{2}{\sqrt{\pi}} \int_{x}^{\infty} e^{-t^2} dt.$$

A simple change of variable verifies that $\text{erf}(-x) = -\text{erf}(x)$:

$$\mathrm{erf}(-x) = \frac{2}{\sqrt{\pi}} \int_0^{-x} \mathrm{e}^{-t^2} \mathrm{d}t = -\frac{2}{\sqrt{\pi}} \int_0^{(-x)} \mathrm{e}^{-t^2} d(-x) = -\mathrm{erf}(x).$$

The CDF of $X \sim \mathcal{N}(0, 0.5)$ is

$$P_X(x) = \frac{1}{\sqrt{2\pi \times 0.5}} \int_{-\infty}^{x} \mathrm{e}^{-t^2/(2\times 0.5)} \mathrm{d}t$$

$$= \begin{cases} \frac{1}{\sqrt{2\pi \times 0.5}} \int_{-\infty}^{0} e^{-t^2} \mathrm{d}t + \frac{1}{\sqrt{\pi}} \int_0^x \mathrm{e}^{-t^2} \mathrm{d}t, \ x \geq 0 \\ \frac{1}{\sqrt{2\pi \times 0.5}} \int_{-\infty}^{0} e^{-t^2} \mathrm{d}t - \frac{1}{\sqrt{\pi}} \int_0^x \mathrm{e}^{-t^2} \mathrm{d}t, \ x < 0 \end{cases}.$$

Since $\mathrm{erf}(-x) = -\mathrm{erf}(x)$, we have

$$P_X(x) = 0.5\{1 + \mathrm{erf}(x)\}, X \sim \mathcal{N}(0, 0.5).$$

Using the equation

$$\mathrm{erf}\left(\frac{x}{\sqrt{2}}\right) = \frac{2}{\sqrt{\pi}} \int_0^{x/\sqrt{2}} \mathrm{e}^{-t^2} \mathrm{d}t = \frac{2}{\sqrt{2\pi}} \int_0^{x} \mathrm{e}^{-t^2/2} \mathrm{d}t,$$

we can prove the following equations that relate the error function to the standard normal pdf:

1. $$\Phi(x) = \frac{1}{\sqrt{2\pi}} \int_{-\infty}^{x} e^{-t^2/2} \mathrm{d}t = 0.5\left\{1 + \mathrm{erf}\left(\frac{x}{\sqrt{2}}\right)\right\} \tag{2.41}$$

2. $$Q(x) = 1 - \Phi(x) = 0.5\left\{1 - \mathrm{erf}\left(\frac{x}{\sqrt{2}}\right)\right\} = 0.5\mathrm{erfc}\left(\frac{x}{\sqrt{2}}\right) = P \tag{2.42}$$

3. $$x = Q^{-1}(P) = \sqrt{2}\mathrm{erfinv}(1 - 2P). \tag{2.43}$$

The proofs are left as an exercise.

The following MATLAB commands implement the computation of (2.41)–(2.43):

> > **erf(x)** % Error function
> > **erfc(x)** % Complementary error function
> > **0.5*(1 + erf(x/sqrt(2)))** % St. Normal P (t < x)
> > **0.5*erfc(x/sqrt(2))** % St. Normal P (t > x)
> > **Qinv = sqrt(2)*erfinv(1–2*P)** % Inverse Q(P).

MATLAB can also calculate the Q function with the command:

> > **Q = qfunc(x)** % Standard normal P (t > x)

and its inverse with the command:

> > **Qinv = qfuncinv(x)** % Inverse of Q(x)

Other packages, such as MAPLE, have similar erf and erfc commands. We can also simply integrate the Gaussian pdf numerically to get probabilities as in the following example. Integration can also be used to get right tail or any other probability.

Example 2.10 Test scores X for an exam are normally distributed with $X \sim \mathcal{N}(83, 64)$. Write an expression for the pdf of the test scores and use MATLAB to obtain the probability of a grade within two standard deviations from the mean.

Solution

The pdf for the test scores is

$$P_X(x) = \frac{1}{\sqrt{128\pi}} \exp\left\{-\frac{(x-83)^2}{128}\right\}.$$

The mean and standard deviation are $m_x = 83$, $\sigma_x = 8$, and the of grades within two standard deviations of the mean lie in the interval $83 \pm 16 = [67, 99]$. We need the probability:

$$\Pr\{67 \le x \le 99\} = \frac{1}{\sqrt{128\pi}} \int_{67}^{99} \exp\left\{-\frac{(x-83)^2}{128}\right\} dx.$$

The following MATLAB commands evaluate the integral:

> > **fun = @(x) exp(-(x-83).^2/128)./sqrt(128*pi);**
> > **integral(fun,83–16,83 + 16) % within 2 sigma**

ans =
0.9545

This shows that for a normal distribution, the two-sigma range has a 95% probability. ■

2.5 Multiple Random Variables

Multivariate distributions govern the joint probability of a set or vector of random variables. For two variables, we use a bivariate distribution. In the discrete case, the joint probability of two variables is a 2-dimensional array:

Fig. 2.9 Region *A* in the $(\boldsymbol{x}_1, \boldsymbol{x}_2)$-plane

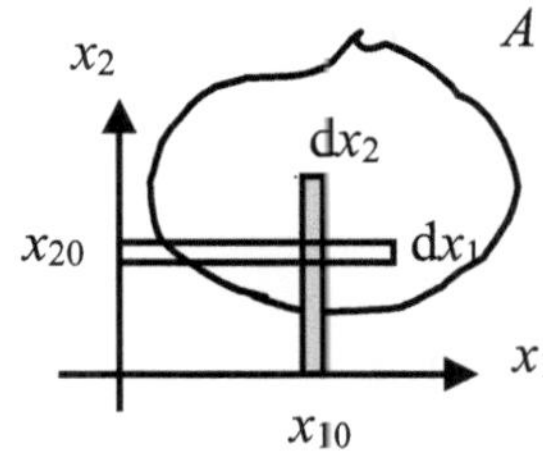

$$p_{XY}(x_i, y_j) = P(X = x_i, Y = y_j). \tag{2.44}$$

For continuous variables, the multivariate distribution function for a $n \times 1$ random vector $\boldsymbol{X}$ is

$$P_{1,\ldots,n}(\boldsymbol{X}) = P(\boldsymbol{X} \le \boldsymbol{x}), \ -\infty < x_i < \infty, i = 1, \ldots, n. \tag{2.45}$$

The probability of $\boldsymbol{X}$ in a region A is

$$P(\boldsymbol{X} \in \mathrm{A}) = \iint\limits_{\Omega} \ldots \int p_{1,\ldots,n}(\boldsymbol{x}) \mathrm{d}\boldsymbol{x}. \tag{2.46}$$

In the bivariate case (Fig. 2.9), A is a subset of the (x_1, x_2)-plane and the probability of X in A is

$$P(X \in A) = \iint\limits_{A} p_{12}(\boldsymbol{x}) \mathrm{d}\boldsymbol{x}. \tag{2.47}$$

2.5.1 Marginal Distributions

The marginal probability governing one of the two discrete variables is obtained by adding a column or row of the joint probability matrix. The marginal pdf $p_i(x_i)$ for a continuous random variable x_i is obtained from the joint pdf $p_{1,\ldots,n}(\boldsymbol{x})$ by integrating w.r.t. all variables with the exception of x_i :

$$p_i(x_i) = \int\limits_{-\infty}^{\infty} \ldots \int\limits_{-\infty}^{\infty} p_{1,\ldots,n}(\boldsymbol{x}) \mathrm{d}x_1 \ldots \mathrm{d}x_{i-1} \mathrm{d}x_{i+1} \ldots \mathrm{d}x_n. \tag{2.48}$$

Example 2.11 Find the marginal pdf for x_1 if the joint pdf for x_1, x_2 is.

$$p_{12}(x_1, x_2) = \begin{cases} x_1^2 + 2x_2^2, & x_1, x_2 \in [0, 1] \\ 0, & \text{elsewhere} \end{cases}.$$

Solution

Integrating w.r.t. x_2 gives the marginal pdf

$$p_1(x_1) = \int_{-\infty}^{\infty} p_{12}(x_1, x_2)\mathrm{d}x_2$$
$$= \int_0^1 (x_1^2 + 2x_2^2)\mathrm{d}x_2 = x_1^2 + \frac{2}{3}.$$

By the axioms of probability, integration with respect to all variables must give 1. Check! ■

2.5.2 *Conditional Probability Density*

The conditional probability for random variables is defined as

$$P(x_1|x_2) = P(x_1, x_2)/P(x_2).$$

For a continuous random variable, the probability of a random variable X_i having values in an infinitesimal interval around x_i is $p_i(x_i)\mathrm{d}x_i$. Substituting in terms of the pdfs gives

$$p_{1|2}(x_1|x_2)\mathrm{d}x_1 = \frac{p_{1,2}(x_1, x_2)\mathrm{d}x_1\mathrm{d}x_2}{p_2(x_2)\mathrm{d}x_2}.$$

Since the equality holds for all perturbations, we have the conditional density

$$p_{1|2}(x_1|x_2) = \frac{p_{1,2}(x_1, x_2)}{p_2(x_2)}. \tag{2.49}$$

More generally, the probability of X_1 given $X_i = x_i, i = 2, \ldots, n$, gives

$$p_{1|2,\ldots,n}(x_1|x_2, \ldots, x_n)\mathrm{d}x_1 = \frac{p_{1,2,\ldots,n}(x_1, x_2, \ldots, x_n)\mathrm{d}x_1\mathrm{d}x_2 \ldots \mathrm{d}x_n}{p_{2,\ldots,n}(x_2, \ldots, x_n)\mathrm{d}x_2 \ldots \mathrm{d}x_n}.$$

Hence, the conditional density of X_1 given $X_i = x_i, i = 2, \ldots, n$, is

$$p_{1|2,\ldots,n}(x_1|x_2,\ldots,x_n) = \frac{p_{1,2,\ldots,n}(x_1,x_2,\ldots,x_n)}{p_{2,\ldots,n}(x_2,\ldots,x_n)}. \tag{2.50}$$

In general, the conditional density is a ratio of a joint density and a marginal density similarly to the conditional probability.

The joint density can be written in terms of the conditional density as

$$p_{1,2,\ldots,n}(x_1,x_2,\ldots,x_n) = p_{1|2,\ldots,n}(x_1|x_2,\ldots,x_n).p_{2,\ldots,n}(x_2,\ldots,x_n). \tag{2.51}$$

For mutually independent random variables, the probability formula gives

$$\begin{aligned} P(x_1,x_2) &= p_{12}(x_1,x_2)\mathrm{d}x_1\mathrm{d}x_2 = P(x_1)P(x_2) \\ &= [p_1(x_1)\mathrm{d}x_1][p_2(x_2)\mathrm{d}x_2]. \end{aligned}$$

The pdfs satisfy

$$p_{12}(x_1,x_2) = p_1(x_1)p_2(x_2). \tag{2.52}$$

Substituting in the formula for the conditional density gives

$$p_{1|2}(x_1|x_2) = \frac{p_{1,2}(x_1,x_2)}{p_2(x_2)} = \frac{p_1(x_1)p_2(x_2)}{p_2(x_2)}. \tag{2.53}$$

For mutually independent random variables x_1 and x_2, the conditional pdf is

$$p_{1|2}(x_1) = p_1(x_1) \tag{2.54}$$

Independent random variables are uncorrelated, i.e., they satisfy $E\{X_1X_2\} = E\{X_1\}E\{X_2\}$. To prove this fact, we find the expectation of the product

$$\begin{aligned} E\{X_1X_2\} &= \int_{-\infty}^{\infty}\int_{-\infty}^{\infty} x_1x_2p_{X_1X_2}(x_1,x_2)\mathrm{d}x_1\mathrm{d}x_2 \\ &= \int_{-\infty}^{\infty} x_1p_{X_1}(x_1)\mathrm{d}x_1 \int_{-\infty}^{\infty} x_2p_{X_2}(x_2)\mathrm{d}x_2 = E\{X_1\}E\{X_2\}. \end{aligned}$$

However, uncorrelated random variables may or may not be independent. Hence, independence is sufficient but is not necessary for the variance of the sum to be equal to the sum of the variances of individual terms (2.28). Uncorrelated variables X and Y are independent if and only if any two functions of the random variables, $f(X)$ and $g(Y)$, are uncorrelated.

2.5.2.1 Law of Iterated Expectations

The conditional expectation of a random variable x is its expected value given a known value of another random variable y, i.e., $E\{x|y\}$. The conditional expectation varies randomly with the variable y and has an expected value that is obtained by averaging over the ensemble of y values. We therefore have the following result.

Theorem 2.3 Law of Iterated Expectations *The expected value of the conditional expectation of a random variable x given a random variable y, averaged over the ensemble of y values is the expectation of x.*

$$E\{E\{x|y\}\} = E\{x\}. \tag{2.55}$$

Proof The conditional expectation of x.

$$E\{x|y\} = \int_{-\infty}^{\infty} x p_{x|y}(x|y)\mathrm{d}x.$$

Taking the expectation w.r.t. y gives

$$E\{E\{x|y\}\} = \int_{-\infty}^{\infty}\left[\int_{-\infty}^{\infty} x p_{x|y}(x|y)\mathrm{d}x\right] p_y(y)\mathrm{d}y.$$

We interchange the order of integration to write

$$\begin{aligned} E\{E\{x|y\}\} &= \int_{-\infty}^{\infty} x\left[\int_{-\infty}^{\infty} p_{xy}(x,y)\mathrm{d}y\right]\mathrm{d}x \\ &= \int_{-\infty}^{\infty} x p_x(x)\mathrm{d}x = E\{x\}. \end{aligned}$$

■

2.5.2.2 Bayes Rule for Random Variables

Using the definition of the conditional density, we write

$$p_{1,2}(x_1, x_2) = p_{2|1}(x_2|x_1)p_1(x_1)$$

$$p_{1|2}(x_1|x_2) = \frac{p_{2|1}(x_2|x_1)p_1(x_1)}{p_2(x_2)}.$$

Generalizing to n variables

$$\begin{aligned} p_{1,\ldots,n}(x_1, x_2, \ldots, x_n) &= p_{2,\ldots,n|1}(x_2, \ldots, x_n|x_1)p_1(x_1) \\ &= p_{1|2,\ldots,n}(x_1|x_2, \ldots, x_n)p_{2,\ldots,n}(x_2, \ldots, x_n) \end{aligned}$$

gives the Bayes rule

$$p_{1|2,\ldots,n}(x_1|x_2, \ldots, x_n) = \frac{p_{2,\ldots,n|1}(x_2, \ldots, x_n|x_1)p_1(x_1)}{p_{2,\ldots,n}(x_2, \ldots, x_n)}. \tag{2.56}$$

Example 2.12 Given two random variables x and y, find the pdf of their sum in terms of their two pdfs.

Solution

The probability of the sum in an infinitesimal interval is

$$P(z < Z < z + \mathrm{d}z) = p_Z(z)\mathrm{d}z = \int_{-\infty}^{\infty} [p_{XY}(x, y)\mathrm{d}y]\mathrm{d}x.$$

Use independence of x and y to replace the joint pdf with the product of the marginal pdfs

$$p_Z(z)\mathrm{d}z = \int_{-\infty}^{\infty} [p_X(x)p_Y(y)\mathrm{d}y]\mathrm{d}x.$$

At any fixed x, $dy = dz$, $y = z - x$, as depicted in Fig. 2.10

$$p_Z(z)\mathrm{d}z = \left\{\int_{-\infty}^{\infty} p_X(x)p_Y(z - x)\mathrm{d}x\right\}\mathrm{d}z.$$

Eliminating the differential gives the pdf of the sum

$$p_Z(z) = \int_{-\infty}^{\infty} p_X(x)p_Y(z - x)\mathrm{d}x.$$

The pdf is a convolution integral, and by the convolution theorem of the Fourier transform, we have

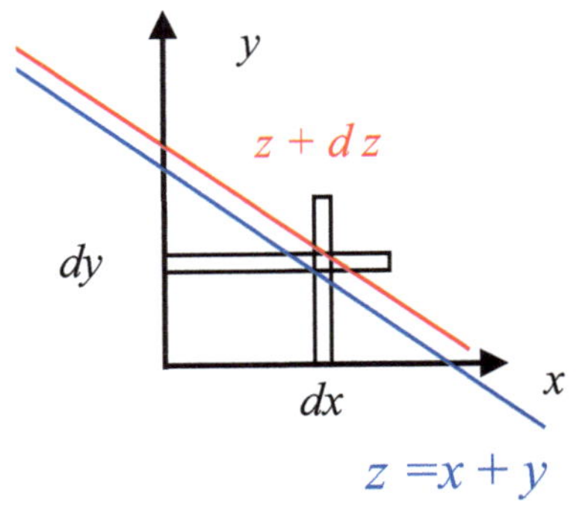

Fig. 2.10 Sum of independent random variables

$$\mathcal{F}\{p_Z(z)\} = \mathcal{F}\{p_X(x)\}\mathcal{F}\{p_Y(y)\}.$$

If x and y are not independent, the integrand is $p(x, z - x)$ and the pdf of the sum is not a convolution integral. ■

2.5.2.3 Correlation Coefficient

The correlation coefficient is a normalized measure of correlation between X and Y defined as

$$\rho_{XY} = \frac{\mathrm{cov}(X, Y)}{\sqrt{\mathrm{var}(X)\mathrm{var}(Y)}}. \tag{2.57}$$

The numerator is given by

$$\begin{aligned} cov(X, Y) &= E\{(X - m_x)(Y - m_y)\} \\ &= E\{XY\} - m_x E\{Y\} - E\{X\}m_y + m_x m_y. \end{aligned}$$

We now have

$$\mathrm{cov}(X, Y) = E\{XY\} - m_x m_{y.} \tag{2.58}$$

This reduces to the variance property for $X = Y$.

The coefficient is called the correlation in statistics and economics literature. It measures linear relationship only (see De Groot p. 253 and Ex. 4.6.4) and gives misleading results for nonlinear relations.

Theorem 2.4 *The correlation coefficient is zero for uncorrelated random variables and has values between* -1 *and* 1.

$$-1 \leq \rho_{XY} \leq 1. \tag{2.59}$$

Proof We first show that for uncorrelated random variables X and Y, the correlation coefficient is zero. We use the formula for the covariance.

$$\mathrm{cov}(X, Y) = E\{XY\} - m_x m_y = E\{X\}E\{Y\} - m_x m_y = 0.$$

For uncorrelated X and Y, $\rho_{XY} = 0$.

For $Y = X$, the covariance is equal to the variance

$$\mathrm{cov}(X, X) = E\{(X - m_x)^2\} = \mathrm{Var}(X),$$

and the correlation coefficient is unity

$$\rho_{XY} = \frac{\mathrm{cov}(X, X)}{\sqrt{\mathrm{Var}(X)\mathrm{Var}(X)}} = 1.$$

For $Y = -X$, the covariance is equal to minus the variance

$$\mathrm{cov}(X, -X) = E\{-(X - m_x)^2\} = -\mathrm{Var}(X),$$

and the means is

$$E\{-X\} = -E\{X\} = -m_x.$$

Thus, we have

$$-X - E\{-X\} = -[X - m_x],$$

and the correlation coefficient is -1

$$\rho_{XY} = \frac{\mathrm{cov}(X, -X)}{\sqrt{\mathrm{Var}(X)\mathrm{Var}(X)}} = -1.$$

To show the bounds on the correlation coefficient, we recall that $E\{X\} > 0$ for $X > 0$ and consider the inequality:

$$0 < E\{[a(X - m_x) + (Y - m_Y)]^2\}.$$

Expanding gives

$$0 < a^2\sigma_X^2 + 2a.\mathrm{cov}(X, Y) + \sigma_Y^2, \forall a.$$

The quadratic in a has no real roots if the discriminant is negative or zero:

$$cov^2(X, Y) - \sigma_X^2\sigma_Y^2 \leq 0.$$

Zero discriminant correspond to equal roots with $X = \pm Y$, with strict inequality otherwise:

$$\text{cov}^2(X, Y)/(\sigma_X^2 \sigma_Y^2) = \rho_{XY}^2 \leq 1.$$

The inequality becomes an equality for $X = \pm Y$ and $\text{cov}(X, Y) = \pm \text{Var}(X)$. ■

Random variables are said to be orthogonal if the expectation of their product is zero:

$$E\{XY\} = 0, \text{ real scalar.} \tag{2.60}$$

For the vector case, orthogonality of two vectors implies that each entry of the first vector is orthogonal to each entry of the second vector:

$$E\{\boldsymbol{X}\boldsymbol{Y}^T\} = [\boldsymbol{0}]. \tag{2.61}$$

2.6 Correlation R_X and Covariance C_X

The generalization of the 2nd moment to the vector case is the correlation:

$$R_X = E\{\boldsymbol{x}\boldsymbol{x}^T\}. \tag{2.62}$$

The covariance matrix generalizes the variance to the vector case:

$$C_X = E\{(\boldsymbol{x} - \boldsymbol{m}_x)(\boldsymbol{x} - \boldsymbol{m}_x)^T\} = [E\{(x_i - m_i)(x_j - m_j)\}]. \tag{2.63}$$

The matrix is identical to the correlation for $\boldsymbol{x}$ zero-mean. The covariance matrix can be written in terms of variances and correlation coefficients as

$$C = \begin{bmatrix} \sigma_1^2 & \rho_{12}\sigma_1\sigma_2 & \cdots & \rho_{1n}\sigma_1\sigma_n \\ \rho_{12}\sigma_1\sigma_2 & \sigma_2^2 & \cdots & \rho_{2n}\sigma_2\sigma_n \\ \vdots & \vdots & \ddots & \vdots \\ \rho_{1n}\sigma_1\sigma_n & \rho_{2n}\sigma_2\sigma_n & \cdots & \sigma_n^2 \end{bmatrix}. \tag{2.64}$$

The matrix is clearly diagonal for uncorrelated variables

$$C = \text{diag}\{\sigma_1^2, \ldots, \sigma_n^2\}. \tag{2.65}$$

For orthogonal variables with one or both zero-mean, the cross-correlation is zero

$$cov(X, Y) = E\{(X - m_x)(Y - m_y)\}$$

$$
\begin{aligned}
&= E\{XY\} - m_x m_y \\
&= -m_x m_y m_x \\
&= 0 \text{ or } m_y = 0 \Rightarrow \text{cov}(X, Y) = 0.
\end{aligned}
$$

Theorem 2.5 *The covariance matrix is positive definite for linearly independent random variables and positive semidefinite otherwise.*

$$C_x > 0\, or\; C_x \geq 0. \tag{2.66}$$

Proof For any real vector $\forall z \in \mathcal{R}^n$, we have the inequality

$$
\begin{aligned}
0 \leq E\left\{\left[z^T(\boldsymbol{x} - \boldsymbol{m}_x)\right]^2\right\} &= z^T E\left\{(\boldsymbol{x} - \boldsymbol{m}_x)(\boldsymbol{x} - \boldsymbol{m}_x)^T\right\} z \\
&= z^T C_x z.
\end{aligned}
$$

This is a positive semidefinite quadratic form

$$z^T C_x z \geq 0, \forall z \in \mathcal{R}^n$$

and hence $C_x \geq 0$(linearly dependent x_i) or $C_x > 0$. ■

The matrix is positive semidefinite if the random variables are not linearly independent. Linear dependence implies that $\exists z \neq \mathbf{0}\; s.t. z^T(\boldsymbol{x} - \boldsymbol{m}_x) = 0$, and if the variables are linearly independent, then no such z exists.

The covariance of two vectors is called the cross-covariance and is given by:

$$
\begin{aligned}
C_{xz} &= E\left\{(\boldsymbol{x} - \boldsymbol{m}_x)(z - \boldsymbol{m}_z)^T\right\} \\
&= E\left\{\boldsymbol{x} z^T\right\} - \boldsymbol{m}_x \boldsymbol{m}_z^T.
\end{aligned} \tag{2.67}
$$

If $\boldsymbol{x} = z$, we have the covariance relation

$$
\begin{aligned}
C_z &= E\left\{(z - \boldsymbol{m}_z)(z - \boldsymbol{m}_z)^T\right\} \\
&= E\left\{z z^T\right\} - \boldsymbol{m}_z \boldsymbol{m}_z^T.
\end{aligned} \tag{2.68}
$$

The proof is similar to the proof of the scalar case and is left as an exercise.

2.7 Multivariate Normal Distribution

The multivariate normal distribution is a generalization of normal distribution to n linearly independent random variables $\{x_1, \ldots x_n\}$. The joint pdf is given by

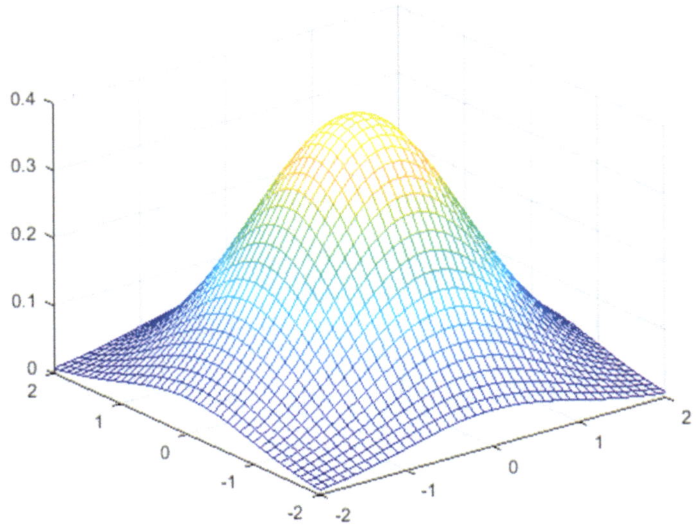

Fig. 2.11 Bivariate normal distribution

$$p_X(\boldsymbol{x}) = \frac{1}{[2\pi]^{n/2}\sqrt{\det(C_x)}} \exp\left\{-\frac{1}{2}(\boldsymbol{x}-\boldsymbol{m}_x)^T C_x^{-1}(\boldsymbol{x}-\boldsymbol{m}_x)\right\}$$
$$\boldsymbol{x} = \begin{bmatrix} x_1 \ldots x_n \end{bmatrix}^T \tag{2.69}$$

with mean $\boldsymbol{m}_x$ and covariance matrix C_x.The pdf reduces to the univariate normal distribution for $n = 1$ with a scalar mean and the covariance matrix replaced by the variance. A plot of a bivariate normal pdf is shown in Fig. 2.11.

$$p_{X_1X_2}(x_1, x_2)$$

The following properties make the multivariable normal distribution particularly useful.

2.7.1 Properties of Multivariate Normal

a. The pdf $\mathcal{N}(\boldsymbol{m}_x, C_x)$ is completely defined by $\boldsymbol{m}_x$ and C_x, i.e., first- and second-order statistics.
b. If the joint pdf of uncorrelated random variable is normal, then they are independent.
c. For a jointly normal pdf, all marginal and conditional pdfs are normal. However, marginal normal pdfs do not guarantee a jointly normal pdf.
d. Linear transformation of normal vector gives a normal vector

Property a is obvious from the expression for the pdf. We prove Properties b and part of c next and prove Property d in Sect. 2.8.3. We leave the proof of the marginal density part of c as an exercise.

Property b If Gaussian random variables x_i are mutually uncorrelated, then they are also mutually independent.

Proof If x_i are mutually uncorrelated, $\rho_{ij} = 0,\ i, j = 1, \ldots, n$, we have the diagonal covariance matrix.

$$C_x = \operatorname{diag}\{\sigma_1^2, \sigma_2^2, \ldots, \sigma_n^2\}, \det(C_x) = \prod_{i=1}^{n} \sigma_i^2.$$

Recall the quadratic form for the diagonal case is just the weighted sum of the diagonal terms. Thus, the exponent of the pdf becomes

$$(\boldsymbol{x} - \boldsymbol{m}_x)^T C_x^{-1} (\boldsymbol{x} - \boldsymbol{m}_x) = \sum_{i=1}^{n} \left(\frac{x_i - m_i}{\sigma_i}\right)^2.$$

Because exponents add for multiplication, we have

$$\begin{aligned} p_X(\boldsymbol{x}) &= \frac{1}{[2\pi]^{\frac{n}{2}} \sqrt{\prod_{i=1}^{n} \sigma_i^2}} \exp\left\{-\frac{1}{2} \sum_{i=1}^{n} \left(\frac{x_i - m_i}{\sigma_i}\right)^2\right\} \\ &= \prod_{i=1}^{n} \frac{1}{\left[2\pi\sigma_i^2\right]^{1/2}} \exp\left\{-\frac{1}{2} \left(\frac{x_i - m_i}{\sigma_i}\right)^2\right\} = \prod_{i=1}^{n} p_{X_i}(x_i). \end{aligned}$$

■

To prove that the conditional density is Gaussian (Property c), we need the joint density of two Gaussian random vectors. We stack the two vectors to form the vector $\boldsymbol{y} = \operatorname{col}\{\boldsymbol{x}, \boldsymbol{z}\}$ with mean

$$E\{\boldsymbol{y}\} = E\left\{\begin{bmatrix} \boldsymbol{x} \\ \boldsymbol{z} \end{bmatrix}\right\} = \begin{bmatrix} E\{\boldsymbol{x}\} \\ E\{\boldsymbol{z}\} \end{bmatrix} = \begin{bmatrix} \boldsymbol{m}_x \\ \boldsymbol{m}_z \end{bmatrix}$$

and covariance

$$C_y = E\left\{\begin{bmatrix} \boldsymbol{x} - \boldsymbol{m}_x \\ \boldsymbol{z} - \boldsymbol{m}_z \end{bmatrix} \left[(\boldsymbol{x} - \boldsymbol{m}_x)^T \ (\boldsymbol{z} - \boldsymbol{m}_z)^T \right]\right\} = \begin{bmatrix} C_x & C_{xz} \\ C_{zx} & C_z \end{bmatrix},$$

where

$$C_x = E\{(\boldsymbol{x} - \boldsymbol{m}_x)(\boldsymbol{x} - \boldsymbol{m}_x)^T\}$$

$$C_z = E\{(z - m_z)(z - m_z)^T\}$$

$$C_{xz} = E\{(x - m_x)(z - m_z)^T\}$$

$$C_{zx} = E\{(z - m_z)(x - m_x)^T\} = C_{xz}^T.$$

Using the inverse of a partitioned matrix, the inverse of the covariance matrix is

$$\begin{aligned}
C_y^{-1} &= \begin{bmatrix} C_x & C_{xz} \\ C_{zx} & C_z \end{bmatrix}^{-1} = \begin{bmatrix} \mathrm{A} & \mathrm{B} \\ \mathrm{B}^T & \mathrm{C} \end{bmatrix} \\
\mathrm{A} &= \left(C_x - C_{xz} C_z^{-1} C_{zx}\right)^{-1} = C_x^{-1} + C_x^{-1} C_{xz} \mathrm{C} C_{zx} C_x^{-1} = \mathrm{A}^T \\
\mathrm{B} &= -\mathrm{A} C_{xz} C_z^{-1} = -C_x^{-1} C_{xz} \mathrm{C} \\
\mathrm{C} &= \left(C_z - C_{zx} C_x^{-1} C_{xz}\right)^{-1} = C_z^{-1} + C_z^{-1} C_{zx} \mathrm{A} C_{xz} C_z^{-1} = \mathrm{C}^T.
\end{aligned} \tag{2.70}$$

If the joint pdf is Gaussian, then it is given by

$$p_y(y) = \frac{1}{[2\pi]^{(n+m)/2}\sqrt{\det(C_y)}} \exp\left\{-\frac{1}{2}(y - m_y)^T C_y^{-1}(y - m_y)\right\}.$$

Using the rule for the determinant of the partitioned matrix, we obtain the determinant

$$\det(C_y) = \det(C_x)\det\left(C_z - C_{zx} C_x^{-1} C_{xz}\right) = \det(C_z)\det\left(C_x - C_{xz} C_z^{-1} C_{zx}\right). \tag{2.71}$$

The above results allow us to characterize the conditional pdf of jointly Gaussian vectors $\boldsymbol{x}$ and $\boldsymbol{z}$. In Chap. 9, this result is fundamental in the derivation of estimators of $\boldsymbol{x}$ using noisy measurements $\boldsymbol{z}$ where we make use of the following theorem.

Theorem 2.6 *For* $\boldsymbol{x}$, $\boldsymbol{z}$ *jointly Gaussian, the conditional pdf for* $\boldsymbol{x}$ *given* $\boldsymbol{z}$ *is Gaussian.*

$$p_{x|z}(\boldsymbol{x}|\boldsymbol{z}) = \frac{1}{[2\pi]^{n/2}\sqrt{\det(C_{x|z})}} \exp\left\{-\frac{1}{2}(\boldsymbol{x} - \boldsymbol{m}_{x|z})^T C_{x|z}^{-1}(\boldsymbol{x} - \boldsymbol{m}_{x|z})\right\} \tag{2.72}$$

with mean

$$\boldsymbol{m}_{x|z} = E\{\boldsymbol{x}|\boldsymbol{z}\} = m_x + C_{xz} C_z^{-1}(\boldsymbol{z} - \boldsymbol{m}_z) \tag{2.73}$$

and covariance

$$C_{x|z} = C_x - C_{xz} C_z^{-1} C_{zx}. \tag{2.74}$$

Proof The conditional pdf is

$$p(\boldsymbol{x}|\boldsymbol{z}) = p_{xz}(\boldsymbol{x}, \boldsymbol{z})/p_z(\boldsymbol{z}).$$

Substituting for the pdfs gives the exponent

$$(\boldsymbol{y} - \boldsymbol{m}_y)^T \begin{bmatrix} \mathrm{A} & \mathrm{B} \\ \mathrm{B}^T & \mathrm{C} - C_y^{-1} \end{bmatrix} (\boldsymbol{y} - \boldsymbol{m}_y) = .$$

Expand the quadratic to obtain

$$(\boldsymbol{x} - \boldsymbol{m}_x)^T \mathrm{A}(\boldsymbol{x} - \boldsymbol{m}_x) + 2(\boldsymbol{x} - \boldsymbol{m}_x)^T \mathrm{B}(\boldsymbol{z} - \boldsymbol{m}_z) + (\boldsymbol{z} - \boldsymbol{m}_z)(C - C_y^{-1})(\boldsymbol{z} - \boldsymbol{m}_z)^T.$$

Substituting for A, B, C, followed by some messy algebra shows that the exponent is equal to

$$(\boldsymbol{x} - \boldsymbol{m}_{x|z})^T C_{x|z}^{-1} (\boldsymbol{x} - \boldsymbol{m}_{x|z})$$

with

$$\boldsymbol{m}_{x|z} = E\{\boldsymbol{x}|\boldsymbol{z}\} = m_x + C_{xz} C_z^{-1} (\boldsymbol{z} - \boldsymbol{m}_z)$$

$$C_{x|z} = C_x - C_{xz} C_z^{-1} C_{zx}.$$

Using (2.71),the conditional pdf includes the reciprocal of the square root of the term

$$[2\pi]^n \frac{\det(C_y)}{\det(C_z)} = [2\pi]^n \frac{\det(C_z)\det(C_x - C_{xz}C_z^{-1}C_{zx})}{\det(C_z)} = [2\pi]^n \det(C_{x|z}).$$

■

Note that if $\boldsymbol{z}$ is a random vector, then $E\{\boldsymbol{x}|\boldsymbol{z}\}$ is also random and is characterized by the following theorem.

Theorem 2.7 *If $\boldsymbol{x}$ and $\boldsymbol{z}$ are jointly Gaussian, then $E\{\boldsymbol{x}|\boldsymbol{z}\}$ is a Gaussian affine transformation of $\boldsymbol{z}$.*

Proof Follows from the expression (2.73). ■

In addition to its excellent mathematical properties, the normal distribution provides a good approximation for the random behavior of many physical phenomena. This is in part due to a property of the sum of independent random variables. Recall that for n independent random variables

$$X_i, i = 1, \ldots, n, E\{X_i\} = m_i, \mathrm{Var}\{X_i\} = \sigma_i^2,$$

the pdf of the sum

$$Z = \sum_{i=1}^{n} X_i$$

is the convolution of their pdfs with mean

$$E\{Z\} = m_z = \sum_{i=1}^{n} m_i$$

and variance

$$Var\{Z\} = \sigma_z^2 = \sum_{i=1}^{n} \sigma_i^2.$$

A property of convolutions is that the convolution of a large number of positive functions is approximately in the bell shape of the normal distribution. This leads to the following theorem.

Theorem 2.8 Central Limit Theorem: Z is asymptotically Gaussian.

$$\underset{n \to \infty}{Lim}\, P_Z(z) = \mathcal{N}\left(m_z, \sigma_z^2\right). \tag{2.75}$$

This property is known as convergence in distribution and is discussed in the appendix on stochastic convergence. For a finite number of terms, the pdf is close to normal after adding a relatively small number of independent variables.

2.8 Transformation of Random Variables

2.8.1 *Linear Transformation*

Consider linear transformation of a random variable X with mean and variance $\left(m_x, \sigma_x^2\right)$ to a random variable

$$Y = aX + b. \tag{2.76}$$

Taking the expectation shows that the mean of Y is

$$m_y = E\{ax + b\} = am_x + b.$$

The variance of Y is

$$\sigma_y^2 = E\left\{\left(y - m_y\right)^2\right\} = a^2 E\left\{(x - m_x)^2\right\} = a^2\sigma_x^2. \tag{2.77}$$

Next, we consider a random vector $\boldsymbol{x}$ of mean $\boldsymbol{m}_x$ and covariance matrix C_{xx}. We derive expression for the mean and covariance of a random vector

$$\boldsymbol{y} = A\boldsymbol{x} + \boldsymbol{b} \tag{2.78}$$

in terms of the mean and covariance of $\boldsymbol{x}$. The mean of $\boldsymbol{y}$ is given by

$$E\{\boldsymbol{y}\} = E\{A\boldsymbol{x} + \boldsymbol{b}\} = A\boldsymbol{m}_x + \boldsymbol{b}. \tag{2.79}$$

The covariance of $\boldsymbol{y}$ is

$$C_{yy} = E\left\{\left(\boldsymbol{y} - \boldsymbol{m}_y\right)\left(\boldsymbol{y} - \boldsymbol{m}_y\right)^T\right\} = E\left\{A(\boldsymbol{x} - \boldsymbol{m}_x)(\boldsymbol{x} - \boldsymbol{m}_x)^T A^T\right\} = AC_{xx}A^T. \tag{2.80}$$

2.8.2 Diagonalizing Transformation

Since C_x is symmetric positive definite, it has the form $C_x = V\Lambda V^T$ where V is a matrix of eigenvectors and $\Lambda = \text{diag}\{\lambda_1, \ldots, \lambda_n\}$ is a diagonal matrix of real eigenvalues λ_i for C_x symmetric. The linear transformation with $A = V^T = V^{-1}$ gives

$$\boldsymbol{y} = V^T(\boldsymbol{x} - \boldsymbol{m}_x) \tag{2.81}$$

with the diagonal covariance

$$C_y = AC_xA^T = V^T\left(V\Lambda V^T\right)V = \Lambda. \tag{2.82}$$

Example 2.13 A sensor measures a voltage in a circuit with additive Gaussian noise with pdf $\mathcal{N}\left(0, \sigma^2\right)$. Find the mean and variance of the measured voltage V_m for a voltage V if the sensor has a bias b and a sensor gain K.

Solution

The output of the sensor is

$$V_m = KV + b.$$

The mean of V_m is

$$m_s = a \times 0 + b = b.$$

The variance of V_m is

$$\sigma_s^2 = a^2\sigma^2.$$

■

2.8.3 Nonlinear Transformation

The equations for the mean and covariance that result from a linear transformation are useful but they do not provide the pdf of the transformed variable, and are only useful if the transformation is linear. This section shows how to find the pdf that results from transforming a variable with known pdf. The results include linear transformation as a special case. We begin by examining the special case of one–one transformation and then provide results for the general nonlinear transformation.

Consider the mapping of a random variable X as in Fig. 2.12

$$Y = g(X). \tag{2.83}$$

Subject to constraints on $g(.)$, we characterize the random variable Y. For any set A, the probability of Y satisfies

$$P(Y \in A) = P(g(X) \in A). \tag{2.84}$$

Using the inverse mapping

$$Y = g^{-1}(X), \tag{2.85}$$

we rewrite the probability as

$$P(Y \in A) = P\big(X \in g^{-1}(A)\big). \tag{2.86}$$

For a discrete random variable X, the probability of $Y = y$ is obtained by adding the probabilities of all values of X that satisfy $g(x) \le y$. This gives the CDF of Y

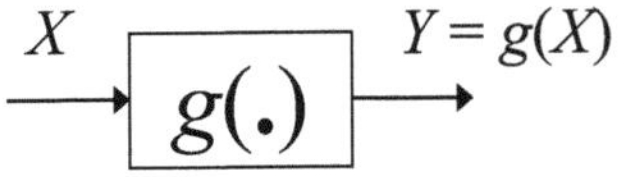

Fig. 2.12 Nonlinear mapping

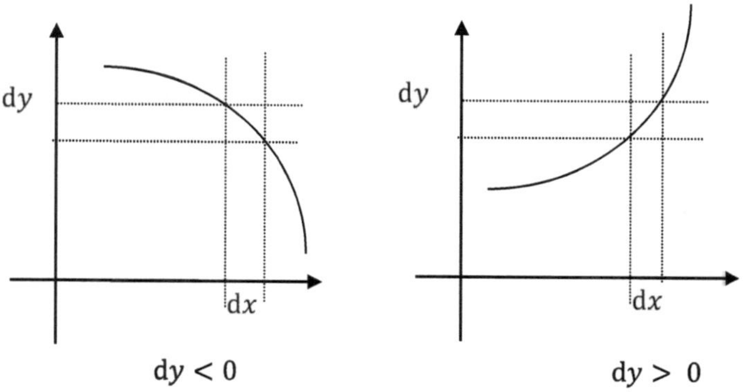

Fig. 2.13 Change dy with dx for monotone functions

$$P_Y(y) = P(Y = y) = \sum_{x \in g^{-1}(y)} p_X(x). \tag{2.87}$$

For a continuous random variable X, the CDF of Y is obtained by integrating over all values of X that satisfy $y \leq g(x)$. This gives the CDF of Y

$$P_Y(y) = \int_{x, g(x) \leq y} p_X(x) \mathrm{d}x. \tag{2.88}$$

If $g(x)$ is monotone as in Fig. 2.13, then the change in Y with X depends on whether it is monotonically increasing or monotonically decreasing. For a monotonically increasing function $g(x)$, Y increases with X, and the CDF of Y is obtained from the CDf of X :

$$P_Y(y) = P_X\left(x = g^{-1}(y)\right). \tag{2.89}$$

If $g(x)$ is monotonically decreasing, then the probability of $Y \leq y$ is equal to the probability of $X \geq g^{-1}(y)$. The CDF of Y is therefore

$$P_Y(y) = 1 - P_X\left(x = g^{-1}(y)\right). \tag{2.90}$$

The CDF expressions allow us to derive the following theorem.

Theorem 2.9 *If a random variable X with pdf $p_X(x)$ is mapped by a monotone function $g(x)$ to a random variable* **Y**, *then pdf of* **Y** *is.*

$$p_Y(y) = \begin{cases} \sum_{i=1}^{l} p_X\left(g^{-1}(Y)\right)\left|\frac{\partial X}{\partial Y}\right|, & X = g^{-1}(Y) \\ 0, & \text{elsewhere} \end{cases}. \tag{2.91}$$

Proof We differentiate (2.89) and (2.90) to obtain.

$$p_Y(y) = \begin{cases} p_X\left(g^{-1}(y)\right)\frac{\partial x}{\partial y}, & g(x)\text{increasing} \\ -p_X\left(g^{-1}(y)\right)\frac{\partial x}{\partial y}, & g(x)\text{decreasing} \end{cases}.$$

The pdf is zero for Y values that do not correspond to a feasible value of X. The results obtained are equivalent to

$$p_Y(y) = \begin{cases} p_X\left(g^{-1}(y)\right)\left|\frac{\partial x}{\partial y}\right|, & x = g^{-1}(y) \\ 0, & \text{elsewhere} \end{cases}.$$

If l solutions x_i exists for the same value y, we have

$$p_Y(y) = \begin{cases} \sum\limits_{i=1}^{l} p_X\left(g^{-1}(Y)\right)\left|\frac{\partial X_i}{\partial Y}\right|, & X_i = g^{-1}(Y) \\ 0, & \text{elsewhere} \end{cases}.$$

(2.91) Generalization to the vector case requires some basic calculus that is omitted here. ■

Based on Theorem 2.9, with generalization to the vector case, we have the following transformation procedure:

Procedure

Solve for $\boldsymbol{x}$.

(a) If no solution exists, then $p_Y(\boldsymbol{y}) = 0$
(b) If one or more solutions $\boldsymbol{x}_i(\boldsymbol{y}), i = 1, \ldots, l$, exist

$$\begin{aligned} p_Y &= \sum_{i=1}^{l} p_X(\boldsymbol{x}_i)\left|\det\left(\partial \boldsymbol{x}/\partial \boldsymbol{y}|_{x=x_i}\right)\right| \\ &= \sum_{i=1}^{l} p_X(\boldsymbol{x}_i)/\left|\det\left(\partial \boldsymbol{y}/\partial \boldsymbol{x}|_{x=x_i}\right)\right|. \end{aligned} \tag{2.92}$$

Example 2.14 Drag or air resistance $\boldsymbol{f}$ is related to velocity $\boldsymbol{v}$ by the formula

$$f(v) = kv^2,$$

where $k > 0$ is a constant. Find the pdf for the random variable F for random velocity $V \sim \mathcal{N}\left(0, \sigma_v^2\right)$.

Fig. 2.14 Square law plot

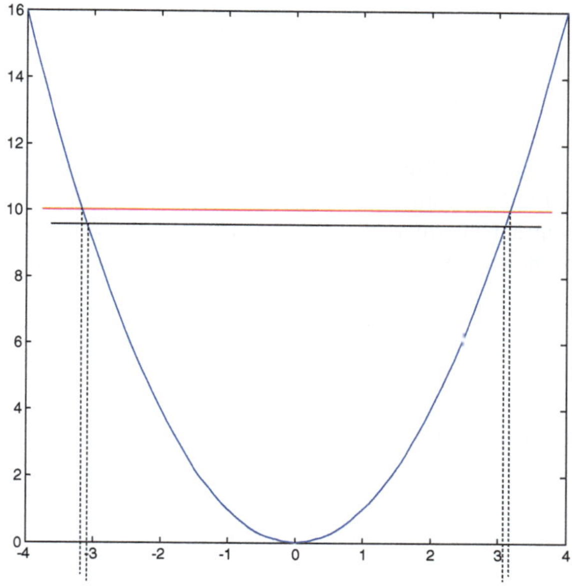

Solution

This is a nonlinear transformation as shown in Fig. 2.14 and solving for V gives two roots

$$f = kv^2, v = \pm\left(\frac{f}{k}\right)^{1/2}.$$

We find the derivative

$$\left|\frac{\partial v}{\partial f}\right| = \frac{1}{2\sqrt{kf}}.$$

From Fig. 2.14, we see that we need to add two terms to obtain the pdf of F. From the symmetry, the two terms are equal, and the pdf is

$$\begin{aligned} p_F(f) &= 2 \times \frac{1}{2\sqrt{2\pi k|f|\sigma_x^2}} \exp\left\{-\frac{\left(\sqrt{f/k}\right)^2}{2\sigma_x^2}\right\} \\ &= \frac{1}{\sqrt{2\pi}}\left(\frac{f}{2k\sigma_x^2}\right)^{-1/2} \exp\left\{-\frac{f}{2k\sigma_x^2}\right\}. \end{aligned}$$

Negative values of F have zero probability. The pdf of F is

$$p_F(f) = \begin{cases} \frac{1}{\sqrt{2\pi}}\left(\frac{f}{2k\sigma_x^2}\right)^{-1/2} \exp\left\{-\frac{f}{2k\sigma_x^2}\right\}, & f \geq 0 \\ 0, & f < 0 \end{cases}.$$

■

The example shows that normality is lost with nonlinear transformation of a normal random variable. The distribution of the force is a one degree of freedom central Chi-square distribution. The general form for n degrees of freedom is the pdf of the sum of n i.i.d. random variables governed by a standard normal distribution

$$Y = \sum_{i=1}^{n} X_i^2, X_i \sim \mathrm{N}(0, 1)$$

and is given by

$$p_Y(y) = \begin{cases} \frac{1}{2^{n/2}\Gamma(n/2)} y^{(n/2)-1} \mathrm{e}^{-y/2}, & y \geq 0 \\ 0, & y < 0 \end{cases},$$

where $\Gamma(.)$ is the Gamma function

$$\Gamma(u) = \int_0^\infty t^{u-1}\mathrm{e}^{-t}\mathrm{d}t = (u-1)!, \quad u \text{ integer}$$

$$\Gamma(1/2) = \sqrt{\pi}.$$

If the i.i.d. variables are not zero-mean, $X_i \sim \mathcal{N}(m_i, 1)$, we have the n degrees of freedom noncentral Chi-square distribution:

$$p_Y(y) = \begin{cases} \frac{1}{2}\left(\frac{y}{\lambda}\right)^{\frac{(n-2)}{4}} \mathrm{e}^{-\frac{y+\lambda}{2}} I_{n/2-1}\left(\sqrt{\lambda y}\right), & y \geq 0 \\ 0, & y < 0 \end{cases}$$

$$\lambda = \sum_{i=1}^{n} m_i^2, \; I_r(u) = \frac{(u/2)^r}{\pi^{1/2}\Gamma(r+1/2)} \int_0^\pi \mathrm{e}^{u\cos(t)} \sin^{2r}(t)\mathrm{d}t.$$

Example 2.15 Two particles are moving independently at random from the origin with one moving along the horizontal axis and the other moving along the vertical axis. Find the pdf of the distance between the two particles if their motions are governed by normal distributions with variance σ^2 and means m_1 and m_2, respectively.

Solution

If the horizontal motion is X_1 and the vertical motion is X_2, then the distance Y between the two particles is

$$Y = \sqrt{X_1^2 + X_2^2}, \, ind.X_i \sim \mathcal{N}\left(m_i, \sigma^2\right), i = 1, 2.$$

It can be shown that Y is governed by the Rician distribution

$$p_Y(y) = \begin{cases} \frac{y}{\sigma^2} \exp\left(-\frac{y^2+\alpha^2}{2\sigma^2}\right) I_0\left(\frac{\alpha y}{\sigma^2}\right), & y \geq 0 \\ 0, & y < 0 \end{cases}$$

$$\alpha^2 = m_1^2 + m_2^2$$

$$I_0(u) = \frac{1}{\pi} \int_0^{\pi} \mathrm{e}^{u \cos(t)} \mathrm{d}t.$$

■

Example 2.16 Linear Transformation of Multivariate Normal Vector Show that normality is preserved under linear transformation of a multivariate normal random vector $\boldsymbol{x} \sim \mathcal{N}(\boldsymbol{m}_x, C_x)$:

$$\boldsymbol{y} = A\boldsymbol{x} + \boldsymbol{b} \Leftrightarrow \boldsymbol{x} = A^{-1}(\boldsymbol{y} - \boldsymbol{b}). \tag{2.93}$$

Solution

The pdf of $\boldsymbol{x}$ is

$$p_X(\boldsymbol{x}) = \frac{1}{[2\pi]^{n/2}\sqrt{\det(C_x)}} \exp\left\{-\frac{1}{2}(\boldsymbol{x} - \boldsymbol{m}_x)^T C_x^{-1} (\boldsymbol{x} - \boldsymbol{m}_x)\right\}.$$

From (2.78) to (2.80), we see that for the linear transformation results in

$$\boldsymbol{m}_y = A\boldsymbol{m}_x + \boldsymbol{b}, \quad C_y = AC_xA^T.$$

We must show that $\boldsymbol{y} \sim \mathcal{N}\left(\boldsymbol{m}_y, C_y\right)$; in other words, we must show that the pdf of $\boldsymbol{y}$ is

$$p_Y(\boldsymbol{y}) = \frac{1}{[2\pi]^{n/2}\sqrt{\det\left(C_y\right)}} \exp\left\{-\frac{1}{2}\left(\boldsymbol{y} - \boldsymbol{m}_y\right)^T C_y^{-1} \left(\boldsymbol{y} - \boldsymbol{m}_y\right)\right\}.$$

By differentiating the expression

$$x(y) = A^{-1}(y - b),$$

we obtain the Jacobian determinant

$$\left|\det\left[\frac{\partial x}{\partial y}\right]\right| = \left|\det\left[A^{-1}\right]\right|.$$

We recall the determinant properties

$$\det\left[A^{-1}\right] = \frac{1}{\det[A]}$$

$$\det[A] = \det\left[A^T\right]$$

$$\det[ABC] = \det[A]\det[B]\det[C].$$

The transformation results in multiplying the term $[2\pi]^{n/2}\sqrt{\det(C_x)}$ by $|\det[A]| = \sqrt{\det[A]\det\left[A^T\right]}$ to obtain

$$[2\pi]^{n/2}\sqrt{\det[A]\det(C_x)\det\left[A^T\right]} = [2\pi]^{n/2}\sqrt{\det\left(C_y\right)}.$$

We next substitute for x in terms of y in the exponent. The exponent includes the term

$$x - m_x = A^{-1}(y - b - Am_x) = A^{-1}\left(y - m_y\right)$$

and can be written as

$$(x - m_x)^T C_x^{-1}(x - m_x) = \left(y - m_y\right)^T A^{-T} C_x^{-1} A^{-1}\left(y - m_y\right).$$

We observer that

$$A^{-T} C_x^{-1} A^{-1} = \left(AC_x A^T\right)^{-1} = C_y^{-1}.$$

The exponent can therefore be expressed as

$$(x - m_x)^T C_x^{-1}(x - m_x) = \left(y - m_y\right)^T \underbrace{\left(AC_x A^T\right)^{-1}}_{C_y^{-1}}\left(y - m_y\right).$$

The pdf of y is

$$
\begin{aligned}
p_Y(\boldsymbol{y}) &= p_X\big[\boldsymbol{x}(\boldsymbol{y})\big]\left|\det\left(\frac{\partial \boldsymbol{x}}{\partial \boldsymbol{y}}\right)\right| \\
&= \frac{1}{[2\pi]^{n/2}\sqrt{\det(C_y)}}\exp\left\{-\frac{1}{2}\big(\boldsymbol{y}-\boldsymbol{m}_y\big)^T C_y^{-1}\big(\boldsymbol{y}-\boldsymbol{m}_y\big)\right\}.
\end{aligned}
\tag{2.94}
$$

In the special case where the transformation matrix $A = S^T$, $S =$ the matrix of eigenvectors of C_x, and $\boldsymbol{b} = S^T\boldsymbol{m}_x$ then by (2.81)–(2.82), $\boldsymbol{y} = S^T(\boldsymbol{x} - \boldsymbol{m}_x)$, $C_y = \Lambda$, and $\boldsymbol{y} \sim \mathcal{N}(0, I_n)$. ■

Example 2.17 Coordinate Transformation.

For zero-mean Gaussian cartesian coordinates (x, y) with variance σ^2, find the joint and marginal pdfs for the corresponding polar coordinates (r, θ).

Solution

The polar coordinates are

$$r = \sqrt{x^2 + y^2}, \theta = \tan^{-1}(y/x)$$

$$x = r\cos(\theta),\ y = r\sin(\theta).$$

The joint pdf for the Cartesian coordinates x and y is

$$p_{XY}(x, y) = \frac{1}{2\pi\sigma^2}\exp\left\{-\frac{x^2 + y^2}{2\sigma^2}\right\}.$$

Transformation to the new coordinates changes the infinitesimal element as shown in Fig. 2.15 and gives the pdf

$$p_Z(z) = p_X[x(z)]\left|\det\left[\frac{\partial x}{\partial z}\right]\right|$$

$$
p_{R\Theta}(r, \theta) = \begin{cases} \frac{r}{2\pi\sigma^2}\exp\left(-\frac{r^2}{2\sigma^2}\right), & r \geq 0, \theta \in [0, 2\pi] \\ 0, & \text{elsewhere} \end{cases}.
\tag{2.95}
$$

Integrate w.r.t. r and change variable of integration to get

$$
\begin{aligned}
\int_0^\infty p_{R\Theta}(r, \theta)dr &= \int_0^\infty \frac{r}{2\pi\sigma^2}\exp\left(-\frac{r^2}{2\sigma^2}\right)dr \\
&= \frac{1}{2\pi}\int_0^\infty \left(\frac{r}{\sigma}\right)\exp\left\{-\frac{1}{2}\left(\frac{r}{\sigma}\right)^2\right\}d\left(\frac{r}{\sigma}\right) = \frac{1}{2\pi}.
\end{aligned}
$$

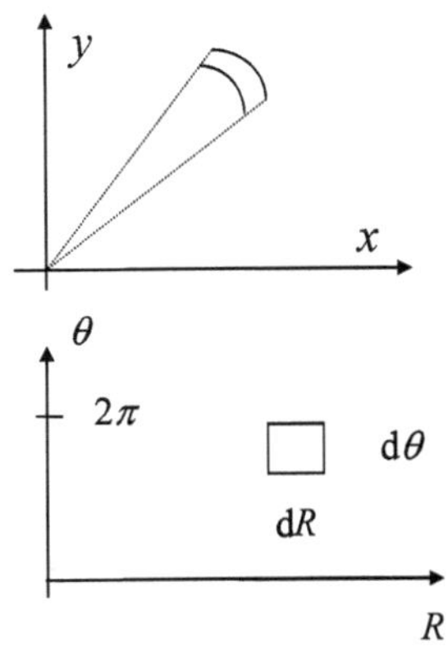

Fig. 2.15 Joint pdf for polar coordinates

The marginal pdf for θ is uniform $U[0, 2\pi]$

$$p_{\Theta}(\theta) = \begin{cases} \frac{1}{2\pi}, & \theta \in [0, 2\pi] \\ 0, & \text{elsewhere} \end{cases}. \tag{2.96}$$

Integrate w.r.t. θ to obtain the marginal distribution for r

$$\int_0^{2\pi} p_{R\Theta}(r, \theta) d\theta = \int_0^{2\pi} \frac{r}{2\pi\sigma^2} \exp\left(-\frac{r^2}{2\sigma^2}\right) d\theta$$

This gives the Rayleigh distribution of Fig. 2.16.

$$p_R(r) = \begin{cases} \frac{r}{\sigma^2} \exp\left(-\frac{r^2}{2\sigma^2}\right), & r > 0 \\ 0, & \text{elsewhere} \end{cases}$$

■

2.9 Pseudorandom Number Generators

Pseudorandom number generators are not actually random. They only appear random and sufficiently resemble random numbers to be useful in simulations or experiments that require random inputs. This makes them essential in many applications in a variety of fields including physics, engineering, biology, etc. They are essential for Monte Carlo simulation that investigate system behavior over a range of parameter values or initial conditions by simulation over a sample of randomly selected values.

Pseudorandom numbers are generated using a deterministic algorithm that generates a sequence of numbers that appear random and are repeated. Hence, a long sequence is desirable to minimize repetition. In addition, the sequence generated depends on the starting point in the sequence or the seed. MATLAB random number

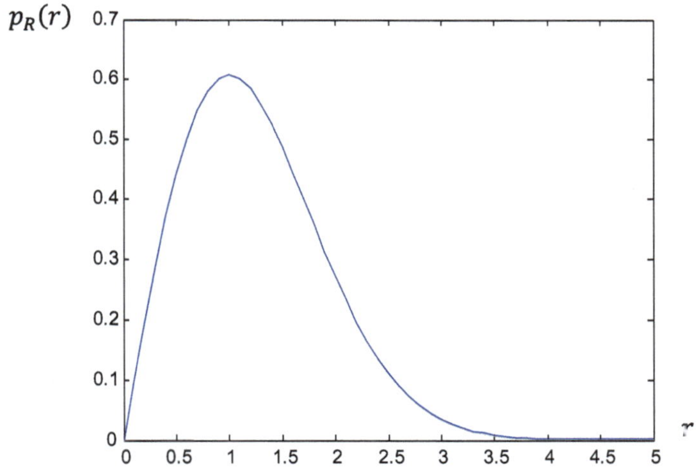

Fig. 2.16 The Rayleigh distribution

generators allow the user to select the generator algorithm and change the seed with the command.

> > **rng(seed, 'Name') % seed = nonnegative integer seed**

where **Name** is the name of the generator and **seed** is a nonnegative integer. The current settings are obtained with the command.

> > **rng.**

To reinitialize the random sequence every time that the command **rng** is called, we use.

> > **rng('shuffle').**

The following theorem shows that many distributions can be generated using a uniform number generator.

Theorem 2.10 *If the random variable* $\boldsymbol{U} \sim \mathcal{U}(\mathbf{0}, \mathbf{1})$ *then, for any continuous distribution function* $\boldsymbol{F}$, *the random variable* $\boldsymbol{X} = \boldsymbol{F}^{-1}(\boldsymbol{U}) \sim \boldsymbol{F}$.

Proof Assume that F is monotone increasing to simplify the proof (one–one). The CDF, shown in Fig. 2.17, is the integral.

$$P_U(u) = \int_0^u 1du = u, \quad U \sim \mathcal{U}(0, 1)$$

$$\begin{aligned} P(X \leq x) &= P\left[P^{-1}(U) \leq x\right] \\ &= P[U \leq P(x)] \end{aligned}$$

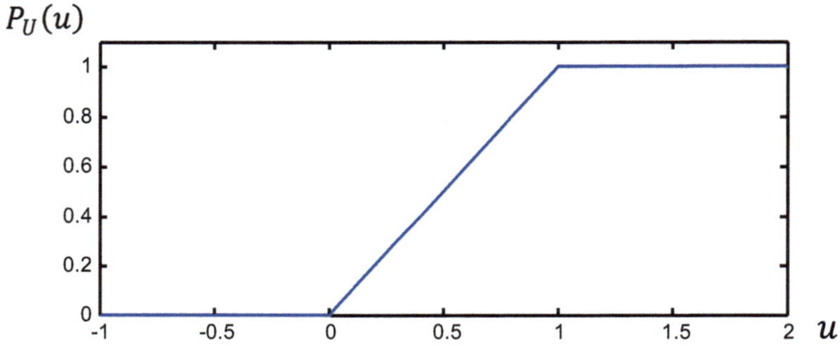

Fig. 2.17 Uniform cumulative distribution function

$$= \int_0^{P(x)} 1 \mathrm{d}u$$
$$= P(x).$$

■

MATLAB has many ways of generating pseudorandom numbers but some of the commands require the Statistics and Machine Learning Toolbox. The following two commands do not. The first generates uniform numbers X with mean $E\{X\} = 0.5$ and varaince $\sigma_x^2 = 1/3$.

> > **rand** % Uniform distribution over [0,1].

The second generates standard normal X with mean $E\{X\} = 0$ and varaince $\sigma_x^2 = 1$.

> > **randn** % Standard normal N(0,1).

By shifting and scaling, we can change the mean and variance of the distribution.

> > **y = sigmay*randn + ybar.**

For $E\{X\} = 0, \sigma_X = 1$, we can generate Y with mean m_Y and variance σ_y^2. To show this, we find the mean and variance of $Y = m_Y + \sigma_Y X$

$$E\{Y\} = m_Y + \sigma_Y E\{X\} = m_Y$$

$$E\left\{(Y - m_Y)^2\right\} = E\left\{\sigma_Y^2 X^2\right\} = \sigma_{Y.}^2$$

Thus, we can use $X \sim \mathcal{N}(0, 1)$ to obtain $Y \sim \mathcal{N}\left(m_Y, \sigma_Y^2\right)$.

With the MATLAB Statistics and Machine Learning Toolbox, we have many more options. Commands in the form.

> > **xxxrnd(parameters, [m,n]),**

where parameters is a list of parameters, m is the number of rows and n is the number of columns generated. The label xxx is the name of distribution that can be **chi2rnd (Chi-square), binornd (binomial), poissrnd (Poisson), nbin (negative binomial), logn (log normal), exprnd(exponential), etc.**

MATLAB also allows us to generate multivariable normal data with mean $\boldsymbol{m}_x$ and covariance matrix C_x with the command **mvnrnd(mx,Cx,N)**.

MATLAB also allows us to create a distribution with specified parameters with the command **makedist**. The command requires selecting the name of the distribution from the list of options that MATLAB provides, and specifying its parameters. If the parameter values are not specified, MATLAB uses a standard set of parameters for the distribution.

Example 2.18 Generate the following using two different MATLAB commands:

(a) A 2×1 Gaussian vector whose entries have unity mean and standard deviation 2.
(b) An exponential random number with mean $\lambda = 2$.

Solution

(A) Use the MATLAB commands.

```
>> mean=1;sigma=2;m=2;n=1;
>> Z = normrnd(mean, sigma, m,n)
     Z =
        0.1349
        -2.3312
Alternatively, we use the command
>> makedist('Normal','mu',1,'sigma',2)
     ans =
      NormalDistribution
      Normal distribution
         mu = 1
        sigma = 2
```

(b) Use the MATLAB commands

```
>> lambda=2;
>> Z = exprnd(lambda)
     Z =
        2.5567
Alternatively, we use the command
>> makedist('Exponential','mu',2)
   ans =
    ExponentialDistribution
    Exponential distribution
     mu = 2
```

■

It is also possible to generate a truncated distribution starting with a known distribution created by MATLAB with the command **truncate**. For example, we could specify a normal distribution with known mean and variance, then truncate the distribution to limit its support to the interval $[\boldsymbol{a}, \boldsymbol{b}]$.

Example 2.19 Generate a normal distribution with mean 1 and standard deviation 2, then truncate it to create a truncated normal distribution with support [0, 5]. Use the truncated distribution to generate a 4 by 1 random vector.

Solution

```
> > pd = makedist('Normal','mu',1,'sigma',2);
> > qd = truncate(pd,0,5).
qd =
 NormalDistribution.
 Normal distribution.
     mu = 1.
   sigma = 2.
Truncated to the interval [0, 5].
> > x = random(qd,4,1).

x =

   0.9790
   1.2230
   1.7982
   4.8471
```

■

The command **random** uses the name of the distribution as a parameter.

> > random('Name', parameters, [m,n]).

where parameters is a list of parameters, m is the number of rows and n is the number of columns generated. The **Name** can be **Normal, Binomial, Poisson, Chai square, Exponential, Uniform, Rician, etc.**

Example 2.20
Generate the following using the MATLAB command **random.**

(a) A 2 × 3 Gaussian matrix whose entries have unity mean and variance 4.
(b) A 1 × 3 vector whose entries are random number with parameter $\lambda = 2$.

Solution

(a) Use the MATLAB commands

>> mean=1; variance=4; m=2; n=3;
>> rn = random('Normal', mean, variance, m, n)
rn =

$$\begin{matrix} 0.6174 & 2.1776 & 3.8573 \\ -2.3294 & -4.3447 & 7.4942 \end{matrix}$$

(b) Use the MATLAB commands

>> min=1.1;max=2;
>> random('Uniform',min,max,1,3) % 1 by 3 array with uniform pdf
ans =

$$\begin{matrix} 1.9551 & 1.3080 & 1.6462 \end{matrix}$$

■

Example 2.21 Generate 10 vectors from a multivariate normal distribution with mean $\boldsymbol{m_x} = \begin{bmatrix} 1 \ 2 \ 1 \ 2 \ 1 \end{bmatrix}^{\mathrm{T}}$ and covariance matrix.

$$C_x = \begin{bmatrix} 1 & 0.1 & 0.2 & 0.3 & 0.2 \\ 0.1 & 2 & 0.1 & 0.1 & 0.1 \\ 0.2 & 0.1 & 2 & 0.1 & 0.1 \\ 0.3 & 0.1 & 0.1 & 1 & 0.1 \\ 0.2 & 0.1 & 0.1 & 0.1 & 1 \end{bmatrix}$$

using the MATLAB command **mvrnd**

Solution

```
N=10;
mx=[1,2,1,2,1];
Cx=[1,.1,.2,.3,.2;.1,2,.1,.1,.1;.2,.1,2,.1,.1;.3,.1,.1,1,.1;
.2,.1,.1,.1,1];
X = mvnrnd(mx,Cx,N)
```

X =

$$\begin{matrix} 0.1363 & 0.3773 & -0.0956 & 3.0221 & -0.3267 \\ 1.0774 & 2.0537 & 2.0638 & 2.3224 & 1.1612 \\ -0.2141 & 2.6580 & 0.5193 & 1.8464 & 1.4935 \\ -0.1135 & 3.4413 & 2.0827 & 3.2561 & 3.4609 \\ 0.9932 & 4.1777 & 0.0163 & 1.2882 & 0.3734 \\ 2.5326 & 2.2745 & -0.6502 & 3.0902 & 1.4638 \\ 0.2303 & -0.1811 & -1.2282 & 2.4525 & 0.6548 \\ 1.3714 & 0.9900 & 1.7151 & 1.8555 & -0.8450 \\ 0.7744 & 0.4799 & 0.6466 & 2.0803 & 0.4670 \\ 2.1174 & 5.4275 & 1.0825 & 1.3364 & -0.4480 \end{matrix}$$

■

For more on random number generators, the reader is referred to:

(1) The Mathworks Random Number page.
www.mathworks.com/discovery/random-number.html?s_tid=srchtitle

(2) Agner Fog's library of programs for generating uniform and nonuniform distributions.
http://www.agner.org/random/

2.9.1 True Random Number Generators

There are physical systems that can generate truly random number. We provide two examples:

(1) A true random number generator built from carbon nanotubes: This is a static random-access memory (SRAM) cell that uses fluctuations in thermal noise to generate random bits.
https://spectrum.ieee.org/nanoclast/computing/hardware/a-true-random-number-generator-built-from-carbon-nanotubes-promises-better-security-for-flexible-electronics
(2) A memristor (a nonlinear passive component relating electric charge and magnetic flux linkage) true random number generator.
http://spectrum.ieee.org/semiconductors/memory/a-memristor-true-random number-generator/?utm_source=techalert&utm_medium=email&utm_campaign=071912

2.10 The Method of Moments

In our discussion of moments, we do not discuss how estimates of the moments can be obtained from data. This topic is discussed extensively in later chapters. In this section, we discuss parameter estimation using a classical approach that is based on the topics reviewed in this chapter. One of the oldest methods to estimate parameters of a distribution is to calculate estimates of the moments of the distribution then equate the estimates of expression for the moments in terms of the parameters. The resulting set of equations is then solved for the parameters. This is known as the method of moments. The procedure is as follows:

(1) Collect the data $\{x_i, i = 1, 2, \ldots, N\}$.
(2) Obtain estimates of m moments of a distribution using the expressions

$$\widehat{E\{X^k\}} = \frac{1}{N}\sum_{i=1}^{N} x_i^k, k = 1, 2, \ldots, m$$

where m is the number of unknown parameters $\boldsymbol{\theta} = \{\{\theta_i, i = 1, 2, \ldots, m\}\}$.

(3) Write m expressions relating the moments to the unknown parameters and replace the moments with their estimates to obtain m equations:

$$E\{X^k\} = f_k(\boldsymbol{\theta}) \approx \widehat{E\{X^k\}}, k = 1, 2, \ldots, m.$$

(4) Solve the m equations for the unknown parameters.

The method of moments can provide acceptable estimates in some cases but in others the estimates can be unacceptable. In cases where the method provides a good approximation, estimates obtained using the method of moments can be used as an initial guess for numerical algorithms to obtain other estimates.

Example 2.22 Use the method of moments to obtain expressions for the parameters of a Gamma distribution $\boldsymbol{Gamma}(\boldsymbol{\alpha}, \boldsymbol{\beta})$ using the data $\{x_i, i = 1, 2, \ldots, N\}$.

Solution

For two parameter we calculate the sample mean $\overline{X}$ and the sample mean-square $\overline{X^2}$

$$\widehat{E\{X^k\}} = \frac{1}{N}\sum_{i=1}^{N} x_i^k, k = 1, 2.$$

We obtain the equations

$$E\{X\} = \alpha\beta \approx \overline{X}$$

$$E\{X^2\} = \sigma^2 + E\{X\}^2 = \alpha\beta^2 + (\alpha\beta)^2 \approx \overline{X^2}.$$

Substituting the first equation in the second gives

$$\beta\overline{X} + \overline{X}^2 \approx \overline{X^2}.$$

We solve for the parameters

$$\hat{\beta} = \frac{\overline{X^2} - \overline{X}^2}{\overline{X}}$$

$$\hat{\alpha} = \frac{\overline{X}}{\hat{\beta}} = \frac{\overline{X}^2}{\overline{X^2} - \overline{X}^2}.$$

■

Problems

2.1. The time to failure of a Lithium-ion car battery T is governed by the exponential distribution

$$P_T(t) = 1 - \exp(-\lambda t), t \geq 0$$

where λ is a positive constant of dimension $yrs.^{-1}$

- Show that the battery fails on average after λ^{-1} years.
- If $\lambda = 0.1\ yrs.^{-1}$, find the probability that the battery will last more than 12 years.

2.2. A component of an electronic instrument has a time to failure governed by the pdf

$$p(t) = \begin{cases} 0.95, & 0 < t \leq 1 \\ 0.95\mathrm{e}^{-\alpha(t-1)}, & 1 < t \end{cases}$$

(a) Find the value of α

(b) Find the mean time to failure (MTTF) for the component.

2.3. Prove the properties of the expectation of Eqs. (2.12)–(2.17).

2.4. Entropy is a measure of the uncertainty of a random variable and is defined as the average amount of information associated with its occurrence

$$H(u) = E\{I(u)\} = E\left\{\log \frac{1}{P(u)}\right\}$$

where $I(u)$ denotes the information associated with the event u. The definition can be extended to two random variables using the joint density and the mutual information

$$H(u, y) = E\{I(u, y)\} = E\left\{\log \frac{1}{P(u, y)}\right\}$$

We can also define a conditional entropy for u given that y has occurred as

$$H(u|y) = E\{I(u|y)\} = E\left\{\log \frac{1}{P(u|y)}\right\}$$

Prove that the joint entropy is given by the sum

$$H(u, y) = H(u|y) + H(y)$$

2.5. Find the mean and variance of the output of an amplifier at time t_0 with gain k and saturation level K with Gaussian noise input $u(t_0) \sim \mathcal{N}(0, \sigma^2)$.

2.6. Write an expression for the expected value of the integral

$$g(x) = \int_{-a(x)}^{a(x)} f(u)du$$

if the pdf of the random variable x is $p_X(x)$ and $a(.)$ and $f(.)$ are continuous functions of their arguments. Evaluate the integral in the case of uniformly distributed $x \sim U[0, \pi]$, $f(u) = u^2$, and $a(x) = \sin(x)$

2.7. Consider the random process $z = \boldsymbol{a}^T \boldsymbol{x}$ where $\boldsymbol{a}^T = \begin{bmatrix} 2 & 3 \end{bmatrix}$ and $\boldsymbol{x} = \begin{bmatrix} x_1 & x_2 \end{bmatrix}^T$ is a zero-mean random vector with uncorrelated entries x_1 and x_2. The autocorrelation function of x_1 is

$$R_1(\tau) = 2e^{-2|\tau|}$$

and x_2 is zero-mean unity white noise.

(a) Find the autocorrelation function of the vector z.
(b) Find the power spectral density function of z.

2.8. Use the transformation approach to find the pdf of $Z = aX + bY$ where X and Y are mutually independent random variables and $a, b \in \mathcal{R}$.

a. Let $Z_1 = aX + bY, Z_2 = X$
b. Find the joint pdf of Z_1 and Z_2
c. Integrate w.r.t. Z_2 to find the pdf of Z_1

Show that setting for $a = b = 1$ yields that the pdf of the sum of two independent random variables as the convolution of the pdfs of the two variables.

2.9. Obtain the mean and variance of the variable

$$y = \boldsymbol{a}^T \boldsymbol{x} = \sum_{i=1}^{n} a_i x_i$$

$$E\{x_i\} = m_i, Var\{x_i\} = \sigma_i^2, i = 1, 2, \ldots, N$$

Use the result to prove Eq. (2.28)

2.10. Let $\boldsymbol{X} = \{X_1, X_2, X_3\}$ be a set of i.i.d. random variables with joint pdf

$$p_X(\boldsymbol{x}) = \prod_{i=1}^{3} p_{X_i}(x_i)$$

Determine the joint pdf of $\boldsymbol{Y} = \{Y_1, Y_2, Y_3\}$ where.
Y_1 = smallest random variable of $\{X_1, X_2, X_3\}$.
Y_2 = middle random variable of $\{X_1, X_2, X_3\}$.
Y_3 = largest random variable of $\{X_1, X_2, X_3\}$.
Hint: List all permutations of $\{X_1, X_2, X_3\}$, then use the transformation of Theorem 2.9, adding over the permutations.

2.11. Show that the pmf of the sum of two discrete independent random variables $X \in \mathcal{X}$ and $Y \in \mathcal{Y}$ is given by the convolution summation

$$P_Z(z) = \sum_{x \in \mathcal{X}, z-x \in \mathcal{Y}} P_X(x) P_Y(z-x)$$

2.12. For the exponential distribution of Problem 2.1, show that the total time to failure for n batteries used in succession

$$p_T(t_i) = \lambda e^{-\lambda t_i}, t_i \geq 0, i = 1, 2, \ldots, n$$

is governed by the Erlang distribution

$$p_T(t) = \frac{\lambda^n t^{n-1}}{(n-1)!} e^{-\lambda t}, t \geq 0, \lambda > 0$$

Show that the mean and variance are

$$E\{T\} = \frac{n}{\lambda}, \text{Var}\{T\} = \frac{n}{\lambda^2}$$

Hint: Recall that the Gamma function is given by

$$\Gamma(n) = \int_0^\infty x^{n-1} \mathrm{e}^{-x} \mathrm{d}x = (n-1)!, n \text{ integer}$$

2.13. If $x \sim \mathcal{N}(m_x, \sigma_x^2)$, find the mean, mean square, and variance of the log-normal random variable $y = Ae^{bx}$.

2.14. Obtain the pdf of $y = Ae^{bx}$ if $X \sim \mathcal{N}(m_x, \sigma_x^2)$.

2.15. Show that the distribution of the product of N independent random variables $\{X_i, i = 1, 2, \ldots, N\}$ converges in distribution to a log-normal distribution.
Hint: Recall the central limit theorem.

2.16. Find the mean and variance of the Gaussian mixture

$$p_X(x, \lambda) = (1-\lambda) \frac{\mathrm{e}^{-x^2/(2\sigma_1^2)}}{\sqrt{2\pi\sigma_1^2}} + \lambda \frac{\mathrm{e}^{-x^2/(2\sigma_2^2)}}{\sqrt{2\pi\sigma_2^2}}, \quad 0 < \lambda < 1$$

2.17. In some applications, the pdf governing a random variable can best be approximated by the sum of known pdfs

$$p_X(x) = \sum_{i=1}^{l} w_i p_i(x)$$

where w_i are constant weights and $p_i(x)$ are pdfs. Show that:

(a) The sum of the weights w_i must be equal to unity.
(b) The moments of the pdf are given by the weighted sum of the moments of the pdfs $p_i(x), i = 1, 2, \ldots, l$, with weights w_i.
(c) The variance of the pdf is given by

$$\text{var}(x) = \sum_{i=1}^{l} w_i \sigma_i^2 + \sum_{i=1}^{l} w_i (m_i - m_x)^2$$

where $\left(m_i, \sigma_i^2\right)$ are the (mean, variance) of $p_i(x), i = 1, 2, \ldots, l$, and $E\{x\} = m_x$

2.18. Prove that the expectation of a random variable X conditioned on a random variable Y is given by

$$E\{\text{var}(X|Y)\} = \text{var}(X) - \text{var}(E\{(X|Y)\})$$

2.19. Prove that $\log E\{X\} \geq E\{\log(x)\}$.

Hint: The function $\log(x)$ is a concave function of x, i.e., $-\log(x)$ is a convex function of x.

2.20. The moment generating function of the joint pdf of random variables $\{x_i, i = 1, \ldots, n\}$ is defined as

$$M_x(\boldsymbol{x}) = E\left\{\exp\left(\boldsymbol{s}^T \boldsymbol{x}\right)\right\}$$

$$\boldsymbol{x} = \begin{bmatrix} x_1 \ x_2 \ \ldots \ x_n \end{bmatrix}^T, \boldsymbol{s} = \begin{bmatrix} s_1 \ s_2 \ \ldots \ s_n \end{bmatrix}^T$$

Show that

$$E\left\{\prod_{i=1}^{n} x_i^{j_i}\right\} = \left.\frac{\partial^{j_1+j_2\ldots+j_n} M_x(\boldsymbol{s})}{\partial^{j_1} s_1 \partial^{j_2} s_2 \ldots \partial^{j_n} s_n}\right|_{\boldsymbol{s}=\boldsymbol{0}}$$

2.21. Show that the moment generating function of the sum of mutually independent random variables $\{x_i, i = 1, \ldots, n\}$ is equal to the product of their moment generating functions.

2.22. Show that the moment generating function for a vector of jointly Gaussian random variables $\boldsymbol{x} = \begin{bmatrix} x_1 \ x_2 \ \ldots \ x_n \end{bmatrix}^T, \boldsymbol{x} \sim \mathcal{N}(\boldsymbol{m}_x, C_x)$ is

$$M_x(t) = E\{\exp(t^T x)\} = \exp\left(t^T m_x + \frac{1}{2} t^T C_x t\right)$$

where $t = \left[t_1 \; t_2 \; \ldots \; t_n \right]^T$.

2.23. Using the moment generating results of Problems 2.12 and 2.13, show that for zero-mean Gaussian random variables $\{x_i, i = 1, 2, 3, 4\}$

$$E\{x_1x_2x_3x_4\} = \frac{\partial^4 M_x(t)}{\partial t_1 \partial t_2 \partial t_3 \partial t_4}_{t=0} = C_{12}C_{34} + C_{13}C_{24} + C_{14}C_{23}$$

where $C_{ij} = E\{x_i x_j\}$.

2.24. Show that for a uniformly distributed random variable $X \sim U[a, b]$ the expectation.

$E\{(X - c)^i (X - c)^j\}, i + j$ odd, $c = (a + b)/2$, is always zero.

2.25. Show that the expectation of $E\{(X - m_x)^i (X - m_x)^j\}, i + j$ odd, is always zero if X is a random variable whose distribution $p_x(x)$ is symmetric about its mean.

2.26. Show that, for any random variable X governed by a symmetric distribution $p_x(x)$ with zero-mean, the expectation of X^i is zero for i odd. The prove that the variables X^i and X^j are uncorrelated but not independent for $i + j$ odd integer, i, j integer.

2.27. Show that the expectations of all the odd terms in the Taylor series expansion of the function $y = h(x)$ of a random variable X are zero if X is a random variable whose distribution $p_x(x)$ is symmetric about its mean.

2.28. The distribution of the reciprocal of a random variable is known in statistics as the inverse distribution. For a random variable X with pdf $p_X(x)$, find the pdf of the random variable $Y = 1/X$.

(a) Using the properties of the CDF.
(b) Using the formula for the transformation of a random variable.

2.29. Use the result of Problem 13 to find the distribution of $Y = 1/X$, if $p_X(x) = U(a, b)$, the find the mean and variance of Y.

2.30. A random vector $x = \left[x_1 \; x_2 \right]^T$ is translated by the constant vector $d = \left[d_1 \; d_2 \right]^T$ and rotated by a constant angle θ. Using MAPLE, determine the distribution of the translated and rotated vector z if the coordinates of x are governed by a bivariate normal distribution with zero-mean and covariance matrix.

$$C_x = \begin{bmatrix} \sigma_1^2 & \rho_{12}\sigma_1\sigma_2 \\ \rho_{12}\sigma_1\sigma_2 & \sigma_2^2 \end{bmatrix}$$

Hint: The rotation matrix is given by

$$R = \begin{bmatrix} \cos(\theta) & -\sin(\theta) \\ \sin(\theta) & \cos(\theta) \end{bmatrix}$$

2.31. Given a random vector $x \in \mathcal{R}^n$ of mean m_x and covariance matrix C_x.

(a) Find the expected value of the quadratic form $(x - m_x)^T Q(x - m_x)$, $Q \in \mathcal{R}^{n\times n}$
(b) Find the expected value of the quadratic form $x^T Qx$, $Q \in \mathcal{R}^{n\times n}$

Hint: Use the properties of the trace of a matrix.

2.32 Given mutually independent random vectors $x \in \mathcal{R}^n$ and $y \in \mathcal{R}^m$. Use the transformation approach to find the pdf of $z = x + y$.
2.33. Show that the sum of two independent uniform random variables $U[0, a], U[0, b], b > a$ is a trapezoidal random variable with support $[0, a + b]$.
2.34. Show that if two variables x and z are jointly Gaussian then the marginal density $p_X(x)$ is Gaussian.
2.35. Show that for $X \sim \mathcal{N}(0, 1)$ and $f(x)$ continuous

$$E\{Xf(x)\} = E\left\{\frac{df(X)}{dX}\right\}$$

2.36. Show that for $X \sim \mathcal{N}\left(m_x, \sigma^2\right)$ and $f(x)$ continuous and bounded then

$$E\left\{\frac{df(X)}{dX}\right\} = \frac{1}{\sigma^2}E\{(X - m_x)f(x)\}$$

2.37. Find the conditional pdf $p_{XY}(x, y|x + y)$ of two random variables $X \sim \mathcal{N}\left(m_x, \sigma_x^2\right)$ and $Y \sim \mathcal{N}\left(m_y, \sigma_y^2\right)$.
2.38. Show that the central moments of the normal distribution are given by.

(a) $E\left\{(X - \mu)^k\right\} = \begin{cases} 0, & \text{odd} \\ \sigma^k(k - 1)!!, & k \text{ even} \end{cases}$

(b) $E\left\{|X - \mu|^k\right\} = \sigma^k(k - 1)!!\begin{cases} \sqrt{\frac{2}{\pi}}, & k \text{ odd} \\ 1, & k \text{ even} \end{cases} = \sqrt{\frac{2^k}{\pi}}\sigma^k\Gamma\left(\frac{k+1}{2}\right)$

where $(k)!!$ denotes the double factorial, which is the factorial including only the product of all numbers that have the same parity (odd or even) as k.

2.39. Use the result of Problem 2.29 to show that the even moments of the standard normal distribution $\mathcal{N}(0, 1)$ are given by

$$E\left\{X^{2k}\right\} = \frac{(2k)!}{2^k k!}, k = 1, 2, \ldots$$

2.40. Show that for the Gaussian vectors $\boldsymbol{x} \in \mathcal{R}^n$ and $\boldsymbol{y} \in \mathcal{R}^m$, the vector $\tilde{\boldsymbol{x}} = \boldsymbol{x} - E\{\boldsymbol{x}|\boldsymbol{y}\}$ is statistically independent of any vector $\boldsymbol{z} = A\boldsymbol{y} + \boldsymbol{b}$, $A \in \mathcal{R}^{l\times m}$, $\boldsymbol{b} \in \mathcal{R}^l$.

2.41 Show that for the Gaussian vectors $\boldsymbol{x} \in \mathcal{R}^n$, $\boldsymbol{y} \in \mathcal{R}^m$ and $\boldsymbol{z} \in \mathcal{R}^l$, if $\boldsymbol{y}$ and $\boldsymbol{z}$ are independent, then $E\{\boldsymbol{x}|\boldsymbol{y}, \boldsymbol{z}\} = E\{\boldsymbol{x}|\boldsymbol{y}\} + E\{\boldsymbol{x}|\boldsymbol{z}\} - \boldsymbol{m}_x$, then use the result of Problem 2.39 to show that $E\{\boldsymbol{x}|\boldsymbol{y}, \tilde{\boldsymbol{z}}\} = E\{\boldsymbol{x}|\boldsymbol{y}\} + E\{\boldsymbol{x}|\tilde{\boldsymbol{z}}\} - \boldsymbol{m}_x$, $\tilde{\boldsymbol{z}} = \boldsymbol{z} - E\{\boldsymbol{x}|\boldsymbol{y}\}$.

2.42. Use the results of Problems 2.40 and 2.41 to show that $E\{\boldsymbol{x}|\boldsymbol{y}, \boldsymbol{z}\} = E\{\boldsymbol{x}|\boldsymbol{y}, \tilde{\boldsymbol{z}}\}$, where $\tilde{z} = z - E\{z|y\}$.

2.43. Show that the product of Gaussian random variables is distributed as the sum of two χ^2 random variables, which is not a χ^2 distribution.

2.44. Finite mixtures of Gaussian distributions provide models for a variety of physical phenomena. Write a MATLAB program to generate and plot the sum

$$p(x) = \sum_{i=1}^{2} w_i p(\mu_i, 1)$$

for $\mu_1 = 0$ and $\mu_2 = 2, 4$, (a) with equal weights $w_1 = w_2 = 0.5$, and (b) with unequal weights $w_1 = 0.25$, $w_2 = 0.75$

Hint: Use the MATLAB command **normpdf** to generate the data in the range [− 4, 6]

Bibliography

1. Anderson, B. D. O., & Moore, J. B. (1979). *Optimal Filtering*. Prentice Hall.
2. Brown, R. G., & Hwang, P. Y. C. (2012). *Introduction to random signals and applied Kalman filtering*. Wiley.
3. Casella, G., & Berger, R. L. (1990). *Statistical inference*. Duxbury Press.
4. De Groot, M. H., & Schervish, M. J. (2002). *Probability & statistics*. Addison-Wesley.
5. Gray, R. M., & Davisson, L. D. (1986). *Random processes: A mathematical approach for engineers*. Prentice Hall.
6. Hogg, R. V., McKean, J. W., & Craig, A. T. (2005). *Introduction to mathematical statistics*. Prentice Hall.
7. Kay, S. M. (1998). *Fundamentals of statistical signal processing: detection theory*. Prentice Hall.
8. Mendel, J. M. (1995). *Lessons in estimation theory for signal processing, communications, and control*. Prentice Hall PTR, Englewood Cliffs.
9. Papoulis, A., & Pillai, S. U. (2002). *Probability, random variables, and stochastic processes* (4th ed.). McGraw Hill.
10. Stark & Woods. (2002). *Probability and random processes*. Prentice Hall.

Chapter 3
Random Signals

3.1 Random Processes

Random processes are functions that map the elements of the sample space $\mathcal{S}$ to the set of continuous-time functions $\mathcal{C}$ as depicted in Fig. 3.1. Thus, a random process is a function of time and an elementary event in the sample space $X(t, s)$. As shown in Fig. 3.2, for a fixed time point t^* the random process is a random variable $X(t^*, s)$, and for a fixed outcome of the random experiment s^* it is an ordinary time function $x(t, s^*)$. Random processes may have no deterministic structure as in the example shown in Fig. 3.3 or may have a deterministic structure as in the following example.

Example 3.1: Random Binary Signal Consider a random sequence of rectangular pulses of Fig. 3.4 where:

i. The pulses are of fixed duration T.
ii. The pulse amplitude is $+\ 1$ or $-\ 1$ with equal probability.
iii. The amplitudes of different pulses are statistically independent.
iv. The initial positive switching time (start time) D for the sequence is uniformly distributed in the interval $[0, T]$, i.e., $D \sim U[0, T]$.

Note that a positive, or negative, interval of duration greater than T can occur if two or more consecutive pulses have the same amplitude.

The signal can be expressed mathematically in the form

$$X(t) = \sum_{k=-\infty}^{\infty} A_k p(t - kT - D). \tag{3.1}$$

$p(t)$ unit amplitude pulse of duration T.
A_k binary r.v. in $\{-1, 1\}$ = amplitude of kth pulse.
D random start time, uniformly distributed in $[0, T]$. ■

M. S. Fadali, *Introduction to Random Signals, Estimation Theory, and Kalman Filtering*, https://doi.org/10.1007/978-981-99-8063-5_3

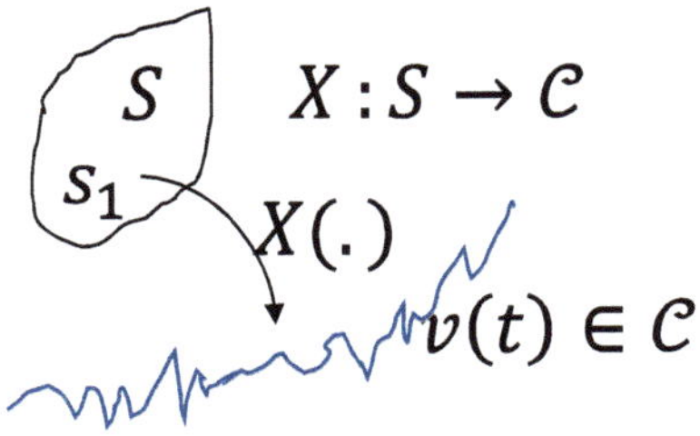

Fig. 3.1 Random signal X maps the sample space S to the set of continuous functions C

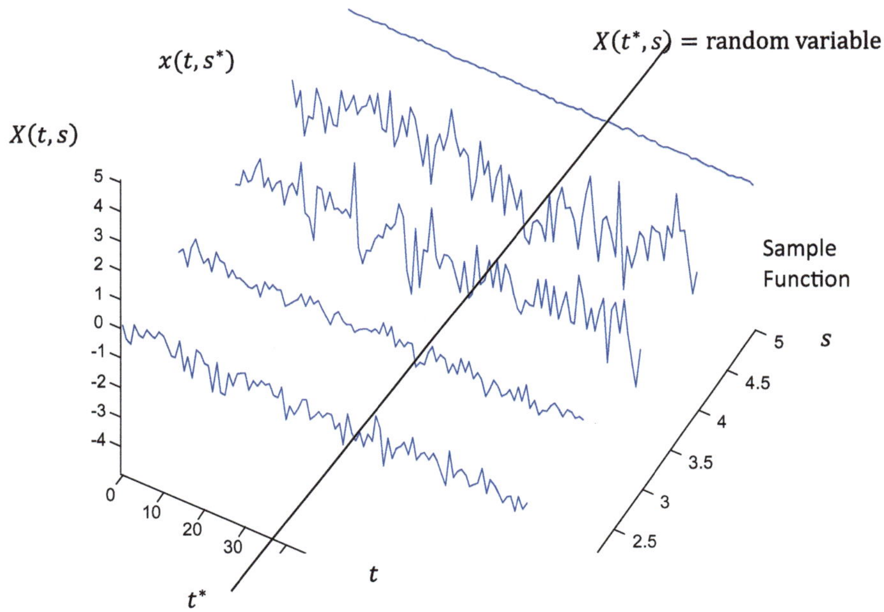

Fig. 3.2 Random process

The following is an example of a discrete random process that does not have a deterministic structure.

Example 3.2: Random Walk The process defines a random walk along a straight path starting at the origin with equal probability for a step forward and a step backward of length L. The mean value of the ith step is

$$E\{X_i\} = (1/2)L + (1/2)(-L) = 0,$$

and the variance is

$$\sigma^2 = E\{X_i^2\} = (1/2)L^2 + (1/2)(-L)^2 = L^2.$$

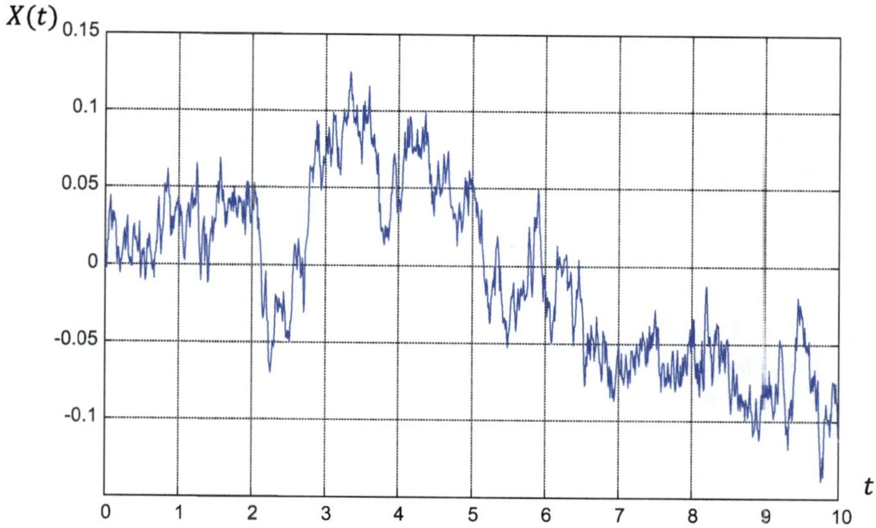

Fig. 3.3 Random signal with no deterministic structure

Fig. 3.4 Random binary signal

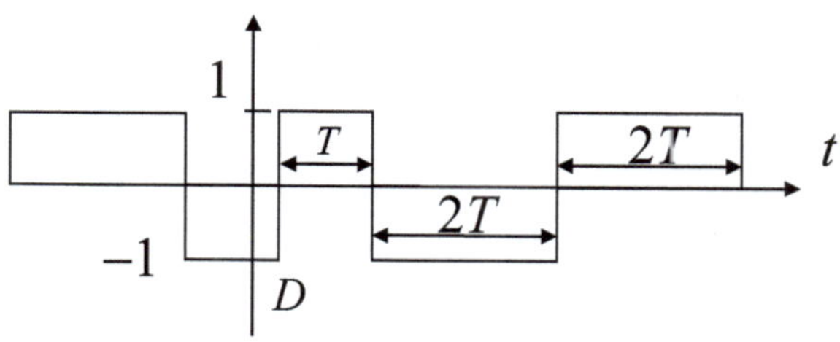

The distance traveled after n steps is

$$X(n) = \sum_{i=1}^{n} X_i.$$

The mean and mean square of the distance traveled after n steps are

$$E\{X(n)\} = 0,$$
$$E\{X^2(n)\} = nE\{X_i^2\} = n\sigma^2 = nL^2.$$

Because the mean distance traveled is zero, its variance is equal to its mean square. ∎

3.1.1 Joint Densities

As depicted in Fig. 3.2, a random signal includes a random variable at each time point. Hence, a joint density is defined for each set of two or more of these random variables. For a random signal $X(t)$, the joint density of its values at two distinct time points t_1 and t_2 is

$$p_{X_i X_j}\left(x_i, x_j\right), \quad i \neq j, X_i = X(t_i). \tag{3.2}$$

The joint density is a measure of joint probability for values of the signal at two time points that specifies how fast the random signal $X(t)$ changes with time. Higher-order densities relating more time points provide more information but are hard to compute.

3.1.1.1 Statistically Independent Random Signals

Using the properties of random variables, we can define mutual independence of random signals as mutual independence for the two sets of random variables associated with two random signals. Two random signals $X(t)$ and $Y(t)$ are independent if for any choice of time points t_i and t_j, $i, j = 1, 2, \ldots$, including $t_i \neq t_j$.

$$p_{X_1 X_2 \ldots Y_1 Y_2 \ldots} = p_{X_1 X_2 \ldots} p_{Y_1 Y_2 \ldots}, \tag{3.3}$$

where $X_i = X(t_i), \quad Y_j = Y\left(t_j\right)$.

3.1.2 Gaussian Random Process

A Gaussian random process is one where all marginal and joint density functions of any order are Gaussian. Recall that if the joint density is Gaussian, then all marginal densities and all conditional densities are also Gaussian.

Because linear transformation of a Gaussian process gives another Gaussian process, the response of a linear system to a Gaussian process is Gaussian. Figure 3.3 shows a process generated by filtering Gaussian white noise with a filter of transfer function $1/(s + 1)$. The response of linear systems to a random input is discussed in Chap. 4.

3.2 Autocorrelation

The autocorrelation is defined as the expectation of the product of the values of the random process at two different time points

$$R_{XX}(t_1, t_2) = E\{X(t_1)X(t_2)\}. \tag{3.4}$$

Recall that for fixed t the random process is a random variable. Hence, the autocorrelation is calculated using the joint pdf of $X(t_1)$ and $X(t_2)$. For the continuous case, the autocorrelation is

$$R_{XX}(t_1, t_2) = \int_{-\infty}^{\infty}\int_{-\infty}^{\infty} x_1 x_2 p_{X_1 X_2}\big(x_{1,}, t_1, x_2, t_2\big) dx_1 dx_2. \tag{3.5}$$

For the discrete case, the autocorrelation is

$$R_{XX}(k_1, k_2) = \sum_i \sum_j x_1(i) x_2(j) p_{ij}(k_1, k_2). \tag{3.6}$$

Clearly, autocorrelation is an *ensemble* average, i.e., the average over the population, using the *joint* density function.

Similarly, we define the autocovariance as

$$C_{XX}(t_1, t_2) = E\{[X(t_1) - E\{X(t_1)\}][X(t_2) - E\{X(t_2)\}]\}. \tag{3.7}$$

Expanding the definition of the autocovariance gives

$$C_{XX}(t_1, t_2) = E\{X(t_1)X(t_2)\} - E\{X(t_1)\}E\{X(t_2)\}. \tag{3.8}$$

This is equal to the autocorrelation for zero-mean.

To extend the definition of autocorrelation to a complex random vector $\boldsymbol{x}$, we use the formula

$$R_{XX}(t_1, t_2) = E\big\{\boldsymbol{x}(t_1)\boldsymbol{x}^*(t_2)\big\}. \tag{3.9}$$

For the autocovariance, we have the formula

$$C_{XX}(t_1, t_2) = E\big\{[\boldsymbol{x}(t_1) - E\{\boldsymbol{x}(t_1)\}][\boldsymbol{x}(t_2) - E\{\boldsymbol{x}(t_2)\}]^*\big\}. \tag{3.10}$$

Both the autocorrelation and the autocovariance are second-order statistics and, while providing important information, do not completely characterize a random signal. Hence, it is possible for two completely different signals to have the same covariance and the same autocorrelation.

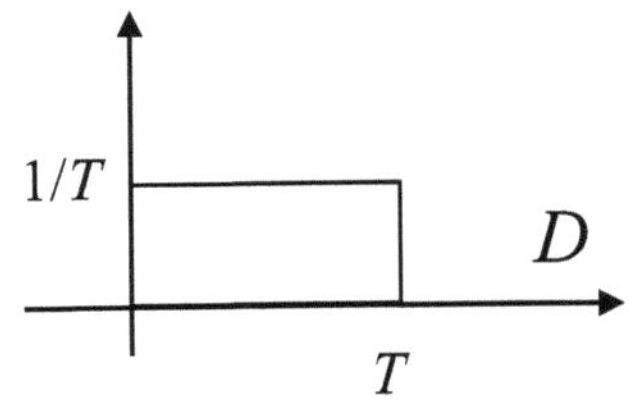

Fig. 3.5 Uniform distribution of first positive switching time

Example 3.3: Autocorrelation of Binary Signal Determine the autocorrelation function of the random binary signal of Example 3.1.

Solution

Consider two time points t_1 and $t_2, 0 < t_1 < t_2 < T$. If the random first positive switching time $D \sim U[0, T]$ is in the interval $[t_1, t_2]$ and t_1 and t_2 are not in the same pulse. Using the properties of the uniform distribution of Fig. 3.5, the probability of D in the interval $[t_1, t_2]$ is

$$P(t_1 < D < t_2) = \frac{t_2 - t_1}{T}, 0 < t_1 < t_2 < T.$$

The point D is outside the interval $[t_1, t_2]$ if either $0 < D < t_1$or $t_2 < D < T$. The probability of D outside the interval $[t_1, t_2]$ is

$$\begin{aligned} P(0 < D < t_1 \text{ or } t_2 < D < T) &= 1 - P(t_1 < D < t_2) \\ &= 1 - \frac{t_2 - t_1}{T}. \end{aligned}$$

Interchanging t_1 and t_2 gives the probabilities for $0 < t_2 < t_1 < T$.

To evaluate the autocorrelation, we need to examine the product of amplitudes of the signal at different time points. If the two time points are in the same pulse $(0 < D < t_1$ or $t_2 < D < T)$, then the sign of the amplitudes is the same and product is always unity. If the two time points are in two consecutive pulses $(t_1 < D < t_2)$, then the signs of the amplitudes may or may not be the same and the product of amplitudes is ± 1. Thus, the product of amplitudes is

$$X(t_1)X(t_2) = \begin{cases} 1, & 0 < D < t_1 \text{ or } t_2 < D < T \\ \pm 1 & t_1 < D < t_2 \end{cases}.$$

The probability of unity product if t_1 is in the same pulse as t_2 is

$$\text{P}\{X(t_1)X(t_2) = 1 | 0 < D < t_1 \text{ or } t_2 < D < T\} = 1.$$

If t_1 is not in the same pulse as t_2, then $+ 1$ and $- 1$ are equally probable

$$P\{X(t_1)X(t_2) = 1|t_1 < D < t_2\} = P\{X(t_1)X(t_2) = -1|t_1 < D < t_2\} = 0.5.$$

We have three mutually exclusive possibilities, and we can add to get the autocorrelation for $0 < t_1 < t_2 < T$. The results for $0 < t_2 < t_1 < T$ are obtained by interchanging t_1 and t_2, and the autocorrelation for both cases is given by

$$\begin{aligned} R_{XX}(t_1, t_2) &= E\{X(t_1)X(t_2)\} \\ &= \sum_{3\,\text{cases}} X(t_1)X(t_2)P(X(t_1)X(t_2)) \\ &= (1)[1 - |t_2 - t_1|/T] + (1)(0.5)|t_2 - t_1|/T + (-1)(0.5)|t_2 - t_1|/T \\ &= 1 - |t_2 - t_1|/T. \end{aligned}$$

For $|t_2 - t_1| > T$, the amplitudes are independent, and therefore uncorrelated, and the autocorrelation is

$$R_{XX}(t_1, t_2) = E\{X(t_1)X(t_2)\} = E\{X(t_1)\}E\{X(t_2)\} = 0.$$

The autocorrelation of the random binary signal is

$$\begin{aligned} R_{XX}(t_1, t_2) &= E\{X(t_1)X(t_2)\} \\ &= \begin{cases} 1 - \frac{|t_2 - t_1|}{T}, & |t_2 - t_1| \leq T \\ 0, & |t_2 - t_1| > T \end{cases}. \end{aligned}$$

Figure 3.6 shows the autocorrelation function of the random binary signal. We observe that the autocorrelation is a function of the width of the interval $|t_2 - t_1|$.

■

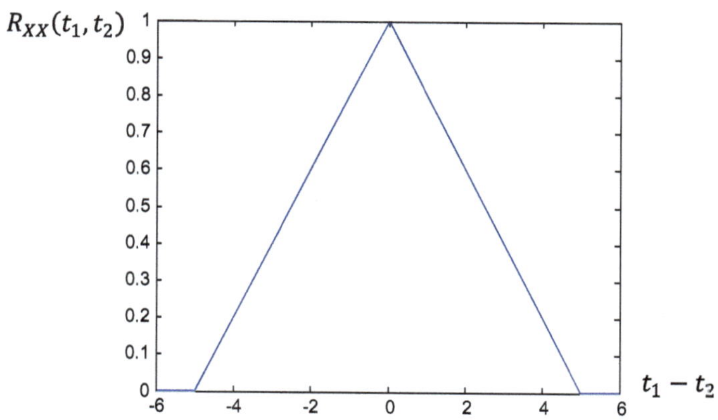

Fig. 3.6 Triangular autocorrelation function

3.3 Stationary Random Process

Stationarity extends the concept of time invariance to random signals. As in the case of time invariance, stationarity is only approximately possible in practice. Two definitions are used for stationarity.

- **Strictly stationary random process**

  Strictly stationary random signals have pdfs that are unaffected by any time shift, i.e., for

$$\begin{aligned} X_i &= X(t_i), i = 1, 2, \ldots, k, \\ X_i' &= X(t_i + \tau), i = 1, 2, \ldots, k, \end{aligned}$$

  the marginal and joint pdfs $p_{X_i}, p_{X_i X_j}, p_{X_i \ldots X_k}$ are the same. Strict stationary implies wide-sense stationary.
- **Wide-sense stationary (WSS) random process**

  Wide-sense stationary random processes have a constant mean and a shift-invariant autocorrelation function:

  – Stationarity of the mean

$$m = E\{X(t)\}, m \text{ constant.} \tag{3.11}$$

  – Stationarity of the autocorrelation

$$R_{XX}(\tau) = E\{X(t)X(t+\tau)\}. \tag{3.12}$$

For discrete time, we replace t by the discrete time k and rewrite the autocorrelation as

$$R_{XX}(m) = E\{X(k)X(k+m)\}. \tag{3.13}$$

Example 3.4 Show that the random process

$$Y(t) = X + \cos(t), X \sim N(0, 9)$$

is nonstationary.

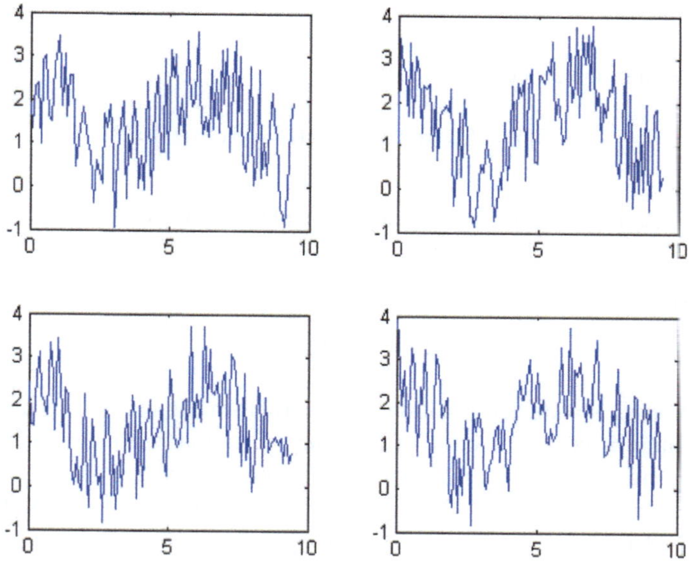

Fig. 3.7 Nonstationary signal

Solution

The mean of $Y(t)$ is

$$E\{Y(t)\} = E\{X + \cos(t)\} = \cos(t).$$

The mean is a function of time, and the process is nonstationary in the mean. A plot of four realizations of the process is shown in Fig. 3.7.

The autocorrelation of the process is

$$E\{Y(t_1).Y(t_2)\} = E\{X^2\} + \cos(t_1)\cos(t_2) = 9 + \frac{1}{2}[\cos(t_1 - t_2) + \cos(t_1 + t_2)].$$

The autocorrelation is a function of $t_1 - t_2$ and $t_1 + t_2$, and the process is not stationary in the autocorrelation. ∎

Example 3.5 A random process $Y(t)$ has values in the following set with equal probability

$$Y(t) \in \{\pm 5\sin(t), \pm 5\cos(t)\}.$$

The process has only four possible equally probable joint outcomes: two positive sines, two positive cosines, two negative sines, or two negative cosines.

(a) Find the mean $E\{Y(t)\}$ and the autocorrelation $R_Y(t + \tau, t)$ of the process.
(b) Explain why the process is wide-sense stationary but not strictly stationary.

Solution

(a) The mean of the process is

$$E\{Y(t)\} = 0.25[\pm 5\sin(t) \pm 5\cos(t)] = 0.$$

The autocorrelation for four equally probable joint outcomes is

$$\begin{aligned} R_Y(t+\tau, t) &= \frac{1}{4}[2 \times 25\sin(t)\sin(t+\tau) + 2 \times 25\cos(t)\cos(t+\tau)] \\ &= 12.5\cos(\tau) = R_Y(\tau). \end{aligned}$$

(b) The process is wide-sense stationary because its mean is constant and its autocorrelation is only a function of the time separation τ. To show that it is not strict sense stationary, consider two time points, 0 and $\pi/4$. At $t = 0$, the process has the values $Y(t) \in \{0, \pm 5\}$; while at $t = \pi/4$, it has the values $Y(t) \in \left\{\pm 5/\sqrt{2}\right\}$. Thus, even though their mean is the same, the first-order distributions are different at the two time points with

$$p_{Y(0)}(y) = 0.5\delta(y) + 0.25\delta(y-5) - 0.25\delta(y+5),$$

$$p_{Y(\pi/4)}(y) = 0.5\delta\left(y - 5/\sqrt{2}\right) + 0.5\delta\left(y + 5/\sqrt{2}\right).$$

The process is not strictly stationary. The result is not surprising because completely different distributions can have the same mean but different higher moments. ■

3.4 Ergodic Random Processes

Ergodicity is when the time average of a single realization is equal to the ensemble average. Ergodicity, like stationarity, is an idealization but although physical random signals are not ergodic, many of them are approximately ergodic. Ergodic signals tend to "look random" and random signals that have a deterministic structure are typically not ergodic. Ergodicity allows us to use a single recording of a time signal to obtain its statistical properties, including all moments, autocorrelation, etc. Because averaging will only give a single value for any property, stationarity is clearly a necessary condition for ergodicity. However, stationarity is not sufficient for ergodicity.

There are two important ergodicity properties that we state in terms of a sample realization $x_A(t)$:

i. **Ergodicity in Mean**: The time average is equal to the mean

$$m_{x_A} = \mathcal{L}im_{T\to\infty} \frac{1}{T} \int_0^T x_A(t)\mathrm{d}t = E\{X(t)\}. \tag{3.14}$$

Ergodicity in the mean requires the covariance to satisfy $\mathrm{Cov}_{XX}(\tau) \to 0$, as $\tau \to \infty$. For zero-mean, the autocorrelations satisfies $R(\tau) \to 0$, as $\tau \to \infty$.

ii. **Ergodicity in Autocorrelation**: The time autocorrelation is equal to the autocorrelation

$$\begin{aligned}\mathscr{R}_{X_A}(\tau) &= \mathcal{L}im_{T\to\infty} \frac{1}{T} \int_0^T x_A(t)x_A(t+\tau)\mathrm{d}t \\ &= E\{X(t)X(t+\tau)\} = R_{XX}(\tau).\end{aligned} \tag{3.15}$$

Ergodicity in autocorrelation requires the 4th moment to satisfy $\mathrm{Cov}_{ZZ}(\tau_1) \to 0$ as $\tau_1 \to \infty$, $Z = X(t)X(t+\tau)$.

The ergodicity conditions are derived in (Shanmugan and Breipohl, p. 170–178).

The expression for the time autocorrelation in the discrete-time case is more useful in practice

$$\mathscr{R}_{X_A}(n) = \frac{1}{N-|m|} \sum_{k=0}^{N-|m|-1} x_A(k)x_A(k+|m|),\, m \in [-(N-1), (N-1)]. \tag{3.16}$$

Example 3.6: Stationary not Ergodic Show that a random constant $X(t) = A$ for all t with random amplitude $A \sim \mathcal{N}(0, \sigma^2)$ is stationary but not ergodic.

Solution

The constant is Gaussian and is completely characterized by its mean and variance, which are fixed for all t. Hence, it is stationary. If we obtain a sample realization with amplitude $A_1 \neq 0$, then the sample mean is $A_1 \neq 0$, the population mean. Therefore, the process is not ergodic. ■

Example 3.7 Obtain the autocorrelation of the random process with a deterministic structure $X(t) = A\sin(\omega_0 t)$ with $A \sim \mathcal{N}(0, \sigma^2)$, ω_0 constant, then use a sample realization is $X_A(t) = A_1 \sin(\omega_0 t)$, with fixed amplitude A_1 to obtain the time autocorrelation. Compare the time autocorrelation and the autocorrelation and explain why the process is not ergodic.

Solution

The time autocorrelation is the integral

$$\mathscr{R}_{X_A}(\tau) = \underset{T\to\infty}{\mathcal{L}im}\frac{1}{T}\int_0^T X_A(t)X_A(t+\tau)\mathrm{d}t = \underset{T\to\infty}{\mathcal{L}im}\frac{1}{T}\int_0^T A_1^2\sin(\omega_0 t)\sin[\omega_0(t+\tau)]\mathrm{d}t$$

$$= \underset{T\to\infty}{\mathcal{L}im}\frac{A_1^2}{2T}\int_0^T \{\cos(\omega_0\tau) - \cos[\omega_0(2t+\tau)]\}\mathrm{d}t.$$

The integral of the second term is

$$\int_0^T \cos[\omega_0(2t+\tau)]\mathrm{d}t = \int_0^T \{\cos(\omega_0\tau)\cos(2\omega_0 t) - \sin(\omega_0\tau)\sin(2\omega_0 t)\}\mathrm{d}t.$$

The integral of the term is finite and division by T then taking the limit gives zero. Thus, the time autocorrelation is

$$\mathscr{R}_{X_A}(\tau) = \frac{A_1^2}{2}\cos(\omega_0\tau).$$

The autocorrelation is

$$R_{XX}(t_1,t_2) = E\{X(t_1)X(t_2)\} = E\{A^2\}\sin(\omega_0 t_1)\sin(\omega_0 t_2)$$
$$= \frac{\sigma^2}{2}\{\cos(\omega_0[t_1-t_2]) - \cos(\omega_0[t_1+t_2])\}.$$

The autocorrelation is not a function of the time shift only and is not equal to time autocorrelation. Therefore, the process is not ergodic. ■

3.5 Properties of Autocorrelation

We present useful properties of the autocorrelation function that we use throughout the text. Several of the properties apply to the stationary case only, and we mostly assume real scalar processes.

(a) The mean square value

From the definition of the autocorrelation

$$R_{XX}(t,t) = E\{X^2(t)\} = R_{XX}(t). \tag{3.17}$$

For a wide-sense stationary process X, we have

$$R_{XX}(t) = R_{XX}(0), \forall t. \tag{3.18}$$

(b) The peak value of the autocorrelation

Consider a wide-sense stationary process X. We assume, w.l.o.g., that the process is zero-mean since the mean can be subtracted if nonzero. The autocorrelation coefficient for the process is

$$\rho = \frac{E\{X(t)X(t+\tau)\}}{\sqrt{E\{X^2(t)\}E\{X^2(t+\tau)\}}} = \frac{R_X(\tau)}{R_X(0)}.$$

It is shown in Chap. 2 that the magnitude of the correlation coefficient is bounded as

$$|\rho| = \left|\frac{R_X(\tau)}{R_X(0)}\right| \leq 1.$$

Because the mean square $R_X(0)$ is always positive, we have

$$|R_X(\tau)| \leq R_X(0). \tag{3.19}$$

Note that adding a mean m to X would add m^2 to $R_X(\tau)$ but would not change the location of the peak.

(c) Autocorrelation is an even function

$$\begin{aligned} R_{XX}(t_1, t_2) &= E\{X(t_1)X(t_2)\} = E\{X(t_2)X(t_1)\} \\ &= R_{XX}(t_2, t_1). \end{aligned} \tag{3.20}$$

For a real wide-sense stationary process X, we have

$$R_{XX}(t_2 - t_1) = R_{XX}(t_1 - t_2). \tag{3.21}$$

In other words, if $\tau = t_2 - t_1$, then $R_{XX}(\tau)$ is an even function of τ. This was the case for the autocorrelation function of Example 3.3.

For the more general case of a complex random vector, the autocorrelation changes as follows:

$$R_{XX}(t_1, t_2) = E\{\boldsymbol{x}(t_1)\boldsymbol{x}^*(t_2)\} = E\{\boldsymbol{x}(t_2)\boldsymbol{x}^*(t_1)\}^* = R_{XX}(t_2, t_1)^*, \tag{3.22}$$

where $*$ denotes the conjugate transpose and reduces to the transpose for real $\boldsymbol{x}$.

For a stationary random vector,

$$R_{XX}(t_2 - t_1) = R_{XX}(t_1 - t_2)^*. \tag{3.23}$$

(d) Autocorrelation of the Sum of Random Processes

The autocorrelation of the sum of two wide-sense stationary random processes is

$$E\{[X(t)+Y(t)][X(t+\tau)+Y(t+\tau)]\} = R_{XX}(\tau)+R_{XY}(\tau)+R_{YX}(\tau)+R_{YY}(\tau). \tag{3.24}$$

If the two processes are orthogonal, then $R_{XY}(\tau)=0$, $R_{YX}(\tau)=0$, and the autocorrelation of the sum $Z(t)=X(t)+Y(t)$ is the sum of autocorrelations

$$R_{ZZ}(\tau) = R_{XX}(\tau)+R_{YY}(\tau). \tag{3.25}$$

The formula can be extended to the sum of a number of mutually orthogonal random processes.

(e) DC power and autocorrelation

For a real wide-sense stationary process X, the square of the mean is the DC power of the signal. The reason for this terminology is explained in Sect. 3.7. We write the random process as the sum of a zero-mean, ergodic process $X_{zm}(t)$, and a constant mean m_x and evaluate the autocorrelation

$$\begin{aligned} R_{XX}(\tau) &= E\{X(t)X(t+\tau)\} = E\{[X_{zm}(t)+m_x][X_{zm}(t+\tau)+m_x]\} \\ &= R_{X_{zm}X_{zm}}(\tau)+m_x^2. \end{aligned}$$

For an ergodic process, the covariance, which is equal to the autocorrelation for a zero-mean process, goes to zero as $\tau \to \infty$ (see Sect. 3.4)

$$R_{XX}(\tau) \to 0 \text{ as } \tau \to \infty. \tag{3.26}$$

In general, for an ergodic process X the autocorrelation satisfies

$$R_{XX}(\tau) \to m_x^2 \text{ as } \tau \to \infty. \tag{3.27}$$

(f) Periodic Component

For a real wide-sense stationary process X with a periodic component, the autocorrelation R has a periodic component of the same period but provides no information about the phase. To show this, we consider a random process from (Cooper and McGillem, p. 218). They consider a zero-mean random process X that is the sum of a nonperiodic random process X_{np} and an uncorrelated X_p

$$X_p = A\cos(\omega_0 t+\theta), \theta \sim U[0, 2\pi].$$

Note that any periodic component can be expanded as a Fourier series with each term similar to the periodic component considered here. The autocorrelation of X is

$$R_{XX}(\tau) = E\{X(t)X(t+\tau)\} = E\big\{\big[X_{np}(t) + X_p(t)\big]\big[X_{np}(t+\tau) + X_p(t+\tau)\big]\big\}.$$

Because X is the sum of orthogonal terms, its autocorrelation is the sum of autocorrelations

$$R_{XX}(\tau) = R_{X_{np}X_{np}}(\tau) + R_{X_pX_p}(\tau).$$

The autocorrelation of the periodic component is

$$\begin{aligned} R_{X_pX_p}(\tau) &= E\big\{X_p(t)X_p(t+\tau)\big\} = A^2E\{\cos(\omega_0 t + \theta)\cos(\omega_0 t + \omega_0\tau + \theta)\} \\ &= \big(A^2/2\big)E\{\cos(\omega_0\tau) + \cos(\omega_0[2t+\tau] + 2\theta)\}. \end{aligned}$$

The second term can be expanded as

$$\cos(\omega_0[2t+\tau] + 2\theta) = \cos(\omega_0[2t+\tau])\cos(2\theta) - \sin(\omega_0[2t+\tau])\sin(2\theta).$$

For $\theta \sim U[0, 2\pi]$, we have

$$E\{\cos(2\theta)\} = (2\pi)^{-1}\int_0^{2\pi}\cos(2\theta)\mathrm{d}\theta = 0,$$

$$E\{\sin(2\theta)\} = (2\pi)^{-1}\int_0^{2\pi}\sin(2\theta)\mathrm{d}\theta = 0.$$

The autocorrelation of the periodic component is

$$R_{X_pX_p}(\tau) = \big(A^2/2\big)\cos(\omega_0\tau).$$

The autocorrelation of X is

$$R_{XX}(\tau) = R_{X_{np}X_{np}}(\tau) + \big(A^2/2\big)\cos(\omega_0\tau).$$

We observe that the autocorrelation has a periodic component of the same period as the periodic component of X but provide no information about the phase of X.

3.6 Cross-Correlation Function

The cross-correlation is the function

$$R_{XY}(t_1, t_2) = E\{X(t_1)Y(t_2)\}. \tag{3.28}$$

For wide-sense stationary random process, the cross-correlation is a function of the separation τ

$$R_{XY}(\tau) = E\{X(t)Y(t+\tau)\}. \tag{3.29}$$

We list properties of the cross-correlation and leave their proofs as an exercise:

I. Skew symmentry $R_{XY}(\tau) = R_{YX}(-\tau)$, (3.30)

II. $|R_{XY}(\tau)| \leq \sqrt{R_{XX}(0)R_{YY}(0)}$, $|\rho| \leq 1$, (3.31)

III. $|R_{XY}(\tau)| \leq \frac{1}{2}[R_{XX}(0) + R_{YY}(0)]$ (3.32)

IV. $R_{XY}(\tau) = \begin{cases} 0, & \text{orthogonal} \\ m_X m_Y, & \text{uncorrelated} \end{cases}.$ (3.33)

3.6.1 Time Delay Estimation

The cross-correlation function can be used to estimate the time delay associated with the transmission and reflection of a signal. This allows the calculation of the distance of the object from which the signal is reflected and has applications in radar and sonar. If a signal $X(t)$ it transmitted, as depicted in Fig. 3.8, then the cross-correlation of the transmitted signal and the reflected noisy signal is

$$R_{XY}(\tau) = E\{X(t)[X(t+\tau-T_d) + n(t)]\} = R_{XX}(\tau - T_d), \tag{3.34}$$

where $n(t)$ is a noise signal that is assumed zero-mean and uncorrelated with the transmitted signal. The cross-correlation is in this case a delayed autocorrelation and the location of its peak gives the value of the time delay.

Example 3.8: Cross-correlation Obtain the cross-correlation of the two random signals

$$X(t) = A\sin(\omega_0 t + \theta), \theta \sim U[0, 2\pi],$$
$$Y(t) = B\cos(\omega_0 t + \theta), \theta \sim U[0, 2\pi].$$

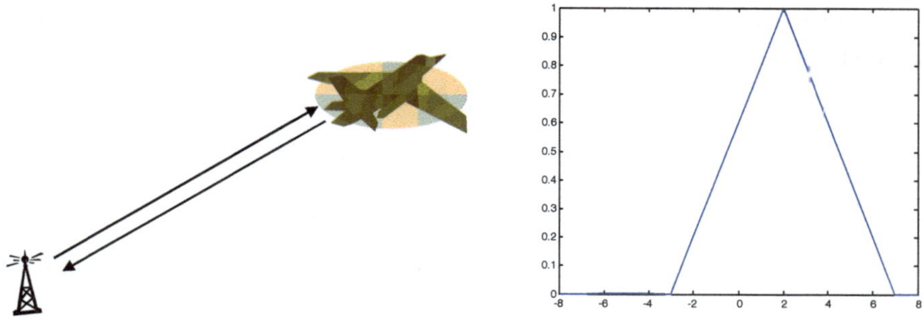

Fig. 3.8 Time delay estimation

Solution

The cross-correlation is

$$
\begin{aligned}
R_{XY}(\tau) &= E\{X(t)Y(t+\tau)\} = AB.E\{\sin(\omega_0 t+\theta)\cos(\omega_0(t+\tau)+\theta)\} \\
&= \frac{AB}{2}.E\{\sin(\omega_0(2t+\tau)+2\theta)+\sin(\omega_0\tau)\}.
\end{aligned}
$$

The expected value of the first term in the above expression is zero and the cross-correlation is

$$
R_{XY}(\tau) = \frac{AB}{2}\sin(\omega_0\tau).
$$

We observe that, unlike the autocorrelation, the cross-correlation is zero at $\tau = 0$ and its maximum is at $\tau = 2\pi/\omega_0$ and its multiples. ■

3.7 Power Spectral Density Function (PSD)

For wide-sense stationary stochastic processes, the power spectral density function (PSD) is defined as the Fourier transform of the autocorrelation

$$
S_{XX}(j\omega) = \mathfrak{F}\{R_{XX}(\tau)\} = \int_{-\infty}^{\infty} R_{XX}(\tau)\mathrm{e}^{-j\omega\tau}\mathrm{d}\tau. \tag{3.35}
$$

Alternatively, the PSD can be defined as the two-sided Laplace transform

$$
S_{XX}(s) = \mathscr{L}_2\{R_{XX}(\tau)\} = \int_{-\infty}^{\infty} R_{XX}(\tau)\mathrm{e}^{-s\tau}\mathrm{d}s. \tag{3.36}
$$

The two-sided Laplace transform must be used because the autocorrelation function is noncausal. The inverse transform

$$R_{XX}(\tau) = \frac{1}{2\pi}\int_{-\infty}^{\infty} S_{XX}(j\omega)e^{j\omega\tau}d\omega$$

$$R_{XX}(\tau) = \frac{1}{2\pi j}\int_{-j\infty}^{j\infty} S_{XX}(s)e^{s\tau}ds \tag{3.37}$$

is known as the *Wiener–Khinchine relation.* To obtain the autocorrelation from the PSD, we inverse transform. As with the one-sided transform, this is accomplished by (1) partial fraction expansion, then table-look up. When solving by hand with a high-order denominator, it is easier to obtain the solution by a partial fraction expansion in terms of s^2 rather than s because the PSD is a function of s^2.

Setting $\tau = 0$ in the Wiener–Khinchine relation gives the mean square value in terms of the PSD

$$R_{XX}(0) = E\{X^2(t)\} = \frac{1}{2\pi}\int_{-\infty}^{\infty} S_{XX}(j\omega)d\omega = \frac{1}{2\pi j}\int_{-j\infty}^{j\infty} S_{XX}(s)ds. \tag{3.38}$$

For discrete-time random signals, we use the discrete-time Fourier transform to obtain the PSD

$$S_{XX}(e^{j\omega}) = \mathfrak{F}\{R_{XX}(\tau)\} = \sum_{k=-\infty}^{\infty} R_{XX}(k)e^{-j\omega k}, \omega \in [-\pi, \pi]. \tag{3.39}$$

Because this is the transform of a discrete-time signal, it is a periodic function of frequency. Because the autocorrelation is even, we can write PSD as

$$S_{XX}(e^{j\omega}) = R_{XX}(0) + \sum_{k=1}^{\infty} R_{XX}(k)\left(e^{j\omega k} + e^{-j\omega k}\right), \omega \in [-\pi, \pi].$$

Substituting for the complex exponentials gives the expression

$$S_{XX}(e^{j\omega}) = R_{XX}(0) + 2\sum_{k=1}^{\infty} R_{XX}(k)\cos(\omega k), \omega \in [-\pi, \pi]. \tag{3.40}$$

For the *Wiener–Khinchine relation*, we invert the discrete-time transform to obtain the autocorrelation

$$R_{XX}(k) = \frac{1}{2\pi} \int_{-\pi}^{\pi} S_{XX}\left(e^{j\omega}\right) e^{j\omega k} d\omega. \tag{3.41}$$

Setting $\tau = 0$ in (3.38) gives the mean square value in terms of the PSD

$$R_{XX}(0) = E\left\{X^2(t)\right\} = \frac{1}{2\pi} \int_{-\pi}^{\pi} S_{XX}(j\omega) d\omega. \tag{3.42}$$

We can also use the noncausal or two-sided z-transform to define the PSD function

$$S_{XX}(z) = \mathcal{Z}_2\{R_{XX}(k)\} = \sum_{k=-\infty}^{\infty} R_{XX}(k) z^{-k}. \tag{3.43}$$

While the Laplace version of the formula is useful because we have a table of integrals for rational PSD functions, the z-transform version for the discrete-time cases is less useful.

Example 3.9 Find the PSD of a process whose autocorrelation is

$$R_{XX}(\tau) = \sigma^2 e^{-\beta|\tau|}.$$

Solution

Using the transform tables, we inverse transform to obtain

$$S_{XX}(j\omega) = \mathfrak{F}\{R_{XX}(\tau)\} = \frac{2\sigma^2\beta}{\omega^2 + \beta^2},$$

$$S_{XX}(s) = \mathcal{L}_2\{R_{XX}(\tau)\} = \frac{2\sigma^2\beta}{-s^2 + \beta^2}.$$

The s-domain expression can always be obtained from the frequency domain expression by substituting s for $j\omega$, or $-s^2$ for ω^2. ■

3.7.1 *Properties of the Power Spectral Density (PSD)*

The following are properties of the PSD of a wide-sense stationary process:

(1) The PSD is a real and even function of frequency.

From the properties of the Fourier transform, the Fourier transform of real function $f(\tau)$ has an even real part and an odd imaginary part

$$\mathfrak{F}\{f(\tau)\} = \text{even} + j\text{odd}.$$

If the function $f(\tau)$ is both real and even, then its Fourier transform $\mathfrak{F}\{f(\tau)\}$ is even. Because the autocorrelation is real and even for real random processes, we have

$$S_{XX}(j\omega) = S_{XX}\left(\omega^2\right) = \text{real, even}.$$

For the discrete case, Equation (3.40) shows that the PSD is even because it is expressed in terms of the even cosine function. It is also well known that the discrete-time Fourier transform of real and even function, such as the autocorrelation, is real and even (see Oppenheim, et. al., page 391).

(2) The PSD is nonnegative: since its integral gives the energy over BW.

(3) Power in a Finite Band.

The PSD is a measure of the distribution of power over a range of frequencies. For the frequency range $\omega \in [\omega_1, \omega_2]$, the power is

$$\text{Power}(\omega_1, \omega_2) = \frac{1}{2\pi}\int\limits_{-\omega_2}^{-\omega_1} S_{XX}(j\omega)\mathrm{d}\omega + \frac{1}{2\pi}\int\limits_{\omega_1}^{\omega_2} S_{XX}(j\omega)\mathrm{d}\omega. \tag{3.44}$$

The above properties apply to both continuous-time and discrete-time processes.

Example 3.10 For the PSD function

$$S_{XX}(j\omega) = \frac{2\sigma^2\beta}{\omega^2 + \beta^2},$$

find the power for the frequency range $\omega \in [\omega_1, \omega_2]$.

Solution

For the frequency range $\omega \in [\omega_1, \omega_2]$, using the symmetry of the PSD, we have

$$\text{Power}(\omega_1, \omega_2) = 2 \times \frac{1}{2\pi}\int\limits_{\omega_1}^{\omega_2} \frac{2\sigma^2\beta}{\omega^2 + \beta^2} d\omega = \left(\frac{2\sigma^2\beta}{\pi}\right)\frac{1}{\beta}\tan^{-1}\left(\frac{\omega}{\beta}\right)\Bigg]_{\omega_1}^{\omega_2}.$$

Using Integration tables gives

$$\text{Power}(\omega_1, \omega_2) = \frac{2\sigma^2}{\pi}\left[\tan^{-1}\left(\frac{\omega_2}{\beta}\right) - \tan^{-1}\left(\frac{\omega_1}{\beta}\right)\right].$$

■

Example 3.11 Find the mean square value and the variance of the process of Example 3.10.

Solution

Using the same integral as in Example 3.10, the mean square is

$$E\left\{X^2(t)\right\} = \frac{1}{2\pi}\int_{-\infty}^{\infty} S_{XX}(j\omega)\mathrm{d}\omega = \frac{1}{2\pi j}\int_{-j\infty}^{j\infty} S_{XX}(s)\mathrm{d}s = \left(\frac{2\sigma^2\beta}{2\pi}\right)\frac{1}{\beta}\tan^{-1}\left(\frac{\omega}{\beta}\right)\Bigg]_{-\infty}^{\infty} = \sigma^2,$$

where we use the identity $\tan^{-1}(\infty) = \pi/2$. Since the process is zero-mean, the variance is equal to the mean square. ■

3.7.2 Cross-Spectral Density Function

The Fourier transform of the cross-correlation gives the cross-spectral density

$$S_{XY}(j\omega) = \mathfrak{F}\{R_{XY}(\tau)\} = \int_{-\infty}^{\infty} R_{XY}(\tau)e^{-j\omega\tau}\mathrm{d}\tau. \tag{3.45}$$

The cross-spectral density is clearly zero for orthogonal signals since the cross-correlation is identically zero.

Recall the skew symmetry of the cross-correlation $R_{XY}(\tau) = R_{YX}(-\tau)$ and rewrite the cross-spectral density as

$$S_{XY}(j\omega) = -\int_{\infty}^{-\infty} R_{YX}(-\tau)\mathrm{e}^{j\omega(-\tau)}d(-\tau).$$

Change of variable of integration to $\lambda = -\tau$ then observe that the RHS is $S_{XY}(j\omega)$ with j replaced by $-j$, i.e., its conjugate

$$S_{XY}(j\omega) = S_{YX}^*(j\omega). \tag{3.46}$$

3.7.2.1 Coherence Function

The coherence is a measure of the relation between two signals

$$\gamma_{XY}^2(j\omega) = \frac{|S_{XY}(j\omega)|^2}{S_{XX}(j\omega)S_{YY}(j\omega)} \leq 1, \forall\omega. \tag{3.47}$$

If the two signals are the ergodic stimulus and response of a linear system, the coherence provides a measure of causality. This has been extensively applied in neuroscience where the association of a response with a particular stimulus is often difficult. The coherence function also widely used in acoustics.

From the definition of orthogonality, the coherence is zero for two orthogonal signals. For identical signals, it is clearly unity. Hence, the coherence has the extreme values

$$\gamma_{XY}^2(j\omega) = \begin{cases} 1, & X(t) = Y(t) \\ 0, & \text{orthogonal } X(t) \text{ and } Y(t) \end{cases}. \tag{3.48}$$

Spectral Density of Sum

$$Z(t) = X(t) + Y(t).$$

We recall that the autocorrelation is

$$R_{ZZ}(\tau) = R_{XX}(\tau) + R_{XY}(\tau) + R_{YX}(\tau) + R_{YY}(\tau).$$

Fourier transforming gives

$$S_{ZZ}(j\omega) = S_{XX}(j\omega) + S_{XY}(j\omega) + S_{YX}(j\omega) + S_{YY}(j\omega). \tag{3.49}$$

For orthogonal random process, the cross-correlation and the cross-spectra are zero $S_{XY}(j\omega) = 0$ and the spectral density of the sum is

$$S_{ZZ}(j\omega) = S_{XX}(j\omega) + S_{YY}(j\omega). \tag{3.50}$$

If X and Y uncorrelated and either X or Y is zero-mean, then the two processes are orthogonal.

3.8 Spectral Factorization

The PSD is defined for stationary processes and, in most applications, we deal with "lumped processes" whose PSD can be modeled as a ratio of polynomials. Several applications in random signal processing require the factorization of the spectral density function.

3.8.1 Continuous-Time Processes

It was shown in Sect. 3.7.1 that if the PSD is a rational function, then it is an even function of frequency, i.e., $S(\omega) = S(-\omega)$, and can be written as $S(\omega^2)$ or $S(-s^2)$. Consequently, if s_i is a root of its numerator or denominator, then $-s_i$ is also a root and $S(s)$ can always be factorized as

$$S(s) = \frac{N(s)N(-s)}{D(s)D(-s)} = L(-s)L(s). \tag{3.51}$$

For real process, all roots of $L(s)$ are real or complex conjugate. Complex zeros on the imaginary axis must have even multiplicity for the PSD to be positive. The properties of the autocorrelation, the inverse transform of $S(s)$, imply that $S(s)$ has no imaginary axis poles.

It is simple to factorize the PSD using the following procedure.

Procedure 1. PSD Factorization for continuous-time processes

1. Find the roots of the numerator and denominator polynomials of the spectral density function.
2. Separate the roots in the RHP from those in the LHP.
3. Construct two functions, one including all RHP roots $L(-s)$ and another including all LHP roots $L(s)$.

The following example applies Procedure 1 to a simple example.

Example 3.12 Obtain the spectral factorization of $S(s)$.

$$S(s) = \frac{-s^2+1}{s^4-2s^2+4}.$$

Solution

For spectral factorization, we obtain the roots of the numerator and denominator and separate the roots in the RHP form those in the left half plane. The numerator roots are $\{\pm 1\}$ and the denominator roots are $\{\pm 1.2247 \pm j\ 0.7071\}$. The spectral factorization is

$$S(s) = L(-s)L(s) = \frac{-s+1}{s^2 - 2.4495s + 2} \times \frac{s+1}{s^2 + 2.4495s + 2}.$$

■

In many problems, we encounter two special cases where spectral factorization is particularly easy:

(1) Quadratic: $-a^2s^2 + b^2 = (-as + b)(as + b)$.
(2) Quartic: $s^4 - as^2 + b^2 = \left(s^2 + b - cs\right)\left(s^2 + b + cs\right), c = \sqrt{a + 2b}$.

The expression for the quartic is obtained using the factorization

$$s^4 - as^2 + b^2 = \left(s^2 + b\right)^2 - (a + 2b)s^2.$$

The following examples show how simple spectral factorization is for the quadratic and quartic.

Example 3.13 Obtain the spectral factorization of $S(s)$.

$$S(s) = \frac{-s^2 + 1}{-s^2 + 4}.$$

Solution

$$S(s) = L(-s)L(s) = \left[\frac{-s+1}{-s+2}\right]\left[\frac{s+1}{s+2}\right].$$

■

Example 3.14 Obtain the spectral factorization of $S(s)$, then obtain the mean square value of the process

$$S(s) = \frac{-s^2 + 1}{s^4 - 2s^2 + 9}.$$

Solution

For this example, the coefficient is

$$a = 2, b = \sqrt{9} = 3.$$

We calculate

$$c = \sqrt{a + 2b} = \sqrt{2 + 2 \times 3} = \sqrt{8} = 2\sqrt{2}$$

and obtain the spectral factorization

$$S(s) = L(-s)L(s) = \left[\frac{-s+1}{s^2 - 2\sqrt{2}s + 3}\right]\left[\frac{s+1}{s^2 + 2\sqrt{2}s + 3}\right].$$

The mean square value is

$$E\left\{X^2(t)\right\} = \frac{1}{2\pi j}\int_{-j\infty}^{j\infty} S_{XX}(s)\mathrm{d}s = \frac{1}{2\pi j}\int_{-j\infty}^{j\infty}\left[\frac{-s+1}{s^2 - 2\sqrt{2}s + 3}\right]\left[\frac{s+1}{s^2 + 2\sqrt{2}s + 3}\right]\mathrm{d}s.$$

Using the table of integrals of Appendix I gives

$$E\left\{X^2(t)\right\} = \frac{a_0^2 b_2 + a_1^2 b_0}{2b_0 b_1 b_2} = \frac{1^2(1) + 1^2(3)}{2(3)\left(2\sqrt{2}\right)(1)} = \frac{1}{3\sqrt{2}}.$$

■

3.8.2 Discrete-Time Processes

Substituting z for $e^{j\omega}$ in (3.40) gives

$$S(z) = R(0) + \sum_{m=1}^{\infty} R(m)(z^m + z^{-m}).$$

Because $S(z)$ is a function of $z^m + z^{-m}$, substituting $S(z) = S\left(z^{-1}\right)$, and if $S(z)$ has a root at z_i, then it also has a root at $1/z_i$. This gives the spectral factorization

$$S(z) = \frac{N(z)N\left(z^{-1}\right)}{D(z)D\left(z^{-1}\right)} = r^2 L(z)L\left(z^{-1}\right), \tag{3.52}$$

where r is a normalizing constant such that $L(\infty) = 1$.

It is simple to factorize the PSD using the following procedure.

Autoregressive (AR) processes are random processes with $L(z)$ in the form

$$L(z) = \frac{b_0}{z^n + a_{n-1}z^{n-1} + \ldots + a_1 z + a_0}. \tag{3.53}$$

This corresponds to a process x_k governed by the difference equation

$$x_{k+n} = b_0 u_k - a_{n-1}x_{k+n-1} - \cdots - a_1 x_{k+1} - a_0 x_k,$$

with noise process u_k. Moving average (MA) processes are random processes with $L(z)$ in the form

$$L(z) = \frac{b_m z^m + b_{m-1} z^{m-1} + \cdots + b_1 z + b_0}{z^m}. \tag{3.54}$$

This corresponds to a process x_k governed by the difference equation

$$x_{k+m} = b_0 u_k + b_1 u_{k+1} \ldots + b_{m-1} u_{k+m-1} + b_m u_{k+m}.$$

Autoregressive moving average (ARMA) processes are random processes with $L(z)$ in the form

$$L(z) = \frac{b_m z^m + b_{m-1} z^{m-1} + \cdots + b_1 z + b_0}{z^n + a_{n-1} z^{n-1} + \cdots + a_1 z + a_0}. \tag{3.55}$$

This corresponds to a process x_k governed by the difference equation

$$\begin{aligned} x_{k+n} &= b_0 u_k + b_1 u_{k+1} \ldots + b_{m-1} u_{k+m-1} \\ &\quad + b_m u_{k+m} - a_{n-1} x_{k+n-1} - \cdots - a_1 x_{k+1} - a_0 x_k. \end{aligned} \tag{3.56}$$

Procedure 2. PSD Factorization for discrete-time processes

(1) Find the roots of the numerator and denominator polynomials of the spectral density function.
(2) Separate the roots inside the unit circle from those outside the unit circle.
(3) Construct two functions, one including all roots outside the unit circle and another including all roots inside the unit circle.

The following example applies Procedure 2 to a simple example.

Example 3.15 Obtain the spectral factorization of $S(z)$ for the AR process

$$S(z) = \frac{b^2}{(z-a)\left(z^{-1}-a\right)}, \quad |a| < 1.$$

Solution

For this simple example, the factorization is obviously

$$S(z) = L(z)L\left(z^{-1}\right) = \left(\frac{bz}{z-a}\right)\left(\frac{bz^{-1}}{z^{-1}-a}\right).$$

■

3.9 Examples of Stochastic Processes

This section presents several important stochastic processes that are important in many applications. The first is an idealized process known as white noise. Band-limited white noise is a more physically plausible stochastic process. Next, we present the Gauss–Markov process that is Gaussian but belong to a class known as Markovian processes. Finally, we compare the telegraph wave to the Gauss–Markov process.

Example 3.16: White Noise White noise is a stationary random process with a constant SDF.

$$S_W(j\omega) = A.$$

By the Wiener–Khinchine relation, the autocorrelation is the impulse

$$R_W(\tau) = A\delta(\tau).$$

Figure 3.10 shows the SDF and autocorrelation of white noise. The name white is because of the spectrum covers all frequencies analogously to white light.▪

White noise is clearly not physically realizable because its infinite bandwidth requires infinite energy and because a Dirac delta is not physically realizable. However, it is an important mathematical tool in the analysis of random signals. A more physically plausible model is band-limited white noise.

Example 3.17: Band-limited White Noise A band-limited base-band random signal has a constant PSD function over a finite bandwidth

$$S_{bw}(j\omega) = \begin{cases} A, & |\omega| \le 2\pi W \\ 0, & |\omega| > 2\pi W \end{cases}.$$

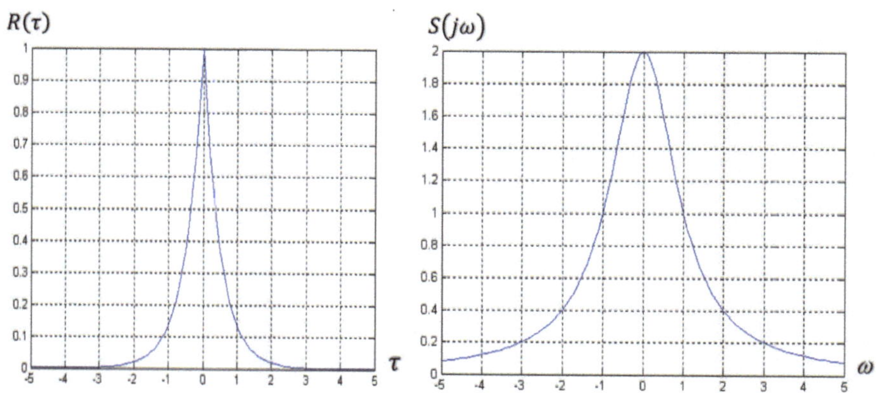

Fig. 3.9 Autocorrelation and PSD

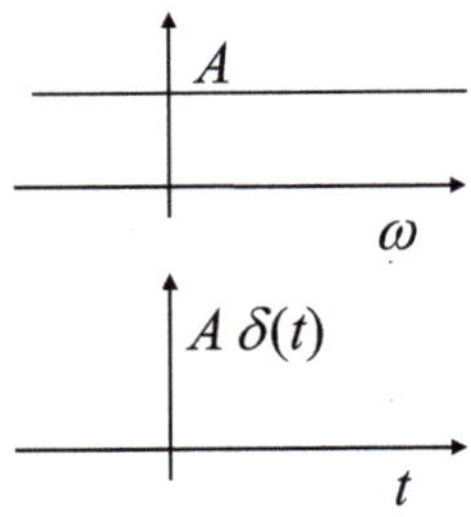

Fig. 3.10 White noise

Although the model is more physically plausible than white noise, it is still not physically realizable because of the abrupt changes in its PSD (Fig. 3.11).

$$R_{bw}(\tau) = 2WA\frac{\sin(2\pi W\tau)}{2\pi W\tau}.$$

Example 3.18 BW centered around $f = W_0$

$$S_{bw}(j\omega) = \begin{cases} A, & ||\omega| - 2\pi W_0| \leq \pi\Delta W \\ 0, & ||\omega| - 2\pi W_0| > \pi\Delta W \end{cases}.$$

Using the frequency shifting theorem of the inverse Fourier transform, we shift by $\pm W_0$ and replace W in the expression for band-limited white noise with $\Delta W/2$ (Figs. 3.12 and 3.13).

$$R_{bw}(\tau) = \Delta WA\frac{\sin(2\pi\Delta W\tau/2)}{\pi\Delta W\tau}\left[e^{j2\pi W_0\tau} + e^{-j2\pi W_0\tau}\right],$$

$$R_{bw}(\tau) = 2\Delta WA\frac{\sin(\pi\Delta W\tau)}{\pi\Delta W\tau}\cos(2\pi W_0\tau).$$

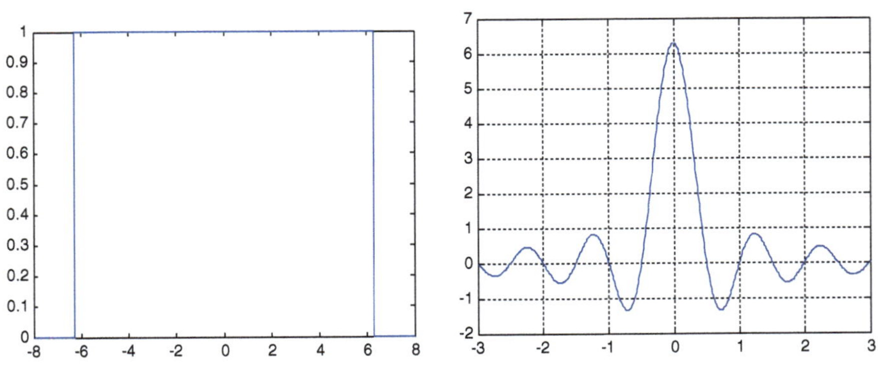

Fig. 3.11 Autocorrelation and power spectral density of band-limited white noise

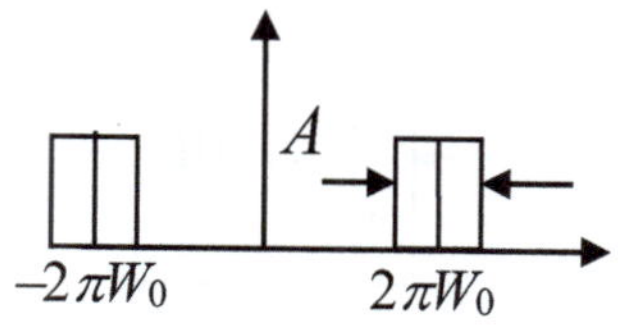

Fig. 3.12 PSD of band-limited white noise

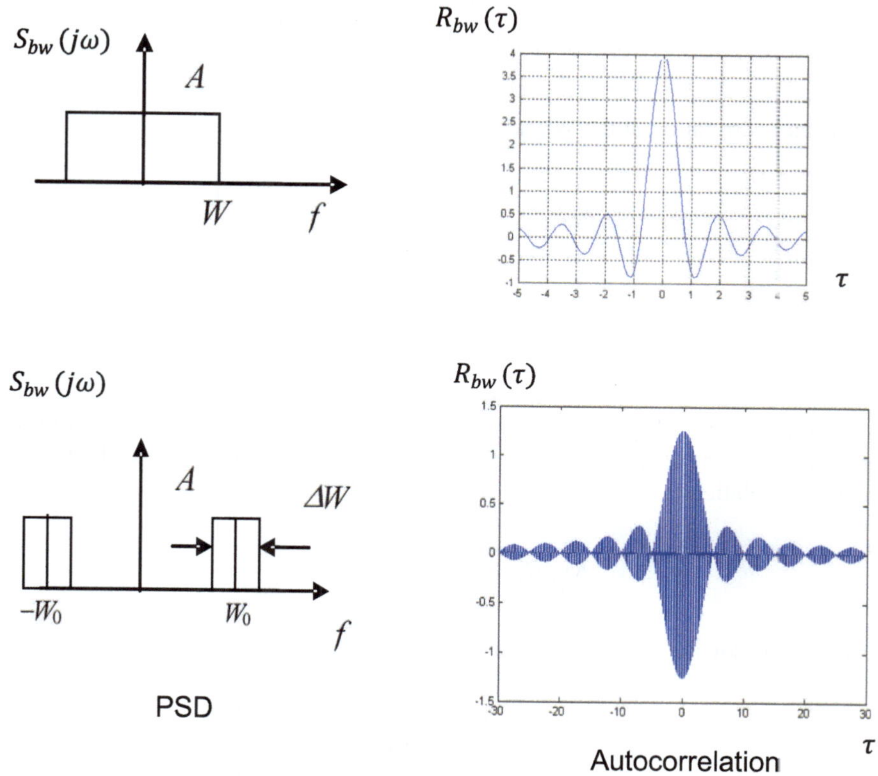

Fig. 3.13 Autocorrelation and PSD of band-limited white noise

Example 3.19: Wiener Process or Brownian Motion The Wiener process is the continuous analog of random walk. Like the random walk, it a zero-mean process that accumulates input changes. For a continuous process, we replace the summation of the random walk with integration

$$X(t) = \int_0^t F(u)\mathrm{d}u.$$

The expectation of the process is (Fig. 3.14)

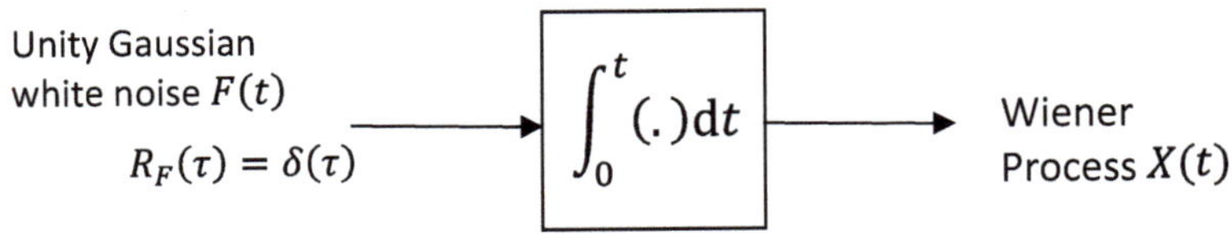

Fig. 3.14 Wiener process

$$E\{X(t)\} = E\left\{\int_0^t F(u)\mathrm{d}u\right\} = 0.$$

The autocorrelation of the Wiener process is

$$\begin{aligned} E\{X(t_1)X(t_2)\} &= E\left\{\int_0^{t_1}\int_0^{t_2} F(u)F(v)\mathrm{d}u\mathrm{d}v\right\} \\ &= \int_0^{t_1}\int_0^{t_2} \delta(u-v)\mathrm{d}u\mathrm{d}v. \end{aligned}$$

We observe from Fig. 3.15 that the integral is more easily evaluated if we first integrate along the longer side of the rectangle. Evaluating the integral shows that the autocorrelation is

$$R_{XX}(t_1, t_2) = \begin{cases} t_2, t_1 \geq t_2 \\ t_1, t_1 < t_2 \end{cases}.$$

The mean square of the process is

$$\begin{aligned} E\{X^2(t)\} &= E\left\{\int_0^{t}\int_0^{t} F(u)F(v)\mathrm{d}u\mathrm{d}v\right\} \\ &= \int_0^{t}\int_0^{t} \delta(u-v)\mathrm{d}u\mathrm{d}v = t. \end{aligned}$$

■

Fig. 3.15 Area of integration to evaluate the autocorrelation of the Wiener process

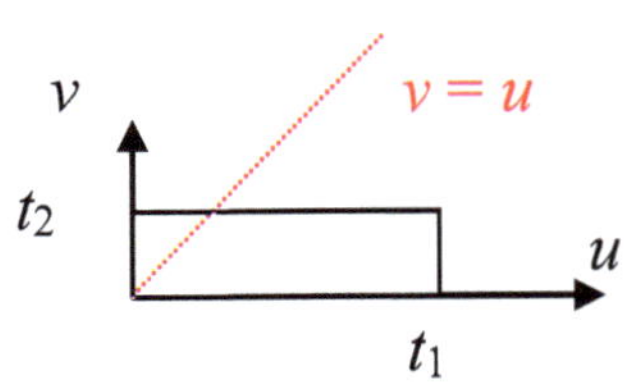

Example 3.20: Narrowband Process A narrow band random process is a process whose PSD is zero outside a small range around a high frequency ω_c. The process is given by

$$N(t) = X(t)\cos(\omega_c t) - Y(t)\sin(\omega_c t),$$

where $X(t)$ and $Y(t)$ are i.i.d. zero-mean Gaussian random processes. Use the Cartesian to polar transformation to obtain the pdf of the signal in terms of the coordinates (R, θ) that are Rayleigh-Uniform (see Chap. 1) (Fig. 3.16)

$$N(t) = R(t)\cos(\omega_c t + \theta(t)),\ \omega_c \gg \Delta\omega.$$

Solution

Because $N(t)$ is a linear combination of Gaussian processes, it is a Gaussian process, and it completely characterized by its mean

$$E\{N(t)\} = E\{X(t)\}\cos(\omega_c t) - E\{Y(t)\}\sin(\omega_c t) = 0,$$

and its mean square

$$E\{N^2(t)\} = E\{X^2(t)\}\cos^2(\omega_c t) + E\{Y^2(t)\}\sin^2(\omega_c t).$$

Because $X(t)$ and $Y(t)$ are zero-mean, their variance is equal to their mean square, which yields

$$E\{N^2(t)\} = \sigma^2.$$

The pdf of the process is

$$p_N(n) = \frac{1}{\sqrt{2\pi\sigma^2}}\exp\left(-\frac{n^2}{2\sigma^2}\right).$$

■

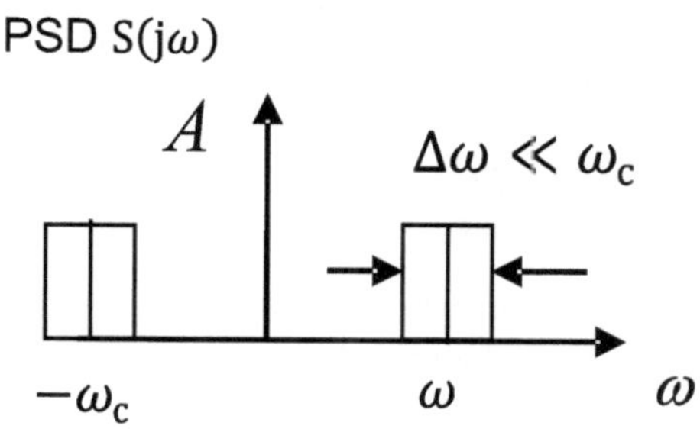

Fig. 3.16 PSD of narrow-band process

3.9.1 Markov Processes

Some random processes have a short memory such that

$$P\{\boldsymbol{x}(t_n) \le x_n | \boldsymbol{x}(t), t \le t_{n-1}\} = P\{\boldsymbol{x}(t_n) \le \boldsymbol{x}_n | \boldsymbol{x}(t_{n-1})\}, t_{n-1} < t_n. \tag{3.57}$$

This is the Markov property where a process can randomly be in one of a finite or infinite set of states at time t_n that depends only on the previous state $\boldsymbol{x}(t_{n-1})$, independently of the behavior of $\boldsymbol{x}(t)$ prior to t_{n-1}. A familiar model that is Markovian is the linear state equation

$$\boldsymbol{x}(t_{k+1}) = A\boldsymbol{x}(t_k) + B\boldsymbol{w}(t_k), \tag{3.58}$$

where $\boldsymbol{x}(t_k)$ is the state vector and $\boldsymbol{w}(t_k)$ is a random input vector at time t_k. The model is clearly Markovian because $\boldsymbol{x}(t_{k+1})$ is independent of the history of the process. In particular, the process is first-order Markov because it depends on only one past value. Higher-order Markov processes that depend on more values can be defined. For example, a process that depends on the two past values is second-order Markov.

Example 3.21: Gauss–Markov Process An ergodic zero-mean Gaussian and Markov process whose PSD is

$$S_{XX}(s) = \frac{2\sigma^2\beta}{-s^2 + \beta^2}.$$

Find the autocorrelation function of the process.

Solution

Using the Wiener–Khinchine relation and the two-sided Laplace transform tables, we obtain the autocorrelation function

$$R_{XX}(\tau) = \sigma^2 e^{-\beta|\tau|}.$$

Figure 3.9 shows the autocorrelation and PSD of a Gauss–Markov process. ■

The Gauss–Markov process is both Gaussian and Markov. Because of its excellent mathematical properties and the fact that it approximately represents many naturally occurring random processes, it is of prime importance.

Example 3.22: PDF of Gauss–Markov Process For a Gauss–Markov process with time constant 0.2 s and variance 4:

(a) Write an expression for the density function governing the process at any time.
(b) Find the joint pdf for the process at times 0.5 s and 0.7 s.

Solution

(a) The process is zero-mean since $R_{XX}(\tau) \to 0$ as $\tau \to \infty$, it is Gaussian, and its variance is 4, which makes the pdf

$$p_X(x) = \frac{1}{\sqrt{2\pi\sigma^2}} e^{-x^2/2\sigma^2} = \frac{1}{\sqrt{8\pi}} e^{-x^2/8}.$$

(b) For a time constant of 0.2 s, we have

$$\beta = \frac{1}{0.2} = 5s^{-1}.$$

The autocorrelation of the process is

$$R(\tau) = \sigma^2 e^{-\beta|\tau|} = 4e^{-5|\tau|}.$$

We consider the vector

$$\boldsymbol{x} = \begin{bmatrix} x_1 \; x_2 \end{bmatrix}^T = \begin{bmatrix} X(0.5) \; X(0.7) \end{bmatrix}^\mathrm{T}.$$

The autocorrelation for the time separation between the entries is

$$R(0.7 - 0.5) = 4e^{-5|0.2|} = 4e^{-1}.$$

The covariance matrix for the vector is

$$C_x = \begin{bmatrix} \sigma^2 & R(0.2) \\ R(0.2) & \sigma^2 \end{bmatrix} = 4\begin{bmatrix} 1 & e^{-1} \\ e^{-1} & 1 \end{bmatrix}.$$

The determinant of the covariance matrix is $|C_x| = 16\left(1 - e^{-2}\right)$, and its inverse is

$$C_x^{-1} = \frac{1}{4\left(1 - e^{-2}\right)} \begin{bmatrix} 1 & -e^{-1} \\ -e^{-1} & 1 \end{bmatrix}.$$

The joint multivariate Gaussian pdf for the process at times 0.5 s and 0.7 s is

$$p_{X_1,X_2}(x_1, x_2) = \frac{1}{2\pi|C_x|^{1/2}} e^{-\boldsymbol{x}^T C_x^{-1} \boldsymbol{x}} = \frac{1}{8\pi\sqrt{1 - e^{-2}}} \exp\left(\frac{x_1^2 + x_2^2 - 2e^{-1} x_1 x_2}{4\left(1 - e^{-2}\right)}\right).$$

■

Example 3.23: Random Telegraph Wave The random telegraph wave is a binary voltage wave of amplitude ± 1. The initial state can be 1 or −1 with equal probability

$$P(X(0) = 1) = P(X(0) = 1) = 0.5.$$

Switching is governed by a Poisson distribution and the probability of k switches in time interval T is

$$P(k) = (aT)^k \mathrm{e}^{-aT}/k!,$$

where a is the average number of switches per unit time. (Papoulis and Pillai) show that the autocorrelation of the process is given by

$$R_{XX}(\tau) = \mathrm{e}^{-2a|\tau|},$$

i.e., the process has the same autocorrelation as the Gauss Markov process. Compare the process to the Gauss Markov and explain why they have the same autocorrelation but are otherwise quite different.

Solution

A complete analysis of the random telegraph wave, as presented in (Papoulis and Pillai) is not required here. The mean of the telegraph wave is

$$E\{X(t)\} = (1)(0.5) + (-1)(0.5) = 0.$$

This is equal to the mean of the ergodic Gauss–Markov process because

$$\lim_{\tau \to \infty} R_{XX}(\tau) = 0.$$

The mean square of the telegraph wave is

$$E\left\{X^2(t)\right\} = (1)^2(0.5) + (-1)^2(0.5) = 1.$$

The two waveforms have the same first- and second-order statistics.

However, while the random telegraph wave can only have values of $\pm$ 1 and is governed by a Poisson process, the values of the Gauss–Markov process are governed by a Gaussian pdf. The two processes are clearly quite different. The example shows that completely different random processes can have the same autocorrelation and PSD. ■

Problems

3.1 Generalize the random walk process so that the probability of a forward step of length L is p and the probability of backward step is $(1 - p)$. Determine the mean and variance of the process and the mean and variance of the distance traveled after n steps.

3.2 For the random walk process of Problem 3.1, determine the probability of advancing by l steps, then show that the return to the starting positions can only occur after and even number of steps.

3.3 Prove the properties of the cross-correlation of wide-sense stationary random processes.

I. Skew symmetry $R_{XY}(\tau) = R_{YX}(-\tau)$.
II. $|R_{XY}(\tau)| \le \sqrt{R_{XX}(0)R_{YY}(0)}$, $|\rho| \le 1$.
III. $|R_{XY}(\tau)| \le \frac{1}{2}[R_{XX}(0) + R_{YY}(0)]$.
IV. $R_{XY}(\tau) = \begin{cases} 0, & \text{orthogonal} \\ m_X m_Y, & \text{uncorrelated} \end{cases}$.

3.4 A discrete-time random signal sent across a communication channel consists of values selected randomly with i.i.d. probability

$$P(x(k) = 1) = p, P(x(k) = -1) = 1 - p, p < 1.$$

Find the mean, mean square, variance, autocorrelation, covariance, the PSD of the signal.

3.5 Find the mean and autocorrelation of the random signal with constant amplitude

$$X(t) = A\cos(\omega t), \omega \sim U(-\pi, \pi).$$

Is the signal stationary (a) in the mean? (b) in the autocorrelation?

3.6 Show that the random process

$$Z(t) = X\cos(\omega t) - Y\sin(\omega t)$$
$$X \sim \mathcal{N}(0, \sigma^2), Y \sim \mathcal{N}(0, \sigma^2)$$

is wide-sense stationary if and only if X and Y are orthogonal.

3.7 An engineer received the following functions and was told that they are estimates of PSDs for a physical process. For each function, decide whether it is a valid PSD for a physical process.

(a) $S(\omega^2) = 5\sin(\omega^2)$.
(b) $S(\omega^2) = 5\left|\sin(\omega^2)\right|\exp(-\omega)$.
(c) $S(\omega^2) = 5\left|\sin(\omega^2)\right|$.

3.8 Show that the following function cannot be a PSD

$$S(\omega) = \begin{cases} \cos(\omega), & \omega \in [-\pi/2, \pi/2] \\ 0, & \text{elsewhere} \end{cases}.$$

Hint: Inverse Fourier transform and check the properties of the corresponding function against those of the autocorrelation.

3.9 If a signal z(t) is obtained by multiplying two independent random signals x(t) × y(t), show that its autocorrelation is given by the product of their autocorrelations

$$R_z(\tau) = R_x(\tau) R_y(\tau).$$

3.10 Find the spectral density function of a signal whose autocorrelation is

$$R(\tau) = \mathrm{e}^{-\beta|\tau|} \cos(\omega_0 \tau),\ \beta, \omega_0 \in \mathcal{R}^+,\ \tau \in \mathcal{R}.$$

3.11 Explain why the following function cannot be an autocorrelation

$$R(\tau) = \mathrm{e}^{-\beta|\tau|} \cos(\omega_0 \tau + \theta),\ \beta, \omega_0 \in \mathcal{R}^+,\ \theta, \tau \in \mathcal{R}.$$

3.12 Find the mean, variance, and autocorrelation of the amplitude-modulated signal

$$x(t) = A\cos(\omega t + \theta),\ A \sim N\left(m, \sigma^2\right),\ \omega, \theta \in \mathcal{R}.$$

3.13 Find the mean, variance, and autocorrelation of the phase-modulated signal

$$x(t) = A\cos(\omega t + \theta),\ \theta \sim U[-\pi/2, \pi/2],\ \omega,\ A \in \mathcal{R}.$$

3.14 Using the results of a single experiment, an engineer was able to approximately determine the autocorrelation of a random process.

(a) What is the property of the process that allowed the engineer to obtain the autocorrelation using a single recording and what is the expression he used?
(b) For the following functions, which could be the correct autocorrelation function he obtained? What is the mean and what is the mean square value of the process? Justify your answer.

$$\text{(i)}\quad R_{XX}(\tau) = \sin(\tau),\quad \text{(ii)}\quad R_{XX}(\tau) = \cos(0.5\tau)\mathrm{e}^{-2|\tau|}.$$

3.15 Find the spectral density function of a signal whose autocorrelation is

$$R(\tau) = \mathrm{e}^{-\beta|\tau|} \cos(\omega_0 |\tau| + \theta),\ \beta, \omega_0 \in \mathcal{R}^+,\ \tau \in \mathcal{R},\ \theta = 2l\pi,\ l = 1, 2, 3, \ldots.$$

3.16 The spectral density function of a random signal is

$$S(\omega^2) = \frac{1}{\sqrt{2\pi\beta^2}} \mathrm{e}^{-\omega^2/2\beta^2}$$

(a) Find the autocorrelation of the signal.

(b) Find the mean of the signal.
(c) Find the mean square of the signal using two different methods.

3.17 The spectral density of a discrete-time stationary Gaussian process is given by

$$S(z) = \frac{1}{1.0904 - 0.306 * z + 0.02z^2 - 0.306/z + 0.02/z^2}.$$

(a) Determine the transfer function of the shaping filter for the process
(b) Find the autocorrelation of the process

3.18 For a Gauss–Markov process with time constant 0.25 s and variance 10, find the third-order pdf $p_x(\boldsymbol{x})$

$$\boldsymbol{x} = \begin{bmatrix} x_1 \; x_2 \; x_3 \end{bmatrix}^T = \begin{bmatrix} x(0) \; x(1) \; x(2) \end{bmatrix}^T.$$

3.19 Show that if a process is Markov, then it is also reverse Markov, i.e., it satisfies

$$P_X(x_k | x_{k+1}, x_{k+2}, \ldots, x_{k+m}) = P_X(x_k | x_{k+1}).$$

3.20 A Poisson process with rate $\lambda \geq 0$ counts the number $N(t)$ of randomly occurring events in an interval of length t. The number of events of any interval of length t is governed by the Poisson model

$$P(N(t+s) - N(s) = n) = \mathrm{e}^{-\lambda t} \frac{(\lambda t)^n}{n!}, n = 0, 1, 2, \ldots.$$

(a) Show that the probability of n events in a period of length t is given by

$$P(N(t) = n) = \mathrm{e}^{-\lambda t} \frac{(\lambda t)^n}{n!}, n = 0, 1, 2, \ldots.$$

(b) Find the mean, variance, and mean square number of events in a time interval of length t
(c) Show that the numbers of events in any two nonoverlapping intervals are mutually independent.
(d) Find the autocorrelation of $N(t)$.

3.21 Let $Z(t) = \alpha X(t) + \beta Y(t), \alpha, \beta \in R$ be a random signal given by the linear combination of two independent wide-sense stationary random signals $(X(t), Y(t))$ with mean $(m_X(t), m_Y(t))$ and autocorrelation $(R_X(\tau), R_Y(\tau))$.

a. Find the mean and autocorrelation Of $Z(t)$.
b. Show that $Z(t)$ is wide-sense stationary.

3.22 Consider the random process

$$x_k = \frac{x_{k_0}}{\alpha k + 1}, \alpha > 0, x_{k_0} \sim N\left(m_x, \sigma_x^2\right).$$

Show that the process is Gauss–Markov.

Appendix 3.1 Brief Review of the Two-Sided Z-Transform

This section provides a brief review of the two-sided z-transform.

The two-sided z-Transform of sequence

$$\{G(k)\} = \{\ldots, G(-1), G(0), \ldots, G(i), \ldots\}$$

is defined as

$$\boldsymbol{G}(z) = \cdots + G(-1)z^1 + G(0) + \cdots + G(i)z^{-i} + \cdots,$$

where z is a time advance operator. It is also possible to define the z-transform using the Laplace transform, the delay theorem for Laplace transforms, and with the definition of $z = \mathrm{e}^{sT}$, where T is the sampling period. With the second definition, the z-transform inherits some properties of the Laplace transform. In particular, the z-transform is a linear transform.

- Linear:

$$\mathcal{Z}\{a\boldsymbol{f}(k) + b\boldsymbol{g}(k)\} = a\boldsymbol{F}(z) + b\boldsymbol{G}(z).$$

Convolution Theorem

The z-transform of the convolution of two sequences is the product of their transforms

$$\mathcal{Z}\left\{\sum_{i=-\infty}^{\infty} G(k-i)\boldsymbol{f}(i)\right\} = \boldsymbol{G}(z)\boldsymbol{F}(z),$$

$$\{\boldsymbol{f}(k)\} = \{\ldots, \boldsymbol{f}(-1), \boldsymbol{f}(0), \ldots, \boldsymbol{f}(i), \ldots\}.$$

Response of DT System

The response of linear system to any input sequence $\{f(k)\}$ is the convolution summation of the input and the impulse response sequence

$$x(k) = \sum_{i=-\infty}^{\infty} G(i)f(k-i) = \sum_{i=-\infty}^{\infty} G(k-i)f(i).$$

By the convolution theorem, the z-transform of the response is the product

$$X(z) = G(z)F(z), G(z) = \sum_{i=-\infty}^{\infty} G(i)z^{-i}.$$

The impulse response sequence and the z-transfer function are z-transform pairs

$$\underset{\text{impulse response}}{G(i)} \quad \overset{z}{\leftrightarrow} \quad \underset{\text{transfer function}}{G(z)}.$$

Even Function

For an even function $G(n) = G(-n)$, the z-transform satisfies

$$G(z) = \sum_{m=-\infty}^{\infty} G(m)z^{-m} = \sum_{m=-\infty}^{\infty} G(-m)z^{-m}.$$

Substituting $l = -m$ gives

$$G(z) = \sum_{l=-\infty}^{\infty} G(l)z^{l} = G\left(z^{-1}\right),$$

that is,

$$G(z) = G\left(z^{-1}\right).$$

Hence, poles are located symmetrically w.r.t. the unit circle.

Bibliography

1. Brown, R. G., & Hwang, P. Y. C. (1997). *Introduction to random signals and applied Kalman filtering* (3ed.). Wiley.
2. Cooper, G. R., & McGillem, C. D. (1999). *Probabilistic methods of signal and system analysis.* Oxford Univ. Press.

3. Moon, T. K., & Sterling, W. C. (2000). *Mathematical methods and algorithms for signal processing*. Prentice Hall, Upper Saddle River
4. Oppenheim, A. V., & Verghese, G. C. (2016). *Mathematical methods and algorithms for signal processing*. Pearson.
5. Oppenheim, A. V., Willsky, A. S., & Nawab, H. (1997). *Signals and systems*. Prentice-Hall.
6. Papoulis, A., & Pillai, S. U. (2002). *Signals, systems and inference*. Pearson.
7. Peebles, P. Z. (1993). *Probability, random variables, and random signal principles* (pp. 253–261). McGraw Hill.
8. Shanmugan, K. S., & Breipohl, A. M. (1988). *Detection, estimation & data analysis* (pp. 132–134). Wiley.
9. Stark, H., & Woods, J. W. (2002). *Probability and random processes with applications to signal processing*. Prentice-Hall.

Chapter 4
Linear System Response to Random Inputs

4.1 Calculus for Random Signals

Dynamic systems are described by integro-differential equations, and their analysis requires integration and differentiation. In calculus, integrals and derivatives are defined as limits. These limits do not exist for all realizations of random signals and traditional calculus fails. The calculus for random signals must use convergence that is defined in an averaged sense as discussed in Appendix F. Convergence can be in law, probability, q^{th} mean for some integer q, or almost sure. The most commonly used convergence to develop calculus is mean square convergence, which we use to develop mean square calculus.

A complete discussion of mean square calculus is beyond the scope of this text. Furthermore, it is possible to perform all the operations needed for system analysis as in traditional calculus while using mean square calculus. The discussion presented here is only intended to make the reader aware of the difficulties associated with calculus for random signals by comparing its definitions to traditional calculus.

4.1.1 Continuity

For deterministic signals, we have the following definition.

Definition 4.1 Continuity of deterministic functions.

A real function $f: \mathcal{E} \to \mathcal{R}$ is continuous on a subset of the real line $\mathcal{E}$ if for any point $t_0 \in \mathcal{E}$

$$\mathcal{L}im_{t \to t_0} x(t) = x(t_0). \tag{4.1}$$

M. S. Fadali, *Introduction to Random Signals, Estimation Theory, and Kalman Filtering*, https://doi.org/10.1007/978-981-99-8063-5_4

It is quite common to have realizations of a random signal that do not satisfy the above definition. The next definition is more appropriate for random signals.

Definition 4.2 Continuity of random functions
A random function $f\colon \mathcal{E} \to \mathcal{R}$ is continuous on a subset of the real line $\mathcal{E}$ if for any point $t_0 \in \mathcal{E}$

$$\lim x(t) = x(t_0), \tag{4.2}$$

i.e.,

$$\mathop{\mathcal{L}im}_{t \to t_0} E\left\{|X(t) - X(t_0)|^2\right\} = 0. \tag{4.3}$$

The following theorem provides a condition for mean square continuity.

Theorem 4.1 Continuity Theorem
A wide-sense stationary process $X(t)$ is mean square continuous if its autocorrelation function $R_{XX}(\tau)$ is continuous at $\tau = 0$.

Proof Expand the quadratic and use autocorrelation definition to obtain

$$E\left\{|X(t+\tau) - X(t)|^2\right\} = R_{XX}(0) + R_{XX}(0) - 2R_{XX}(\tau).$$

Take the limit as $\tau \to \tau \to$

$$\mathop{\mathcal{L}im}_{\tau \to 0} E\left\{|X(t+\tau) - X(t)|^2\right\} = 2\left\{R_{XX}(0) - \mathop{\mathcal{L}im}_{\tau \to 0} R_{XX}(\tau)\right\}.$$

If $R_{XX}(\tau)$ is continuous at $\tau = 0$, we have

$$\mathop{\mathcal{L}im}_{\tau \to 0} E\left\{|X(t+\tau) - X(t)|^2\right\} = 0.$$

The following result allows us to exchange limit and expectation with the mild condition of finite variance ([3], p. 162). ■

Theorem 4.2 Continuity Theorem (Shanmugan & Breipohl [3], Stark & Woods)
For a mean square continuous wide-sense stationary process with finite variance and any continuous function $g(.)$

$$\mathop{\mathcal{L}im}_{t \to t_0} E\{g(X(t))\} = E\{g(X(t_0))\}. \tag{4.4}$$

Starting with the definition of the derivative

$$\dot{x}(t) = \underset{\epsilon \to 0}{\mathcal{L}im} \frac{x(t+) - x(t)}{\epsilon} \tag{4.5}$$

leads to the following definition.

Definition 4.3 Mean square derivative
For a finite variance stationary process, the mean square derivative is

$$\dot{X}(t) = \lim_{\epsilon \to 0} \frac{X(t+\epsilon) - X(t)}{\epsilon}, \tag{4.6}$$

$$\underset{\epsilon \to 0}{\mathcal{L}im}\, E\left\{ \left| \dot{X}(t) - \frac{X(t+\epsilon) - X(t)}{\epsilon} \right|^2 \right\} = 0. \tag{4.7}$$

Stark & Woods, p. 490, show that the limit exists if $\partial R_{XX}(t_1, t_2)/\partial t_1 \partial t_2$ exists at $t = t_1 = t_2$.

Similarly, Stark & Woods, p. 502, define the mean square integral in terms of a sum

$$I_n = \sum_{i=1}^{n} X(t_i)\Delta t_i. \tag{4.8}$$

Definition 4.4 Mean square integral

$$I = \int_{T_1}^{T_2} X(t)\mathrm{d}t = \mathrm{l.i.m.} I_n. \tag{4.9}$$

The integral exists when

$$\lim_{n \to \infty} E\{|I - I_n|^2\} = 0. \tag{4.10}$$

It can be shown that the mean square integral exists for any random process with finite variance. In addition, because integration is a linear operation, the expectation of the integral is equal to the integral of the expectation.

With the definitions of the derivative, integral, and continuity presented in this section and with the ability to interchange the order of integration or differentiation and expectation, we can proceed with the analysis of systems with random inputs. We can handle the equations as we do the more familiar calculus of deterministic signals, with the understanding that the operations are as defined in this section.

4.2 Response to Random Input

The response to a given realization of a random process does not characterize the system response to the process. We must therefore assess the statistical properties of the response in terms of its moments, autocorrelation, and power spectral density.

If the input to a linear system is Gaussian, the response is also Gaussian. Recall that differentiation and integration are linear operations. In Chap. 2, it was shown that linear transformations of a Gaussian random variable give another Gaussian process with a different mean and covariance matrix. Thus, all first-order densities for the output of a linear system are Gaussian if the input is Gaussian.

Two types of analysis of the random response are possible:

1. Stationary steady-state analysis:
    - The analysis is applicable to a wide-sense stationary random input.
    - The system dynamics must be linear time-invariant.
    - In practice, the results are valid after a sufficiently long period after which the transients become negligible. The length of the period depends on the dynamics of the system with slower dynamics requiring a longer period for the transients to die out.
    - Because the response is wide-sense stationary, we can use transforms in the analysis.
2. Nonstationary transient analysis:
    - Only time domain analysis is possible, and transforms cannot be used.
    - The system dynamics can be unstable, time-varying, and even nonlinear.

We consider both types of analysis, first for continuous-time random processes, then for discrete-time random processes.

4.3 Continuous-Time (CT) Random Signals

The output of a continuous-time system is the convolution of its impulse response and the input. For a time-varying system with impulse response matrix $G(t, \tau)$, the response is

$$\boldsymbol{y}(t) = \int_{-\infty}^{\infty} G(t, \tau)\boldsymbol{u}(\tau)\mathrm{d}\tau. \tag{4.11}$$

We first obtain the mean of the output response for both the stationary and nonstationary cases.

4.3.1 Mean Response

The mean of the output is

$$E\{y(t)\} = E\left\{\int_{-\infty}^{\infty} G(t,\tau)u(\tau)\mathrm{d}\tau\right\}.$$

Interchanging the order of expectation and integration gives

$$E\{y(t)\} = \int_{-\infty}^{\infty} G(t,\tau)E\{u(\tau)\}\mathrm{d}\tau.$$

For a process that is stationary in the mean with mean $\boldsymbol{m}_u$, we have

$$E\{y(t)\} = \left[\int_{-\infty}^{\infty} G(t,\tau)\mathrm{d}\tau\right]\boldsymbol{m}_u. \tag{4.12}$$

The integral is bounded if the linear system is bounded-input-bounded-output (BIBO) stable. For a BIBO stable linear time-invariant (LTI) system in the steady state, the expectation is

$$E\{y(t)\} = \left[\int_{-\infty}^{\infty} G(\tau)\mathrm{d}\tau\right]\boldsymbol{m}_u. \tag{4.13}$$

The integral in the expression for the mean is the DC gain

$$\int_{-\infty}^{\infty} G(\tau)\mathrm{d}\tau = \left.\int_{-\infty}^{\infty} G(\tau)\mathrm{e}^{-j\omega\tau}\mathrm{d}\tau\right]_{\omega\to 0}.$$

Hence, the mean of the output is the product of the mean of the input and the DC gain

$$\boldsymbol{m}_y = \boldsymbol{G}(j\omega)]_{\omega\to 0}\boldsymbol{m}_u. \tag{4.14}$$

This result is intuitive because one would expect the constant mean to be scaled by the DC gain.

4.3.2 *Stationary Steady-State Analysis for Continuous-Time Systems*

For a stable system in the steady state with a wide-sense stationary input, the autocorrelation of the output of a linear system is given by the following theorem.

Theorem 4.3 *The autocorrelation of the output of a stable LTI system with impulse response matrix $G(\tau)$ due to a wide-sense stationary input $\boldsymbol{u}(t)$ with autocorrelation $R_{uu}(\tau)$ is*

$$R_{yy}(\tau) = G(-\tau) * R_{uu}(\tau) * G^T(\tau). \tag{4.15}$$

The PSD of the output is

$$S_{yy}(s) = \boldsymbol{G}(-s)S_{uu}(s)\boldsymbol{G}^T(s). \tag{4.16}$$

Proof For a BIBO stable system, the response of the system remains bounded and is given by the convolution integral. Substituting the convolution integral in the expression for the autocorrelation gives

$$\begin{aligned} R_{yy}(\tau) &= E\left\{\boldsymbol{y}(t)\boldsymbol{y}^T(t+\tau)\right\} \\ &= E\left\{\int_{-\infty}^{\infty} G(\eta)\boldsymbol{u}(t-\eta)\mathrm{d}\eta\left[\int_{-\infty}^{\infty} G(\upsilon)\boldsymbol{u}(t+\tau-\xi)\mathrm{d}\xi\right]^T\right\}. \end{aligned}$$

Interchanging the order of expectation and integration gives

$$\begin{aligned} R_{yy}(\tau) &= \int_{-\infty}^{\infty}\int_{-\infty}^{\infty} G(\eta)E\left\{\boldsymbol{u}(t-\eta)\boldsymbol{u}^T(t+\tau-\xi)\right\}G^T(\xi)\mathrm{d}\eta\mathrm{d}\xi \\ &= \int_{-\infty}^{\infty}\int_{-\infty}^{\infty} G(\eta)R_{uu}(\tau+\eta-\xi)G^T(\xi)\mathrm{d}\eta\mathrm{d}\xi. \end{aligned}$$

We change the variable of integration to $\gamma = \eta + \tau$ and change order of integration

$$\begin{aligned} R_{yy}(\tau) &= \int_{-\infty}^{\infty}\int_{-\infty}^{\infty} G(\gamma-\tau)R_{uu}(\gamma-\xi)G^T(\xi)\mathrm{d}\gamma\mathrm{d}\xi \\ &= \int_{-\infty}^{\infty} G(\gamma-\tau)\left[\int_{-\infty}^{\infty} R_{uu}(\gamma-\xi)G^T(\xi)\mathrm{d}\xi\right]\mathrm{d}\gamma. \end{aligned}$$

We rewrite the expression as

$$R_{yy}(\tau) = \int_{-\infty}^{\infty} G(-[\tau - \gamma]) \left[\underbrace{\int_{-\infty}^{\infty} R_{uu}(\gamma - \xi) G^T(\xi) \mathrm{d}\xi}_{R_{uu}(\gamma) * G^T(\gamma)} \right] \mathrm{d}\gamma.$$

Because the second integral is with respect to ξ, it is a convolution that yields a function of γ. The second convolution integral gives the result

$$R_{yy}(\tau) = G(-\tau) * R_{uu}(\tau) * G^T(\tau).$$

We use the definition of Laplace transform then changing the variable of integration to $z = -t$

$$\mathcal{L}\{G(-t)\} = \int_0^{\infty} G(-t) e^{-st} dt = \int_0^{\infty} G(z) e^{-(-s)z} dz = \boldsymbol{G}(-s).$$

Laplace transforming gives the PSD

$$S_{yy}(s) = \boldsymbol{G}(-s) S_{uu}(s) \boldsymbol{G}^T(s).$$

■

For the single–input–single–output (SISO) case, the PSD of the output simplifies to

$$S_{yy}(s) = G(-s) G(s) S_{uu}(s). \tag{4.17}$$

In the frequency domain, the input–output relation simplifies to

$$Y(j\omega) = G(j\omega) U(j\omega),$$

and the PSD of the output is (Fig. 4.1).

$$S_{yy}(j\omega) = |G(j\omega)|^2 S_{uu}(j\omega). \tag{4.18}$$

Example 4.1 Obtain the PSD and mean square value of the output of the low-pass filter with transfer function

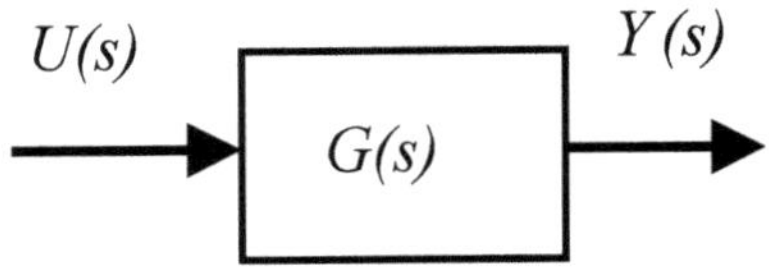

Fig. 4.1 Block diagram

$$G(s) = \frac{1}{Ts+1},$$

due to a Gauss–Markov random input with autocorrelation

$$R_{uu}(\tau) = \sigma^2 \mathrm{e}^{-\beta|\tau|}.$$

Solution

We first obtain the PSD of the Gauss–Markov process and its spectral factorization

$$S_{uu}(s) = \frac{2\sigma^2\beta}{-s^2+\beta^2} = \frac{\sqrt{2\sigma^2\beta}}{-s+\beta} \times \frac{\sqrt{2\sigma^2\beta}}{s+\beta}.$$

The PSD of the output is

$$\begin{aligned} S_{yy}(s) &= G(-s)G(s)S_{uu}(s) \\ &= \frac{1}{-Ts+1} \times \frac{1}{Ts+1} \times \frac{\sqrt{2\sigma^2\beta}}{-s+\beta} \times \frac{\sqrt{2\sigma^2\beta}}{s+\beta} \\ &= \frac{\sqrt{2\sigma^2\beta}}{(-Ts+1)(-s+\beta)} \times \frac{\sqrt{2\sigma^2\beta}}{(Ts+1)(s+\beta)}. \end{aligned}$$

The mean square value of the output is

$$R_{yy}(0) = \frac{1}{2\pi j} \int_{-j\infty}^{j\infty} S_{yy}(s)\mathrm{d}s = \frac{1}{2\pi j} \int_{-j\infty}^{j\infty} L(s)L(-s)\mathrm{d}s.$$

The causal factor of the PSD is

$$L(s) = \frac{\sqrt{2\sigma^2\beta}}{(s+\beta)(Ts+1)} = \frac{\sqrt{2\sigma^2\beta}}{Ts^2+(\beta T+1)s+\beta}.$$

Using the integration table for two-sided Laplace transforms, we have

$$I_2 = \frac{c_0^2+0}{2d_0d_1} = \frac{2\sigma^2\beta}{2\beta(1+\beta\mathrm{T})}.$$

The mean square value of the output is

$$E\{y^2(t)\} = R_{yy}(0) = \frac{\sigma^2}{\beta T + 1}.$$

Example 4.2 Obtain the PSD and mean square value of the output of the low-pass filter with transfer function

$$G(s) = \frac{1}{Ts + 1}$$

due to white noise then due to band-limited white noise (BLWN) with PSD

$$S_{uu}(j\omega) = \begin{cases} A, & |\omega| \le \omega_c \\ 0, & |\omega| > \omega_c \end{cases}$$

and compare the results.

Solution

The PSD of the system for white noise input with PSD $S_{uu}(j\omega) = A$ is

$$S_{yy}(j\omega) = |\boldsymbol{G}(j\omega)|^2 S_{uu}(j\omega) = \frac{A}{(\omega T)^2 + 1}.$$

The mean square value of the output is

$$\begin{aligned} E\{y^2\} &= \frac{1}{2\pi}\int_{-\infty}^{\infty} \frac{A}{(\omega T)^2 + 1} \mathrm{d}\omega \\ &= \frac{A}{2\pi T}\tan^{-1}(\omega T)\Big]_{-\infty}^{\infty} = \frac{A}{2T}. \end{aligned}$$

With BLWN, $G(.)$ input, the PSD of the output is

$$S_{yy}(j\omega) = |G(j\omega)|^2 S_{uu}(j\omega) = \begin{cases} \frac{A}{(\omega T)^2+1}, & |\omega| \le \omega_c \\ 0, & |\omega| > \omega_c \end{cases},$$

and its mean square value is

$$E\{y^2\} = \frac{1}{2\pi}\int_{-\omega_c}^{\omega_c} \frac{A}{(\omega T)^2 + 1} \mathrm{d}\omega = \frac{A}{\pi T}\tan^{-1}(\omega_c T).$$

For a much larger signal bandwidth than the system bandwidth $\omega_c \gg \omega_{BW} = 1/T, \omega_c T \gg 1$, and $\tan^{-1}(\omega_c T)$ is approximately unity. The mean square output for BLWN is approximately equal to the mean square output for white noise

$$E\{y^2\} \approx A/(2T).$$

Clearly, band-limited white noise is a more physically plausible noise model than white noise since the latter implies a signal of infinite energy. However, using white noise makes the system analysis much simpler. The example shows that when dealing with a low-pass system the results of using band-limited white noise are approximately the same as when using white noise. Thus, it is possible to use white noise in system analysis and avoid the complications of using band-limited white noise.

The reader may observe that even band-limited white noise is not physically plausible because of the abrupt changes in its PSD. A better representation of physical noise is one where the PSD has no abrupt changes, and the bandwidth is limited. By comparison to white noise, it is called **colored noise**. We show that we can obtain an equivalent band-limited noise to a noise model where the PSD does not have abrupt changes (Fig. 4.2).

We model a noise process as the response of a physical filter with transfer function $G(s)$ to a white noise input with $S_{uu}(s) = A$. The PSD of the output is

$$S_{yy}(s) = AG(s)G(-s),$$

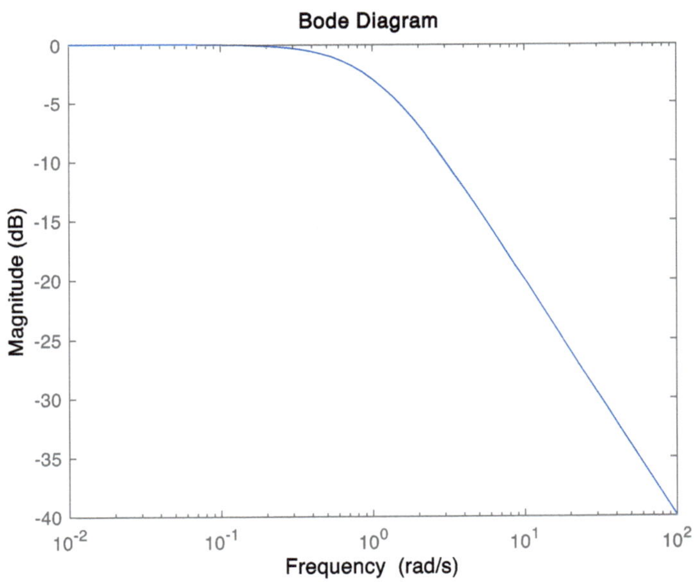

Fig. 4.2 Frequency response of LPF

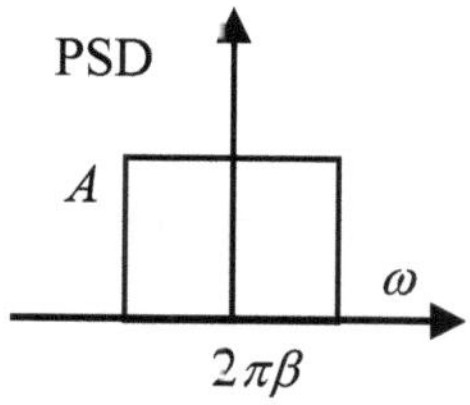

Fig. 4.3 PSD of BLWN with bandwidth β

and its mean square value is

$$E\{y^2\}\big]_{\text{physical filter}} = \frac{1}{2\pi j}\int_{-j\infty}^{j\infty} AG(s)G(-s)\mathrm{d}s.$$

To approximate the physical noise with ideal band-limited white noise (BLWN) of bandwidth β, we equate the mean square values. The mean square value for BLWN is

$$E\{y^2\}\big]_{\text{BLWN}} = \frac{1}{2\pi}\int_{-2\pi\beta}^{2\pi\beta} A\mathrm{d}\omega = 2A\beta.$$

Equating means square values gives the bandwidth of the equivalent BLWN process (Fig. 4.3).

$$\beta = \frac{1}{2\times 2\pi j}\int_{-j\infty}^{j\infty} G(s)G(-s)\mathrm{d}s. \tag{4.19}$$

Example 4.3 Find the equivalent BLWN for a noise process with PSD

$$S_{yy}(s) = \frac{1}{(T_1 s+1)(T_2 s+1)} = \frac{1}{T_1T_2s^2+(T_1+T_2)s+1}.$$

Solution

$$\beta = \frac{1}{2\times 2\pi j}\int_{-j\infty}^{j\infty} G(s)G(-s)\mathrm{d}s.$$

Using the complex integration of Appendix A

$$I_2 = \frac{c_0^2}{2d_0d_1} = \frac{1}{2(T_1 + T_2)}.$$

The bandwidth of the BLWN is

$$\beta = \frac{1}{4(T_1 + T_2)}.$$

4.3.3 *Shaping (Innovations) Filter*

Although physical noise is never white and only approximately stationary over a limited time span, it can often be modeled as wide-sense stationary colored noise that is the output of a linear filter. The filter that yields colored noise with unity white noise input is called the shaping or innovations filter. We use spectral factorization to obtain the filter transfer function

$$S_{yy}(s) = L(-s)L(s).$$

Using (4.17), the spectral density of the output of the filter is

$$S_{yy}(s) = G(-s)G(s) \times 1.$$

Thus, the transfer function of the shaping filter is the causal factor of the PSD (Fig. 4.4).

$$G(s) = L(s). \tag{4.20}$$

Example 4.4 **Gauss–Markov process.**
Obtain the shaping filter for the Gauss–Markov process.

Solution

The Gauss–Markov process has the PSD

$$S_{yy}(s) = \frac{2\sigma^2\beta}{-s^2 + \beta^2}.$$

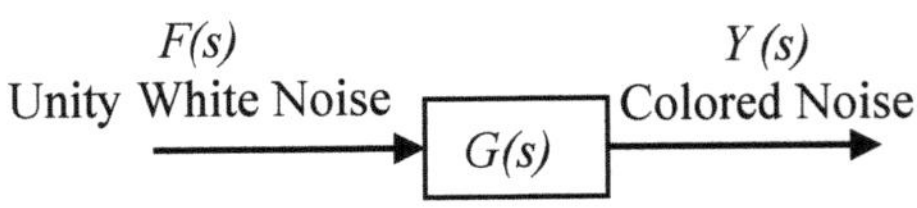

Fig. 4.4 Shaping filter

Spectral factorization of the PSD gives

$$S_{yy}(s) = L(-s)L(s)$$
$$= \frac{\sqrt{2\sigma^2\beta}}{-s+\beta} \times \frac{\sqrt{2\sigma^2\beta}}{s+\beta}.$$

The transfer function of the shaping filter is that causal factor

$$G(s) = \frac{\sqrt{2\sigma^2\beta}}{s+\beta}.$$

4.4 Nonstationary Analysis for Continuous-Time Systems

This section discusses the response of linear system to random initial conditions and random inputs. The state-space model of linear time-invariant system is given by

$$\begin{aligned} \dot{\boldsymbol{x}}(t) &= A\boldsymbol{x}(t) + B\boldsymbol{u}(t), \\ \boldsymbol{y}(t) &= C\boldsymbol{x}(t), \end{aligned} \tag{4.21}$$

where $\boldsymbol{x} \in \mathcal{R}^n$ is the state vector, $\boldsymbol{u}(t) \in \mathcal{R}^m$ is the input vector, and $\boldsymbol{y}(t) \in \mathcal{R}^l$ is the output vector. For linear systems where the principle of superposition applies and the responses due to the initial conditions and input can be obtained separately, the total response is the sum of the two responses

$$\boldsymbol{x}(t) = \mathrm{e}^{At}\boldsymbol{x}(t_0) + \int_0^t \mathrm{e}^{A(t-\tau)}B\boldsymbol{u}(\tau)\mathrm{d}\tau, \tag{4.22}$$

where $\boldsymbol{x}(0)$ is the initial condition vector, and

$$\mathrm{e}^{At} = \sum_{i=0}^{\infty} \frac{(At)^i}{i!} \tag{4.23}$$

is the matrix exponential. The matrix exponential is the inverse transform

$$\mathrm{e}^{At} = \mathcal{L}^{-1}\{[sI_n - A]^{-1}\}. \tag{4.24}$$

The output of the system is

$$y(t) = Ce^{At}x(0) + C\int_0^t e^{A(t-\tau)}Bu(\tau)d\tau. \tag{4.25}$$

The first term in the output is the zero-input response due to the initial conditions, and the second term is the zero-state response due to the input. To calculate the autocorrelation, we assume that the two responses are orthogonal so that the autocorrelation of the sum is the sum of the autocorrelations. If responses are correlated, then we must add the cross-correlation terms.

4.4.1 Zero-Input Response

The zero-input response of the system is its response to initial conditions with the input equal to zero

$$y(t) = Ce^{At}x(0). \tag{4.26}$$

For random initial conditions and a stable system in the steady state, the autocorrelation of the output is the expectation

$$\begin{aligned} R_{yy}(t_1, t_2) &= E\{y(t_1)y^T(t_2)\} \\ &= E\left\{Ce^{At_1}x(0)x^T(0)e^{A^Tt_2}C^T\right\} \\ &= Ce^{At_1}E\{x(0)x^T(0)\}e^{A^Tt_2}C^T. \end{aligned}$$

The autocorrelation of the zero-input response is

$$R_{yy}(t_1, t_2) = Ce^{At_1}R_{xx}(0)e^{A^Tt_2}C^T, \tag{4.27}$$

where $R_{xx}(0) = E\{x(0)x(0)^T\}$ is the mean square of the process.

Example 4.5 For initial conditions with mean square $R_{xx}(0) = I_2$, find the spectral density function and autocorrelation of the zero-input response of the system

$$\begin{aligned} \begin{bmatrix} \dot{x}_1 \\ \dot{x}_2 \end{bmatrix} &= \begin{bmatrix} 0 & 1 \\ -8 & -6 \end{bmatrix}\begin{bmatrix} x_1 \\ x_2 \end{bmatrix} + \begin{bmatrix} 0 \\ 1 \end{bmatrix}u, \\ y &= \begin{bmatrix} 1 & 0 \end{bmatrix}\begin{bmatrix} x_1 \\ x_2 \end{bmatrix}. \end{aligned}$$

Solution

The state-transition matrix of the system is

$$e^{At} = \begin{bmatrix} 2 & 0.5 \\ -4 & -1 \end{bmatrix} e^{-2t} + \begin{bmatrix} -1 & -0.5 \\ 4 & 2 \end{bmatrix} e^{-4t},$$

and the product with the output matrix is

$$Ce^{At} = \begin{bmatrix} 2 \; 0.5 \end{bmatrix} e^{-2t} + \begin{bmatrix} -1 \; -0.5 \end{bmatrix} e^{-4t}.$$

The autocorrelation of the zero-input response is

$$\begin{aligned} R_{yy}(t_1, t_2) &= \left(\begin{bmatrix} 2 \; 0.5 \end{bmatrix} e^{-2t_1} + \begin{bmatrix} -1 \; -0.5 \end{bmatrix} e^{-4t_1} \right) \\ & I_2 \left(\begin{bmatrix} 2 \\ 0.5 \end{bmatrix} e^{-2t_2} + \begin{bmatrix} -1 \\ -0.5 \end{bmatrix} e^{-4t_2} \right) \\ &= 4.25 e^{-2(t_1+t_2)} - 2.25 e^{-2(t_1+2t_2)} - 2.25 e^{-2(2t_1+t_2)} + 1.25 e^{-4(t_1+t_2)}. \end{aligned}$$

4.4.2 *Forced (Zero-State) Response MIMO Time-Varying Case*

The zero-state response of a linear system with $\boldsymbol{u}(t)$ input and state-transition matrix $\phi(t, \tau)$ is the integral

$$\boldsymbol{x}(t) = \int_0^t \phi(t, \tau) B(\tau) \boldsymbol{u}(\tau) \mathrm{d}\tau. \tag{4.28}$$

The autocorrelation is the expectation.

$$\begin{aligned} R_{XX}(t_1, t_2) &= E\left\{ \boldsymbol{x}(t_1) \boldsymbol{x}^T(t_2) \right\} \\ &= \int_0^{t_1} \int_0^{t_2} \phi(t_1, \eta) B(\eta) E\left\{ \boldsymbol{u}(\eta) \boldsymbol{u}^T(\xi) \right\}. \\ & B(\xi)^T \phi(t_2, \xi)^T \mathrm{d}\eta \mathrm{d}\xi \end{aligned}$$

In terms of the autocorrelation of the input, the autocorrelation of the zero-state response is

$$R_{XX}(t_1, t_2) = \int_0^{t_1} \int_0^{t_2} \phi(t, \eta) B(\eta) R_{uu}(\eta, \xi) B(\xi)^T \phi(t, \xi)^T \mathrm{d}\eta \mathrm{d}\xi. \tag{4.29}$$

The mean square zero-state response is obtained with $t_1 = t_2$

$$R_{XX}(t) = \int_0^t \int_0^t \phi(t,\eta)B(\eta)R_{\boldsymbol{uu}}(\eta,\xi)B(\xi)^T\phi(t,\xi)^T \mathrm{d}\eta \mathrm{d}\xi. \tag{4.30}$$

The zero-state output response for a single-input time-invariant system is

$$X(t) = \int_0^t \phi(t-\eta)Bu(\eta)\mathrm{d}\eta.$$

Hence, with a wide-sense stationary random input, we have the autocorrelation

$$R_{XX}(t_1,t_2) = \int_0^{t_1} \int_0^{t_2} \phi(t_1-\eta)BR_{uu}(\eta-\xi)B^T\phi(t_2-\xi)^T \mathrm{d}\eta \mathrm{d}\xi, \tag{4.31}$$

and the mean square

$$R_{XX}(0) = \int_0^t \int_0^t \phi(t-\eta)BR_{uu}(\eta-\xi)B^T\phi(t-\xi)^T \mathrm{d}\eta. \tag{4.32}$$

To obtain the expressions for the output of the system, we multiply by the output matrix to obtain

$$R_{yy}(t_1,t_2) = E\{\boldsymbol{y}(t_1)\boldsymbol{y}^T(t_2)\} = \int_0^{t_1} \int_0^{t_2} C\phi(t_1,\eta)B(\eta)$$
$$R_{uu}(\eta,\xi)B(\tau)^T\phi(t_2,\xi)^T C^T \mathrm{d}\eta,$$

that is

$$R_{yy}(t_1,t_2) = \int_0^{t_1} \int_0^{t_2} G(t_1,\eta)R_{uu}(\eta,\xi)G^T(t_2,\xi)\mathrm{d}\eta \mathrm{d}\xi, \tag{4.33}$$

where $G(t,t_0) = C\phi(t,t_0)B(t)$ is the impulse response matrix. For a wide-sense stationary input and a linear time-invariant system, we have the autocorrelation

$$R_{yy}(t_1,t_2) = \int_0^{t_1} \int_0^{t_2} G(t_1-\eta)R_{uu}(\eta-\xi)G^T(t_2-\xi)\mathrm{d}\eta \mathrm{d}\xi. \tag{4.34}$$

For a SISO time-invariant system with wide-sense stationary random input the autocorrelation is

$$R_{yy}(t_1, t_2) = E\{y(t_1)y(t_2)\} = \int_0^{t_1}\int_0^{t_2} g(\eta)g(\xi)R_{uu}(\eta - \xi + t_2 - t_1)\mathrm{d}\eta\mathrm{d}\xi \tag{4.35}$$

The mean square zero-state response is

$$R_{yy}(t) = E\{y^2(t)\} = \int_0^{t}\int_0^{t} g(\eta)g(\xi)R_{uu}(\eta - \xi)\mathrm{d}\eta\mathrm{d}\xi. \tag{4.36}$$

For white noise inputs, the autocorrelation is a Dirac delta and has the sifting property. Figure 4.5 shows that integrating the delta function is more convenient along in the longer direction so that the limit of the remaining integral is $\min(t_1, t_2)$ (see the discussion of the Wiener process in Chap. 3). The autocorrelation becomes

$$R_{yy}(t_1, t_2) = \int_0^{\min(t_1,t_2)} G(t_1, \eta)G^T(t_2, \eta + t_2 - t_1)\mathrm{d}\eta. \tag{4.37}$$

For the time-invariant case, this becomes

$$R_{yy}(t_1, t_2) = \int_0^{\min(t_1,t_2)} G(\eta)G^T(\eta + t_2 - t_1)\mathrm{d}\eta. \tag{4.38}$$

To obtain the autocorrelation for the state vector, the transfer function is calculated with the output matrix equal to the identity, that is, $G(t) = e^{At}B$.

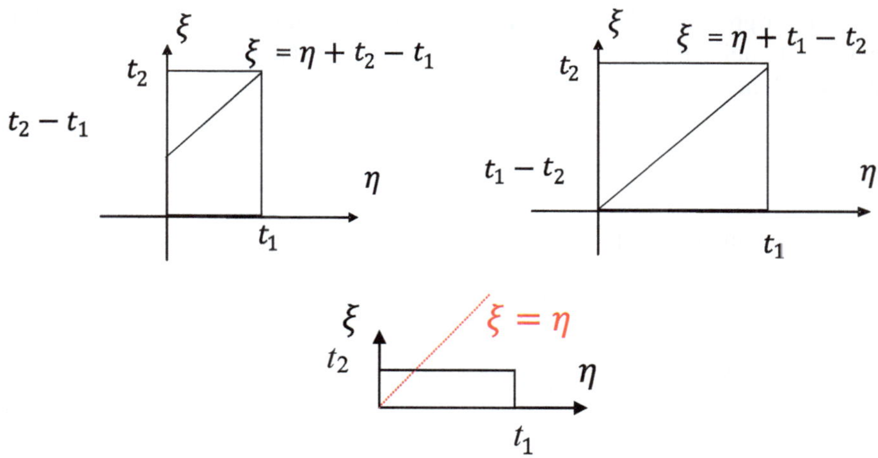

Fig. 4.5 Region of integration

Example 4.6 For unity white noise input, find the mean square value of the zero-state response of the system

$$\begin{bmatrix} \dot{x}_1 \\ \dot{x}_2 \end{bmatrix} = \begin{bmatrix} 0 & 1 \\ -8 & -6 \end{bmatrix}\begin{bmatrix} x_1 \\ x_2 \end{bmatrix} + \begin{bmatrix} 0 \\ 1 \end{bmatrix}u,$$
$$y = \begin{bmatrix} 1 & 0 \end{bmatrix}\begin{bmatrix} x_1 \\ x_2 \end{bmatrix}.$$

Solution

The impulse response of the system is

$$\begin{aligned} g(t) = Ce^{At}B &= \begin{bmatrix} 1 & 0 \end{bmatrix}\left(\begin{bmatrix} 2 & 0.5 \\ -4 & -1 \end{bmatrix}\mathrm{e}^{-2t} + \begin{bmatrix} -1 & -0.5 \\ 4 & 2 \end{bmatrix}\mathrm{e}^{-4t}\right)\begin{bmatrix} 0 \\ 1 \end{bmatrix} \\ &= 0.5\left(\mathrm{e}^{-2t} - \mathrm{e}^{-4t}\right). \end{aligned}$$

Using (4.38), we evaluate the autocorrelation

$$\begin{aligned} R_{yy}(t) &= \int_0^t 0.25\left(\mathrm{e}^{-2\eta} - \mathrm{e}^{-4\eta}\right)^2 \mathrm{d}\eta \\ &= 0.25\int_0^t \left(\mathrm{e}^{-4\eta} - 2\mathrm{e}^{-6\eta} + \mathrm{e}^{-8\eta}\right)\mathrm{d}\eta \\ &= 0.1771 - 0.0625\mathrm{e}^{-4t} - 0.0833\mathrm{e}^{-6t} - 0.03125\mathrm{e}^{-8t}. \end{aligned}$$

Example 4.7 Obtain the autocorrelation of the Wiener process.

Solution

The Wiener process is the output of an integrator with unity white noise input. For unity white noise input, we have the autocorrelation $R_{uu}(t) = \delta(t)$. The transfer function of the integrator is

$$G(s) = \frac{1}{s}.$$

Hence, its impulse response is

$$g(t) = 1,\ t > 0,$$
$$R_{yy}(t_1, t_2) = \int_0^{t_2}\int_0^{t_1} g(\eta)g(\xi)R_{uu}(\eta - \xi + t_2 - t_1)\mathrm{d}\eta\mathrm{d}\xi$$

$$= \int_0^{t_2}\int_0^{t_1} (1)(1)\delta(\eta - \xi + t_2 - t_1)\mathrm{d}\eta\mathrm{d}\xi$$

$$= \begin{cases} \int_0^{t_1} \mathrm{d}\eta, \; t_2 \geq t_1 \\ \int_0^{t_2} \mathrm{d}\xi, \; t_1 > t_2 \end{cases} = \min(t_1, t_2).$$

4.4.3 Covariance Computation

For a time-invariant linear system with wide-sense stationary noise input, the response due to the noise is the zero-state response of Sect. 4.4.2. In the steady state, the autocorrelation of the response with white noise input with autocorrelation $R(\tau) = W\delta(\tau)$ is

$$R_{xx}(\tau) = \int_0^t \phi(t-\eta)BWB^T\phi(t-\eta)^T\mathrm{d}\eta.$$

The following procedure, due to Van Loan, computes the autocorrelation matrix.

Procedure:

1. Form the $2n \times 2n$ matrix

$$M = \left[\begin{array}{c|c} M_{11} & M_{12} \\ 0 & M_{22} \end{array}\right] = \left[\begin{array}{c|c} -A & BWB^T \\ 0 & A^T \end{array}\right].$$

2. Obtain the matrix exponential

$$e^{Mt} = \left[\begin{array}{c|c} e^{-At} & e^{-At}R_{xx}(t) \\ 0 & e^{A^Tt} \end{array}\right].$$

3. Transpose the lower right corner to obtain the state-transition matrix

$$e^{At} = \left[e^{A^Tt}\right]^T.$$

4. Multiply the result of step 3 by the upper right corner.

$$R_{xx}(t) = e^{At} \times e^{-At}R_{xx}(t).$$

The proof of the formulas used in the procedure starts with the residual matrix that is the Laplace transform of the matrix exponential e^{Mt}

$$[sI_{2n} - M]^{-1} = \left[\begin{array}{c|c} sI_n + A & -BWB^T \\ 0 & sI_n - A^T \end{array}\right]^{-1}$$
$$= \left[\begin{array}{c|c} (sI_n + A)^{-1} & (sI_n + A)^{-1}BWB^T\left(sI_n - A^T\right)^{-1} \\ 0 & \left(sI_n - A^T\right)^{-1} \end{array}\right].$$

Inverse Laplace transform using the convolution theorem

$$e^{Mt} = \left[\begin{array}{c|c} e^{-At} & \int_0^t e^{-A\eta}BWB^Te^{A^T(t-\eta)}d\eta \\ 0 & e^{A^Tt} \end{array}\right].$$

Premultiplying the upper right-hand corner by $e^{A(t-t)} = e^{At} \times e^{-At} = I_n$

$$e^{At} \times e^{-At}\int_0^t e^{A(t-\eta)}BWB^Te^{A^T(t-\eta)}d\eta = e^{-At} \times \int_0^t e^{A(t-\eta)}BWB^Te^{A^T(t-\eta)}d\eta$$
$$= e^{-At} \times R_{xx}(t).$$

Multiplying by e^{At} gives the desired autocorrelation.

For any known numerical value of t, the autocorrelation $R_{xx}(t)$ can be calculated using MATLAB with the matrix exponential command.

>> eMT = expm(M * T).

The command **exp** is inappropriate in this case and will only give the matrix $\left[e^{m_{ij}T}\right]$, where m_{ij} is the ijth entry of M.

Example 4.8 Obtain the discrete-time model corresponding the to the state equation with zero-mean Gaussian white noise input $u(t)$ whose PSD $W = 1$ with a sampling period $\Delta t = 0.1s$

$$\dot{x}(t) = \begin{bmatrix} 0 & 1 \\ -8 & -6 \end{bmatrix} x(t) + \begin{bmatrix} 0 \\ 1 \end{bmatrix} u(t).$$

Solution

We use the Van Loan procedure. We start with the product

$$BWB^T = \begin{bmatrix} 0 \\ 1 \end{bmatrix} 1 [0\ 1] = \begin{bmatrix} 0\ 0 \\ 0\ 1 \end{bmatrix}$$

then form the matrix

$$M = \left[\begin{array}{c|c} -A & BWB^T \\ 0 & A^T \end{array}\right] = \left[\begin{array}{c|c} -\begin{bmatrix} 0 & 1 \\ -8 & -6 \end{bmatrix} & \begin{bmatrix} 0\,0 \\ 0\,1 \end{bmatrix} \\ \begin{bmatrix} 0\,0 \\ 0\,0 \end{bmatrix} & \begin{bmatrix} 0 & -8 \\ 1 & -6 \end{bmatrix} \end{array}\right].$$

We calculate the matrix exponential

$$e^{Mt} = \left[\begin{array}{cc|cc} 0.9510 & -0.1352 & -2\times10^{-4} & -0.0051 \\ 1.0817 & 1.7622 & -0.0051 & 0.3987 \\ \multicolumn{2}{c|}{0\;0} & 0.9671 & -0.5936 \\ \multicolumn{2}{c|}{0\;0} & 0.0742 & 0.5219 \end{array}\right].$$

We obtain the state-transition matrix by transposing the lower right submatrix

$$\phi = e^{A\Delta t} = \left[e^{A^T\Delta t}\right]^T = \begin{bmatrix} 0.9671 & 0.0742 \\ -0.5396 & 0.5219 \end{bmatrix}$$

then multiply by the upper right submatrix to obtain the noise covariance matrix

$$Q = \phi \times \text{Upper right submatrix} = 10^{-3}\begin{bmatrix} 0.2 & 2.8 \\ 2.8 & 57.0 \end{bmatrix}.$$

The following MATLAB commands perform the above operations:

```
>> A = [0,1; – 8, – 6]; B = [0;1];
>> M = [–A,B*B';zeros(2,2),A'];
>> eM = expm(M*0.1);
>> phi = eM(3:4,3:4)'
phi =
   0.9671 0.0742
 – 0.5936 0.5219
>> Q = phi*eM(1:2,3:4)
Q =
   0.0002 0.0028
   0.0028 0.0570
```

4.5 Discrete-Time (DT) Random Signals

The output of a discrete-time system is the convolution summation of its impulse response and the input. For a time-varying system with impulse response matrix $G(k, m)$, the response is

$$y(m) = \sum_{k=-\infty}^{\infty} G(k, m)u(k). \tag{4.39}$$

We first obtain the mean of the output response for both the stationary and nonstationary case.

4.5.1 Mean Response

The output of a discrete-time systems is given by the convolution of the input $\boldsymbol{u}(i)$ and the impulse response matrix

$$E\{y(k)\} = E\left\{\sum_{k=-\infty}^{\infty} G(k, m)\boldsymbol{u}(k)\right\} = \sum_{k=-\infty}^{\infty} G(k, m)E\{\boldsymbol{u}(k-i)\}. \tag{4.40}$$

For a process that is stationary in the mean with mean $\boldsymbol{m}_u$, we have

$$E\{y(k)\} = \left[\sum_{k=-\infty}^{\infty} G(k, m)\right]\boldsymbol{m}_u. \tag{4.41}$$

For a BIBO stable system, the summation is finite. For a linear time-invariant system the impulse response sequence is

$$\{G(k)\} = \{..., G(-1), G(0), ..., G(i), ...\},$$

The formula has a physical interpretation in terms of the frequency response $\boldsymbol{G}\left(e^{j\omega T}\right)$

$$\boldsymbol{m}_y = \left[\sum_{i=-\infty}^{\infty} G(i)\right]\boldsymbol{m}_u = \left[\sum_{i=-\infty}^{\infty} G(i)z^{-i}\right]_{z=1}\boldsymbol{m}_u,$$

that is,

$$\boldsymbol{m}_y = \boldsymbol{G}(1)\boldsymbol{m}_u. \tag{4.42}$$

Thus, as in the continuous-time case, the mean is scaled by the DC gain.

4.5.2 Stationary Steady-State Analysis for Discrete-Time Systems

For a stable system in the steady state with a wide-sense stationary input, the autocorrelation of the output of a LTI system is given by the following theorem.

Theorem 4.4 *The autocorrelation of the output of a stable LTI system with impulse response matrix $G(m)$ due to a wide-sense stationary input $\boldsymbol{u}(k)$ with autocorrelation $R_{uu}(m)$ is*

$$R_{yy}(m) = G(-m) * R_{uu}(m) * G^T(m). \tag{4.43}$$

The PSD of the output is

$$S_{yy}(z) = \boldsymbol{G}(z^{-1}) S_{uu}(z) \boldsymbol{G}^T(z). \tag{4.44}$$

Proof For a BIBO stable system, the response of the system remains bounded and is given by the convolution summation. Substituting the convolution summation in the expression for the autocorrelation gives

$$\begin{aligned} E\{\boldsymbol{y}(k)\boldsymbol{y}^T(k+m)\} &= E\left\{\left[\sum_{i=-\infty}^{\infty} G(i)\boldsymbol{u}(k-i)\right]\left[\sum_{j=-\infty}^{\infty} G(j)\boldsymbol{u}(k+m-j)\right]^T\right\} \\ &= \sum_{i=-\infty}^{\infty}\sum_{j=-\infty}^{\infty} G(i)E\{\boldsymbol{u}(k-i)\boldsymbol{u}^T(k+m-j)\}G^T(j) \\ &= \sum_{i=-\infty}^{\infty}\sum_{j=-\infty}^{\infty} G(i)R_{uu}(m+i-j)G^T(j). \end{aligned}$$

Substituting $i = l - m$, or equivalently $l = m + i$, gives the result

$$R_{yy}(m) = \sum_{l=-\infty}^{\infty} G(-(m-l)) \underbrace{\sum_{j=-\infty}^{\infty} R_{uu}(l-j)G^T(j)}_{R_{uu}(l)*G^T(l)} = G(-m) * R_{uu}(m) * G^T(m).$$

To obtain the PSD of the output, we z-transform the autocorrelation and use the convolution theorem. We first obtain the z-transform of $G(-k)$. For the sequence $\{G(k)\} = \{\ldots, G(-1), G(0), \ldots, G(i), \ldots\}$, the z-transform is $\boldsymbol{G}(z) = \cdots + G(-1)z^1 + G(0) + \cdots + G(i)z^{-i} + \cdots$. Expanding $G(-n)$ as

$$\{G(-k)\} = \{..., G(1), G(0), ..., G(-i), ...\}$$

shows that its z-transform is

$$Z\{G(-k)\} = \cdots + G(1)z^{1} + G(0) + \cdots + G(-i)z^{-i} + \cdots = \boldsymbol{G}\left(z^{-1}\right).$$

By the convolution theorem, the PSD is the product of z-transform

$$S_{yy}(z) = \boldsymbol{G}\left(z^{-1}\right)S_{uu}(z)\boldsymbol{G}^{T}(z).$$

■

Using (4.43) gives the mean square of the output

$$E\left\{\boldsymbol{y}(k)\boldsymbol{y}^{T}(k+m)\right\}\Big]_{m=0} = R_{yy}(0) = \sum_{i=-\infty}^{\infty}\sum_{j=-\infty}^{\infty} G(i)R_{uu}(i-j)G^{T}(j). \quad (4.45)$$

For the SISO case, the autocorrelation simplifies to

$$R_{yy}(m) = \sum_{i=-\infty}^{\infty}\sum_{j=-\infty}^{\infty} g(i)g(j)R_{uu}(m+i-j), \quad (4.46)$$

and the PSD is

$$S_{yy}(z) = \boldsymbol{G}\left(z^{-1}\right)\boldsymbol{G}(z)S_{uu}(z). \quad (4.47)$$

Substituting $z = e^{j\omega T}$ for sampling period T gives the frequency domain expression

$$S_{yy}\left(e^{j\omega T}\right) = \left|\boldsymbol{G}\left(e^{j\omega T}\right)\right|^{2} S_{uu}\left(e^{j\omega T}\right). \quad (4.48)$$

Example 4.9 For the transfer function,

$$\boldsymbol{G}(z) = \frac{bz}{z-a}, \ |a| < 1$$

Determine the PSD of the output due to a unity white noise sequence.

Solution

The PSD of the output is given by

$$S_{yy}(z) = \boldsymbol{G}\left(z^{-1}\right)\boldsymbol{G}(z)S_{uu}(z).$$

For unity white noise input, the PSD of the input is $S_{uu}(z) = 1$, and the PSD of the output is

$$S_{yy}(z) = \frac{b^2}{(z-a)\left(z^{-1}-a\right)},\ |a| < 1.$$

4.5.2.1 Cross-Correlation

Like the continuous-time case, the cross-correlation of a wide-sense stationary input and the output of a linear time-invariant system with impulse response matrix $G(k)$ is

$$R_{yu}(m) = E\{\boldsymbol{y}(k)\boldsymbol{u}^T(k+m)\} = E\left\{\left[\sum_{i=-\infty}^{\infty} G(i)\boldsymbol{u}(k-i)\right]\boldsymbol{u}^T(k+n)\right\}.$$

Changing the order of expectation and summation

$$R_{yu}(m) = \sum_{i=-\infty}^{\infty} G(i)E\{\boldsymbol{u}(k-i)\boldsymbol{u}^T(k+m)\}$$

gives an expression for the autocorrelation of the output in terms of the autocorrelation of the input and the impulse response matrix

$$R_{yu}(m) = \sum_{i=-\infty}^{\infty} G(i)R_{uu}(m+i). \tag{4.49}$$

If the input and output are interchanged, we have

$$\begin{aligned} R_{uy}(m) &= E\{\boldsymbol{u}(k)\boldsymbol{y}^T(k+m)\} \\ &= E\left\{\boldsymbol{u}(k)\left[\sum_{i=-\infty}^{\infty} G(i)\boldsymbol{u}(k+m-i)\right]^T\right\} \\ &= \sum_{i=-\infty}^{\infty} E\{\boldsymbol{u}(k)\boldsymbol{u}^T(k+m-i)\}G^T(i). \end{aligned}$$

Thus, reversing the order shifts and transposes the autocorrelation resulting in the convolution summation

$$R_{uy}(m) = \sum_{i=-\infty}^{\infty} R_{uu}(m-i)G^T(i). \tag{4.50}$$

Using the convolution theorem, z-transforming gives the cross-spectral density

$$S_{uy}(z) = S_{uu}(z)\boldsymbol{G}^T(z). \tag{4.51}$$

For a SISO system, we have the transfer function

$$\boldsymbol{G}(z) = \frac{S_{uy}(z)}{S_{uu}(z)}. \tag{4.52}$$

For unity white noise input the denominator $S_{uu} = 1$, and the transfer function is the cross-correlation

$$G(z) = S_{uy}(z). \tag{4.53}$$

This provides us with a simple approach to evaluate the transfer function of a linear time-invariant system. In practice, we use band-limited white noise with a bandwidth much larger than the system bandwidth to approximate white noise for system identification.

Example 4.10 Find the transfer function of an electronic device if the cross-correlation of its output and unity band-limited white noise input with bandwidth much larger than the system bandwidth is

$$S_{uy}(z) = \frac{bz}{z-a},\ |a| < 1.$$

Solution

For unity white noise input, we have the PSD $S_{uu}(z) = 1$ and

$$\boldsymbol{G}(z) = S_{uy}(z) = \frac{bz}{z-a},\ |a| < 1.$$

4.5.2.2 The Shaping (Innovations) Filter

Physical noise is never white and is only approximately stationary over a limited time span but can often be modeled as wide-sense stationary colored noise. To simplify the analysis, we use a linear filter with white noise input to generate the colored noise as shown in Fig. 4.6. The filter that yields colored noise with white noise input is called the **shaping filter** or **innovations filter**. We use spectral factorization to obtain the filter transfer function

$$S_{yy}(z) = L\left(z^{-1}\right)L(z).$$

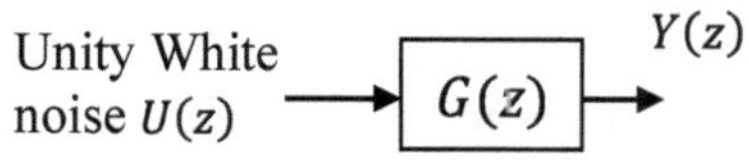

Fig. 4.6 Transfer function of the shaping filter for colored noise with PSD S_{yy}

Using (4.44), the spectral density of the output of the filter is

$$S_{yy}(s) = G\left(z^{-1}\right)G(z) \times 1.$$

Thus, the transfer function of the shaping filter is the causal factor of the PSD

$$\boldsymbol{G}(z) = L(z). \tag{4.54}$$

Example 4.11

$$S_{yy}(z) = \frac{b^2}{(z-a)\left(z^{-1}-a\right)}, \ |a| < 1.$$

Find the transfer function of the shaping filter for colored noise with PSD S_{yy}.

Solution

$$\begin{aligned} S_{yy}(z) &= b^2 L(z) L\left(z^{-1}\right) \\ &= \left(\frac{bz}{z-a}\right)\left(\frac{bz^{-1}}{z^{-1}-a}\right), \\ \boldsymbol{G}(z) &= \frac{bz}{z-a}. \end{aligned}$$

4.5.3 Nonstationary Analysis for Discrete-Time Systems

A discrete-time system is governed by the state-space model

$$\begin{aligned} \boldsymbol{x}(k+1) &= \phi(k+1,k)\boldsymbol{x}(k) + \Gamma(k)\boldsymbol{u}(k), \\ \boldsymbol{y}(k) &= C(k)\boldsymbol{x}(k), \end{aligned}$$

where $\boldsymbol{x}(k) \in \mathcal{R}^n$ is the state vector, $\boldsymbol{u}(k) \in \mathcal{R}^m$ is the input vector, and $\boldsymbol{y}(k) \in \mathcal{R}^l$ is the output vector. The matrix $\phi(k+1,k)$ is the state-transition matrix at time k, and the matrix $\Gamma(k)$ is the input matrix at time k.

The solution of the discrete-time state equation is

$$\boldsymbol{x}(k) = \phi(k,0)\boldsymbol{x}(0) + \sum_{i=0}^{k-1} \phi(k,i+1)\Gamma(i)\boldsymbol{u}(i),$$

where $\boldsymbol{x}(0)$ is the initial condition vector. The first term is the zero-input response due to the initial conditions, and the second term is the zero-state response due to the input. To calculate the autocorrelation, we assume that the two responses are orthogonal so that the autocorrelation of the sum is the sum of the autocorrelations. If the two responses are correlated, we must add the cross-correlation terms.

The state-transition matrix $\phi(k, k_0)$ satisfies

$$\phi(k, k_0) = \phi(k, l)\phi(l, k_0), l \in [k_0, k]$$
$$\phi(k, k_0) = I_n.$$

Hence, the output of the system is

$$\boldsymbol{y}(k) = C\phi(k, 0)\boldsymbol{x}(0) + \sum_{i=0}^{k-1} C\phi(k, i+1)\Gamma(i)\boldsymbol{u}(i). \tag{4.55}$$

For physical systems, the system dynamics is continuous-time but a discrete - time model is needed for signal processing, or communication. Thus, discrete-time models often represent a continuous-time system at sampling points. For a linear time-invariant system, the state-transition matrix is the matrix exponential of the continuous-time state matrix A. If sampling is nonuniform, the matrix is

$$\phi(k+1, k) = \mathrm{e}^{A\Delta t_k},$$

where $\Delta t_k = t_{k+1} - t_k$ is the sampling period at time k and the system is time-varying. For uniform sampling with a LTI system and sampling period Δt, we have the constant matrix

$$\phi = \mathrm{e}^{A\Delta t}. \tag{4.56}$$

The input matrix of the discrete-time system in terms of the state matrix A and the input matrix B of the continuous-time system is

$$\Gamma(k) = \int_{t_k}^{t_{k+1}} \mathrm{e}^{A(t_{k+1}-\tau)} B \mathrm{d}\tau. \tag{4.57}$$

For uniform sampling with a linear time-invariant system and sampling period Δt, we have the constant matrix

$$\Gamma(k) = \int_0^{\Delta t} \mathrm{e}^{A\tau} B \mathrm{d}\tau. \tag{4.58}$$

4.5.3.1 Zero-Input Response

The zero-input response of discrete-time system is given by

$$\boldsymbol{y}(k) = C\phi(k,0)\boldsymbol{x}(0). \tag{4.59}$$

To obtain its autocorrelation with random initial conditions, we substitute for the output and take the expectation

$$\begin{aligned} R_{yy}(k_1,k_2) &= E\{\boldsymbol{y}(k_1)\boldsymbol{y}^T(k_2)\} \\ &= E\{[C\phi(k_1,0)\boldsymbol{x}(0)][C\phi(k_2,0)\boldsymbol{x}(0)]^T\} \\ &= E\{C\phi(k_1,0)\boldsymbol{x}(0)\boldsymbol{x}(0)^T\phi(k_2,0)^T C^T\}. \end{aligned}$$

In terms of the mean square, the autocorrelation is

$$R_{yy}(k_1,k_2) = C\phi(k_1,0)R_{xx}(0)\phi(k_2,0)^T C^T. \tag{4.60}$$

For a linear time-invariant system with uniform sampling, the autocorrelation becomes

$$R_{yy}(m) = C\mathrm{e}^{Ak\Delta t}R_{xx}(0)\mathrm{e}^{A^T(k+m)\Delta t}C^T. \tag{4.61}$$

Example 4.12 For initial conditions with mean square $R_{xx}(0) = I_2$, find the spectral density function and autocorrelation of the zero-input response of the discrete-time system with sampling period Δt

$$\begin{aligned} \begin{bmatrix} \dot{x}_1 \\ \dot{x}_2 \end{bmatrix} &= \begin{bmatrix} 0 & 1 \\ -8 & -6 \end{bmatrix}\begin{bmatrix} x_1 \\ x_2 \end{bmatrix} + \begin{bmatrix} 0 \\ 1 \end{bmatrix}u, \\ y &= [1\ 0]\begin{bmatrix} x_1 \\ x_2 \end{bmatrix}. \end{aligned}$$

Solution

The state-transition matrix of the system is

$$\mathrm{e}^{A\Delta t} = \begin{bmatrix} 2 & 0.5 \\ -4 & -1 \end{bmatrix}\mathrm{e}^{-2\Delta t} + \begin{bmatrix} -1 & -0.5 \\ 4 & 2 \end{bmatrix}\mathrm{e}^{-4\Delta t}.$$

We evaluate the product with the output matrix

$$\begin{aligned} C\mathrm{e}^{Ak\Delta t} &= [1\ 0]\left(\begin{bmatrix} 2 & 0.5 \\ -4 & -1 \end{bmatrix}\mathrm{e}^{-2k\Delta t} + \begin{bmatrix} -1 & -0.5 \\ 4 & 2 \end{bmatrix}\mathrm{e}^{-4k\Delta t}\right) \\ &= [2\ 0.5]\mathrm{e}^{-2k\Delta t} - [1\ 0.5]\mathrm{e}^{-4k\Delta t}. \end{aligned}$$

The autocorrelation is

$$\begin{aligned}
R_{yy}(m) &= Ce^{Ak\Delta t}R_{xx}(0)e^{A^T(k+m)\Delta t}C^T \\
&= \left(\begin{bmatrix} 2 & 0.5 \end{bmatrix}e^{-2k\Delta t} - \begin{bmatrix} 1 & 0.5 \end{bmatrix}e^{-4k\Delta t}\right) \\
&\quad I_2\left(\begin{bmatrix} 2 \\ 0.5 \end{bmatrix}e^{-2(k+m)\Delta t} - \begin{bmatrix} 1 \\ 0.5 \end{bmatrix}e^{-4(k+m)\Delta t}\right) \\
&= 4.25e^{-2(2k+m)\Delta t} + 1.25e^{-4(2k+m)\Delta t} \\
&\quad - 2.25e^{-2(3k+2m)\Delta t} - 2.25e^{-2(3k+m)\Delta t}.
\end{aligned}$$

4.5.3.2 Zero-State Response

The zero-state response is the convolution summation

$$\boldsymbol{y}(t) = \sum_{i=0}^{k-1} C\phi(k, i+1)\Gamma(i)\boldsymbol{u}(i).$$

To obtain its autocorrelation, we substitute and take the expectation

$$\begin{aligned}
R_{yy}(k_1, k_2) &= E\left\{\boldsymbol{y}(k_1)\boldsymbol{y}^T(k_2)\right\} \\
&= E\left\{\left[\sum_{i=0}^{k_1-1} C\phi(k_1, i+1)\Gamma(i)\boldsymbol{u}(i)\right]\left[\sum_{j=0}^{k_2-1} C\phi(k_2, j+1)\Gamma(j)\boldsymbol{u}(j)\right]^T\right\} \\
&= E\left\{\sum_{i=0}^{k_1-1}\sum_{j=0}^{k_2-1} C\phi(k_1, i+1)\Gamma(i)\boldsymbol{u}(i)\boldsymbol{u}(j)^T\Gamma(j)^T\phi(k_2, j+1)^T C^T\right\} \\
&= \sum_{i=0}^{k_1-1}\sum_{j=0}^{k_2-1} C\phi(k_1, i+1)\Gamma(i)E\{\,\boldsymbol{u}(i)\boldsymbol{u}(j)^T)\}\Gamma(j)^T\phi(k_2, j+1)^T C^T.
\end{aligned}$$

The autocorrelation of the zero-input response is

$$R_{yy}(k_1, k_2) = \sum_{i=0}^{k_1-1}\sum_{j=0}^{k_2-1} C\phi(k_1, i+1)\Gamma(i)R_{uu}(i-j)\Gamma(j)^T\phi(k_2, j+1)^T C^T. \tag{4.62}$$

In the LTI case, we have

$$R_{yy}(m) = E\left\{\boldsymbol{y}(k)\boldsymbol{y}^T(k+m)\right\}$$

$$= \sum_{i=0}^{k-1} \sum_{j=0}^{k+m-1} C\phi(k-i-1)\Gamma R_{uu}(i-j)\Gamma^T \phi(k+m-j-1)^T C^T. \tag{4.63}$$

For uniform sampling, the autocorrelation becomes

$$\begin{aligned} R_{yy}(m) &= E\{y(k)y^T(k+m)\} \\ &= \sum_{i=0}^{k-1} \sum_{j=0}^{k+m-1} C\mathrm{e}^{A(k-i-1)\Delta t}\Gamma R_{uu}(i-j)\Gamma^T \mathrm{e}^{A^T(k+m-j-1)\Delta t} C^T. \end{aligned} \tag{4.64}$$

For a white noise process with autocorrelation matrix $R_{uu}(k) = S\delta(k)$, the autocorrelation reduces to the single summation

$$R_{yy}(m) = E\{y(k)y^T(k+m)\} = \sum_{i=0}^{k-1} C\mathrm{e}^{A(k-i-1)\Delta t}\Gamma S\Gamma^T \mathrm{e}^{A^T(k+m-i-1)\Delta t} C^T. \tag{4.65}$$

Example 4.13 For white noise input with PSD $S_{uu} = 1$, find the spectral density function and autocorrelation of the zero-input response of the discrete-time system with sampling period Δt

$$\begin{bmatrix} x_1(k+1) \\ x_2(k+1) \end{bmatrix} = \left(\begin{bmatrix} 2 & 0.5 \\ -4 & -1 \end{bmatrix} \mathrm{e}^{-2\Delta t} + \begin{bmatrix} -1 & -0.5 \\ 4 & 2 \end{bmatrix} \mathrm{e}^{-4\Delta t} \right) \begin{bmatrix} x_1(k) \\ x_2(k) \end{bmatrix} + \begin{bmatrix} 0 \\ 1 \end{bmatrix} u(k),$$

$$y = \begin{bmatrix} 1 & 0 \end{bmatrix} \begin{bmatrix} x_1 \\ x_2 \end{bmatrix}.$$

Solution

The impulse response matrix is

$$\begin{aligned} C\mathrm{e}^{Ak\Delta t}B &= \begin{bmatrix} 1 & 0 \end{bmatrix} \left(\begin{bmatrix} 2 & 0.5 \\ -4 & -1 \end{bmatrix} \mathrm{e}^{-2k\Delta t} + \begin{bmatrix} -1 & -0.5 \\ 4 & 2 \end{bmatrix} \mathrm{e}^{-4k\Delta t} \right) \begin{bmatrix} 0 \\ 1 \end{bmatrix} \\ &= 0.5\left(\mathrm{e}^{-2k\Delta t} - \mathrm{e}^{-4k\Delta t}\right). \end{aligned}$$

The autocorrelation of the output is

$$\begin{aligned} R_{yy}(m) &= \sum_{i=0}^{k-1} C\mathrm{e}^{A(k-i-1)\Delta t}\Gamma S\Gamma^T \mathrm{e}^{A^T(k+m-i-1)\Delta t} C^T \\ &= 0.25 \sum_{i=0}^{k-1} \left(\mathrm{e}^{-2(k-i-1)\Delta t} - \mathrm{e}^{-4(k-i-1)\Delta t}\right)\left(\mathrm{e}^{-2(k+m-i-1)\Delta t} - \mathrm{e}^{-4(k+m-i-1)\Delta t}\right) \\ &= 0.25 \sum_{i=0}^{k-1} \left(\mathrm{e}^{-2(2k+m-i-1)\Delta t} + \mathrm{e}^{-4(2k+m-i-1)\Delta t} - \mathrm{e}^{-2(3k+2m-i-1)\Delta t} - \mathrm{e}^{-2(3k+m-i-1)\Delta t}\right). \end{aligned}$$

Problems

4.1. For a system with the given transfer function in the steady state and with unity white noise input

$$\boldsymbol{G}(s) = \begin{bmatrix} \frac{1}{Ts+1} \\ \frac{1}{2Ts+1} \end{bmatrix}.$$

(a) Find the power spectral density of the output vector.
(b) Find the mean square value of the output vector.

4.2 Find the PSD and mean square output for the second-order Butterworth filter with unity white noise input

$$G(s) = \frac{1}{s^2 + \sqrt{2}s + 1}.$$

4.3 Consider the random process of Problem 2.7 with $z = \boldsymbol{a}^T \boldsymbol{x}$ where $\boldsymbol{a}^T = \begin{bmatrix} 2 \; 3 \end{bmatrix}$ and $\boldsymbol{x} = \begin{bmatrix} x_1 \; x_2 \end{bmatrix}^T$ is a zero-mean random vector with uncorrelated entries x_1 and x_2. The autocorrelation function of x_1 is

$$R_1(\tau) = 2e^{-2|\tau|},$$

and x_2 is zero-mean unity white noise. Find the filter transfer function that yields z as output for unity white noise input.

4.4 For a noise process with the power spectral density

$$S_{XX}(s) = \frac{-s^2 + c^2}{\left(-s^2 + a^2\right)\left(-s^2 + b^2\right)}.$$

(a) Find the shaping filter whose output with unity white noise is the given noise process.
(b) Find the autocorrelation function of the noise process.
(c) Find the steady-state mean square value.
(d) Unity white noise input.

4.5 Consider the shaping filter of Problem 4.2 (Use MAPLE when necessary)

(a) Obtain the corresponding differential equation.
(b) Obtain the corresponding impulse response.
(c) Find the mean square response due to an initial condition $x(0) \sim \mathcal{N}(0, 2), \dot{x}(0) = 0$.
(d) Find the mean square response due to unity white noise input.

(e) Find the total response due to the initial condition of Part c and the input of Part d.
(f) Find the steady-state mean square response and compare the result to your answer from Part e.

4.6 The equation of motion for an overdamped and asymptotically stable (real poles in the LHP) spring-mass-damper system is

$$m\ddot{y} + c\dot{y} + ky = f,$$

where m = mass, k = spring constant, c = damper constant, y = displacement of the mass, $f(t)$ = unity white Gaussian force with variance σ^2:

(a) Find the mean, mean square, and variance of the displacement at time t.
(b) Find the mean, mean square, and variance of the displacement in the steady state using spectral density of the applied force.
(c) Show that the answers of (a) are identical to those of (b) in the limit as $t \to \infty$.

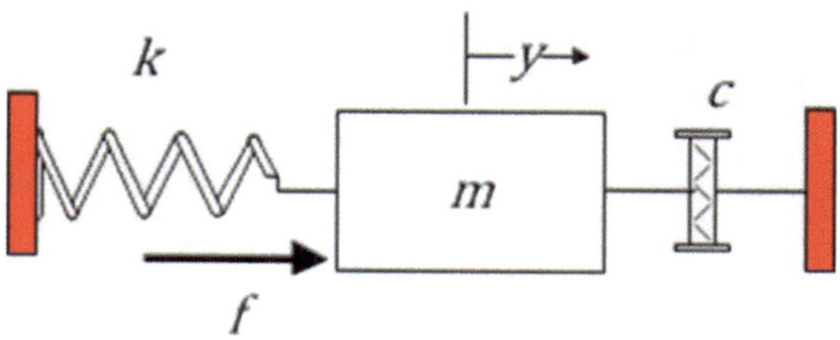

4.7 For the spring-mass-damper system with $m = 1$ kg, $k = 0.1$ kg/m, $c = 0.05$ kg/m/s, and $f(t)$ = Gauss–Markov force with variance $\sigma^2 = 1$ and time constant 0.5 s:

(d) Find the mean, mean square, and variance of the displacement at time t.
(e) Find the mean, mean square, and variance of the displacement in the steady state using spectral density of the applied force.
(f) Show that the answers of (a) are identical to those of (b) in the limit as $t \to \infty$.

4.8 Find the mean and variance of the output in the steady state due to a disturbance input for the block diagram with disturbance Gaussian white noise input in the steady state

$$d(t) \sim \mathcal{N}(0, \sigma^2),\ E\{d(t)d(t+\tau)\} = R(\tau) = W\delta(\tau)$$

and transfer functions

$$G(s) = \frac{K_p}{Ts+1}, \; C(s) = K_c \frac{s+a}{s}.$$

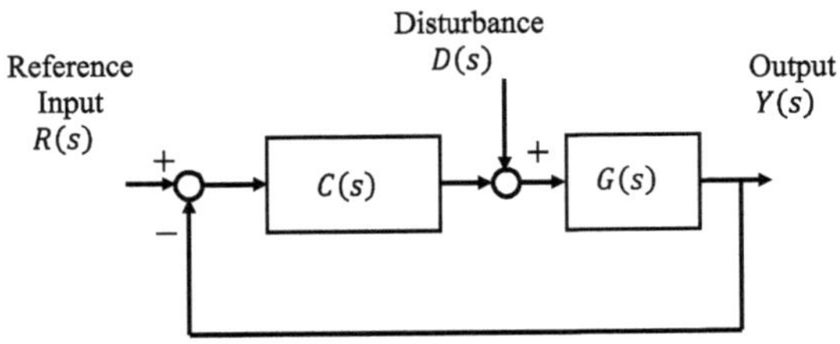

4.9 Obtain a state-space model for a random constant a.

4.10 Obtain a state-space model for a random ramp with slope a_1 and intercept a_0.

4.11 Obtain a state-space model for the random harmonic motion

$$\ddot{y}(t) + \omega_0^2 y(t) = 0$$

with random initial conditions $y(0)$ and $\dot{y}(0)$

4.12 The state equation for drug ingestion in the bloodstream is[1]

$$\dot{\boldsymbol{x}} = \begin{bmatrix} -K_1 & 0 \\ K_1 & -K_2 \end{bmatrix} \boldsymbol{x} + \begin{bmatrix} 1 \\ 0 \end{bmatrix} u.$$

(a) Find the autocorrelation of the zero-input response due to random initial conditions with an autocorrelation $R_{xx}(0) = \sigma_x^2 I_2$.
(b) Find the mean square state vector due to unity white Gaussian noise input.
(c) Find the total mean square output due to the initial conditions of (a) and the input of (b). Assume that the two responses are orthogonal.

4.13 Obtain a time-varying state-space model for the random harmonic motion with zero-state matrix.

4.14 Show that the transfer function of the RC circuit is

$$G(s) = \frac{1/RT}{s+3/T}, \; T = 2RC.$$

Then, for zero-mean Gaussian input $v_{in} \sim \mathbf{N}(0, \sigma^2)$, obtain.

(a) The mean, mean square, and variance of the output at time t.
(b) The mean, mean square, and variance of the output in the steady state with and without using the results of Part a.

[1] N.H. McClamroch, *State Models of Dynamic Systems: A Case Study Approach*, Springer-Verlag, 1980.

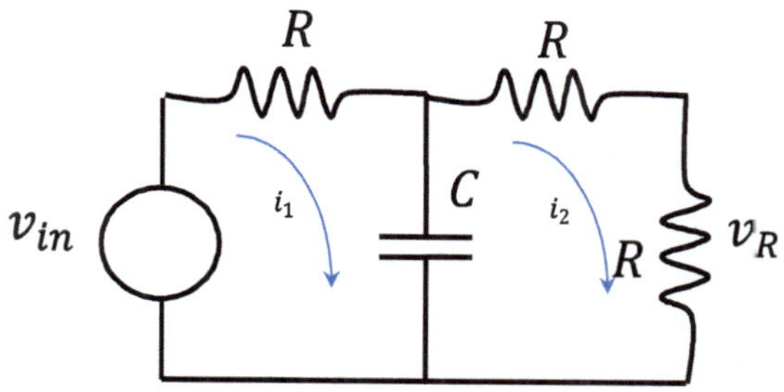

4.15 For the Gauss–Markov process with variance σ^2 and time constant β^{-1}

(a) Obtain a state-space model whose output is the process with additive noise.
(b) With a sampling period Δt, find the discrete state-space model with zero-mean white Gaussian process noise $u(t)$ with covariance $q\delta(t)$ and zero-mean white Gaussian measurement noise with covariance r using the Van Loan procedure.
(c) Repeat part (b) using basic principles.

4.16 Find the discrete state equation and noise covariance matrix for a unit point mass subject to a zero-mean unity Gaussian white noise force (a) using the Van Loan process (b) using basic principles.

4.17 Simulate the filter with transfer function

$$G(s) = \frac{1}{s+1}$$

with the sinusoidal signal $z(t)$ with additive white noise $n(t)$

$$z(t) = A\cos(t) + n(t)$$

with $A = 1$, and with $R_n(\tau) = 0.05$. Obtain plots of the input and the output of the filter.

4.18 Find the autocorrelation of the noisy sinusoidal signal of Problem 4.15. Why is it not possible to obtain the PSD of this signal?

4.19 Find the spectral density function and the autocorrelation of the output of a FIR filter with unity white noise input if the filter has the z-transfer function

$$H(z) = K\sum_{i=0}^{N} z^{-i}.$$

4.20 For the z-transfer function of discrete-time linear time-invariant system given by

$$G(z) = \frac{z}{z - a}$$

with the Gaussian noise sequence whose autocorrelation is given by $R_{uu}(n) = \sigma^2 e^{-\beta|n|}$, determine the following

(a) The spectral density function of the output.
(b) The autocorrelation function of the output.
(c) The mean square of the output.
(d) The change in the output if the discrete-time system is described by the difference equation

$$y(k+1) - ay(k) = u(k+1).$$

4.21 Consider the state equation

$$x_{k+1} = a_k x_k + b_k w_k,$$

where the noise term is governed by the recursion

$$w_{k+1} = c w_k + d u_k,$$

$$u_k \sim \mathcal{N}(0, q),\ E\{u_k u_j\} = 0,\ j \neq k.$$

(a) Verify that w_k is not white noise using the solution of the difference equation governing it.
(b) Write the transfer function of the shaping filter for the noise w_k
(c) Write a state equation for the two given equations with white noise input.

4.22 Consider the single input, multiple output system with transfer function

$$g(z) = \begin{bmatrix} \frac{z}{(z-0.1)(z-0.2)} & \frac{z-0.5}{(z-0.2)(z-0.4)} \end{bmatrix}^T.$$

Obtain the PSD of the output with unity Gaussian white noise input.

4.23 A mechanical spring-mass-damper system has constant mass and damper constant, but its spring constant can vary randomly. The equation of motion of the system is

$$\ddot{y}(t) + \dot{y}(t) + ky(t) = u(t),$$

$$k = \overline{k} + k_r,\ E\{k\} = \overline{k} = 0.25,\ k_r random.$$

(a) Obtain an expression for the deterministic component of the system impulse response $h(t)$ with $k = \overline{k}$.

(b) Show that a series solution for any input $u(t)$ is

$$y(t) = \sum_{i=0}^{\infty} y_i(t)$$

of the impulse response of the system is given by the convolution

$$\begin{aligned} y_i(t) &= -k_r h(t) * y_{i-1}(t) \\ &= (-k_r)^i h(t) * h(t) * \ldots * h(t) * u(t), i \\ &= 1, 2, \ldots, \\ y_0(t) &= h(t) * u(t). \end{aligned}$$

(c) How do the statistics of the input and the spring constant affect the different terms of the series solution assuming that k_r and $u(t)$ are mutually independent.

Hint: Substitute the series solution in the equation of motion, then prove the result by induction.

Bibliography

1. Brown, R. G., & Hwang, P. Y. C. (2012). *Introduction to random signals and applied Kalman filtering*. Wiley.
2. Papoulis, A., & Pillai, S. U. (2002). *Probability, random variables and stochastic processes*. McGraw Hill
3. Shanmugan, K. S., & Breipohl, A. M. (1988). *Detection, estimation & data analysis*. Wiley.
4. Söderström, T. (1994). *Discrete-time stochastic systems: Estimation and control*. Prentice Hall.

Chapter 5
Estimation and Estimator Properties

Mathematical modeling is an important part of many fields of engineering and science. Mathematical models are used to simulate physical systems and test their behavior under different conditions. These mathematical models depend on model parameters that must be evaluated from measurements collected by scientists or engineers from physical systems operating under the conditions of validity of the mathematical model.

For example, an aerospace engineer modeling a rocket collects measurements of position, velocity, and acceleration that together with the equation of motion allow him to evaluate the parameters of the rocket's model. An electronic engineer measures voltage and current at the input and output of an electronic device and uses his measurements to obtain an input–output model of the device. A wildlife biologist collects population data for species in a particular environment to develop a predator–prey model for wildlife management.

The measurements in all the above applications include measurement errors that must be considered when evaluating the model parameters. The measurement errors can be deterministic, random, or both, depending on the application. The presence of errors in the measurements makes it necessary to collect far more measurements than the number of parameters to be evaluated because the measurement redundancy corrects for the errors in each measurement.

The two most popular approaches for parameter estimation are the minimum square error or least-squares approach and the maximum-likelihood approach. They are the subjects of later chapters. In this chapter, we discuss properties of estimators that allow us to assess their quality.

Because parameter estimates are based on noisy measurements, the estimates themselves are random. Even in cases where the noise distribution is known, the distribution of the parameter estimates can be quite complicated. However, to assess

M. S. Fadali, *Introduction to Random Signals, Estimation Theory, and Kalman Filtering*, https://doi.org/10.1007/978-981-99-8063-5_5

the quality of an estimate it is essential to learn more about its distribution and how its distribution is influenced by the size of the data sample used in estimation. In some cases, it is not possible to obtain this information for finite sample size, but much can be learned by taking the limit as the sample size goes to infinity. Properties evaluated by taking the limit are known as **large sample properties**. These properties allow us to evaluate the effect of increasing sample size on the quality of the estimator. Properties that can be evaluated without taking the limit are valid for any sample size and are known as **small sample properties**. We discuss small sample properties next.

5.1 Small Sample Properties

The random estimator of a random parameter.

5.1.1 Unbiased Estimators

An unbiased estimator is an estimator that is correct "on the average," i.e., if the estimator $\hat{\boldsymbol{\theta}}$ of a parameter vector $\boldsymbol{\theta}$ is unbiased, then the estimation error satisfies

$$Bias\left(\hat{\boldsymbol{\theta}}\right) = \boldsymbol{E}\left\{\hat{\boldsymbol{\theta}} - \boldsymbol{\theta}\right\} = \mathbf{0}. \tag{5.1}$$

For a random parameter vector $\boldsymbol{\theta}$, this gives the condition

$$E\left\{\hat{\boldsymbol{\theta}}\right\} = E\{\boldsymbol{\theta}\}, \tag{5.2}$$

and for a constant vector, the condition reduces to

$$E\left\{\hat{\boldsymbol{\theta}}\right\} = \boldsymbol{\theta}. \tag{5.3}$$

To show that an estimator is unbiased, we simply find the expected value of the estimator as in the following example.

Example 5.1 Show that the sample mean is an unbiased estimator of the mean.

Solution

Consider a sample of N independent identically distributed (i.i.d.) measurements $\{z_i, i = 1, \ldots, N\}$ from a population of mean $\boldsymbol{m}_z$. The parameter to be estimated is $\boldsymbol{\theta} = \boldsymbol{m}_z$, and the estimator is

$$\hat{\theta} = \overline{z} = \frac{1}{N} \sum_{i=1}^{N} z_i.$$

To show that the estimator is unbiased, we take the expectation

$$E\left\{\hat{\theta}\right\} = \frac{1}{N} \sum_{i=1}^{N} E\{z_i\} = m_z = \theta.$$

5.1.2 Efficiency

Efficiency is a measure of the spread of the distribution governing the estimator, which in turn determines its performance. If the distribution is wide and the estimator is unbiased, or has a small bias, then the probability of obtaining an estimate that is far from the correct parameter value becomes higher. To see this, assume for simplicity that the random estimator is governed by a normal distribution. Then, as Fig. 5.1 shows, the area under the normal distribution curve for a given interval around the mean decreases as the variance increases. In other words, the probability of obtaining an estimate that is close to the correct value becomes smaller as the variance or spread of the distribution increases.

Clearly, a good estimator is one that has a small mean squared error value. The means squared error is

$$\text{MSE}\left(\hat{\theta}\right) = E\left\{\left(\hat{\theta} - \theta\right)^2\right\}. \tag{5.4}$$

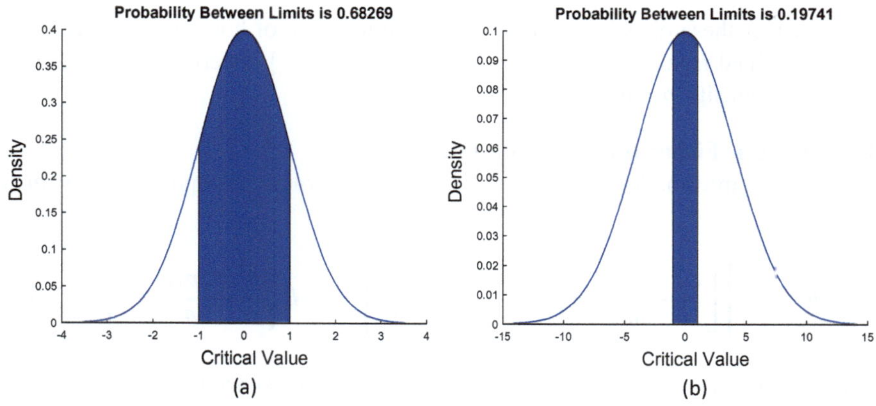

Fig. 5.1 Area under zero-mean normal distribution curves in the interval [−1, 1] around the means (**a**), standard normal (**b**) $\mathcal{N}(0, 16)$

For an unbiased estimator, the mean squared error has the following property.

Fact: The mean squared error of an unbiased estimator is equal to its variance.

Proof We expand the expression for the mean squared error and substitute for the mean square value in terms of the variance and the mean to obtain

$$\begin{aligned} E\left\{\left(\hat{\theta}-\theta\right)^2\right\} &= E\left\{\hat{\theta}^2\right\} - 2\theta E\left\{\hat{\theta}\right\} + \theta^2 \\ &= \sigma_{\hat{\theta}}^2 + E\left\{\hat{\theta}\right\}^2 - 2\theta E\left\{\hat{\theta}\right\} + \theta^2. \end{aligned}$$

Combining terms to complete the square, we have

$$\mathrm{MSE}\left(\hat{\theta}\right) = \sigma_{\hat{\theta}}^2 + \left(E\left\{\hat{\theta}\right\} - \theta\right)^2 = \mathrm{var}\left(\hat{\theta}\right) + \mathrm{bias}\left(\hat{\theta}\right)^2.$$

For an unbiased estimator, we have $E\left\{\hat{\theta}\right\} = \theta$ and $\mathrm{MSE}\left(\hat{\theta}\right) = \sigma_{\hat{\theta}}^2$. ■

Example 5.1 shows that the sample mean is an unbiased estimator. Hence, the variance of the sample mean is equal to its square error.

In summary, an unbiased estimator is an estimator whose distribution has a mean equal to the true value of the estimated parameter and whose variance is equal to the mean squared estimation error. The standard deviation of the distribution of the estimator $\sigma_{\hat{\theta}}$ is known as the **standard error**.

Given the importance of the variance of the estimator, one may wonder:

(a) Is there a limit below which this variance cannot be reduced regardless of the estimator used?
(b) How is this limit affected by the distribution of the random data?
(c) How is this limit affected by the quality of the data used in estimation?

To answer these questions, we first define a measure of how much information can be obtained from a set of data $\boldsymbol{z} = col\{z_1, z_2, \ldots, z_k\}$ governed by a particular probability density $p(\boldsymbol{z}|\boldsymbol{\theta}) = p(\boldsymbol{z})$.

Definition 5.1 Fisher Information Matrix

The Fisher information matrix for the data set $\boldsymbol{z}$ governed by the pdf $p(\boldsymbol{z}|\boldsymbol{\theta}) = p(\boldsymbol{z})$ is

$$I(\boldsymbol{\theta}) = E\left\{\left[\frac{\partial \ln[p(\boldsymbol{z}|\boldsymbol{\theta})]}{\partial \boldsymbol{\theta}}\right]\left[\frac{\partial \ln[p(\boldsymbol{z}|\boldsymbol{\theta})]}{\partial \boldsymbol{\theta}}\right]^T\right\} = -E\left\{\frac{\partial^2 \ln[p(\boldsymbol{z}|\boldsymbol{\theta})]}{\partial \boldsymbol{\theta}^2}\right\}. \tag{5.5}$$

The partial derivative $\partial \ln[p(\boldsymbol{z}|\boldsymbol{\theta})]/\partial \boldsymbol{\theta}$ is known as the **score function**.

For a scalar parameter θ, the information matrix reduces to the scalar

$$
\begin{aligned}
I(\theta) &= E\left\{\left[\frac{\partial \ln[p(z|\boldsymbol{\theta})]}{\partial \boldsymbol{\theta}}\right]^2\right\} \\
&= \int_{-\infty}^{\infty} \left[\frac{\partial \boldsymbol{p}(z|\boldsymbol{\theta})}{\partial \boldsymbol{\theta}}\right]^2 \frac{1}{\boldsymbol{p}(z;\boldsymbol{\theta})} \boldsymbol{d}z, \quad \boldsymbol{d}z = dz_1 dz_2 \dots dz_N \\
&= -E\left\{\frac{\partial^2 \ln[p(z|\boldsymbol{\theta})]}{\partial \boldsymbol{\theta}^2}\right\}.
\end{aligned} \tag{5.6}
$$

The proofs of the above equalities are left as an exercise.

Example 5.2 Find the score function and the Fisher information for the binomial distribution

$$
p(x|p) = \binom{N}{x} p^x (1-p)^{N-x}, x = 0, 1, 2, \dots, \quad E\{x\} = Np.
$$

Solution

The log of the pdf is

$$
\ln[p(x)] = \ln\binom{N}{x} + x\ln(p) + (N-x)\ln(1-p).
$$

The score function is

$$
\frac{\partial \ln[p(x)]}{\partial p} = \frac{x}{p} - \frac{N-x}{1-p}.
$$

The second derivative is

$$
\frac{\partial^2 \ln[p(x)]}{\partial p^2} = -\frac{x}{p^2} - \frac{N-x}{(1-p)^2}.
$$

We obtain the Fisher information

$$
\begin{aligned}
I(p) &= -E\left\{\frac{\partial^2 \ln[p(x)]}{\partial p^2}\right\} = E\left\{\frac{x}{p^2} + \frac{N-x}{(1-p)^2}\right\} \\
&= \frac{Np}{p^2} + \frac{N-Np}{(1-p)^2} \\
&= \frac{N}{p} + \frac{N}{1-p} = \frac{N}{p(1-p)}.
\end{aligned}
$$

Because the joint p.d.f. for N measurements $\{x_i, i = 1, \ldots, N\}$ is the product of the density functions $\prod_{i=1}^{N} p(x_i)$, the corresponding natural log is the sum of logs

$$\sum_{i=1}^{N} \ln(p(x_i)).$$

Hence, both the score function and the Fisher information corresponding to N measurements are scaled by N.

The inverse of the information matrix provides a lower bound below which the variance of the estimator cannot be reduced. The lower bound, known as the Cramer–Rao lower bound (CRLB), is discussed next.

Theorem 5.1 Cramer–Rao Inequality

The variance of an unbiased estimator $\hat{\boldsymbol{\theta}}$ of a deterministic $\boldsymbol{\theta}$ is bounded below by

$$\begin{gathered} Var\left(\hat{\boldsymbol{\theta}}\right) = E\left\{\tilde{\boldsymbol{\theta}}\tilde{\boldsymbol{\theta}}^T\right\} \geq I(\boldsymbol{\theta})^{-1} \\ \tilde{\boldsymbol{\theta}} = \hat{\boldsymbol{\theta}} - \boldsymbol{\theta}. \end{gathered} \tag{5.7}$$

For a scalar parameter

$$E\left\{\tilde{\theta}^2\right\} \geq I(\theta)^{-1}. \tag{5.8}$$

The proof of the inequality requires that the derivatives exist and be absolutely integrable (see [3]).

The efficiency of an estimator is measured by how close it approaches the CRLB.

Definition 5.2 Efficiency

The efficiency of an estimator is given by

$$Var\left(\hat{\boldsymbol{\theta}}\right)^{-1} I(\theta)^{-1}. \tag{5.9}$$

An efficient estimator is one whose efficiency is equal to unity and its variance is equal to the CRLB.

Example 5.3 Show that the sample mean is an efficient estimator of the mean of a multivariate normal distribution.

Solution

The log of the pdf for the distribution is

$$-\ln[p(z)] = \frac{1}{2\sigma^2}(z - m_z\mathbf{1})^T(z - m_z\mathbf{1})$$
$$+\frac{N}{2}\ln\left(2\pi\sigma^2\right) = \frac{1}{2\sigma^2}\left(z^Tz - 2m_z\mathbf{1}^Tz + m_z^2\mathbf{1}^T\mathbf{1}\right) + \cdots.$$

The score function for the distribution is

$$\frac{\partial\ln[p(z)]}{\partial p} = \frac{1}{\sigma^2}\left(-\mathbf{1}^Tz + m_z\mathbf{1}^T\mathbf{1}\right) = \frac{N}{\sigma^2}(m_z - \overline{z}).$$

The Fisher information for the distribution is

$$-E\left\{\partial^2\ln[p(z)]/\partial\theta^2\right\} = \frac{2N}{2\sigma^2} = \frac{N}{\sigma^2}, \mathbf{1}^T\mathbf{1} = N.$$

Hence, the Cramer–Rao lower bound is

$$\frac{\sigma^2}{N} = E\left\{\tilde{\theta}^2(k)\right\}.$$

The variance of the sample mean is

$$E\left\{\tilde{\theta}^2(k)\right\} = var\left\{\hat{\theta}^2\right\} = \frac{\sigma_z^2}{N},$$
$$\tilde{\theta}(N) = \theta - \hat{\theta}(N).$$

Thus, the variance of the sample mean is equal to the Cramer–Rao lower bound, and the sample mean is an efficient estimator.

The next theorem provides a necessary and sufficient condition for an efficient estimator.

Theorem 5.2 *The variance of an estimator is equal to the CRLB if and only if the score function satisfies*

$$\frac{\partial\ln[p(z)]}{\partial\boldsymbol{\theta}} = C(\boldsymbol{\theta})\left(\hat{\boldsymbol{\theta}} - \boldsymbol{\theta}\right), \tag{5.10}$$

where $C(\boldsymbol{\theta})$ is independent of the data z.

The proof is in Appendix 3 of [1].

Example 5.4 Use Theorem 5.2 to show that the sample mean is an efficient estimator for normal data with distribution $\mathcal{N}\left(m_z, \sigma^2\right)$.

Solution

The log of the pdf for the distribution is

$$
\begin{aligned}
-\ln[p(z)] &= \frac{1}{2\sigma^2}(z - m_z\mathbf{1})^T(z - m_z\mathbf{1}) + \frac{N}{2}\ln\left(2\pi\sigma^2\right) \\
&= \frac{1}{2\sigma^2}\left(z^T z - 2m_z\mathbf{1}^T z + m_z^2\mathbf{1}^T\mathbf{1}\right) + \cdots .
\end{aligned}
$$

The score function for the distribution is

$$
\frac{\partial \ln[p(z)]}{\partial m_z} = \frac{1}{\sigma^2}\left(-\mathbf{1}^T z + m_z\mathbf{1}^T\mathbf{1}\right) = -\frac{N}{\sigma^2}(\overline{z} - m_z).
$$

This is the in the form of Theorem 5.2 with $C(\boldsymbol{\theta}) = -N/\sigma^2$ independent of the data and the sample mean is an efficient estimator or the mean.

Efficient Unbiased Estimator

An unbiased estimator $\hat{\theta}^*(N)$ is more efficient than any other unbiased estimator if $\sigma^2_{\hat{\theta}^*} \leq \sigma^2_{\hat{\theta}}, \forall N$. When comparing two estimators, the following definition is useful.

Definition 5.3 Relative efficiency of two unbiased estimators

Let $\hat{\boldsymbol{\theta}}_1$ and $\hat{\boldsymbol{\theta}}_2$ be two unbiased estimators of $\boldsymbol{\theta}$ with variances $Var\left(\hat{\boldsymbol{\theta}}_1\right)$ and $Var\left(\hat{\boldsymbol{\theta}}_2\right)$, respectively. Then, the relative efficiency of $\hat{\boldsymbol{\theta}}_1$ with respect to $\hat{\boldsymbol{\theta}}_2$ is the ratio

$$
\frac{Var\left(\hat{\boldsymbol{\theta}}_2\right)}{Var\left(\hat{\boldsymbol{\theta}}_1\right)}, \tag{5.11}
$$

and $\hat{\boldsymbol{\theta}}_1$ is more efficient than $\hat{\boldsymbol{\theta}}_2$ if $Var\left(\hat{\boldsymbol{\theta}}_1\right) < Var\left(\hat{\boldsymbol{\theta}}_2\right)$.

Example 5.5 Sample Mean

Show that the sample mean of N i.i.d. Gaussian measurements $\left\{z_i \sim \mathcal{N}\left(m_z, \sigma^2\right), i = 1, 2, \ldots, N\right\}$ is an efficient estimator of the mean m_z.

Solution

We write the joint pdf of the measurements as

$$
p_X(z) = \frac{1}{[2\pi]^{N/2}\sqrt{\det(C_z)}}\exp\left\{-\frac{1}{2\sigma^2}(z - m_z1)^T(z - m_z1)\right\},
$$

where $z = \left[\,z_1 \ldots z_N\,\right]^T$, $1 = \left[\,1 \ldots 1\,\right]^T$, $C_z = \sigma^2 I_N$ and $\det(C_z) = \sigma^{2N}$.

The CRLB is given by

$$
E\left\{\tilde{\theta}^2(k)\right\} \geq -1/E\left\{\partial^2 \ln[p(z)]/\partial\theta^2\right\}.
$$

We take the log of the pdf and only retain the relevant terms including the parameter $\theta = m_z$

$$
\begin{aligned}
-\ln[p(z)] &= \frac{1}{2\sigma^2}(z - m_z\mathbf{1})^T(z - m_z\mathbf{1}) + \frac{N}{2}\ln\left(2\pi\sigma^2\right) \\
&= \frac{1}{2\sigma^2}\left(z^T z - 2m_z\mathbf{1}^T z + m_z^2 1^T \mathbf{1}\right) + \ldots \\
&= \frac{1}{2\sigma^2}\left(z^T z - 2m_z\mathbf{1}^T z + m_z^2 N\right) + \ldots.
\end{aligned}
$$

The CRLB for the sample mean is the second derivative

$$
-E\left\{\partial^2 \ln[p(z)]/\partial\theta^2\right\} = \frac{2N}{2\sigma^2}.
$$

The Cramer–Rao lower bound is therefore equal to the variance of the sample mean

$$
CRLB = \sigma^2/N = E\left\{\tilde{\theta}^2(k)\right\}.
$$

Thus, the sample mean is an efficient estimator for any variance $\sigma^2 > 0$.

The example shows that for N measurements $\{z_i, i = 1, \ldots, N\}$, the information is multiplied by N and the CRLB is therefore divided by N.

5.2 Large Sample Properties

Large sample properties are derived by taking the limit as the sample size goes to infinity. Hence, we can only assume that they will be valid if the sample size is very large. To emphasize the sample size on which the estimator is based, we will use the notation $\hat{\boldsymbol{\theta}}(N)$ for an estimate based on N data points. In many cases, a desirable estimator property does not hold for a finite sample size but is approximately true for if the sample size is large. We first consider the effect of sample size on bias.

Asymptotically Unbiased Estimators

A biased estimator may sometimes have desirable properties that make it preferable to an unbiased estimator if its bias is small, particularly if the bias goes to zero as the sample size goes to infinity. This leads to the following definition.

Definition 5.4 Asymptotic unbiasedness

An estimator is asymptotically unbiased if satisfies the condition

$$
\mathcal{L}im_{N\to\infty}\, E\left\{\boldsymbol{\theta} - \hat{\boldsymbol{\theta}}(N)\right\} = 0. \tag{5.12}
$$

For $\boldsymbol{\theta}$ random, we have

$$\underset{N\to\infty}{\mathcal{L}im}\; E\left\{\hat{\boldsymbol{\theta}}(N)\right\} = E\{\boldsymbol{\theta}\}, \tag{5.13}$$

and for $\boldsymbol{\theta}$ constant,

$$\underset{N\to\infty}{\mathcal{L}im}\; E\left\{\hat{\boldsymbol{\theta}}(N)\right\} = \boldsymbol{\theta}. \tag{5.14}$$

Example 5.6 Show that the sample variance calculated using i.i.d. measurements $\{x_i, i = 1, \ldots, N\}$,

$$\widehat{\sigma^2} = \frac{1}{N}\sum_{i=1}^{N}(x_i - \overline{x})^2$$

is a biased estimator but is asymptotically unbiased, where $\overline{x}$ is the sample mean.

Solution

The expected value of the estimate is

$$E\left\{\widehat{\sigma^2}\right\} = \frac{1}{N}\sum_{i=1}^{N} E\left\{x_i^2 - 2x_i\overline{x} + \overline{x}^2\right\}.$$

Substituting for the sample mean gives

$$E\left\{\widehat{\sigma^2}\right\} = \frac{1}{N}\left[\sum_{i=1}^{N} E\left\{x_i^2\right\} - 2E\left\{\frac{1}{N}\sum_{i=1}^{N} x_i \sum_{j=1}^{N} x_j\right\} + E\left\{\frac{1}{N^2}\sum_{i=1}^{N}\sum_{j=1}^{N}\sum_{l=1}^{N} x_j x_l\right\}\right]$$

$$= \frac{1}{N^2}\left[N\sum_{i=1}^{N} E\left\{x_i^2\right\} - 2\sum_{i=1}^{N}\sum_{j=1}^{N} E\left\{x_i x_j\right\} + \frac{N}{N}\sum_{i=1}^{N}\sum_{j=1}^{N} E\left\{x_i x_j\right\}\right]$$

$$= \frac{1}{N^2}\left[(N-1)\sum_{i=1}^{N} E\left\{x_i^2\right\} - \frac{1}{N}\sum_{i=1}^{N}\sum_{\substack{j=1\\ i\neq j}}^{N} E\left\{x_i x_j\right\}\right]$$

$$= \frac{1}{N^2}\left[(N-1)\sum_{i=1}^{N} E\left\{x_i^2\right\} - \sum_{i=1}^{N}\sum_{\substack{j=1\\ i\neq j}}^{N} E\{x_i\}E\left\{x_j\right\}\right]$$

$$= \frac{1}{N^2}\left[(N-1)\sum_{i=1}^{N}(\sigma^2+\mu^2) - N(N-1)\mu^2\right]$$

$$= \frac{N-1}{N}\sigma^2.$$

The estimator is clearly biased, but it is asymptotically unbiased because

$$\mathcal{L}im_{N\to\infty} E\left\{\widehat{\sigma^2}(N)\right\} = \sigma^2.$$

5.2.1 Consistent Estimators

Suppose that we have a data sample $\{x_i, i = 1, 2, \ldots, N\}$ that was collected with the purpose of estimating a particular parameter using a given estimator. One would at least hope that adding more data to our sample would result in an improvement in the estimate we obtain. In other words, one would hope that the sequence of estimators corresponding to sample size $N, N+1, N+2, \ldots,$ would get closer to the correct value of the parameter as N increases. Moreover, one would hope that the estimator corresponding to the limit as $N \to \infty$ would yield the correct value. This essential property of estimators is known as **consistency**.

Definition 5.5 Consistency: An estimator is consistent if it converges in probability to the parameter value, i.e.,

An estimator is consistent if it converges in probability to the parameter value, i.e.,

$$\mathcal{L}im_{N\to\infty} P\left[\left\|\boldsymbol{\theta} - \hat{\boldsymbol{\theta}}(N)\right\| \geq \epsilon\right] = 0. \tag{5.15}$$

Example 5.7 Show that the sample mean $\overline{x}$ of i.i.d. measurements $\{x_i, i = 1, \ldots, N\}$, is a consistent estimator of the mean. This is known as the **weak law of large numbers**.

Solution

We recall that for N i.i.d. variables of variance σ^2, the variance of their mean is $\sigma_{\overline{x}}^2 = \sigma^2/N$ and write the probability as

$$P[|\mu - \overline{x}| \geq \epsilon] = P\left[|\mu - \overline{x}| \geq \frac{\epsilon\sqrt{N}}{\sigma} \times \frac{\sigma}{\sqrt{N}}\right].$$

We use the Chebyshev inequality

$$P[|\mu - \overline{x}| \geq k\sigma_{\overline{x}}] \leq \frac{1}{k^2}$$

to write

$$P\left[|\mu - \overline{x}| \geq \frac{\epsilon\sqrt{N}}{\sigma} \times \sigma_{\overline{x}}\right] \leq \frac{\sigma^2}{N\epsilon^2}.$$

In the limit as $N \to \infty$, we have

$$\mathcal{L}im_{N\to\infty} P\left[|\mu - \overline{x}| \geq \frac{\epsilon\sqrt{N}}{\sigma} \times \sigma_{\overline{x}}\right] \leq \mathcal{L}im_{N\to\infty} \frac{\sigma^2}{N\epsilon^2} = 0.$$

5.2.2 Asymptotic Efficiency

Some estimators have a variance that is significantly larger than the CRLB but whose variance decreases to approach the bound in the limit as $N \to \infty$. Such estimators are said to be asymptotically efficient. Asymptotically, efficient estimators are expected to have a low variance if they are based on a large but finite number of data points.

5.2.3 Asymptotic Normality

The distribution governing the estimator is usually quite complex for an estimator obtained using a small sample size. For some estimators, the distribution converges to a normal distribution in the limit as $N \to \infty$. This is clearly desirable because a normal distribution is completely defined by its first two moments. If in addition the estimator is unbiased and efficient, or asymptotically unbiased and efficient, then for large sample size we know the probability of getting a bad estimate is low.

Example 5.8 For an unbiased asymptotically efficient (CRLB $= 0.01$), and asymptotically Gaussian estimator of a parameter whose true value is $m = 2$, determine the probability that the estimator is within 0.1 standard deviation of the correct value.

Solution

We use the MATLAB script and obtain the plot of Fig. 5.2.

```
m=2;sigma=0.1; % Mean, standard deviation

step=0.2*sigma; % Interval width
```

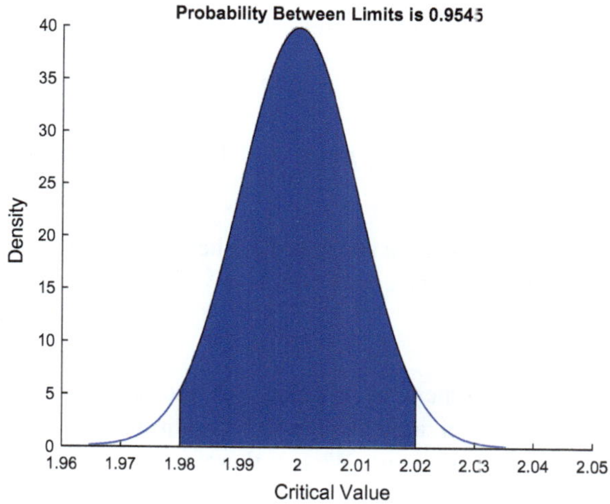

Fig. 5.2 Asymptotic distribution of estimator

p = normspec([m-step,m+step],2,sigma^2,'inside')

p = 0.9545

Example 5.9 Show that the sample mean $\overline{z}$ of i.i.d. measurements $\{z_i, i = 1, \ldots, N\}$, is asymptotically Gaussian.

Solution

We recall that the central limit theorem states that the distribution of the sum of any i.i.d. measurements $\{z_i, i = 1, \ldots, N\}$ converges to a normal distribution. The sample mean is equal to the sum times a constant

$$\overline{z} = \frac{1}{N}\sum_{i=1}^{N} z_i.$$

Because linear transformations preserve Gaussianity, the sample mean must converge to a Gaussian distribution.

5.3 Random Sample

Parameter estimates are obtained using experimental data that is collected through a random experiment. If the measurements are collected from a very large or infinite population, or from a finite population with replacement, then we can assume that the measurements are mutually independent. If the experiment is repeatable then it is reasonable to assume that the measurements are identically distributed. This is why in engineering applications it is common to assume that the measurements are i.i.d.

If the pdf of each measurement is $p_X(x)$, then the joint pdf of i.i.d. measurements $\{x_i, i = 1, 2, \ldots, N\}$ is

$$p_X(x_i, i = 1, 2, \ldots, N) = \prod_{i=1}^{N} p_X(x_i). \tag{5.16}$$

Example 5.10 Give an example where the data in a random sample can and another where it cannot be considered i.i.d.

Solution

The measurement of a voltage at the same node of an electric circuit can be repeated under almost identical conditions. Each measurement is unaffected by earlier measurements. We can therefore assume that voltage measurements $v_i, i = 1, 2, \ldots, N$, are i.i.d.

A fish tank has a finite number of fish of different weights that are assumed to vary randomly. Fish are drawn from the tank one at a time and weighed but not returned to the tank. The weight data $\{W_i, i = 1, 2, \ldots, N\}$ is not identically distributed because each time a fish is drawn and not returned the weight distribution for the fish in the tank changes.

5.3.1 Sufficient Statistics

In many parameter estimation problems, it is possible to avoid saving the complete data set by calculating a single statistic. If a single statistic captures all the information needed to estimate a parameter, it is known as a sufficient statistic. $T(\boldsymbol{x})$ is a sufficient statistic for a parameter θ if any estimate of θ using a sample $\boldsymbol{x}$ depends only on $T(\boldsymbol{x})$. $T(\boldsymbol{x})$ is a sufficient statistic for a parameter θ if any statistical inference about the parameter θ using the sample $\boldsymbol{x}$ only requires knowledge of $T(\boldsymbol{x})$. Thus, if $T(\boldsymbol{x}) = T(\boldsymbol{y})$ then inference based on observing the sample $\boldsymbol{x}$ yields identical results to inference based on observing the sample $\boldsymbol{y}$ and the estimate $\hat{\theta}(\boldsymbol{x})$ using the sample $\boldsymbol{x}$ is equal to the estimate using $\hat{\theta}(\boldsymbol{y})$ using the sample $\boldsymbol{y}$.

Definition 5.6 Sufficient Statistic

$T(\boldsymbol{x})$ is a sufficient statistic for a parameter θ if the conditional distribution $P(\boldsymbol{X} = \boldsymbol{x}|T(\boldsymbol{X}) = T(\boldsymbol{x}))$ of the sample $\boldsymbol{x}$ given $T(\boldsymbol{x})$ does not depend on θ.

The following theorem allows us to identify a sufficient statistic.

Theorem 5.3 *A statistic $T(\boldsymbol{x})$, with an i.i.d. random sample $\boldsymbol{x} = \{x_i, i = 1, 2, \ldots, N\}$ is a sufficient statistic for a parameter θ if and only if the joint pdf can be factored as*

$$l(\boldsymbol{x}|\theta) = a(\boldsymbol{x})b(T(\boldsymbol{x}), \theta)$$

for all random samples $\{x_i, i = 1, 2, \ldots, N\}$.

The conditions of the theorem ensure that the only part of the likelihood function that includes information on the parameter θ depends on $T(\boldsymbol{x})$ only and not on the random sample from which it is calculated.

Example 5.11 Show that the sum of the sample data for an i.i.d. random sample $\{x_i, i = 1, 2, \ldots, N\}$ drawn from a Poisson distribution with parameter λ is a sufficient statistic for the parameter.

Solution

The Poisson pdf is given by

$$p(x) = \frac{\mathrm{e}^{-\lambda}\lambda^x}{x!}.$$

The likelihood for the random sample is given by

$$l(\boldsymbol{x}|\lambda) = \prod_{i=1}^{N} p(x_i|\lambda) = \left(\prod_{i=1}^{N} \frac{1}{x_i!}\right)\mathrm{e}^{-N\lambda}\lambda^{T(\boldsymbol{x})},$$

$$T(\boldsymbol{x}) = \sum_{i=1}^{N} x_i.$$

The likelihood can be factorized as the product of

$$l(\boldsymbol{x}|\lambda) = a(\boldsymbol{x})b(T(\boldsymbol{x}), \lambda),$$

$$a(\boldsymbol{x}) = \left(\prod_{i=1}^{N} \frac{1}{x_i!}\right),$$

$$(T(\boldsymbol{x}), \lambda) = \mathrm{e}^{-N\lambda}\lambda^{T(\boldsymbol{x})}.$$

By Theorem 5.3, the sum $T(\boldsymbol{x})$ is a sufficient statistic for the parameter λ .

5.4 Estimation for the Autocorrelation and the Power Spectral Density

Although examples in this textbook may give the impression that the autocorrelation and power spectral density of noise processes are readily available, this is unfortunately not the case. In fact, both are difficult to estimate in practice. For processes where the form of these functions is unknown, they can be estimated using nonparametric approaches. These are approaches where the functions are estimated from data without the use of theory and with no associated parameters.

5.4.1 Autocorrelation Standard Estimate (ACS)

Assuming an ergodic process, we can use a single realization of a stationary stochastic process to obtain an estimate of its autocorrelation. For a continuous-time recording of duration T, we have the unbiased autocorrelation standard estimate (ACS)

$$\hat{R}_{XX}(\tau) = \frac{1}{T-\tau}\int_{0}^{T-\tau} X_T(t)X_T(t+\tau)\mathrm{d}t, \tau \ll T. \tag{5.17}$$

A more useful formula is the estimate for a discrete-time process

$$\hat{R}_{XX}(k) = \frac{1}{N-k}\sum_{i=0}^{N-k-1} X(i)X(i+k), k \ll N. \tag{5.18}$$

It is easy to prove that the ACS is an unbiased estimator and the proof is left as an exercise.

For large N, a biased version of the estimate with division by N rather than $N-k$ is sometimes used. It can be easily shown that the shown version of the estimator is unbiased and the version using N, or respectively T, is asymptotically unbiased. The proof is left as an exercise.

Because the length of the overlap in the product is $N-k$, or $T-\tau$ in the continuous-time case, the computed estimate is only meaningful for values of k, or τ, that are sufficiently short. The rule of thumb is that we must have k/N or $\tau/T < 0.1$.

To calculate the ACS using MATLAB, we use the command for the cross-correlation.

>> **c=xcorr(x,y, 'unbiased')** % cross-correlation

with the second argument omitted.

>> **r=xcorr(x, 'unbiased')** % autocorrelation

The MATLAB command uses the formula for complex random processes.

$$\hat{R}_{XY}(k) = \begin{cases} \frac{1}{N-k}\sum_{i=0}^{N-k-1} X(i+k)Y^*(i), k \geq 0 \\ \hat{R}^*_{YX}(-k), k < 0 \end{cases}, \tag{5.19}$$

$$\left[Y(i-k)X^*(i)\right]^* = X(i)Y^*(i-k). \tag{5.20}$$

The command exploits the even structure of the autocorrelation. The default divides by N and the option "unbiased" is required to divide by $N-k$.

For a record of length $N+k$, we rewrite the estimator as

$$\hat{R}_{XX}(k)=\frac{1}{N}\sum_{i=0}^{N-1}X(i)X(i+k),\, k\ll N. \tag{5.21}$$

The variance of the ACS is given by

$$\begin{aligned} var\left\{\hat{R}_{XX}(k)\right\} &= E\left\{\left(\hat{R}_{XX}(k)-R_{XX}(k)\right)^{2}\right\}=E\left\{\hat{R}_{XX}^{2}(k)\right\}-R_{XX}^{2}(k) \\ &= \frac{1}{N^{2}}\sum_{i=0}^{N-1}\sum_{j=0}^{N-1}E\{X(i)X(i+k)X(j)X(j+k)\}-R_{XX}^{2}(k). \end{aligned} \tag{5.22}$$

We used $R_{XX}(\tau)$ as the mean because the estimator is unbiased. Restricting the analysis to the Gaussian case, we use the expression

$$\begin{aligned} E\{X_1X_2X_3X_4\} &= E\{X_1X_2\}E\{X_3X_4\}+E\{X_1X_3\}E\{X_2X_4\} \\ &\quad + E\{X_1X_4\}E\{X_2X_3\} \end{aligned} \tag{5.23}$$

to obtain

$$\begin{aligned} &E\{X(i)X(i+k)X(j)X(j+k)\} \\ &= E\{X(i)X(j)\}E\{X(i+k)X(j+k)\} \\ &\quad + E\{X(i)X(j+k)\}E\{X(j)X(i+k)\}. \end{aligned} \tag{5.24}$$

Thus, we simplify the expression for the variance to

$$\begin{aligned} var\left\{\hat{R}_{XX}(k)\right\} &= \frac{1}{N^{2}}\sum_{i=0}^{N-1}\sum_{j=0}^{N-1}R_{XX}^{2}(k) \\ &\quad + R_{XX}^{2}(i-j)+R_{XX}(i-j-k)R_{XX}(i-j+k)-R_{XX}^{2}(k) \\ &= \frac{1}{N^{2}}\sum_{i=0}^{N-1}\sum_{j=0}^{N-1}R_{XX}^{2}(i-j)+R_{XX}(i-j-k)R_{XX}(i-j+k). \end{aligned}$$

Letting $j=i-l$, we have

$$var\left\{\hat{R}_{XX}(k)\right\}=\frac{1}{N^{2}}\sum_{i=0}^{N-1}\sum_{l=i-N+1}^{i}R_{XX}^{2}(l)+R_{XX}(l-k)R_{XX}(l+k).$$

We change the order of summation (see Fig. 5.3).

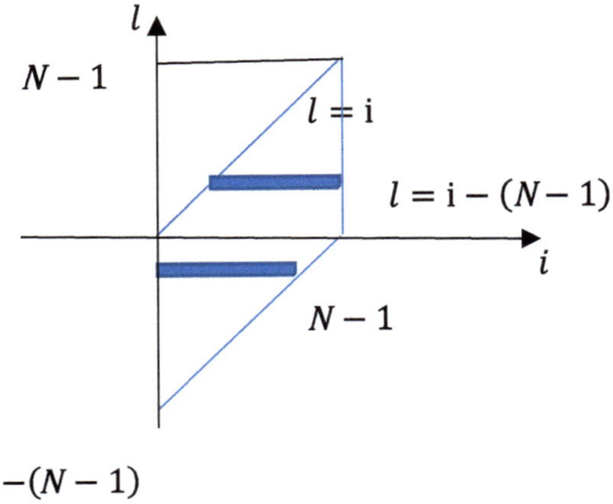

Fig. 5.3 Changing the order of summation

$$var\left\{\hat{R}_{XX}(k)\right\} = \frac{1}{N^2}\sum_{l=-(N-1)}^{0}\sum_{i=0}^{l-(N-1)} R_{XX}^2(l) + R_{XX}(l-k)R_{XX}(l+k)$$
$$+\frac{1}{N^2}\sum_{l=0}^{N-1}\sum_{i=l}^{i=N-1} R_{XX}^2(l) + R_{XX}(l-k)R_{XX}(l+k)$$
$$= \frac{1}{N^2}\sum_{l=-(N-1)}^{N-1}(N-1-l)\left[R_{XX}^2(l) + R_{XX}(l-k)R_{XX}(l+k)\right]$$
$$var\left\{\hat{R}_{XX}(k)\right\} = \frac{1}{N}\sum_{l=-(N-1)}^{N-1}\left(1-\frac{l+1}{N}\right)\left[R_{XX}^2(l) + R_{XX}(l-k)R_{XX}(l+k)\right]. \tag{5.25}$$

For large N, we have

$$var\left\{\hat{R}_{XX}(k)\right\} < \frac{1}{N}\sum_{l=-\infty}^{\infty}\left[R_{XX}^2(l) + R_{XX}(l-k)R_{XX}(l+k)\right]. \tag{5.26}$$

At $k = 0$, using the symmetry of the autocorrelation, we obtain

$$var\left\{\hat{R}_{XX}(0)\right\} < \frac{2}{N}\sum_{l=-\infty}^{\infty} R_{XX}^2(l) < \frac{4}{N}\sum_{l=0}^{\infty} R_{XX}^2(l).$$

The bound is approximately correct for small k. For an ergodic process, $R_{XX}(k)$ approaches zero as $k \to \infty$, and with N large the variance becomes

$$var\left\{\hat{R}_{XX}(k)\right\} \approx \frac{1}{N}\sum_{l=-(N-1)}^{N-1} R_{XX}^2(l) < \frac{1}{N}\sum_{l=-\infty}^{\infty} R_{XX}^2(l),$$

which is half the value for $k = 0$. We conclude that the variance of the estimator has the upper bound

$$var\left\{\hat{R}_{XX}(\tau)\right\} < \frac{4}{N}\sum_{k=0}^{\infty} R_{XX}^2(l). \tag{5.27}$$

Similarly, we can show that for the continuous case

$$var\left\{\hat{R}_{XX}(\tau)\right\} < \frac{4}{T}\int_0^{\infty} R_{XX}^2(\tau)\mathrm{d}\tau. \tag{5.28}$$

The variance goes to zero as the record length goes to infinity but can be large. For an unbiased estimator, the variance is equal to the mean-squared error and the estimator converges in mean square. Convergence in mean square implies convergence in probability. Thus, the estimator is consistent.

For a Gauss–Markov process, the autocorrelation is

$$R_{XX}(\tau) = \sigma^2 \mathrm{e}^{-\beta|\tau|},$$

and the upper bound of the variance gives

$$\begin{aligned} var\left\{\hat{R}_{XX}(\tau)\right\} &< \frac{4}{T}\int_0^{\infty} R_{XX}^2(\tau)\mathrm{d}\tau = \frac{4}{T}\int_0^{\infty} \sigma^4 \mathrm{e}^{-2\beta|\tau|}\mathrm{d}\tau \\ &= -\frac{4\sigma^4}{2\beta T}\mathrm{e}^{-2\beta\tau}\Big]_0^{\infty} = \frac{2\sigma^4}{\beta T}. \end{aligned} \tag{5.29}$$

The upper bound on the variance is in terms of the parameters (σ^2, β), which in practice are only known approximately.

The accuracy of the estimator is defined as

$$\text{Accuracy} = \sqrt{\frac{var\left\{\hat{R}_{XX}(\tau)\right\}}{\sigma^4}}. \tag{5.30}$$

The upper bound on the variance can be used to find the length of the recorded signal T for a specified accuracy of the autocorrelation estimate for a Gauss–Markov process. To guarantee an accuracy α, we satisfy the condition

$$(\text{Accuracy})^2 = \frac{var\left\{\hat{R}_{XX}(\tau)\right\}}{\sigma^4} \le \frac{2}{\beta T} = \alpha^2, \tag{5.31}$$

Hence, the length of the signal must be

$$T = \frac{2}{\alpha^2 \beta}.$$

In practice, the discrete expression is used, and we have

$$var\left\{\hat{R}_{XX}(\tau)\right\} < \frac{4}{N}\sum_{k=0}^{\infty} R_{XX}^2(l) = \frac{4}{N}\sum_{k=0}^{\infty} \sigma^4 e^{-2\beta|l\Delta t|} = \frac{4\sigma^4}{N\left(1 - e^{-2\beta|\Delta t|}\right)}. \quad (5.32)$$

To guarantee an accuracy α, we satisfy the condition.

$$(\text{Accuracy})^2 = \frac{var\left\{\hat{R}_{XX}(\tau)\right\}}{\sigma^4} \le \frac{4}{N\left(1 - e^{-2\beta|\Delta t|}\right)} = \alpha^2,$$
$$N = \frac{4}{\alpha^2\left(1 - e^{-2\beta|\Delta t|}\right)}. \quad (5.33)$$

Example 5.12 **Random Gyro Drift (Adapted from Brown and Hwang** [2]**)**
A gyro has random Gauss–Markov drift with time constant $\beta^{-1} = 10\,\text{h}$. Determine the length of the record needed for the ACS of the autocorrelation of the process to have an accuracy α of (a) 0.1, (b) 0.01.

Solution

Following the Brown and Hwang [2], we first use the continuous-time formula to assess the length of the record required for a given accuracy, then we use the more practical discrete-time formula and compare the results.

The length of the record for accuracy α is

$$T = \frac{2}{\alpha^2 \beta} = \frac{20}{\alpha^2}.$$

Substituting for the accuracy gives

$$T = \begin{cases} 2000\,\text{h}, & \alpha = 0.1 \\ 200{,}000\,\text{h} \approx 23\,\text{years}, & \alpha = 0.01 \end{cases}.$$

The time needed for 10% accuracy is of the order of the mean time to failure of gyro. The example shows that because the ACS has a large variance, a considerable amount of data is needed to estimate the autocorrelation requires huge amounts of data for an accurate estimate.

In practice, the record is sampled, and the discrete formula is used. Using the discrete formula with $\Delta t = 0.01s = 0.01/3600\,\text{h}$, and with $\beta = 0.1\,\text{h}^{-1}$, we have

$$N = \frac{4}{\alpha^2\left(1 - e^{-2\beta|\Delta t|}\right)} = \frac{4}{\alpha^2\left(1 - e^{-5.5556e-07}\right)}.$$

Substituting for the accuracy and rounding up gives

$$N = \begin{cases} 7.2 \times 10^{8} \text{ points}, & \alpha = 0.1 \\ 7.2 \times 10^{10} \text{ points}, & \alpha = 0.01 \end{cases}.$$

Converting to length of recording gives the record length as that obtained with the continuous-time formula.

5.4.2 *Periodogram*

The power spectral density (PSD) is defined as

$$S_{XX}(j\omega) = \mathfrak{F}\{R_{XX}(\tau)\} = \int_{-\infty}^{\infty} R_{XX}(\tau)\mathrm{e}^{-j\omega\tau}\mathrm{d}\tau. \tag{5.34}$$

Because if its random nature, $X(t)$ may violate the Dirichlet conditions, and its Fourier transform may not exist. However, a truncated signal $X_T(t)$ of duration T typically has a Fourier transform. To obtain an estimate of the PSD, we use the periodogram of a realization of $X_T(t)$.

$$\hat{S}(j\omega, T) = \frac{1}{T}|\mathfrak{F}\{X_T(t)\}|^2. \tag{5.35}$$

In practice, a discrete-time signal

$$S(j\omega) = \sum_{k=-\infty}^{\infty} R(k)\mathrm{e}^{-j\omega k} \tag{5.36}$$

is used with the fast Fourier transform

$$\hat{S}(j\omega) = \frac{1}{N}\left|\sum_{k=1}^{N} X(k)\mathrm{e}^{-j\omega k}\right|^2. \tag{5.37}$$

To compute the periodogram, the MATLAB signal processing toolbox has the command.

>> [Pxx,w] = periodogram(x)

We now examine the properties of the periodogram as an estimator of the PSD.

Theorem 5.4 *The periodogram is an asymptotically unbiased estimator of the power spectral density, i.e., the expected value of the periodogram of a realization of $X_T(t)$ approaches its power spectral density in the limit as $T \to \infty$.*

$$\mathcal{L}im_{T\to\infty} E\left\{\frac{1}{T}|\mathfrak{F}\{X_T(t)\}|^2\right\} = S_X(j\omega). \tag{5.38}$$

Proof The expectation of the periodogram is

$$E\left\{\hat{S}(j\omega,T)\right\}=E\left\{\frac{1}{T}|\mathfrak{F}\{X_T(t)\}|^2\right\}$$

$$=\frac{1}{T}E\left\{\int_0^T X(t)\mathrm{e}^{-j\omega t}\mathrm{d}t\times\overbrace{\int_0^T X(s)\mathrm{e}^{j\omega s}\mathrm{d}s}^{\text{conjugate}}\right\}.$$

Exchanging the order of integration and expectation gives

$$E\left\{\hat{S}(j\omega,T)\right\}=\frac{1}{T}\int_0^T\int_0^T E\{X(t)X(s)\}\mathrm{e}^{-j\omega(t-s)}\mathrm{d}t\mathrm{d}s$$

$$=\frac{1}{T}\int_0^T\int_0^T R_{XX}(t-s)\mathrm{e}^{-j\omega(t-s)}\mathrm{d}t\mathrm{d}s.$$

We simplify the integral by changing the variable of integration to $\tau=t-s$, in other words

$$\tau=\begin{cases}t, & s=0\\ t-T, & s=T\end{cases}.$$

The change of variable is depicted in Fig. 5.4. For fixed t, we have the differential $\mathrm{d}\tau=-ds$.

Changing the order of integration gives

$$E\left\{\frac{1}{T}|\mathfrak{F}\{X_T(t)\}|^2\right\}=\frac{1}{T}\int_0^T\left[\int_{t-T}^{t}R_{XX}(\tau)\mathrm{e}^{-j\omega\tau}\mathrm{d}\tau\right]\mathrm{d}t$$

$$=\underbrace{\frac{1}{T}\int_{-T}^{0}\left[\int_0^{T+\tau}R_{XX}(\tau)\mathrm{e}^{-j\omega\tau}\mathrm{d}t\right]\mathrm{d}\tau}_{\tau\leq 0}$$

$$+\underbrace{\frac{1}{T}\int_0^T\left[\int_\tau^T R_{XX}(\tau)\mathrm{e}^{-j\omega\tau}\mathrm{d}t\right]\mathrm{d}\tau}_{\tau>0}.$$

Integration w.r.t. t gives $T+\tau$ for negative τ and $T-\tau$ for positive τ. The two integrals can be combined and written in terms of the absolute value as

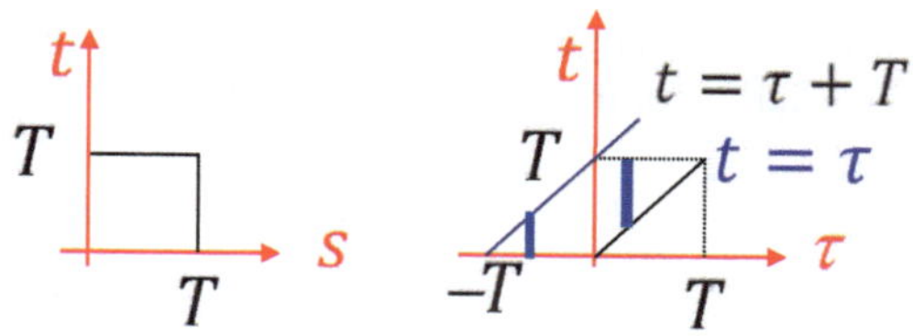

Fig. 5.4 Changing the variable of integration

$$E\left\{\frac{1}{T}|\mathfrak{F}\{X_T(t)\}|^2\right\} = \int_{-T}^{T}\left[1 - \frac{|\tau|}{T}\right]R_{XX}(\tau)\mathrm{e}^{-j\omega\tau}\mathrm{d}\tau.$$

In the limit at $T \to \infty$, the expectation reduces to

$$\underset{T\to\infty}{\mathcal{L}im}\; E\left\{\frac{1}{T}|\mathfrak{F}\{X_T(t)\}|^2\right\} = \int_{-\infty}^{\infty} R_{XX}(\tau)\mathrm{e}^{-j\omega\tau}\mathrm{d}\tau = S_X(j\omega).$$

■

Unfortunately, the periodogram is an inconsistent and inefficient estimator and has a large variance (see Kay [3], p. 97). Using a long record (relative to $X(t)$ variations) does not reduce the variance.

The variance of the periodogram is

$$var\left\{\hat{S}(j\omega, T)\right\} = E\left\{\hat{S}^2(j\omega, T)\right\} - E^2\left\{\hat{S}(j\omega, T)\right\}$$

i. Derive an expression for the mean square.
ii. Derive an expression for the variance

The mean square of the periodogram is

$$\hat{S}(j\omega, T) = \frac{1}{T}|\mathfrak{F}\{XT(t)\}|^2 = \frac{1}{T}\left\{\int_0^T X(t)\mathrm{e}^{-j\omega t}\mathrm{d}t \times \overbrace{\int_0^T X(s)\mathrm{e}^{j\omega s}\mathrm{d}s}^{\text{conjugate}}\right\}.$$

and its expected value is

$$E\left\{\hat{S}^2(j\omega, T)\right\} = \frac{1}{T^2}E\left\{\int_0^T\int_0^T\int_0^T\int_0^T X(t)X(s)X(u)X(v)\mathrm{e}^{-j\omega(t-s+u-v)}\mathrm{d}t\mathrm{d}s\mathrm{d}u\mathrm{d}v\right\}.$$

We interchange expectation and integration and use the property of zero-mean Gaussian random signals (see Problem 2.17)

$$
\begin{aligned}
E\{X(t)X(s)X(u)X(v)\} = {} & R_{XX}(t-s)R_{XX}(u-v) \\
& + R_{XX}(t-u)R_{XX}(s-v) \\
& + R_{XX}(t-v)R_{XX}(s-u).
\end{aligned}
$$

Substituting in the periodogram expression gives

$$
\begin{aligned}
& E\left\{\hat{S}^2(j\omega,T)\right\} \\
& = \frac{1}{T^2}\left\{\int_0^T\int_0^T R_{XX}(t-s)\mathrm{e}^{-j\omega(t-s)}\mathrm{d}t\mathrm{d}s \times \int_0^T\int_0^T R_{XX}(u-v)\mathrm{e}^{-j\omega(u-v)}\mathrm{d}u\mathrm{d}v\right\} \\
& + \frac{1}{T^2}\left\{\int_0^T\int_0^T R_{XX}(t-u)\mathrm{e}^{-j\omega(t+u)}\mathrm{d}t\mathrm{d}u \times \int_0^T\int_0^T R_{XX}(s-v)\mathrm{e}^{j\omega(s+v)}\mathrm{d}s\mathrm{d}v\right\} \\
& + \frac{1}{T^2}\left\{\int_0^T\int_0^T R_{XX}(t-v)\mathrm{e}^{-j\omega(t-v)}\mathrm{d}t\mathrm{d}v \times \int_0^T\int_0^T R_{XX}(u-s)\mathrm{e}^{-j\omega(u-s)}\mathrm{d}s\mathrm{d}u\right\}
\end{aligned}
$$

$$
\begin{aligned}
E\left\{\hat{S}^2(j\omega,T)\right\} = {} & 2E\left\{\hat{S}(j\omega,T)\right\}^2 \\
& + \frac{1}{T^2}\left|\int_0^T\int_0^T R_{XX}(t-u)\mathrm{e}^{-j\omega(t+u)}\mathrm{d}t\mathrm{d}u\right|^2.
\end{aligned}
$$

The variance of the periodogram is

$$
Var\left\{\hat{S}(j\omega,T)\right\} = E\left\{\hat{S}^2(j\omega,T)\right\} - E\left\{\hat{S}(j\omega,T)\right\}^2.
$$

Expanding and substituting gives

$$
\begin{aligned}
Var\left\{\hat{S}(j\omega,T)\right\} = {} & E\left\{\hat{S}(j\omega,T)\right\}^2 + \underbrace{\tfrac{1}{T^2}\left|\int_0^T\int_0^T R_{XX}(t-u)\mathrm{e}^{-j\omega(t+u)}\mathrm{d}t\mathrm{d}u\right|^2}_{\text{nonnegative}} \\
& \geq E\left\{\hat{S}(j\omega,T)\right\}^2.
\end{aligned}
$$

Because the estimator is asymptotically unbiased, we have $\mathcal{L}im_{T\to\infty} E\left\{\hat{S}(j\omega,T)\right\} = S(j\omega)$.

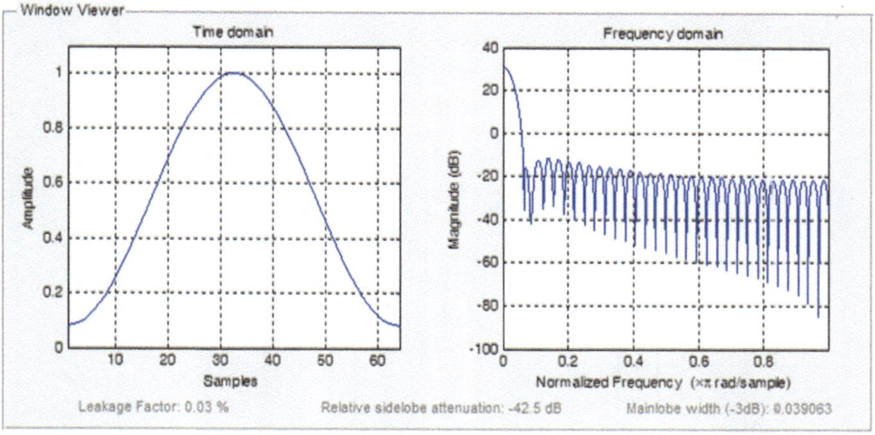

Fig. 5.5 Hamming window and its frequency spectrum

$$\mathcal{L}im_{T\to\infty} \, Var\left\{\hat{S}(j\omega, T)\right\} \geq S^2(j\omega).$$

Thus, the variance of the periodogram does not converge to zero.
We summarize the properties of the periodogram.

- Asymptotically unbiased spectral estimator (good).
- Inconsistent spectral estimator (large variance, and does not converge in probability), i.e., long record does not reduce variance.

There are methods available to reduce the variance, but they increase the bias and require a large number of records:

1. Chop record into short segments of length T/N and average their FFT.

- Loss of frequency resolution (N/T).[1]
- Problem: chopping (multiplication by pulse) is equivalent to frequency convolution with sinc (distortion).

2. Reduce distortion by using windows.

For example, one could use the Hamming window. The Hamming window and its frequency spectrum are shown in Fig. 5.5. The figure is obtained using the MATLAB command.

>> L=64; wvtool(hamming(L))

Window visualization tool

Example 5.13 Use MATLAB to obtain the windowed periodogram of a cosine signal

[1] This follows from the properties of the discrete-time Fourier transform.

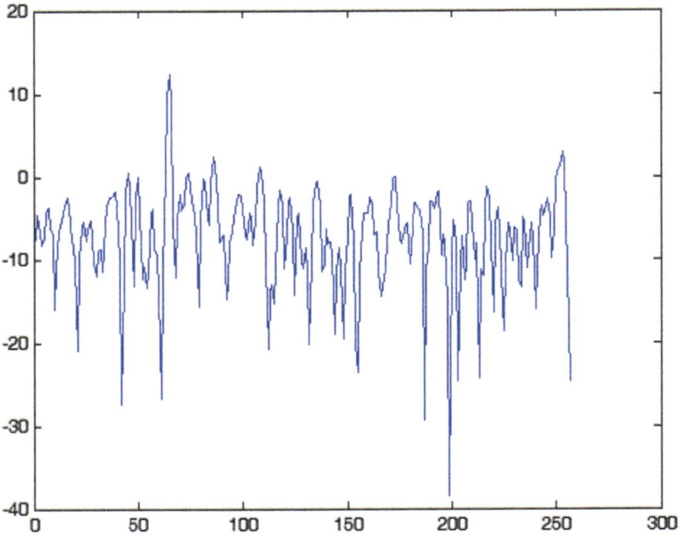

Fig. 5.6 Windowed periodogram

$$x_k = \cos(0.25\pi k) + u_k$$

with the Hamming window, where u_k is a Gaussian white noise sequence with zero-mean and unity variance (Fig. 5.6).

Solution

```
>> n = 0:319;
>>x = cos(pi/4*n)+randn(size(n)); % Data
>> pxx = periodogram(x,hamming(length(x)));
>> plot(10*log10(pxx)) % Plot dBs
>>% Default pxx is the rectangular window.
```

Problems

5.1 Prove that the estimator of the correlation

$$\hat{R}_{XX}(k) = \frac{1}{N-k}\sum_{i=0}^{N-k-1} X(i)X(i+k), \quad k \ll N$$

is unbiased and that the estimator is asymptotically unbiased with division by N instead of $N-k$.

5.2 Prove that the estimator of the correlation

$$\hat{R}_{XX}(\tau) = \frac{1}{T-\tau}\int_{0}^{T-\tau} X_T(t)X_T(t+\tau)\mathrm{d}t, \quad \tau \ll T$$

is unbiased and that the estimator is asymptotically unbiased with division by T instead of $T - \tau$.

5.3 Find the score function, the Fisher information, and the CRLB for the Poisson distribution.

5.4 Find the score function, the Fisher information, and the CRLB for i.i.d. measurements $\{z_i, i = 1, \ldots, N\}$ with mean μ and variance σ^2 governed by the normal distribution

$$p(z) = \frac{1}{\sqrt{2\pi\sigma^2}} \mathrm{e}^{-(z-\mu)^2/2\sigma^2}.$$

5.5 Show that

(i) $I(\theta) = E\left\{\left[\frac{\partial \ln[p(z)]}{\partial \theta}\right]^2\right\} = \int\limits_{-\infty}^{\infty} \left[\frac{\partial p(z)}{\partial \theta}\right]^2 \frac{1}{f(z)} \mathrm{d}z.$

(ii) $I(\theta) = E\left\{\left[\frac{\partial \ln[p(z)]}{\partial \theta}\right]^2\right\} = -E\left\{\frac{\partial^2 \ln[p(z)]}{\partial \theta^2}\right\}.$

5.6 Prove the identities for i.i.d. measurements $\{x_i, i = 1, \ldots, N\}$ with mean μ and variance σ^2.

(i) $(x_i - \overline{x})^2 = (x_i - \mu)^2 - (\overline{x} - \mu)^2 - 2(\overline{x} - \mu)(x_i - \overline{x}).$
(ii) $\sum_{i=1}^{N}(x_i - \overline{x})^2 = \sum_{i=1}^{N}(x_i - \mu)^2 - N(\overline{x} - \mu)^2.$
(iii) $E\left\{\sum_{i=1}^{N}(x_i - \overline{x})^2\right\} = (N - 1)\sigma^2.$

5.7 Using the steps of Example 5.6, show that the sample variance calculated using i.i.d.. measurements $\{x_i, i = 1, \ldots, N\}$.

$$\widehat{\sigma^2} = \frac{1}{N-1}\sum_{i=1}^{N}(x_i - \overline{x})^2$$

is an unbiased estimator of the variance, where $\overline{x}$ is the sample variance.

5.8 Using the results of Problem 5.5, repeat Problem 5.6.

5.9 Prove that the sample variance

$$\widehat{\sigma^2} = \frac{1}{N}\sum_{i=1}^{N}(x_i - \overline{x})^2$$

calculated using i.i.d. measurements $\{x_i, i = 1, \ldots, N\}$ satisfies the variance relation

$$\sigma^2 = E\{x^2\} - E\{x\}^2$$

with each term replaced by its sample mean, i.e.,

$$\widehat{\sigma^2} = \frac{1}{N}\sum_{i=1}^{N} x_i^2 - \overline{x}^2, \quad \overline{x} = \frac{1}{N}\sum_{i=1}^{N} x_i.$$

Use the result to show that the estimator is biased.

5.10 Using the fact that for i.i.d. zero-mean unity variance Gaussian variables $\{z_i, i = 1, \ldots, N\}$, the sum of the squares has the chi-square distribution with N degrees of freedom

$$\sum_{i=1}^{N} z_i^2 \sim \mathcal{X}_N^2.$$

It can be shown that for i.i.d. Gaussian variables $\{x_i, i = 1, \ldots, N\}$.

$$\left(\frac{N-1}{\sigma^2}\right)\widehat{\sigma^2} = \frac{1}{N-1}\sum_{i=1}^{N}(x_i - \overline{x})^2 \sim \mathcal{X}_{N-1}^2.$$

(a) Use the properties of the Chi-square distribution to show that that the estimator is unbiased.
(b) Use the properties to show that the variance of the estimator is

$$Var\left(\widehat{\sigma^2}\right) = \frac{2\sigma^4}{N-1}.$$

(c) Show that the estimator is consistent.
(d) Show that the estimator is asymptotically efficient.

5.11 Using the results of Example 5.2 for the binomial distribution, show that the score function for i.i.d. sample points $\{x_i, i = 1, \ldots, N\}$ satisfies the condition of Theorem 5.2 then verify that the estimator

$$\hat{p} = \frac{1}{nN}\sum_{i=1}^{N} x_i = \frac{\overline{x}}{n}$$

of the parameter p has a variance that is equal to the CRLB.

5.12 Use the following Chebyshev inequality, where c is a real constant to show that convergence of a random variable x in mean square implies convergence in probability

$$P(|X - c| \geq \epsilon) \leq \frac{E\{(X - c)^2\}}{\epsilon^2}.$$

5.13 When randomly testing a product on a production line for defective units, we have n independent Bernoulli trials. The probability of n sound units before a defective unit is encountered is governed by the geometric distribution

$$p(n|\theta) = (1-\theta)^n\theta.$$

Show that if the testing process is repeated N times and the number of consecutive sound units is $n_i, i = 1, 2, \ldots, N$, then the sum $\sum_{i=1}^{N} n_i$ is a sufficient statistic for the parameter θ.

5.14 Repeat Example 5.13 with the Bartlett window for the signal

$$x_k = 0.5\sin(0.2\pi k) + u_k,$$

where u_k is a Gaussian white noise sequence with zero-mean and unity variance.

5.15 The exponential family of distributions is in the form

$$p(\boldsymbol{x}|\boldsymbol{\theta}) = h(\boldsymbol{x})\exp\left(\boldsymbol{c}(\boldsymbol{\theta})^T\boldsymbol{t}(\boldsymbol{x}) - a(\boldsymbol{\theta})\right),$$

where $\boldsymbol{\theta}$ is a vector of parameters, $\boldsymbol{t}(\boldsymbol{x})$ is a vector of sufficient statistics, $\boldsymbol{\theta}$ is a vector of natural parameters, and $\boldsymbol{c}(\boldsymbol{\theta})$ is a vector of parameters. The family includes important distributions that we often encounter in applications including normal, gamma, Poisson, Bernoulli, and multinomial distributions.

(a) Show that the Bernoulli distribution

$$p(x|\theta) = \theta^x(1-\theta)^{1-x}, x \in \{0, 1\}$$

is a member of the exponential family. You will need to write the pmf in terms of a function of the parameter θ.

(b) Show that $a(\boldsymbol{\theta}) = \ln\left\{\int_{-\infty}^{\infty} h(\boldsymbol{x})\exp\left(\boldsymbol{\theta}^T\boldsymbol{t}(\boldsymbol{x})\right)\mathrm{d}\boldsymbol{x}\right\}$.

(c) Obtain the score function for the exponential family.

(d) Use the result of (a) and (b) to write the score function for the Bernoulli distribution.

References

1. Goldberger, A. S. (1964). *Economic theory*. Wiley.
2. Brown, R. G., & Hwang, P. Y. C. (2012). *Introduction to random signals and applied Kalman filtering* (4th ed.) Wiley.

3. Kay, S. M. (1993). *Fundamentals of statistical signal processing: Estimation theory* (Vol. I). Prentice-Hall.

Bibliography

1. Bendat, J. S., & Piersol, A. G. (2011). *Random data: Analysis and measurement procedures.* Wiley.
2. Casella, G., & Berger, R. (1990). *Statistical inference.* Duxbury Press.
3. DeGroot, M. H., & Schervish, M. J. (2012). *Probability and statistics.* Addison-Wesley.
4. Hogg, R. V., McKean, J. W., & Craig, A. T. (2005). *Introduction to mathematical statistics.* Prentice-Hall.
5. Lindsey, J. K. (1996). *Parametric statistical inference.* Clarendon Press.
6. Martinez, W. L., & Martinez, A. R. (2002). *Computational statistics handbook with MATLAB.* Chapman & Hall/CRC.
7. Mendel, J. (1995). *Lessons in estimation theory and signal processing, communications, and control.* Prentice-Hall.
8. Shanmugan & Breipohl. (1988). *Detection, estimation & data analysis* (pp. 132–134). Wiley.
9. Stoica, P., & Moses, R. (1997). *Introduction to spectral analysis.* Prentice Hall.

Chapter 6
Least-Squares Estimation

Least-squares estimation provides a means of determining estimates of model parameters that are optimal in the sense of minimizing the sum of the squares of the estimation errors. Clearly, using the errors in the parameters directly requires knowledge of the parameters we seek to determine and is therefore not feasible. Instead, we rely on errors in estimating measurements that depend on the parameters we seek to estimate. Using physical principles or empirical knowledge of a physical system, we obtain of the form

$$\boldsymbol{y} = \boldsymbol{f}(\boldsymbol{x}, \boldsymbol{\theta}) \tag{6.1}$$

where $\boldsymbol{y}$ is a vector of measurements, $\boldsymbol{x}$ is a vector of system variables, and $\boldsymbol{\theta}$ is a vector of system parameters. The vector $\boldsymbol{f}$ is a vector of functions that represent the behavior of the system and are assumed to be continuous w.r.t. all its arguments. In practice, measurements will include errors or noise and the physical measurements have the form

$$\boldsymbol{z} = \boldsymbol{f}(\boldsymbol{x}, \boldsymbol{\theta}) + \boldsymbol{v} \tag{6.2}$$

where $\boldsymbol{v}$ is a vector of measurement errors or noise.

Example 6.1 Power engineers have measurements of active and reactive power that they use to estimate voltage magnitudes and phases in a power network through node equations. The analysis, known as power flow, results in theoretical equations in the form of (6.1). Because the measurements obtained in practice are noisy, they write an expression for the measurements in the form of (6.2).

The minimum square error problem is as follows.

Problem statement: Given a vector of physical measurements $\boldsymbol{z}$, determine the parameter vector $\boldsymbol{\theta}$ that minimizes the square error $\tilde{\boldsymbol{z}}$ between the measurements and the corresponding values obtained from the mathematical model of the system.

M. S. Fadali, *Introduction to Random Signals, Estimation Theory, and Kalman Filtering*, https://doi.org/10.1007/978-981-99-8063-5_6

In other words, determine the vector $\boldsymbol{\theta}$ such that

$$\min_{\hat{\theta}} J\left(\hat{\theta}\right) = \|\tilde{z}\|_2^2 = \tilde{z}^T\tilde{z} \tag{6.3}$$

where $\tilde{z} = z - \hat{z}$ is the residual vector and $\|\boldsymbol{x}\|_2 = \boldsymbol{x}^*\boldsymbol{x}$ is the 2-norm.

The output of the model, which is an estimate of the measurements, depends on the form of the model selected to model the physical system. It is a function of other measurements $\boldsymbol{x}$ obtained from the physical system and can be written as

$$\hat{z} = f(\boldsymbol{x}, \boldsymbol{\theta}) \tag{6.4}$$

6.1 Linear Model

The nonlinear least-squares problem is generally difficult and is not always solvable. In addition, many engineering models, including nonlinear models, are in the form of a linear function of the model parameters. These models are typically valid for a bounded range of the system variables where the system under study is likely to operate. As long as the model under study is a linear function of its parameter, the model equations can be rewritten in the form

$$\begin{aligned} z(k) &= H(k)\boldsymbol{\theta} + \boldsymbol{v}(k) \\ z(k) &= \left[\, z(k)\ z(k-1)\ \ldots\ z(k-N+1) \,\right]^T \\ \boldsymbol{v}(k) &= \left[\, v(k)\ v(k-1)\ \ldots\ v(k-N+1) \,\right]^T \\ \boldsymbol{\theta} &= \left[\, \theta_1\ \theta_2\ \ldots\ \theta_n \,\right]^T \end{aligned} \tag{6.5}$$

$z(k) = N \times 1$ measurement vector
$H(k) = N \times n$ measurement matrix
$\boldsymbol{v}(k) = N \times 1$ zero-mean measurement noise vector
$\boldsymbol{\theta} = n \times 1$ parameter vector.

The model is referred to as a **linear model** because it is linear in the parameters, excluding the error vector, even though it may arise from a nonlinear equation governing the systems variables. The entries of both the measurement vector and the measurement matrix are measurements obtained from the physical system to be modeled. In the power flow analysis of Example 6.1, as in many other problems, it is standard practice to linearize the nonlinear power flow equations to reduce them to the form of (6.5).

(a) Underdetermined $N < n$ (b) $N = n$ (c) Overdetermined $N > n$

Fig. 6.1 Form of measurement matrix

The assumption of zero-mean noise is not a severe restriction provided that the mean of the noise vector is known. For nonzero-mean noise, we simply subtract the mean from both sides of the linear model and replace the vector z with the vector $z - E\{\boldsymbol{v}\}$.

The linear model falls into one of three cases depicted in Fig. 6.1 depending on the number of measurements N and the number of unknown parameters n:

a. *Fewer measurements than unknowns* $(N < n)$ underdetermined system with infinitely many solutions.
b. *As many measurements as unknowns* $(N = n)$ unique solution (ignoring unknown noise).
c. *More measurements than unknowns* $(N > n)$ overdetermined system, use extra measurements to "filter data" (most common).

For the linear model, the estimate of the measurements is given by the model in terms of the parameter estimate and in the absence of noise.

$$\hat{z} = H\hat{\theta} \tag{6.6}$$

Because some measurements may be more reliable that others, we use a weighting matrix as part of the minimization. Terms given more weight are more important because solutions that do not minimize them do not yield the minimum and are rejected. The weight matrix W is typically a positive definite diagonal matrix $W = \text{diag}\{w_1, w_2, \ldots, w_N\}$. In some applications, the number of the measurements indicates the time at which it was takes so that w_1 is the weight of the first measurement and w_N is the last. Then the later measurements are given more weight if w_i increase with i and vice versa.

Including the weights, and adopting the linear model, we now minimize the performance measure

$$J\left\{\hat{\theta}\right\} = \left[z - H\hat{\theta}\right]^T W\left[z - H\hat{\theta}\right] = \hat{\theta}^T H^T W H\hat{\theta} - 2\hat{\theta}^T H^T W z + z^T W z \tag{6.7}$$

We use calculus to obtain the best estimate in the least-square error sense. We recall the conditions for maximization or minimization using the second-order approximation

$$J(\boldsymbol{x}) = J(\boldsymbol{x}_0) + \left.\frac{\partial J(\boldsymbol{x})}{\partial \boldsymbol{x}}\right]_{\boldsymbol{x}_0}^{T} \Delta\boldsymbol{x} + \frac{1}{2!}\Delta\boldsymbol{x}^T \left.\frac{\partial^2 J(\boldsymbol{x})}{\partial \boldsymbol{x}^2}\right]_{\boldsymbol{x}_0} \Delta\boldsymbol{x} + O\left(\left\|\Delta\boldsymbol{x}^3\right\|\right) \tag{6.8}$$

The series shows that the necessary condition for a minimum or maximum is zero gradient. The sufficient conditions require the Hessian matrix to maintain the same sign for any perturbation vector. Hence, the sufficient for minimum is a positive definite Hessian and the for a maximum a negative definite Hessian. Semidefinite matrices only give a necessary condition.

For our performance measure, the gradient is

$$\frac{\partial J\{\hat{\theta}(k)\}}{\partial \hat{\theta}(k)} = \mathbf{0} = 2H^T W H\hat{\theta} - 2H^T W z \tag{6.9}$$

The necessary Condition for a minimum is

$$H^T W H\hat{\theta}_{\text{WLS}} = H^T W z \tag{6.10}$$

H must be full rank for inverse to exist (assuming positive definite W). The solution is

$$\hat{\theta}_{\text{WLS}} = \left(H^T W H\right)^{-1} H^T W z \tag{6.11}$$

For a scalar weight matrix $W = wI_N$, the scalar w cancels and the solution is the least-squares estimate regardless of the scalar weight. This is to be expected because a scalar weight matrix only scales the performance measure but does not change the location of the minimum. The least-squares solution is

$$\hat{\theta}_{LS} = \left(H^T H\right)^{-1} H^T z \tag{6.12}$$

and is in terms of the pseudoinverse of the matrix H

$$H^{\#} = \left(H^T H\right)^{-1} H^T \tag{6.13}$$

Example 6.2 An engineer recorded the output voltage v and input current i for a nonlinear electronic amplifier as shown in Fig. 6.2. Obtain the parameters of the following model for the amplifier: $v(i) = a_0 + a_1 i + a_3 i^3$.

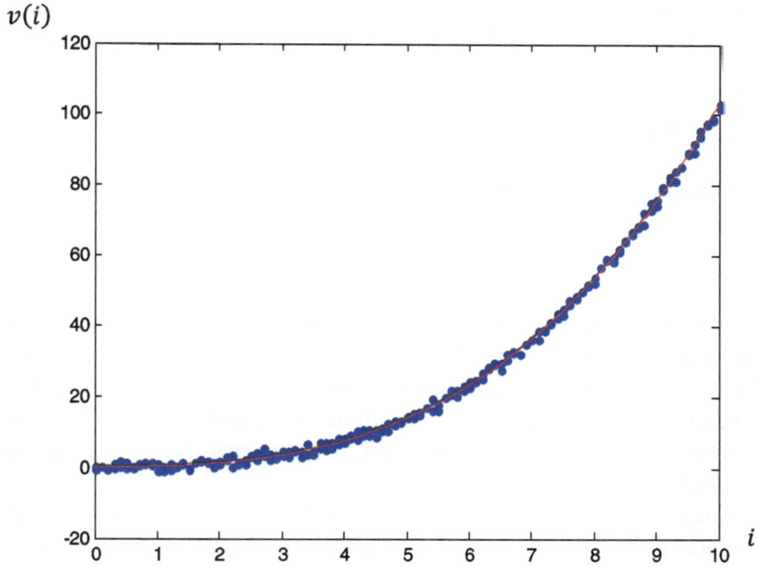

Fig. 6.2 V-I characteristics of amplifier

Solution

Rewrite the formula in vector form

$$v(i) = \begin{bmatrix} 1 \; i \; i^3 \end{bmatrix} \begin{bmatrix} a_0 \\ a_1 \\ a_3 \end{bmatrix}$$

Measuring voltage and current introduces noise w, and we rewrite the formula including an error term

$$z(j) = \begin{bmatrix} 1 \; i(j) \; i(j)^3 \end{bmatrix} \begin{bmatrix} a_0 \\ a_1 \\ a_3 \end{bmatrix} + w(j)$$
$$j = 1, 2, \ldots, N$$

Write a matrix measurement equation including the noise vector $\boldsymbol{w}$

$$z = H\boldsymbol{a} + \boldsymbol{w}$$

$$\begin{bmatrix} z(N) \\ z(N-1) \\ \vdots \\ z(1) \end{bmatrix} = \begin{bmatrix} 1 & i(N) & i^3(N) \\ 1 & i(N-1) & i^3(N-1) \\ \vdots & \vdots & \vdots \\ 1 & i(1) & i^3(1) \end{bmatrix} \begin{bmatrix} a_0 \\ a_1 \\ a_2 \end{bmatrix} + \begin{bmatrix} w(N) \\ w(N-1) \\ \vdots \\ w(1) \end{bmatrix}$$

The least-squares estimate is

$$\hat{\boldsymbol{a}}_{LS} = \left[H^T H\right]^{-1} H^T \boldsymbol{z}$$

The above example demonstrates the use of linear least squares to model a nonlinear input–output relation. Clearly, this approach works for any order polynomial but the number of parameters increases with the order of the polynomial and the inverted matrix becomes ill-conditioned for problems involving more than six parameters. For an exponential relation, it is still possible to use least-squares estimably by taking the natural log of data.

The above solution of the least-squares problem is only valid for a full-rank measurement matrix. If the measurement matrix is not full rank, the solution can be obtained in terms of the Moore–Penrose pseudoinverse of the measurement matrix. The Moore–Penrose pseudoinverse is obtained using the singular value decomposition

$$\begin{aligned} \underbrace{H}_{N\times n} &= \underbrace{U}_{N\times N}\underbrace{\Sigma}_{N\times n}\underbrace{V^*}_{n\times n} \\ \Sigma &= \begin{bmatrix} \Sigma_r & \mathbf{0}_{r\times(n-r)} \\ \mathbf{0}_{(N-r)\times r} & \mathbf{0}_{(N-r)\times(n-r)} \end{bmatrix} \\ \Sigma_r &= \mathrm{diag}\left\{\sigma_1\ \sigma_2\ \ldots\ \sigma_r\right\} \end{aligned} \tag{6.14}$$

where $\sigma_i, i = 1, \ldots, r$, are the singular values of the matrix and can be ordered WLOG as

$$\sigma_1 \geq \sigma_2 \geq \ldots \geq \sigma_r$$

The two matrices U and V are matrices of the left and right singular vectors, respectively. Both are unitary (orthogonal if real), matrices, i.e.,

$$U^{-1} = U^*, V^{-1} = V^* \tag{6.15}$$

where the superscript (*) denotes the conjugate transpose.

Analogously to the inverse of the product of three matrices, the pseudoinverse is given by

$$\underbrace{H^{\#}}_{n\times N} = \underbrace{V}_{n\times n}\underbrace{\Sigma^{\#}}_{n\times N}\underbrace{U^*}_{N\times N} \tag{6.16}$$

$$\Sigma^{\#} = \begin{bmatrix} \Sigma_r^{-1} & \mathbf{0}_{r\times(N-r)} \\ \mathbf{0}_{(n-r)\times r} & \mathbf{0}_{(n-r)\times(N-r)} \end{bmatrix} \tag{6.17}$$

$$\Sigma_r^{-1} = \mathrm{diag}\{\sigma_1^{-1}\ \sigma_2^{-1}\ \ldots\ \sigma_r^{-1}\}$$

It can be shown that the Moore–Penrose pseudoinverse reduces to the pseudoinverse $\left[H^T H\right]^{-1} H^T$ if the matrix is full rank. The following theorem gives the general solution of the least-squares problem.

Theorem 6.1 *For the linear model $z = H\boldsymbol{\theta} + \boldsymbol{v}$, the least-squares estimate of the parameter vector is*

$$\hat{\boldsymbol{\theta}}_{LS}(k) = H^{\#}z$$

Proof Consider the performance measure

$$J\left(\hat{\theta}\right) = \left\|z - H\hat{\theta}\right\|_2^2 = \left[z - H\hat{\theta}\right]^*\left[z - H\hat{\theta}\right]$$

Using the properties of the matrices U and V (6.15), we write

$$\begin{aligned} J\left(\hat{\theta}\right) &= UU^*\left[z - HVV^*\hat{\theta}\right] \\ &= \left\|U^*z - \left(U^*HV\right)V^*\hat{\theta}\right\|_2^2 \end{aligned}$$

Define the vectors

$$\boldsymbol{c} = U^*z = \begin{bmatrix} \boldsymbol{c}_1 \\ \boldsymbol{c}_2 \end{bmatrix}, \quad \boldsymbol{m} = V^*\hat{\theta} = \begin{bmatrix} \boldsymbol{m}_1 \\ \boldsymbol{m}_2 \end{bmatrix}$$

with $r \times 1\boldsymbol{c}_1, \boldsymbol{m}_1, N - r \times 1\boldsymbol{c}_2$, and $n - r \times 1\boldsymbol{m}_2$. ■

We rewrite the performance measure in terms of the vectors as

$$J(\hat{\theta}) = \left\|\boldsymbol{c} - \left(U^*HV\right)\boldsymbol{m}\right\|_2^2$$

Premultiplying H by U^* and postmultipying by V the using (6.15) gives

$$U^* \underbrace{H}_{N\times n} V = U^*U \underbrace{\Sigma}_{N\times n} V^*V = \begin{bmatrix} \Sigma_r & \mathbf{0}_{r\times(n-r)} \\ \mathbf{0}_{(N-r)\times r} & \mathbf{0}_{(N-r)\times(n-r)} \end{bmatrix}$$

We now write the performance measure as

$$J\left(\hat{\theta}\right) = \left\|\boldsymbol{c} - \left(U^*HV\right)\boldsymbol{m}\right\|_2^2 = \left\|\begin{bmatrix} \boldsymbol{c}_1 \\ \hline \boldsymbol{c}_2 \end{bmatrix} - \left[\begin{array}{c|c} \Sigma_r & \mathbf{0}_{r\times(n-r)} \\ \hline \mathbf{0}_{(N-r)\times r} & \mathbf{0}_{(N-r)\times(n-r)} \end{array}\right]\begin{bmatrix} \boldsymbol{m}_1 \\ \hline \boldsymbol{m}_2 \end{bmatrix}\right\|^2$$

There are infinitely many solutions that minimize the performance measure

$$\boldsymbol{m}_1 = \Sigma_r^{-1}\boldsymbol{c}_1, \forall \boldsymbol{m}_2$$

The minimum norm solution corresponds to $\boldsymbol{m}_2 = \boldsymbol{0}$. This gives the expression

$$J\left(\hat{\theta}\right) = \left\| \left[\frac{\boldsymbol{c}_1 - \Sigma_r \boldsymbol{m}_1}{\boldsymbol{c}_2} \right] \right\|^2$$

Recalling the definitions of the vectors $\boldsymbol{m}$ and $\boldsymbol{c}$, we have

$$\begin{aligned}
\underbrace{V^*\hat{\theta}}_{n\times 1} &= \begin{bmatrix} \boldsymbol{m}_1 \\ \boldsymbol{m}_2 \end{bmatrix} = \left[\frac{\Sigma_r^{-1}\boldsymbol{c}_1}{\boldsymbol{0}} \right] \\
&= \left[\begin{array}{c|c} \Sigma_r^{-1} & \boldsymbol{0}_{r\times(N-r)} \\ \hline \boldsymbol{0}_{(n-r)\times r} & \boldsymbol{0}_{(n-r)\times(N-r)} \end{array} \right] \overbrace{U^*\boldsymbol{Z}}^{c} \\
V^*\hat{\theta} &= \left[\begin{array}{c|c} \Sigma_r^{-1} & \boldsymbol{0}_{r\times(N-r)} \\ \hline \boldsymbol{0}_{(n-r)\times r} & \boldsymbol{0}_{(n-r)\times(N-r)} \end{array} \right] U^* z
\end{aligned}$$

Premultiply by V to obtain the least-squares solution

$$\begin{aligned}
\hat{\theta}_{LS} &= \underbrace{V}_{n\times n} \underbrace{\Sigma^{\#}}_{n\times N} \underbrace{U^*}_{N\times N} z = \underbrace{H^{\#}}_{n\times N} z \\
\underbrace{H}_{N\times n} &= \underbrace{U}_{N\times N} \underbrace{\Sigma}_{N\times n} \underbrace{V^*}_{n\times n} \\
\sum_r^{-1} &= \operatorname{diag}\left\{ \sigma_1^{-1}\ \sigma_2^{-1}\ \dots\ \sigma_r^{-1} \right\} \\
\sigma_1 &\geq \sigma_2 \geq \cdots \geq \sigma_r
\end{aligned}$$

When the matrix $H^T H$ is not full rank, the solution obtained using its pseudoinverse is sensitive to perturbations in the matrix H and the vector z. With a large number of measurements, the matrix H is likely to be full rank. Even if the matrix is full rank, the matrix H can be **ill**-conditioned; i.e., small changes in its entries can lead to large changes in the least-squares solution. A measure of conditioning is the **condition number** of the matrix

$$\kappa = \|H\| \left\|H^{-1}\right\| = \frac{\sigma_{max}(H)}{\sigma_{min}(H)}$$

where the 2-norm of the matrix is used. It the ratio is large, then the matrix is ill-conditioned.

MATLAB © commands allow us to obtain the least-squares solution easily. The relevant MATLAB commands are demonstrated in the following example.

Example 6.3 For the linear model of (6.5), use MATLAB to generate a measurement matrix H with entries uniformly distributed in the interval [0, 1] and a normally distributed measurement vector $\boldsymbol{Z}$ with zero-mean and unity variance, then obtain the least-squares estimate of the parameter vector $\boldsymbol{\theta}$. Obtain the condition number of H and discuss the effect of a 1% increase in the entries of H on the estimate and how it is influenced by the condition number.

Solution

Using MATLAB gives

```
>> H = rand(8,3); z = randn(8,1);
```

The singular value decomposition is obtained using the command

```
>> [V,Sigma,U] = svd(H)
```

$V =$

$$\begin{matrix}
-0.3746 & 0.4400 & -0.2654 & -0.4281 & -0.4768 & -0.3038 & -0.2773 & 0.1259 \\
-0.4784 & 0.0442 & -0.1955 & -0.2490 & 0.1125 & 0.3531 & 0.4455 & -0.5773 \\
-0.1903 & -0.5387 & 0.2913 & -0.1595 & -0.1243 & 0.0173 & -0.6193 & -0.4050 \\
-0.4884 & 0.0125 & -0.1834 & 0.8287 & -0.1542 & -0.0670 & -0.1045 & -0.0413 \\
-0.3888 & 0.2102 & 0.1548 & -0.0869 & 0.8034 & -0.1940 & -0.2466 & 0.1710 \\
-0.1100 & 0.3177 & 0.3329 & 0.0287 & -0.1336 & 0.7940 & -0.2251 & 0.2763 \\
-0.3411 & -0.1222 & 0.6976 & -0.0429 & -0.2383 & -0.2685 & 0.4599 & 0.1998 \\
-0.2766 & -0.5950 & -0.3892 & -0.1797 & 0.0265 & 0.1951 & 0.0848 & 0.5828
\end{matrix}$$

sigma =

$$\begin{matrix}
3.3572 & 0 & 0 \\
0 & 0.9102 & 0 \\
0 & 0 & 0.4875 \\
0 & 0 & 0 \\
0 & 0 & 0 \\
0 & 0 & 0 \\
0 & 0 & 0 \\
0 & 0 & 0
\end{matrix}$$

$U =$

$$\begin{matrix}
-0.5098 & 0.1604 & -0.8452 \\
-0.6142 & 0.6200 & 0.4882 \\
-0.6024 & -0.7680 & 0.2176
\end{matrix}$$

To obtain the pseudoinverse, we use the command.
>> pinv(H)

$ans =$

$$\begin{matrix} 0.5946 & 0.4193 & -0.5710 & 0.3943 & -0.1724 & -0.5045 & -1.1792 & 0.6120 \\ 0.1024 & -0.0781 & -0.0405 & -0.0857 & 0.3694 & 0.5699 & 0.6777 & -0.7445 \\ -0.4225 & -0.0387 & 0.6187 & -0.0048 & -0.0385 & -0.0997 & 0.4757 & 0.3780 \end{matrix}$$

The same answer is obtained by evaluating $\left[H^T H\right]^{-1} H^T$.
>> (H'*H)\H'

$ans =$

$$\begin{matrix} 0.5946 & 0.4193 & -0.5710 & 0.3943 & -0.1724 & -0.5045 & -1.1792 & 0.6120 \\ 0.1024 & -0.0781 & -0.0405 & -0.0857 & 0.3694 & 0.5699 & 0.6777 & -0.7445 \\ -0.4225 & -0.0387 & 0.6187 & -0.0048 & -0.0385 & -0.0997 & 0.4757 & 0.3780 \end{matrix}$$

The least-squares estimate is.
>> theta_hat = pinv(H)*z

$$\text{theta} = \begin{matrix} -0.1916 \\ 1.2995 \\ -1.2887 \end{matrix}$$

The condition number of H is.
>> cond(H)

$$\text{ans} = 6.8869$$

It is interesting to examine the effect of a 1% increase in the entries of H on the estimate. Because the condition number is not large, a 1% increase in the entries of H does not cause a big change in the estimate (only a 1% change).

If we construct a new measurement matrix using the matrix H with the first singular value divided by 10^5, the condition number of the matrix becomes 2.7112×10^4. With this large condition number, a 1% increase in the matrix entries results in the changes in the estimates of (−7.3%, 6.3%, 11.7%). This demonstrates the effect of an ill-conditioned measurement matrix on the estimates.

6.2 Properties of the WLS Estimator

Assuming that the constant measurement matrix H is full rank, we derive properties of the WLS estimator of a constant parameter vector $\boldsymbol{\theta}$.

1. The WLS Estimator is Gaussian if the Noise Vector v is Gaussian

Substitute for the measurement in the weighted least-squares estimator

$$\begin{aligned}\hat{\theta}_{\text{WLS}} &= \left[H^T W H\right]^{-1} H^T W z \\ &= \left[H^T W H\right]^{-1} H^T W [H\boldsymbol{\theta} + \boldsymbol{v}]\end{aligned}$$

The estimator is Gaussian because it is an affine transformation of the Gaussian vector $\boldsymbol{v}$.

2. WLS is an unbiased estimator

The expectation of the estimator is

$$\begin{aligned}E\left\{\hat{\theta}_{\text{WLS}}\right\} &= \left[H^T W H\right]^{-1} H^T W E\{z\} \\ &= \left[H^T W H\right]^{-1} H^T W E\{H\theta + \boldsymbol{v}(k)\} \\ &= \left[H^T W H\right]^{-1} H^T W H\theta = \theta\end{aligned}$$

It follows that the residual $\tilde{z} = z - \hat{z}$ is zero-mean

$$E\{\tilde{z}\} = E\left\{H\left[\theta - \hat{\theta}\right] + \boldsymbol{v}\right\} = \mathbf{0}.$$

3. Covariance of LS estimator

Because adding a constant vector to a random vector does not change the latter's covariance. The covariance of the measurement vector is

$$\begin{aligned}\text{Cov}\{z\} &= \text{Cov}\{z - H\boldsymbol{\theta}\} = \text{Cov}\{\boldsymbol{v}\} = R \\ \hat{\theta}_{LS} &= \left[H^T H\right]^{-1} H^T z(k) \\ \text{Cov}\left\{\hat{\theta}_{LS}\right\} &= \left[H^T H\right]^{-1} H^T \text{Cov}\{z\} H \left[H^T H\right]^{-1} \\ &= \left[H^T H\right]^{-1} H^T R H \left[H^T H\right]^{-1}.\end{aligned} \tag{6.18}$$

For i.i.d. uncorrelated measurements $R = \sigma^2 I_N$

$$\text{Cov}\left\{\hat{\theta}_{\text{WLS}}\right\} = \text{Cov}\left\{\hat{\theta}_{LS}(k)\right\} = \sigma^2 \left[H^T H\right]^{-1}. \tag{6.19}$$

As expected, the covariance of the estimator is larger for larger noise variance. In other words, the uncertainty in the estimator is larger if the uncertainty in the measurement is large.

4. Confidence Intervals of LS Estimator

Since the least-squares estimator is Gaussian and unbiased, we can use an estimate of the standard deviation to obtain a confidence interval for the estimator with the desired confidence level. The estimate of the variance for $R = \sigma^2 I_N$ is

$$\hat{\sigma}^2 = \frac{1}{N - n}(z - H\boldsymbol{\theta})^T (z - H\boldsymbol{\theta}) \tag{6.20}$$

Hence, the estimate of the covariance matrix is

$$\text{Cov}\left\{\hat{\theta}_{\text{WLS}}\right\} = \hat{\sigma}^2\left[H^T H\right]^{-1} \tag{6.21}$$

where the diagonal elements are estimates of the variances of the estimators of $\theta_i, i = 1, \ldots, n$.

Thus, if h_{ii}^2 is the ith diagonal entry of $\left[H^T H\right]^{-1}$, we estimate the variance of the estimator of θ_i using

$$Var\left(\hat{\theta}_i\right) = \hat{\sigma}_{\theta_i}^2 = h_{ii}^2 \hat{\sigma}^2 \tag{6.22}$$

The confidence interval for the estimator $\hat{\theta}_i$ is

$$\hat{\theta}_i \pm c_i \hat{\sigma}_{\theta_i}, i = 1, \ldots, n \tag{6.23}$$

where c_i is a constant that is chosen for a specified confidence level. For example, for a 95% confidence level $c_i = 1.96$.

6.3 Best Linear Unbiased Estimator (BLUE)

Because estimators are based on random data samples, they are themselves random variables. The distribution of the estimator can be quite complicated, but we are particularly interested in the mean and variance of this distribution. Modelers prefer estimators that are correct on average; i.e., the mean of their distribution is the actual parameter value. They also prefer estimators that have the minimum variance, i.e., efficient estimators, because a low variance implies that values far from the mean have a low probability

$$J = \min E\left\{\left\|\hat{\boldsymbol{\theta}} - E\left\{\hat{\boldsymbol{\theta}}\right\}\right\|^2\right\} \tag{6.24}$$

Hence, we often seek a minimum variance unbiased estimator (MVUE) that minimizes

$$J = \min E\left\{\left\|\hat{\boldsymbol{\theta}} - \boldsymbol{\theta}\right\|^2\right\}(\text{unbiased}) \tag{6.25}$$

If the estimator is linear,

$$\boldsymbol{z} = H\boldsymbol{\theta} + \boldsymbol{v},$$

then we may not obtain the MVUE. Minimizing the variance gives the best (most efficient) unbiased linear estimator (BLUE).

$$\widehat{\boldsymbol{\theta}}_{\text{BLU}} = F\boldsymbol{z} \tag{6.26}$$

In general, the MVUE may be a nonlinear estimator but for Gaussian statistics BLUE is MVUE because the optimal estimator is linear. In some cases, the performance of BLUE, while suboptimal, may be a good replacement for MVUE. This is particularly useful in cases where MVUE cannot be determined.

Theorem 6.2 *For an unbiased linear estimator $\widehat{\boldsymbol{\theta}} = F\boldsymbol{z}$, the estimator matrix satisfies*

$$FH = I_n \tag{6.27}$$

Proof Substituting from the linear model gives

$$\widehat{\boldsymbol{\theta}} = F\boldsymbol{z} = F[H\boldsymbol{\theta} + \boldsymbol{v}].$$

The expectation of the estimator is

$$E\left\{\widehat{\boldsymbol{\theta}}(k)\right\} = FH\,E\{\boldsymbol{\theta}\} + \mathbf{0} = I_n\boldsymbol{\theta}.$$

Since for an unbiased estimator, the above equation must be valid for any matrix H. Hence, we have the condition

$$FH = I_n$$

■

Theorem 6.3 *For the linear model $\boldsymbol{z} = H\boldsymbol{\theta} + \boldsymbol{v}$, BLUE is weighted least-squares estimator with weighting matrix equal to the inverse of the covariance matrix R of the measurement noise $\boldsymbol{v}$.*

Proof We partition the estimator gain matrix and rewrite the condition for an unbiased estimator $\hat{\boldsymbol{\theta}} = F\mathbf{z}$ **as**

$$FH = I_n = \begin{bmatrix} \boldsymbol{f}_1^T \\ \vdots \\ \boldsymbol{f}_n^T \end{bmatrix} H$$

or equivalently,

$$\boldsymbol{f}_i^T H = \begin{bmatrix} \mathbf{0}_{1\times(i-1)} \ \mathbf{1} \ \mathbf{0}_{1\times(n-i)} \end{bmatrix} = \boldsymbol{e}_i^T$$

where $\boldsymbol{e}_i^T = i$th row of the identity matrix.

Observe that the ith entry of the parameter vector is

$$\theta_i = \boldsymbol{e}_i^T \boldsymbol{\theta} = \boldsymbol{f}_i^T H(k)\boldsymbol{\theta}, i = 1, \ldots, n$$

Minimizing the mean square error implies minimizing the n functions

$$\begin{aligned} E\left\{\left[\theta_i - \hat{\theta}_{i\mathrm{BLU}}\right]^2\right\} &= E\left\{\left[\theta_i - \boldsymbol{f}_i^T \mathbf{z}\right]^2\right\} \\ &= E\left\{\left[\theta_i - \boldsymbol{f}_i^T (H\boldsymbol{\theta} + \boldsymbol{v})\right]^2\right\} = E\left\{\left[\theta_i - \theta_i - \boldsymbol{f}_i^T \boldsymbol{v}\right]^2\right\} \\ &= \boldsymbol{f}_i^T E\left\{\boldsymbol{v}\boldsymbol{v}^T\right\} \boldsymbol{f}_i \\ &= \boldsymbol{f}_i^T R \boldsymbol{f}_i, i = 1, \ldots, n \end{aligned}$$

For an unbiased estimator, we need to incorporate the unbiasedness constraint in the optimization through the use of Lagrange multipliers $\boldsymbol{\lambda} \in \mathcal{R}^{n^2}$, then solve an unconstrained optimization problem in both $\boldsymbol{x}$ and $\boldsymbol{\lambda}$. The performance measure now becomes

$$J\left[\left(\boldsymbol{f}_i, \boldsymbol{\lambda}_i\right), i = 1, \ldots, n\right] = \sum_{i=1}^{n} \boldsymbol{f}_i^T R(k) \boldsymbol{f}_i + \sum_{i=1}^{n} \boldsymbol{\lambda}_i^T \left(H^T(k) \boldsymbol{f}_i - \boldsymbol{e}_i\right)$$

$$= \sum_{i=1}^{n} \left[\boldsymbol{f}_i^T R \boldsymbol{f}_i + \boldsymbol{\lambda}_i^T \left(H^T \boldsymbol{f}_i - \boldsymbol{e}_i\right)\right] = \sum_{i=1}^{n} J_i\left(\boldsymbol{f}_i, \boldsymbol{\lambda}_i\right)$$

$$J_i\left(\boldsymbol{f}_i, \boldsymbol{\lambda}_i\right) = \boldsymbol{f}_i^T R \boldsymbol{f}_i + \boldsymbol{\lambda}_i^T \left(H^T \boldsymbol{f}_i - \boldsymbol{e}_i\right)$$

Differentiate w.r.t. the vector λ_i (assume H full rank)

$$\frac{\partial J_i\left(\boldsymbol{f}_i, \boldsymbol{\lambda}_i\right)}{\partial \boldsymbol{\lambda}_i} = H^T \boldsymbol{f}_i - \boldsymbol{e}_i = 0$$

$$H^T \boldsymbol{f}_i = H^T \left[-\frac{1}{2} R^{-1} H \boldsymbol{\lambda}_i\right] = \boldsymbol{e}_i$$

$$\boldsymbol{\lambda}_i = -2\left(H^T R^{-1} H\right)^{-1} \boldsymbol{e}_i$$

The second necessary condition is

$$\frac{\partial J_i\left(\boldsymbol{f}_i, \boldsymbol{\lambda}_i\right)}{\partial \boldsymbol{f}_i} = R\boldsymbol{f}_i - H\boldsymbol{\lambda}_i = \mathbf{0}.$$

Eliminate λ_i to obtain

$$\boldsymbol{f}_i = -\frac{1}{2} R^{-1} H \boldsymbol{\lambda}_i.$$

Eliminate λ_i to obtain

$$\begin{aligned}\boldsymbol{f}_i &= -\frac{1}{2} R^{-1} H \times -2\left(H^T R^{-1} H\right)^{-1} \boldsymbol{e}_i \\ \boldsymbol{f}_i &= R^{-1} H\left(H^T R^{-1} H\right)^{-1} \boldsymbol{e}_i\end{aligned}$$

The BLUE estimator is

$$\begin{aligned}\boldsymbol{f}_i &= R^{-1} H\left(H^T R^{-1} H\right)^{-1} \boldsymbol{e}_i, i = 1, \ldots, n \\ F_{\text{BLU}}^T &= \left[\boldsymbol{f}_1\ \boldsymbol{f}_2 \ldots \boldsymbol{f}_n\right] \\ &= R^{-1} H\left(H^T R^{-1} H\right)^{-1}\left[\boldsymbol{e}_1\ \boldsymbol{e}_2 \ldots \boldsymbol{e}_n\right] \\ &= R^{-1} H\left(H^T R^{-1} H\right)^{-1} I_n\end{aligned}$$

Transpose and use $\left(A^{-1}\right)^T = \left(A^T\right)^{-1}$ to obtain

$$F_{\text{BLU}} = \left(H^T R^{-1} H\right)^{-1} H^T R^{-1}$$

i.e., we have a WLS estimator with $W = R^{-1}$

R is symmetric (noise correlation = covariance for zero-mean noise), WLS estimator with $W = R^{-1}$.

The following corollary follows from Theorem 6.3. ■

Corollary 6.1 *For the linear model $\boldsymbol{z} = H\boldsymbol{\theta} + \boldsymbol{v}$ with the covariance of the measurement noise vector $R = \sigma_v^2 I_N$, BLUE is a least-squares estimator.*

Proof

$$\begin{aligned}\widehat{\boldsymbol{\theta}}_{\text{BLU}} &= \left[H^T R^{-1} H\right]^{-1} H^T R^{-1} \boldsymbol{z} \\ &= \sigma_v^2\left[H^T H\right]^{-1} \frac{H^T \boldsymbol{z}}{\sigma_v^2} \\ &= \left[H^T H\right]^{-1} H^T \boldsymbol{z} = \widehat{\boldsymbol{\theta}}_{LS}\end{aligned}$$

■

Problems

6.1 Consider the nth order time response polynomial where the model parameters must be estimated using the method of least squares with $x(t)$ as the response of the system at time t

$$x(t) = \sum_{j=0}^{n} a_j \frac{t^j}{j!}, a_j \in \mathcal{R}$$

(a) What are the data that you need to collect for the parameter estimation problem?
(b) Why is it a bad idea to only have n sets of measurements?
(c) Write a scalar equation to represent the given mathematical model as a vector product then modify it to include an error term.
(d) Write a linear model that can be used to solve the parameter estimation problem.
(e) Write an expression for the least-squares estimate.
(f) Write an expression for the BLUE estimate. Explain why this may or may not be the MVUE estimate.
(g) For the given table, find the best least-squares fit for the data with $n = 4$.

Time	Output
0	2.35651858148141
0.05	5.30666114717210
0.10	2.14257069401017
0.15	4.20852521365836
0.20	3.15459561348711
0.25	2.55598182380392
0.30	2.02992411687959
0.35	3.08519623008631

6.2 The equation of motion of a nonlinear spring-mass-damper system is

$$m\ddot{y}(t) + b\dot{y}(t) + c_1 y(t) + c_3 y^3(t) = f(t)$$

where m is the mass, b is the damper constant, and (c_1, c_3) are the spring constants.

(a) What variables must we measure experimentally to collect the data necessary to estimate the parameters m, b, and (c_1, c_3)?
(b) If we have N sets of noisy measurements, write a linear equation that can be used to estimate the parameters of the nonlinear model.
(c) Write an expression for the least-squares estimate of the parameters.

(d) If all the measurements obtained have added Gaussian noise, explain why the noise in the linear model of (b) will not be Gaussian.

6.3 An armature-controlled dc motor is governed by the equation

$$e_a = Ri_a + K_b\omega$$

where e_a is the armature voltage, i_a is the armature current, and ω is the angular velocity of the motor.

(a) Select variables to measure to estimate the motor parameters R and K_b and write an expression for the noisy linear measurement equation.
(b) If N (noisy) measurements are obtained for the variables you chose in (a), write a matrix equation that can be used to obtain a least-squares estimate of the motor parameters.
(c) Write an expression for the estimate in terms of the data (i) for a rank-deficient measurement matrix, (ii) for a full-rank measurement matrix.
(d) Explain why we should use a large number of data points N to estimate two parameters only

6.4 It is possible to improve the numerical properties of the least-squares problem by minimizing the performance measure

$$J = \|z - H\boldsymbol{\theta}\|^2 + \alpha\|\boldsymbol{\theta}\|^2, \alpha > 0$$

Find the optimal estimator corresponding to this performance measure.

6.5 For the matrix $\underbrace{H}_{n\times n} = \underbrace{U}_{n\times n}\underbrace{\Sigma}_{n\times N}\underbrace{V^*}_{N\times N}$ of rank r with the Moore–Penrose pseudoinverse, $\underbrace{V}_{n\times n}\underbrace{\Sigma^\#}_{n\times N}\underbrace{U^*}_{N\times N}$, prove the following properties

i $r = N < n$:

$$H^\# = V\begin{bmatrix} \Sigma_N^{-1} \\ \mathbf{0}_{(n-N)\times N} \end{bmatrix}U^* = H^*\left(HH^*\right)^{-1}$$

ii $N > n = r$:

$$H^\# = V\left[\,\Sigma_n^{-1}\ \mathbf{0}_{n\times(N-n)}\,\right]U^* = \left(H^*H\right)^{-1}H^*$$

iii $r = N = n$: $H^\# = V\Sigma_N^{-1}U^* = H^{-1}$

6.6 Simplify the expression of the least-squares estimator for the case of a linear model where the measurement matrix is normalized and partitioned such that

$$H = \left[\, H_1 \; H_2 \; \ldots \; H_l \,\right], H_i \in \mathcal{R}^{N \times n_i}, \sum_{i=1}^{l} n_i = n$$

$$H_i^T H_j = \mathbf{0}, i \neq j$$

Show that the problem is equivalent to l separate least-squares estimation problems of the form

$$J_i = \|\boldsymbol{z} - H\boldsymbol{\theta}_i\|^2, \;\; i = 1, 2, \ldots, l$$

6.7 The standard formulation of the least-squares estimation problem does not include perturbations in the measurement matrix H. If perturbations in H are considered, the problem is known as **total least squares**. With a perturbation matrix E, the linear model becomes

$$\boldsymbol{z} = [H + E]\boldsymbol{\theta} + \boldsymbol{v}$$

(a) Show that the perturbed linear equation can be written in the form

$$\tilde{C}\left[\frac{\theta}{-1}\right] = \mathbf{0}$$
$$\tilde{C} = C + \Delta C, C = [H|\boldsymbol{z}]$$
$$\Delta C = [E| - \boldsymbol{v}]$$

(b) Show that if the $N \times n$ matrix H has rank r, then by singular value decomposition we can write

$$H = U\Sigma V^* = \sum_{i=1}^{r} \sigma_i \boldsymbol{u}_i \boldsymbol{v}_i^*$$

where the singular values, left singular vectors, and right singular vectors are $\sigma_i, \boldsymbol{u}_i, \boldsymbol{v}_i, i = 1, \ldots, r$.

(c) Show that the closest matrix of rank k to a matrix H of rank $r > k$

$$H = \sum_{i=1}^{r} \sigma_i \boldsymbol{u}_i \boldsymbol{v}_i^*, \sigma_1 \geq \sigma_2 \geq \ldots \geq \sigma_{r-1} \geq \sigma_r$$

is given by

$$H_k = \sum_{i=1}^{k} \sigma_i \boldsymbol{u}_i \boldsymbol{v}_i^*$$

(d) Show that if the nominal matrix C has the singular value decomposition

$$C = U\text{diag}\{\sigma_1, \sigma_2, \ldots, \sigma_n, \sigma_{n+1}\}V^* = \sum_{i=1}^{n+1} -\sigma_{n+1}\boldsymbol{u}_{n+1}\boldsymbol{v}_{n+1}^*$$

$$\sigma_1 \geq \sigma_2 \geq \ldots \geq \sigma_n \geq \sigma_{n+1}$$

then the optimal estimate is given by

$$\hat{\boldsymbol{\theta}} = \frac{\boldsymbol{v}_{n+1}(1:n)}{\boldsymbol{v}_{n+1}(n+1)}$$

with $\boldsymbol{v}_{n+1} = col\{\boldsymbol{v}_{n+1}(1:n), \boldsymbol{v}_{n+1}(n+1)\}$

6.8 Prove the expression for the condition number

$$\kappa = \|H\|\left\|H^{-1}\right\| = \frac{\sigma_{max}(H)}{\sigma_{min}(H)}$$

6.9 Show that for H full rank the condition number of $H^T H$ is $\kappa^2(H)$, where $\kappa(H)$ is the condition number of H.

6.10 Show that the weighted least-squares parameter estimate subject to the linearly independent constraint

$$A\boldsymbol{\theta} = \boldsymbol{b}$$

$A \in \mathcal{R}^{r \times p}, \boldsymbol{b} \in \mathcal{R}^r, r < p$, rank$(A) = r$, is given by

$$\hat{\boldsymbol{\theta}}_{\text{LSc}} = \hat{\boldsymbol{\theta}}_{LS} - \left(H^T H\right)^{-1} A^T \left[A\left(H^T H\right)^{-1} A^T\right]^{-1} \left(A\hat{\boldsymbol{\theta}}_{LS} - b\right)$$

$$\hat{\boldsymbol{\theta}}_{\text{LS}} = \left(H^T H\right)^{-1} H^T \boldsymbol{z}$$

6.11 The linearized equation of motion of a 3-D.O.F. robotic manipulator is given by

$$y(t) = l_1\phi_1(t) + l_2\phi_2(t) + l_3\phi_3(t)$$

where y is the vertical position of the hand, $(l_i, \phi_i(t)), i = 1, 2, 3$, are the (length, angular position) of the ith link.

(a) Select the variables that you need to measure to obtain a least-squares estimate of the link lengths.

(b) If you have N measurements of each of the variables selected in (a), write a linear matrix equation governing the link lengths. Identify the measurement matrix and the parameter vector.

(c) Write an expression for the least-squares estimate and explain why the estimator is unbiased.

6.12 Given the discrete state-space model of a spring-mass-damper system

$$\begin{bmatrix} x_1(k+1) \\ x_2(k+1) \end{bmatrix} = \begin{bmatrix} 0 & 1 \\ -a_0 & -a_1 \end{bmatrix} \begin{bmatrix} x_1(k) \\ x_2(k) \end{bmatrix} + \begin{bmatrix} 0 \\ 1 \end{bmatrix} u(k)$$

$$y(k) = \begin{bmatrix} c_1 & c_2 \end{bmatrix} \begin{bmatrix} x_1(k) \\ x_2(k) \end{bmatrix} + du(k)$$

(a) Select variables to measure so that the system parameters can be estimated.
(b) Write two linear equations that can be used to estimate the parameters using N sets of measurements.
(c) Write matrix equations and expressions of the least-squares estimates of the parameters obtained using noisy measurements with zero-mean Gaussian white additive noise.
(d) Explain why the parameter estimates cannot be equal to the parameter values referring to the properties of the least-squares estimator.
(e) What is the BLUE estimate of the parameters?

6.13 Write a MATLAB program to perform the following

(a) Generate 100 normally distributed random data points with zero-mean and 0.01 variance.
(b) Add to the random data a sampled sinusoidal component of unity amplitude, frequency $f = 2$ Hz.
(c) Use a sampling frequency $f_s = 50$ Hz. Obtain a plot of the combined data z_k and of the deterministic input data u_k versus time with filter parameters $\{1, 2, 3\}$, where

$$z_k = \begin{bmatrix} u_k & u_{k-1} & u_{k-2} \end{bmatrix} \begin{bmatrix} h_0 \\ h_1 \\ h_2 \end{bmatrix} + v_k k = 0, 1, 2, \ldots$$

$$u_k = \begin{cases} \sin(2\pi f k / f_s), & k \geq 0 \\ 0, & k < 0 \end{cases}$$

(d) Find the least-squares estimate of the coefficients of the filter using the data of Part(a) then verify your answer using the MATLAB command **lsqlin**.

Bibliography

1. Kay, S. M. (1993). *Fundamental of statistical signal processing: Estimation theory*. Prentice-Hall.
2. Golub, G. H., & Van Loan, C. F. (1996). *Matrix computations* (3rd ed.). Johns Hopkins University Press.
3. Mendel, J. (1995). *Lessons in estimation theory for signal processing, communications and control*. Prentice Hall.
4. Seber, G. (1977). *Linear regression analysis*. Wiley.
5. Van de Geer, S. A. (2005). Least squares estimation. In *Encyclopedia of statistics in behavioral science*. Wiley.

Chapter 7
The Likelihood Function and Signal Detection

7.1 The Likelihood Function

In any probability model, it is implicitly assumed that the parameters of the probability distribution are known. It is therefore appropriate to write the probability as $P(\boldsymbol{x}|\boldsymbol{\theta})$ where $\boldsymbol{x}$ is a random vector and $\boldsymbol{\theta}$ is a vector of parameters. If the data $\boldsymbol{x} \in \mathcal{R}^n$ is given and the parameter vector $\boldsymbol{\theta}$ is unknown, we can define a function of the parameters given the data as $l(\boldsymbol{\theta}|\boldsymbol{x})$ known as the **likelihood function**. The likelihood is a measure of how likely the parameter values $\boldsymbol{\theta}$ can give rise to the data $\boldsymbol{x}$. The likelihood function is subject to the following principle.

The likelihood principle: Given a set of data $\boldsymbol{x}$, a parameter vector $\boldsymbol{\theta}_1$ is more plausible than a parameter vector $\boldsymbol{\theta}_2$ if it is more probable than $\boldsymbol{\theta}_2$, i.e., $l(\boldsymbol{\theta}_1|\boldsymbol{x}) > l(\boldsymbol{\theta}_2|\boldsymbol{x})$ if $P(\boldsymbol{x}|\boldsymbol{\theta}_1) > P(\boldsymbol{x}|\boldsymbol{\theta}_2)$.

Based on the likelihood principle, we write

$$l(\boldsymbol{\theta}|\boldsymbol{x}) = c\ P(\boldsymbol{x}|\boldsymbol{\theta}), \tag{7.1}$$

where c is a proportionality constant. Because of the proportionality constant, the axioms of probability that apply to $P(\boldsymbol{x}|\boldsymbol{\theta})$ do not apply to the likelihood function and there are no axioms of likelihood. For example, while the probability has values in the interval [0, 1], the likelihood will have values in the interval $[0, c]$.

For a continuous random vector $\boldsymbol{x} \in \mathcal{R}^n$, the probability is

$$P(\boldsymbol{x}|\boldsymbol{\theta}) = p(\boldsymbol{x}|\boldsymbol{\theta})d\boldsymbol{x} = p(\boldsymbol{x}|\boldsymbol{\theta})\mathrm{d}x_1 \ldots \mathrm{d}x_n. \tag{7.2}$$

We express the likelihood function as

$$l(\boldsymbol{\theta}|\boldsymbol{x}) = c\ p(\boldsymbol{x}|\boldsymbol{\theta})\mathrm{d}x_1 \ldots \mathrm{d}x_n = c_1 p(\boldsymbol{x}|\boldsymbol{\theta}), \tag{7.3}$$

M. S. Fadali, *Introduction to Random Signals, Estimation Theory, and Kalman Filtering*, https://doi.org/10.1007/978-981-99-8063-5_7

where we absorb the differentials in the constant multiplier. Hence, the likelihood function is proportional to the pdf.

Because the likelihood function is often used to compare models with different parameter values, the proportionality constant can be set equal to unity.

Example 7.1 Consider a communication system where the sender sends a binary message m where the probability of a 1 is p and the probability of a 0 is $(1 - p)$. Each bit is governed by a Bernoulli distribution and a -bit message is governed by a Binomial distribution.

$$P(m|p) = \frac{n!}{n_1!(n - n_1)!} p^{n_1}(1 - p)^{n-n_1}$$

with n_1 ones and $n - n_1$ zeros. Data is collected to determine the more likely of two hypotheses, a parameter value $p = 1/2$ or a parameter value $p = 1/4$.

To simplify the problem, we assume that the data collected comprises two 2-bit messages with the first message m_1 having $n_1 = 1$ and the second message m_2 having $n_1 = 2$.

For the hypothesis, $p = 1/2$, the binomial probability gives

$$P\left(m_1|p = \frac{1}{2}\right) = \frac{2!}{1!1!}\left(\frac{1}{2}\right)^1\left(\frac{1}{2}\right)^1 = \frac{1}{2}$$

$$P\left(m_2|p = \frac{1}{2}\right) = \frac{2!}{2!0!}\left(\frac{1}{2}\right)^2\left(\frac{1}{2}\right)^0 = \frac{1}{4}$$

and for $p = 1/4$

$$P\left(m_1|p = \frac{1}{4}\right) = \frac{2!}{1!1!}\left(\frac{1}{4}\right)^1\left(\frac{3}{4}\right)^1 = \frac{3}{8}$$

$$P\left(m_2|p = \frac{1}{4}\right) = \frac{2!}{2!0!}\left(\frac{1}{4}\right)^2\left(\frac{3}{4}\right)^0 = \frac{1}{16}.$$

We observe that

$$P\left(m_1|p = \frac{1}{2}\right) > P\left(m_1|p = \frac{1}{4}\right)$$

$$P\left(m_2|p = \frac{1}{2}\right) > P\left(m_2|p = \frac{1}{4}\right).$$

Because likelihood increases with probability, both messages indicate that the data is more likely to have come from a source where the 1 and 0 have equal probability with $p = 1/2$.

7.2 Likelihood Ratio

Given a set of random data $\boldsymbol{x}$ where the structure of the probability model of $\boldsymbol{x}$ is known but its parameters are not, it is often necessary to compare two or more sets of parameter values to choose the most plausible set of values. In the case of choosing between two sets of parameter values, we have the binary hypothesis test

$$H_0\colon \boldsymbol{\theta} = \boldsymbol{\theta}_1$$
$$H_1\colon \boldsymbol{\theta} = \boldsymbol{\theta}_2.$$

It is also possible to extend this to the case of deciding which of multiple hypotheses $H_i, i = 0, 1, \ldots, m$, best fits a given data set. However, this topic is beyond the scope of this text.

When comparing two different hypotheses based on the same data set, the comparison can be based on their likelihoods. This conveniently done by examining their likelihood ratio. For binary hypothesis testing, it is often necessary to choose between two parameter values for a known distribution, $\boldsymbol{\theta}_1$ and $\boldsymbol{\theta}_2$. We use the likelihood ratio to choose between them. For discrete variables, the likelihood ratio is

$$l(\boldsymbol{\theta}_1, \boldsymbol{\theta}_2|\boldsymbol{x}) = \frac{l(\boldsymbol{\theta}_1|\boldsymbol{x})}{l(\boldsymbol{\theta}_2|\boldsymbol{x})} = \frac{P(\boldsymbol{x}|\boldsymbol{\theta}_1)}{P(\boldsymbol{x}|\boldsymbol{\theta}_2)}, \tag{7.4}$$

and for continuous variables, it is

$$\frac{l(\boldsymbol{\theta}_1|\boldsymbol{x})}{l(\boldsymbol{\theta}_2|\boldsymbol{x})} = \frac{p(\boldsymbol{x}|\boldsymbol{\theta}_1)}{p(\boldsymbol{x}|\boldsymbol{\theta}_2)}. \tag{7.5}$$

Another advantage of using the likelihood ration is that it is invariant under transformation of variables. Transformation from variables $\boldsymbol{x}$ to variables $\boldsymbol{y} = \boldsymbol{f}(\boldsymbol{x})$ results in the likelihood ratio

$$\frac{l(\boldsymbol{\theta}_1|\boldsymbol{y})}{l(\boldsymbol{\theta}_2|\boldsymbol{y})} = \frac{p(\boldsymbol{y}|\boldsymbol{\theta}_1)}{p(\boldsymbol{y}|\boldsymbol{\theta}_2)} = \frac{p(\boldsymbol{x}|\boldsymbol{\theta}_1)\big|\det\big[\partial\boldsymbol{x}/\partial\boldsymbol{y}\big]\big|}{p(\boldsymbol{x}|\boldsymbol{\theta}_2)\big|\det\big[\partial\boldsymbol{x}/\partial\boldsymbol{y}\big]\big|} = \frac{p(\boldsymbol{x}|\boldsymbol{\theta}_1)}{p(\boldsymbol{x}|\boldsymbol{\theta}_2)}.$$

Thus, the likelihood ratio is invariant

$$\frac{l(\boldsymbol{\theta}_1|\boldsymbol{y})}{l(\boldsymbol{\theta}_2|\boldsymbol{y})} = \frac{l(\boldsymbol{\theta}_1|\boldsymbol{x})}{l(\boldsymbol{\theta}_2|\boldsymbol{x})}. \tag{7.6}$$

Based on comparing the ratio to an appropriately chosen threshold γ, we choose decision d_1, to accept H_1, or decision d_2 to accept H_2, i.e.,

$$\begin{aligned} d_1 &: l(\boldsymbol{\theta}_1, \boldsymbol{\theta}_2|\boldsymbol{x}) > \gamma \\ d_2 &: l(\boldsymbol{\theta}_1, \boldsymbol{\theta}_2|\boldsymbol{x}) < \gamma \\ d_1 \text{ or } d_2 &: l(\boldsymbol{\theta}_1, \boldsymbol{\theta}_2|\boldsymbol{x}) = \gamma. \end{aligned} \tag{7.7}$$

We summarize the decision as

$$l(\boldsymbol{\theta}_1, \boldsymbol{\theta}_2|\boldsymbol{x}) \lesseqgtr \gamma. \tag{7.8}$$

Likelihood ratios as a means of decision making are the basis for a branch of probability theory known as **stochastic decision theory**. In engineering, the main application of the theory is in the field of signal detection.

Example 7.2 Use the results for the communication system of Example 7.1 for decision making using the likelihood ratio test with a threshold $\boldsymbol{\gamma} = 1$.

Solution

The results for the first message give the ratio

$$l(p_1, p_2|m_1) = \frac{P\left(m_1|p_1 = \frac{1}{2}\right)}{P\left(m_1|p_2 = \frac{1}{4}\right)} = \frac{1/2}{3/8} = \frac{4}{3} > 1$$

$$l(p_1, p_2|m_2) = \frac{P\left(m_2|p_1 = \frac{1}{2}\right)}{P\left(m_2|p_2 = \frac{1}{4}\right)} = \frac{1/4}{1/16} = 4 > 1.$$

The likelihood ratio for both messages is greater than the threshold and the conclusion is that $p_1 = 1/2$ is more likely than $p_2 = 1/4$.

Hypothesis testing is based on random samples, and it is always possible that the test will lead to the wrong conclusion. For example, one could decide that H_0 is true, when H_1 is true. Table 7.1 summarizes all the possible decisions in binary hypothesis testing.

Table 7.1 Hypothesis testing errors

	Truth	
Decision	H_0	H_1
Accept H_1	Type I error (false alarm)	Correct decision (hit)
Accept H_0	Correct decision (proper dismissal)	Type II error (miss)

7.3 Signal Detection

The signal detection problem is to decide based on noisy measurements, with a measure of confidence, if an event has occurred, or to decide which of a number of possible outcomes has occurred. In statistics, this field is call stochastic decision theory. Applications of detection theory include radar/sonar, communications, speech/image processing, biomedicine, fault detection, and seismology. In radar or sonar, the purpose is to decide the presence or absence of a target. This is achieved by transmitting an electromagnetic pulse then testing the received noisy signal. If a pulse is detected, then we can conclude that it was reflected from a target and that a target is present. If no pulse is detected, then we conclude that no target is present.

Detection problems include using noisy measurements to detect (i) known signals in additive noise, (ii) deterministic signals with unknown parameters in additive noise, or (iii) random signals in additive noise. In most cases, it is assumed that the additive noise is white and Gaussian.

We only consider the detection of a known signal using noisy measurements

$$z(k) = s(k) + v(k), k = 0, 1, 2, \ldots \tag{7.9}$$

with white Gaussian additive noise $v(k)$. Deciding if the known signal "s" is present or if the measurement is only noise reduces to a binary hypothesis to test H_0 (no signal), versus H_1 (signal present). The decision D_0 corresponds to noise with no signal while the decision D_1 corresponds to the presence of the known signal. The corresponding likelihoods are, respectively,

$$p(z|H_i), i = 0, 1. \tag{7.10}$$

For hypothesis testing, we use the likelihood ratio

$$l(z) = \frac{p(z|H_1)}{p(z|H_0)} \underset{D_0}{\overset{D_1}{\gtrless}} \gamma \tag{7.11}$$

and compare it to a threshold value γ. The test has four possible outcomes depending on the decision versus the correct hypothesis that we denote by $D_i|H_j, i, j \in \{0, 1\}$. The four outcomes are

1. $D_0|H_0$: correct decision, "proper dismissal."
2. $D_1|H_0$: Type I Error, "false alarm."
3. $D_1|H_1$: correct decision.
4. $D_0|H_1$: Type II Error, "miss."

The terms false alarm and miss come from application to radar or sonar, while the terms Type I and Type II error are the statistical terminology.

If we decompose the space of observations $\mathcal{Z}$ into two disjoint subspaces: $\mathcal{Z} = \mathcal{Z}_0 \oplus \mathcal{Z}_1$, $\mathcal{Z}_0 \cap \mathcal{Z}_1 = \emptyset$, the decomposition is depicted for scalar data with

$$\mathcal{Z}_0 = \{z : z < \gamma\},\ \mathcal{Z}_1 = \{z : z \geq \gamma\} \tag{7.12}$$

in Fig. 7.1. The figure shows the two density functions H_0 to the left and H_1 to the right. The different outcomes and their probabilities are summarized in Table 7.2.

From the axioms of probability, the probabilities satisfy the conditions

$$P_{PD} + P_{FA} = 1. \tag{7.13}$$

The probability of detection is

$$P_D = P(D_1|H_1) = \int\limits_{\mathcal{Z}_1} p(z|H_1)\mathrm{d}z. \tag{7.14}$$

Because we often deal with exponential pdfs, it is more convenient to consider the log-likelihood ratio rather than the likelihood ratio. This does not create any

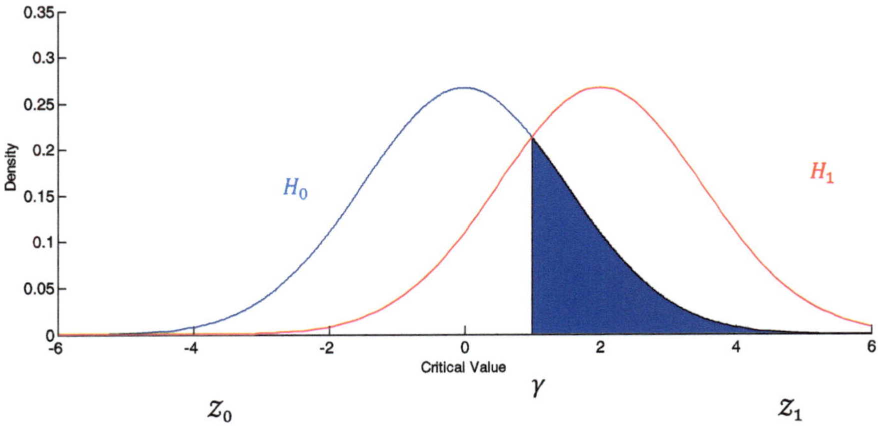

Fig. 7.1 Density functions for binary hypothesis testing

Table 7.2 Outcomes of binary hypothesis testing for signal detection

Name	Decision	Hypothesis	Probability
Proper dismissal	D_0	H_0	$P_{PD} = P(D_0\|H_0) = \int_{\mathcal{Z}_0} p(z\|H_0)\mathrm{d}z$
False alarm	D_1	H_0	$P_{FA} = P(D_1\|H_0) = \int_{\mathcal{Z}_1} p(z\|H_0)\mathrm{d}z$
Detection	D_1	H_1	$P_D = P(D_1\|H_1) = \int_{\mathcal{Z}_1} p(z\|H_1)\mathrm{d}z$
Miss	D_0	H_1	$P_M = P(D_0\|H_1) = \int_{\mathcal{Z}_0} p(z\|H_0)\mathrm{d}z$

problems because both sides of the inequality (7.11) are positive and because log is an order preserving transformation, i.e.,

$$x > y \Leftrightarrow \ln(x) > \ln(y).$$

The log of (7.11) gives the inequality

$$L(z) = \ln[l(z)] = \ln\left[\frac{p(z|H_1)}{p(z|H_0)}\right] \underset{D_0}{\overset{D_1}{\gtrless}} \ln[\gamma]. \tag{7.15}$$

Example 7.3 Given a set of binomially distributed data $\{z_i, i = 1, 2, \ldots, N\}$ with each data point obtained from n Bernoulli trials. Use the likelihood ratio to develop a test with threshold γ to choose between two possible values for p, p_1 and p_2 by testing the two hypotheses

$$\begin{aligned} H_1 &: p = p_1 \\ H_2 &: p = p_2. \end{aligned}$$

Solution

In Chap. 4, we obtained the likelihood

$$p(p_i|z) = \binom{n}{z} p_j^z (1 - p_j)^{n-z}, \; j = 1, 2.$$

For n measurements, we have the likelihood

$$p(\boldsymbol{p}|\boldsymbol{z}) = \prod_{i=1}^{N} \binom{n}{z_i} \times p_j^{\sum_{i=1}^{N} z_i} \times (1 - p_j)^{\sum_{i=1}^{N}(n-z_i)}, \; j = 1, 2.$$

The likelihood ratio is

$$l(p_1, p_2|\boldsymbol{z}) = \frac{p_1^{\sum_{i=1}^{N} z_i} \times (1 - p_1)^{\sum_{i=1}^{N}(n-z_i)}}{p_2^{\sum_{i=1}^{N} z_i} \times (1 - p_2)^{\sum_{i=1}^{N}(n-z_i)}} = \left(\frac{p_1}{p_2}\right)^{\sum_{i=1}^{N} z_i} \left(\frac{1 - p_1}{1 - p_2}\right)^{\sum_{i=1}^{N}(n-z_i)}.$$

The log of the likelihood ratio is

$$\begin{aligned} L(p_1, p_2|\boldsymbol{z}) &= \ln\left(\frac{p_1}{p_2}\right) \sum_{i=1}^{N} z_i + \ln\left(\frac{1 - p_1}{1 - p_2}\right) \sum_{i=1}^{N} (n - z_i) \\ &= \left[\ln\left(\frac{p_1}{p_2}\right) - \ln\left(\frac{1 - p_1}{1 - p_2}\right)\right] N\overline{z} + \ln\left(\frac{1 - p_1}{1 - p_2}\right) nN \end{aligned}$$

where the sample mean is

$$\overline{z} = \frac{1}{N}\sum_{i=1}^{N} z_i.$$

Logarithmic relations give the simplification

$$\ln\left(\frac{p_1}{p_2}\right) - \ln\left(\frac{1-p_1}{1-p_2}\right) = \ln(p_1) - \ln(p_2) - \ln(1-p_1) + \ln(1-p_2)$$
$$= \ln\left(\frac{p_1}{1-p_1}\right) + \ln\left(\frac{1-p_2}{p_2}\right) = \ln\left(\frac{p_1}{p_2}\right) + \ln\left(\frac{1-p_2}{1-p_1}\right) = \ln\left(\frac{1-p_2^{-1}}{1-p_1^{-1}}\right).$$

Assuming WLOG that $p_2 > p_1$, making the above term positive, we write $L(p_1, p_2|z)$ as

$$L(p_1, p_2|z) = \ln\left(\frac{1-p_2^{-1}}{1-p_1^{-1}}\right)N\overline{z} + \ln\left(\frac{1-p_1}{1-p_2}\right)nN.$$

The likelihood ratio test with threshold γ is

$$L(p_1, p_2|z) = \ln\left(\frac{1-p_2^{-1}}{1-p_1^{-1}}\right)N\overline{z} + \ln\left(\frac{1-p_1}{1-p_2}\right)nN \underset{D_0}{\overset{D_1}{\gtrless}} \ln[\gamma].$$

We rewrite the test as

$$\overline{z} \underset{D_0}{\overset{D_1}{\gtrless}} \frac{1}{\ln\left(\frac{1-p_2^{-1}}{1-p_1^{-1}}\right)}\left[\frac{\ln[\gamma]}{N} - \ln\left(\frac{1-p_1}{1-p_2}\right)n\right].$$

Note that γ must be large enough to make the RHS of the inequality positive.

There are many methods to choose the threshold level γ. One of the most popular approaches is the Neyman–Pearson (NP) criterion, defined as follows:

Definition 7.1: NP Criterion

Select the threshold $\boldsymbol{\gamma}$ to maximize the probability of detection P_D (14) such that the probability of false alarm $P_{FA} \leq \alpha$.

From Fig. 7.1, we observe that the probability of detection is maximized subject to the constraint if $P_{FA} = \alpha$.

7.4 Matched Filters

In many applications, including radar and sonar, it is required to detect the presence of a known deterministic signal $s(n), n = 0, 1, \ldots, N-1$ in white Gaussian noise $v(n), n = 0, 1, \ldots, N-1$. This must be achieved while minimizing the probability of a false positive, where no signal exists but the test predicts its existence. Using the NP criterion sets an upper bound on the probability of false alarm but also results in an optimal filter that maximizes the signal-to-noise ratio.

The objective of the filter is to decide between two hypotheses

$$\begin{matrix} H_0\colon z(k) = v(k) \\ H_1\colon z(k) = v(k) + s(k), \end{matrix} \quad k = 0, 1, \ldots, N-1 \tag{7.16}$$

where the noise $v(k) \sim \mathcal{N}\left(0, \sigma^2\right)$ is Gaussian and white with zero-mean and known variance

$$r_{vv}(m) = E\{v(k)v(k+m)\} = \sigma^2\delta(m). \tag{7.17}$$

The decision is based on the likelihood ratio test is

$$l(\mathbf{z}) = \frac{p(\mathbf{z}|H_1)}{p(\mathbf{z}|H_0)} \begin{matrix} D_1 \\ > \\ < \\ D_0 \end{matrix} \gamma \tag{7.18}$$

with the data given by $\mathbf{z} = \left[\, z(0)\ z(1)\ \ldots\ z(N-1)\,\right]^T$.

The likelihood function for the null hypothesis is

$$p(\mathbf{z}|H_0) = \left(2\pi\sigma^2\right)^{-N/2} \exp\left\{-\frac{\mathbf{z}^T\mathbf{z}}{2\sigma^2}\right\} \tag{7.19}$$

and for the alternative hypothesis

$$p(\mathbf{z}|H_1) = \left(2\pi\sigma^2\right)^{-N/2} \exp\left\{-\frac{(\mathbf{z}-\mathbf{s})^T(\mathbf{z}-\mathbf{s})}{2\sigma^2}\right\} \tag{7.20}$$

Substituting for the likelihoods in the likelihood ratio

$$l(\mathbf{z}) = \frac{p(\mathbf{z}|H_1)}{p(\mathbf{z}|H_0)} = \frac{\left(2\pi\sigma^2\right)^{-N/2} \exp\left\{-\frac{(\mathbf{z}-\mathbf{s})^T(\mathbf{z}-\mathbf{s})}{2\sigma^2}\right\}}{\left(2\pi\sigma^2\right)^{-N/2} \exp\left\{-\frac{\mathbf{z}^T\mathbf{z}}{2\sigma^2}\right\}}. \tag{7.21}$$

Thus, the likelihood ratio can be simplified to the form

$$l(\boldsymbol{x}) = \exp\left\{\frac{2\boldsymbol{s}^T\boldsymbol{z} - \boldsymbol{s}^T\boldsymbol{s}}{2\sigma^2}\right\} \underset{D_0}{\overset{D_1}{\gtrless}} \gamma. \tag{7.22}$$

The log-likelihood ratio is

$$L(\boldsymbol{z}) = \ln[l(\boldsymbol{z})] = \frac{2\boldsymbol{s}^T\boldsymbol{z} - \boldsymbol{s}^T\boldsymbol{s}}{2\sigma^2} \underset{D_0}{\overset{D_1}{\gtrless}} \ln[\gamma]. \tag{7.23}$$

Separating known terms from data-dependent terms give the sufficient statistic

$$T(\boldsymbol{z}) = \boldsymbol{s}^T\boldsymbol{z} \underset{D_0}{\overset{D_1}{\gtrless}} \sigma^2\ln[\gamma] + \frac{1}{2}\boldsymbol{s}^T\boldsymbol{s} = \gamma'. \tag{7.24}$$

$T(\boldsymbol{z}) = \boldsymbol{s}^T\boldsymbol{z}$ is the correlation of the data and the known signal, or its replica, and is known as a **replica correlator**. It weights the data samples with the values of the signal so that the larger the signal amplitude the higher the weight and negative amplitudes have negative weights. The replica correlator is depicted in Fig. 7.2. $T(\boldsymbol{z})$ is a **sufficient statistic** in that it contains all the information needed from the data to make a decision.

In terms of the log-likelihood ratio, NP Detection becomes: Choose γ' to maximize P_D and satisfy the false alarm rate constraint

$$P_{FA} = \int_{\mathcal{Z}_1} p(\boldsymbol{z}|H_0)\mathrm{d}\boldsymbol{z} = \alpha. \tag{7.25}$$

The impulse response of the matched filter is obtained in two steps:

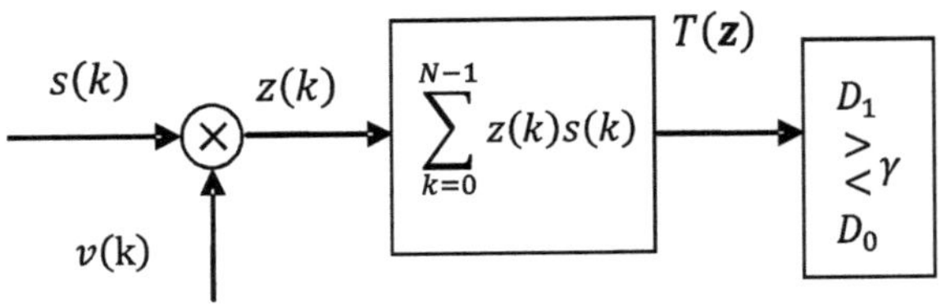

Fig. 7.2 Replica correlator block diagram

i. Flip the signal to obtain $s(-n), n = 0, 1, \ldots, N-1$.
ii. Right shift the signal by $N-1$ to obtain

$$h(n) = \begin{cases} s(N-1-n), n = 0, 1, \ldots, N-1 \\ 0, \text{elsewhere} \end{cases}. \tag{7.26}$$

The output of the matched filter is

$$y(n) = \sum_{k=0}^{n} h(n-k)z(k) = \sum_{k=0}^{n} s(N-1-n+k)z(k), n \geq 0. \tag{7.27}$$

The output of the matched filter can be written in terms of the signal vector and noise vector as

$$y(N-1) = \sum_{k=0}^{N-1} s(k)z(k) = \boldsymbol{s}^T \boldsymbol{z} = \boldsymbol{s}^T (\boldsymbol{s} + \boldsymbol{v}) \tag{7.28}$$

For any finite impulse response (FIR) filter with impulse response $\boldsymbol{h}$, the output of the filter under the hypothesis H_1 is

$$y(N-1) = \sum_{k=0}^{N-1} h(N-1-k)z(k) = \boldsymbol{h}^T \boldsymbol{z} = \boldsymbol{h}^T (\boldsymbol{s} + \boldsymbol{v}). \tag{7.29}$$

The mean of the output under the hypothesis H_1 is

$$E\{y(N-1)|H_1\} = \boldsymbol{h}^T (\boldsymbol{s} + E\{\boldsymbol{v}\}) = \boldsymbol{h}^T \boldsymbol{s}. \tag{7.30}$$

The variance of the output under the hypothesis H_1 is

$$var\{y(N-1)|H_1\} = E\left\{\left(\boldsymbol{h}^T \boldsymbol{v}\right)^2\right\} = \boldsymbol{h}^T E\{\boldsymbol{v}\boldsymbol{v}^T\}\boldsymbol{h} = \sigma^2 \boldsymbol{h}^T \boldsymbol{h}. \tag{7.31}$$

The signal-to-noise ratio (SNR) is defined as

$$\begin{aligned} \eta &= \frac{E\{y(N-1)|H_1\}^2}{var\{y(N-1)|H_1\}} = \frac{\left(\boldsymbol{h}^T \boldsymbol{s}\right)^2}{\sigma^2 \boldsymbol{h}^T \boldsymbol{h}} \\ &= \frac{\left[\sum_{k=0}^{N-1} h(N-1-k)s(k)\right]^2}{\sigma^2 \sum_{k=0}^{N-1} h^2(k)}. \end{aligned} \tag{7.32}$$

Considering all possible FIR filters, we find an upper bound on the SNR.

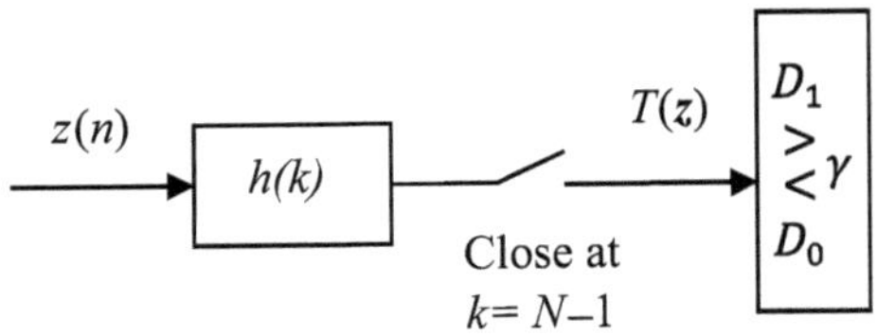

Fig. 7.3 Matched filter block diagram

$$\eta = \frac{\left(\boldsymbol{h}^T \boldsymbol{s}\right)^2}{\sigma^2 \boldsymbol{h}^T \boldsymbol{h}} \leq \frac{\|\boldsymbol{h}\|^2 \|\boldsymbol{s}\|^2}{\sigma^2 \|\boldsymbol{h}\|^2} = \frac{\|\boldsymbol{s}\|^2}{\sigma^2} = \frac{\mathcal{E}}{\sigma^2} \tag{7.33}$$

where $\mathcal{E}$ is the energy of the signal

$$\mathcal{E} = \|\boldsymbol{s}\|^2 = \sum_{k=0}^{N-1} s^2(k). \tag{7.34}$$

The upper bound on the SNR is achieved when the signal and impulse response sequence are colinear and their inner product is equal to the energy of the signal. This gives the maximum SNR.

$$\eta_{\max} = \frac{\mathcal{E}}{\sigma^2}, \tag{7.35}$$

and it is exactly the ratio obtained using the matched filter as shown in Fig. 7.3. The performance of the matched filter improves monotonically with the maximum SNR η.

We next consider the design of a matched filter for a given P_{FA} and obtain the corresponding P_D. We start with the analysis of the statistic $T(z)$ and its distribution. $T(z) = \boldsymbol{s}^T \boldsymbol{z}$ is Gaussian because it is a linear combination of Gaussian random data. Hence, it is completely characterized by its mean and variance. The mean varies depending on the hypothesis. Under H_0, the mean is

$$E\{T(z)|H_0\} = E\left\{\boldsymbol{v}^T \boldsymbol{s}\right\} = E\left\{\boldsymbol{v}^T\right\} s = 0. \tag{7.36}$$

Under H_1, the mean is

$$E\{T(z)|H_1\} = E\left\{(\boldsymbol{s} + \boldsymbol{v})^T \boldsymbol{s}\right\} = \boldsymbol{s}^T \boldsymbol{s} = \mathcal{E}. \tag{7.37}$$

The variance of $T(z) = \boldsymbol{z}^T \boldsymbol{s}$ is the same under H_0 and under H_1 because the perturbation from the mean does not change with the hypothesis. The variance is given by

$$var\{T(z)|H_i\} = E\left\{\left(\boldsymbol{v}^T \boldsymbol{s}\right)^2\right\} = \boldsymbol{s}^T E\left\{\boldsymbol{v}\boldsymbol{v}^T\right\} s = \sigma^2 \boldsymbol{s}^T \boldsymbol{s}. \tag{7.38}$$

In terms of the energy of the signal, the variance is

$$var\{T(z)|H_i\} = \sigma^2\mathcal{E}, i = 0, 1. \tag{7.39}$$

Combining the above results, we have the distribution of $T(\boldsymbol{x})$

$$T(z) \sim \begin{cases} \mathcal{N}(0, \sigma^2\mathcal{E}), H_0 \\ \mathcal{N}(\mathcal{E}, \sigma^2\mathcal{E}), H_1 \end{cases}. \tag{7.40}$$

We normalize T to obtain a standard normal distribution under H_0

$$T'(z) = T(z)/\sqrt{\sigma^2\mathcal{E}}. \tag{7.41}$$

The distribution of the normalized statistic is

$$T'(z) \sim \begin{cases} \mathcal{N}(0, 1), H_0 \\ \mathcal{N}\left(\sqrt{\mathcal{E}/\sigma^2}, 1\right), H_1 \end{cases}. \tag{7.42}$$

The density functions of Fig. 7.4 show that the performance of the matched filter improves as $\mathcal{E}/\sigma^2$ increases (SNR) because the densities move farther apart but have the same shape.

The threshold is calculated as follows:

$$P_{\text{FA}} = P(T > \gamma'|H_0) = P\left(T' > \frac{\gamma'}{\sqrt{\sigma^2\mathcal{E}}}|H_0\right) \tag{7.43}$$

$$T'(z) = \frac{T(z)}{\sqrt{\sigma^2\mathcal{E}}} = Q\left(\frac{\gamma'}{\sqrt{\sigma^2\mathcal{E}}}\right) = \text{right tail probability for } \mathcal{N}(0, 1) \tag{7.44}$$

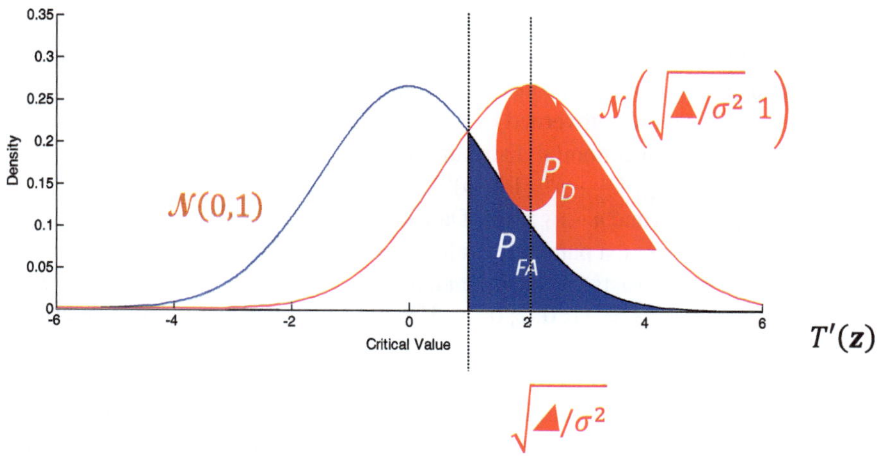

Fig. 7.4 Probability densities of the normalized sufficient statistic $\boldsymbol{T}'(z)$

Solve for γ'

$$\gamma' = \sqrt{\sigma^2\mathcal{E}} \times Q^{-1}(P_{FA}). \tag{7.45}$$

The probability of detection subject to the NP criterion is

$$P_D = P\big(T(z) > \gamma'|H_1\big), T \sim \mathcal{N}\big(\mathcal{E}, \sigma^2\mathcal{E}\big) \tag{7.46}$$

$$\gamma' = \sqrt{\sigma^2\mathcal{E}} \times Q^{-1}(P_{FA}). \tag{7.47}$$

Normalize: standard normal

$$P_D = Q\left(\frac{\gamma' - \mathcal{E}}{\sqrt{\sigma^2\mathcal{E}}}\right) = Q\left\{\frac{\sqrt{\sigma^2\mathcal{E}} \times Q^{-1}(P_{FA}) - \mathcal{E}}{\sqrt{\sigma^2\mathcal{E}}}\right\} \tag{7.48}$$

$$P_D = Q\left\{Q^{-1}(P_{FA}) - \sqrt{\mathcal{E}/\sigma^2}\right\}. \tag{7.49}$$

The matched filter is optimal in the case of a known deterministic signal in Gaussian white noise and is obtained using the NP criterion and maximizing the signal-to-noise ratio. For non-Gaussian Noise.

a. The matched filter detector is not NP optimal but still maximizes SNR.
b. The NP filter is nonlinear, but the linear filter works well for moderate deviations from Gaussian.

Problems

7.1 Prove that for two statistically independent data sets $\boldsymbol{x}_1$ and $\boldsymbol{x}_2$ the likelihoods ratio of the combined data set is equal to the product of their likelihood ratios.

7.2 Assuming that the two data sets are independent, use the results of Problem 7.1 to strengthen the conclusion of Example 7.2.

7.3 Consider a communication system where the sender sends a 5-bit binary message m where the probability of a 1 is p and the probability of a 0 is $(1-p)$. Each bit is governed by a Bernoulli distribution, and a n-bit message is governed by a Binomial distribution.
$P(m|p) = \frac{n!}{n_1!(n-n_1)!}p^{n_1}(1-p)^{n-n_1}$.
with n_1 ones and $n-n_1$ zeros. Data is collected to determine the more likely of two hypotheses, a parameter value $p = 1/2$ or a parameter value $p = 1/4$. The data set consists of 3 independent messages $m_1 = \{1, 1, 1, 0, 0\}$, $m_2 = \{1, 0, 0, 1, 0\}$, $m_3 = \{1, 0, 1, 0, 1\}$. Use the likelihood ratio test to determine the more likely parameter value.

7.4 An engineer tests an electronic circuit on a manufacturing line to determine if it is sound, has a short circuit fault, or has an open circuit fault. The engineer knows that the open circuit fault is twice as likely as the short circuit fault and must determine the probability of the latter using data collected from 100 circuits. The data had 95 sound circuits, 2 with an open circuit fault and 3 with a short circuit fault, and it is known that the probability of a short circuit fault is either 0.02 or 0.1. Determine the more likely of the two probability values. (**Hint**: Use a trinomial pmf).

7.5 A manufacturer purchases a component from two different suppliers whose characteristics depend on a parameter μ. The parameter has values $\mu = \mu_1$ from the first supplier and $\mu = \mu_2$ from the second supplier. Design a test to determine the supplier of a batch of components using a sample set of independent measurements $z(i), i = 1, 2, \ldots, N$, of the parameter μ with additive zero-mean Gaussian noise of known variance σ^2.

7.6 Given a set of measurement vectors $\boldsymbol{x}_i, i = 1, 2, \ldots, N$ of concentrations in a chemical process $\boldsymbol{x} = \begin{bmatrix} x_1 \; x_2 \; \ldots \; x_n \end{bmatrix}^T$, governed by the normal distribution $\mathcal{N}(\boldsymbol{m}_h, C_x)$ when healthy, and the distribution $\mathcal{N}\left(\boldsymbol{m}_f, C_x\right)$ when faulty. Write the equations for the likelihood ratio test to determine the health of the process with test threshold c.

7.7 For the process of Problem 7.6, verify that transforming the variables to a random vector $\boldsymbol{z}$ with unity covariance does not change the likelihood ratio, then write the rate in terms of $\boldsymbol{z}$.

7.8 To decide between the two hypotheses

$$H_0 : \theta = \theta_0$$
$$H_1 : \theta \neq \theta_0$$

we use the generalized likelihood ratio

$$\frac{l(\theta_0)}{\max\limits_{\theta} l(\theta)} \geq \gamma$$

for some constant $\gamma > 0$. To determine whether the parameter $p = 1/2$, an engineer collects 40 messages with 19 messages with 3 ones, 19 messages with 2 ones, 1 message with 4 ones and one message with four zeros. Calculate the generalized likelihood ratio and use a threshold of 1 to determine the answer.

7.9 Generate 100 data points from the normal distribution $\mathcal{N}(1, 9)$, then use the data with the generalized likelihood ratio and a threshold $\gamma = 0.85$ to test the following hypothesis regarding the mean m

$$H_0 : m = 1$$
$$H_1 : m \neq 1$$

7.10 An engineer must determine if noisy measurements include an unknown bias A using N measurements. Assuming that the noise is zero-mean Gaussian with unknown variance, the following generalized likelihood ratio test is used

$$L(\theta, A) = \frac{p\left(\sigma^2, \hat{A}_{ML}\right)}{p\left(\sigma^2\right)} > \gamma$$

where the threshold γ is chosen for a probability of false alarm $PFA = 10^{-4}$, and θ_i, $i = 0, 1$, are the maximum-likelihood estimates of the unknown variance.

(a) Write the equations for the test using the sample variance as an estimate of the variance
(b) Determine the value of the threshold in terms of the variance estimate

7.11 Design a matched filter for a square wave signal of unity amplitude and unity period with zero DC component that switches to one at time zero. Use 100 samples at a sampling rate of 5 Hz and $P_{FA} = 10^{-5}$. Repeat the analysis of Sect. 7.4. on signal detection and evaluate the performance (SNR, P_D) of the filter for zero-mean additive Gaussian white noise with variance (a) 1, (b) 2.

Bibliography

1. Casella, G., & Berger, R. (1990). *Statistical inference.* Duxbury Press.
2. DeGroot, M. H., & Schervish, M. J. (2012). *Probability and statistics.* Addison-Wesley.
3. Goldberger, A. S. (1964). *Economic theory.* J. Wiley.
4. Hippenstiel, R. D. (2002). *Detection theory: Application and digital signal processing.* CRC Press.
5. Hogg, R. V., McKean, J. W., & Craig, A. T. (2005). *Introduction to mathematical statistics.* Prentice-Hall.
6. Kay, S. M. (1993). *Fundamentals of statistical signal processing: Estimation theory* (Vol. I). Prentice-Hall.
7. Lindsey, J. K. (1996). *Parametric statistical inference.* Clarendon Press.
8. Mendel, J. (1995). *Lessons in estimation theory and signal processing, communications, and control.* Prentice-Hall.

Chapter 8
Maximum-Likelihood Estimation

8.1 Maximum-Likelihood Estimator (MLE)

Maximum-likelihood estimates are obtained by maximizing the likelihood function for a given data set with a known distribution. We obtain the estimate in two steps:

1. Obtain the likelihood or log-likelihood for a given data set.
2. Maximize to obtain the estimates.

The maximum-likelihood estimator is optimum in the sense of maximizing the likelihood $l(\theta|z)$ for a particular set of measurements z. Because the natural logarithm is a monotonic transformation, using the log-likelihood $L(\theta|z)$ does not change the solution, and it is often used instead of the likelihood.

To find the conditions for a maximum, we consider the series expansion of the likelihood function

$$L(\boldsymbol{\theta}|\boldsymbol{z}) \approx L\left(\widehat{\boldsymbol{\theta}}_{ML}|\boldsymbol{z}\right) + \left.\frac{\partial L(\boldsymbol{\theta}|\boldsymbol{z})}{\partial \boldsymbol{\theta}}\right]^{T}_{\widehat{\boldsymbol{\theta}}_{ML}} \Delta\boldsymbol{\theta} + \frac{1}{2!}\Delta\boldsymbol{\theta}^{T}\left[\left.\frac{\partial^{2} L(\boldsymbol{\theta}|\boldsymbol{z})}{\partial \boldsymbol{\theta}^{2}}\right]_{\widehat{\boldsymbol{\theta}}_{ML}}\right]\Delta\boldsymbol{\theta},$$
$$\Delta\boldsymbol{\theta} = \boldsymbol{\theta} - \widehat{\boldsymbol{\theta}}_{ML} \tag{8.1}$$

where z is a set of N i.i.d. observations $z(i), i = 1, \ldots, N$, and the derivatives are assumed to exist.

If the maximum exists in the interior of the feasible region for the parameter, the necessary condition for a maximum is

$$\left.\frac{\partial L(\boldsymbol{\theta}|\boldsymbol{z})}{\partial \boldsymbol{\theta}}\right]^{T}_{\widehat{\boldsymbol{\theta}}_{ML}} = \left[\frac{\partial L}{\partial \theta_1}\ \frac{\partial L}{\partial \theta_2} \cdots \frac{\partial L}{\partial \theta_n}\right]^{T}_{\widehat{\boldsymbol{\theta}}_{ML}} = \mathbf{0} \tag{8.2}$$

and the sufficient condition is the negative definite Hessian

M. S. Fadali, *Introduction to Random Signals, Estimation Theory, and Kalman Filtering*, https://doi.org/10.1007/978-981-99-8063-5_8

$$\left.\frac{\partial^2 L(\boldsymbol{\theta}|\boldsymbol{z})}{\partial \boldsymbol{\theta} \partial \boldsymbol{\theta}^T}\right]_{\widehat{\boldsymbol{\theta}}_{ML}} = \left[\frac{\partial^2 L(\boldsymbol{\theta}|\boldsymbol{z})}{\partial \theta_i \partial \theta_j}\right] < 0. \tag{8.3}$$

These conditions work for most of the distributions that are commonly used in engineering. However, if the maximum is on the boundary of the parameter's feasible region, then the conditions are not valid, and boundary must be tested to find the maximum.

In some cases, it is difficult to analytically find the maximum of the likelihood function and a numerical method must be used to find the maximum. Because the likelihood function may be nonconvex, the search must be conducted carefully so that the solution obtained is the global maximum, and not a local maximum.

Example 8.1 Given a data set $z(i), i = 1, \ldots, N$, of normally distributed measurements with unknown mean μ and variance σ^2, find the maximum-likelihood estimators of the parameters μ and σ^2.

Solution

The likelihood function for one measurement is

$$p\left(z(i)|\mu, \sigma^2\right) = \frac{1}{\sqrt{2\pi\sigma^2}} \exp\left\{-\frac{(z(i) - \mu)^2}{2\sigma^2}\right\}.$$

The log-likelihood function is

$$\ln\left[p\left(z(i)|\mu, \sigma^2\right)\right] = -\frac{1}{2}\ln\left[2\pi\sigma^2\right] - \frac{[z(i) - \mu]^2}{2\sigma^2}.$$

For a set of N i.i.d. measurements, the likelihood becomes

$$l\left(\mu, \sigma^2\right) = p\left(z(1), \ldots, z(N)|\mu, \sigma^2\right) = \prod_{i=1}^{N} p\left(z(i)|\mu, \sigma^2\right).$$

This gives the log-likelihood

$$\begin{aligned} L\left(\mu, \sigma^2\right) &= \sum_{i=1}^{N} \ln\left[p\left(z(i)|\mu, \sigma^2\right)\right] \\ &= \sum_{i=1}^{N}\left\{-\frac{1}{2}\ln\left[2\pi\sigma^2\right] - \frac{[z(i) - \mu]^2}{2\sigma^2}\right\} \\ &= -\frac{N}{2}\left(\ln[2\pi] + \ln\left[\sigma^2\right]\right) - \sum_{i=1}^{N} \frac{[z(i) - \mu]^2}{2\sigma^2} \end{aligned}$$

To find maximum-likelihood estimator, we differentiate w.r.t. the parameters to obtain the necessary condition

$$\left.\frac{\partial L(\mu,\sigma^2)}{\partial \boldsymbol{\theta}}\right]_{\widehat{\boldsymbol{\theta}}_{ML}} = \left[\begin{array}{c} \frac{\partial L}{\partial \mu} \\ \frac{\partial L}{\partial \sigma^2} \end{array}\right]_{\widehat{\boldsymbol{\theta}}_{ML}} = \left[\begin{array}{c} \frac{1}{\sigma^2}\sum_{i=1}^{N}[z(i)-\mu] \\ -\frac{N}{2}\times\frac{1}{\sigma^2}+\frac{1}{2\sigma^4}\sum_{i=1}^{N}[z(i)-\mu]^2 \end{array}\right]_{\widehat{\boldsymbol{\theta}}_{ML}} = \mathbf{0}.$$

The maximum-likelihood estimates are

$$\widehat{\mu}_{ML} = \frac{1}{N}\sum_{i=1}^{N} z(i) = \overline{z},\ \widehat{\sigma^2}_{ML} = \frac{1}{N}\sum_{i=1}^{N}[z(i)-\overline{z}]^2.$$

The second derivative provides the sufficient condition:

$$\left.\frac{\partial^2 L(\mu,\sigma^2)}{\partial \theta^2}\right]_{\widehat{\boldsymbol{\theta}}_{ML}} = \left[\begin{array}{cc} -\frac{N}{\sigma^2} & -\frac{1}{\sigma^4}\sum_{i=1}^{N}[z(i)-\mu] \\ -\frac{1}{\sigma^4}\sum_{i=1}^{N}[z(i)-\mu] & \frac{N}{2}\times\frac{1}{\sigma^4}-\frac{1}{\sigma^6}\sum_{i=1}^{N}[z(i)-\mu]^2 \end{array}\right]_{\widehat{\boldsymbol{\theta}}_{ML}}$$

$$= \left[\begin{array}{cc} -\frac{N}{\widehat{\sigma^2}_{ML}} & 0 \\ 0 & -\frac{N}{2\left(\widehat{\sigma^2}_{ML}\right)^2} \end{array}\right] < 0.$$

■

8.2 Properties of Maximum-Likelihood Estimators

The popularity of maximum-likelihood estimators is due to their excellent large sample properties. This section discusses these properties and provides examples to demonstrate their importance.

Theorem 8.1 *ML estimates are*

1. *Consistent.*
2. *Asymptotically Gaussian with mean θ and covariance matrix $J^{-1} = J_i^{-1}/N$* where

$$J_i = -E\left\{\frac{\partial^2 \ln[p(z_i)]}{\partial \theta^2}\right\}, i = 1, \ldots, N \tag{8.4}$$

is the Fisher information matrix.

3. *Asymptotically Efficient.* ■

Example 8.2 Given N i.i.d. measurements from an exponentially distributed population.

$$p(z_i) = \theta \exp(-\theta z_i), i = 1, \ldots N$$

(i) Obtain the Fisher information
(ii) Obtain the Cramer–Rao lower bound (CRLB).

Solution

The likelihood for N measurements is

$$p(\mathbf{z}) = \prod_{i=1}^{N} p(z_i), \; p(z_i)\} = \theta \; \exp(-\theta z_i).$$

The log-likelihood is

$$\ln[p(\mathbf{z})] = \sum_{i=1}^{N} \ln[p(z_i)], \ln[p(z_i)] = \ln[\theta] - \theta z_i.$$

Differentiate to obtain

$$\frac{\partial(\ln[p(z_i)])}{\partial\theta} = \frac{1}{\theta} - z_i.$$

The Fisher information of one measurement is

$$J_i = -E\left\{\frac{\partial^2(\ln[p(z_i)])}{\partial\theta^2}\right\} = \frac{1}{\theta^2}, i = 1, 2, \ldots, N.$$

For N measurements, the Fisher information becomes

$$J = \sum_{i=1}^{N} J_i = N/\theta^2.$$

Hence, the CRLB is

$$\theta^2/N.$$

To obtain the maximum-likelihood estimate, we use the necessary condition

$$\sum_{i=1}^{N} \frac{\partial(\ln[p(z_i)])}{\partial\theta} = \sum_{i=1}^{N}\left[\frac{1}{\theta} - z_i\right] = 0;$$

that is

$$\frac{N}{\widehat{\theta}_{ML}} = \sum_{i=1}^{N} z_i.$$

We solve for the maximum-likelihood estimate

$$\widehat{\theta}_{ML} = \frac{N}{\sum_{i=1}^{N} z_i} = \frac{1}{\overline{z}}.$$

We can show that the expected value of the estimate is

$$E\left\{\widehat{\theta}_{ML}\right\} = \frac{N}{N-1}\theta.$$

The estimate is biased but is asymptotically unbiased. ■

Theorem 8.2 *Any continuous function* $\mathbf{g}(\boldsymbol{\theta}) : \mathcal{R}^n \to\subset \mathcal{R}^r$ *of a consistent estimator is itself a consistent estimator.*

$$\widehat{\mathbf{g}(\boldsymbol{\theta})} = \mathbf{g}\left(\widehat{\boldsymbol{\theta}}\right) \tag{8.5}$$

■

Because maximum-likelihood estimators are consistent, they have the property

$$\widehat{\mathbf{g}(\boldsymbol{\theta})}_{ML} = \mathbf{g}\left(\widehat{\boldsymbol{\theta}}_{ML}\right). \tag{8.6}$$

Example 8.3 Use Theorem 8.2 and the results of Example 8.2 to obtain the MLE of the parameter λ *for* N i.i.d. measurements from an exponentially distributed population.

$$p(z_i) = \lambda^{-1}\exp(-z_i/\lambda), i = 1, \ldots N$$

Solution

Example 8.2 gives the maximum-likelihood estimate

$$\widehat{\theta}_{ML} = \frac{1}{\overline{z}}.$$

We observe that

$$\lambda = \frac{1}{\theta}.$$

Using Theorem 8.2 gives the estimate

$$\widehat{\lambda}_{ML} = \frac{1}{\widehat{\theta}_{ML}} = \overline{z}.$$

■

8.3 Comparison of Estimators

In this section, we consider the MLE when using the linear model

$$z = H\theta + \boldsymbol{v} \tag{8.7}$$

with H deterministic and zero-mean Gaussian white noise $\boldsymbol{v}(k)$ with known covariance matrix R. We show that the estimator is identical to BLUE and inherits its good characteristics.

Because z is a linear transformation of the Gaussian vector $\boldsymbol{v}$, it is also Gaussian, and its conditional Gaussian distribution has the parameters

$$E\{z|\theta\} = H\theta \tag{8.8}$$

$$\text{var}\{z|\theta\} = E\left\{\boldsymbol{v}\boldsymbol{v}^T\right\} = R. \tag{8.9}$$

Next, we show that for the linear model with Gaussian noise the BLUE is the MLE.

Theorem 8.3 *For the linear model with deterministic* $\boldsymbol{H}$ *and multivariate zero-mean Gaussian white noise* $\boldsymbol{v}$, *MLE and BLUE are identical.*

$$\widehat{\theta}_{ML} = \widehat{\theta}_{\text{BLUE}}. \tag{8.10}$$

The estimators are

(i) unbiased, (ii) the most efficient linear estimators, (iii) consistent, and (iv) Gaussian.

Proof The conditional distribution for the measurements gives the likelihood function

$$p(z(k)|\theta) = \frac{1}{\sqrt{(2\pi)^N \det(R(k))}} \exp\left\{-\frac{1}{2}[z(k) - H(k)\theta]^T R^{-1}(k)[z(k) - H(k)\theta]\right\}.$$

Maximizing p is equivalent to minimizing the quadratic in the exponent, that is, solving the equation

$$\frac{d}{d\theta}[z(k) - H(k)\theta]^T R^{-1}(k)[z(k) - H(k)\theta]\bigg]_{\widehat{\theta}_{ML}(k)} = \mathbf{0}.$$

- However, minimizing the quadratic gives BLUE since BLUE is a weighted least-squares estimator with weight matrix $W = R^{-1}$. Hence, the two estimators are identical.

The estimator has the properties of both BLUE and MLE. It follows that the estimators are

- Unbiased since BLUE are unbiased.
- Most efficient since BLUE are most efficient.
- Consistent since MLE is consistent.

Gaussian because they depend linearly on the Gaussian measurement. ■

Example 8.4 Determine the MLE of the parameters of the causal filter with output

$$\begin{aligned} y(k) &= \sum_{i=1}^{n} h(i)u(k-i) = \boldsymbol{u}^T\boldsymbol{h} \\ \boldsymbol{u}^T &= \left[u(k-1)\ u(k-2)\ \ldots\ u(k-n)\right], u(i) = 0, i < 0 \\ \boldsymbol{h}^T &= \left[h(1)\ h(2)\ \ldots\ h(n)\right] \end{aligned}$$

using N noisy measurements $z(i), u(i), i = 1, \ldots, N$, with additive i.i.d. zero-mean Gaussian white noise.

Solution

The noisy measurements are given by

$$\begin{aligned} z(k) &= y(k) + v(k), k = 1, 2, \ldots, N \\ &= [h(k-1)\ h(k-2)\ \ldots\ h(0)]\begin{bmatrix} u(1) \\ u(2) \\ \vdots \\ u(k) \end{bmatrix} + v(k) \\ \boldsymbol{u}^T &= [u(k-1)\ u(k-2)\ u(k-n)\ \ldots], u(i) = 0, i < 0 \\ \boldsymbol{h}^T &= [h(1)\ h(2)\ \ldots\ h(n)]. \end{aligned}$$

Combine equations to obtain the linear model

$$z = H\theta + v$$

$$\begin{bmatrix} z(N) \\ z(N-1) \\ \vdots \\ z(2) \\ z(1) \end{bmatrix} = \begin{bmatrix} u(N-1) & u(N-2) & \cdots & u(N-n) \\ u(N-2) & u(N-3) & \cdots & u(N-n-1) \\ \vdots & \vdots & \ddots & \vdots \\ u(1) & u(0) & 0\ldots & 0 \\ u(0) & 0 & \cdots & 0 \end{bmatrix} \begin{bmatrix} h(1) \\ h(2) \\ \vdots \\ h(n-1) \\ h(n) \end{bmatrix} + \begin{bmatrix} v(N) \\ v(N-1) \\ \vdots \\ v(2) \\ v(1) \end{bmatrix}$$

Because the noise is Gaussian and the model is linear, the maximum-likelihood estimator is the same as the BLUE. The noise is i.i.d., and its covariance matrix is therefore of the form $\sigma^2 I_N$. Hence, BLUE is the least-squares estimate

$$\hat{\theta} = H^{\#} z = \left(H^T H\right)^{-1} H^T z$$

assuming a full-rank measurement matrix. ■

Many of the most important distributions for engineering applications belong to the exponential family

$$p(z|\theta) = h(z)\exp\left(c(\theta)^T t(z) - a(\theta)\right) \tag{8.11}$$

where θ is a vector of natural parameters and $t(z)$ is a vector of sufficient statistics. As shown in Problem 8.12, the maximum-likelihood estimator for the exponential family is the solution of the equation

$$t(z) = E\{t(z|\theta). \tag{8.12}$$

Example 8.5 Show that the Gaussian distribution belongs to the exponential family, then use (8.12) to obtain the ML parameter estimate with N i.i.d. Gaussian measurements

Solution

The likelihood function of N i.i.d. Gaussian measurements

$$\begin{aligned} l\left(\mu,\sigma^2\right) &= p\left(z(1),\ldots,z(N)|\mu,\sigma^2\right) = \prod_{i=1}^{N} p\left(z(i)|\mu,\sigma^2\right) = p\left(z(i)|\mu,\sigma^2\right) \\ &= \frac{1}{(2\pi)^{N/2}\left(\sigma^2\right)^{N/2}} \exp\left\{-\sum_{i=1}^{N} \frac{(z(i)-\mu)^2}{2\sigma^2}\right\}. \end{aligned}$$

The term in the exponent can be expanded and written as

$$\frac{(z(i)-\mu)^2}{2\sigma^2} = \frac{z(i)^2 - 2\mu z(i) + \mu^2}{2\sigma^2} = \left[z(i)^2 \ z(i)\right]\begin{bmatrix} \frac{1}{2\sigma^2} \\ \frac{\mu}{\sigma^2} \end{bmatrix} + \frac{\mu^2}{2\sigma^2}.$$

This allows us to write the log-likelihood in the exponential family form

$$l(\mu, \sigma^2) = \frac{1}{(2\pi)^{N/2}} \exp\left\{ -\sum_{i=1}^{N} \left(\left[z(i)^2 \; z(i) \right] \begin{bmatrix} \frac{1}{2\sigma^2} \\ \frac{\mu}{\sigma^2} \end{bmatrix} + \frac{\mu^2}{2\sigma^2} - \frac{N}{2} \ln(\sigma^2) \right) \right\}.$$

The vector of sufficient statistics is

$$t(z) = \sum_{i=1}^{N} \left[z(i)^2 \; z(i) \right]^T = \left[\sum_{i=1}^{N} z(i)^2 \; \sum_{i=1}^{N} z(i) \right]^T.$$

The expectation of the vector is

$$\begin{aligned} E\left\{ t(z|\theta)^T \right\} &= E\left\{ \left[\sum_{i=1}^{N} z(i)^2 \quad \sum_{i=1}^{N} z(i)|\theta \right] \right\} \\ &= \left[\sum_{i=1}^{N} E\left\{ z(i)^2|\theta \right\} \quad \sum_{i=1}^{N} E\{ z(i)|\theta \} \right] \\ &= N\left[\sigma^2 + \mu_{4\mu}^2 \right] = \sum_{i=1}^{N} \left[z(i)^2 \quad z(i) \right]^T. \end{aligned}$$

This gives the estimate for the mean

$$\widehat{\mu}_{ML} = \frac{1}{N} \sum_{i=1}^{N} z(i) = \overline{z}.$$

The estimate of the variance is

$$\widehat{\sigma}_{ML}^2 = \frac{1}{N} \sum_{i=1}^{N} z(i)^2 - \widehat{\mu}_{ML}^2 = \overline{z^2} - \overline{z}^2 = \frac{1}{N} \sum_{i=1}^{N} [z(i) - \overline{z}]^2.$$

As shown in Example 8.1, the estimates are the sample mean and sample variance, respectively. ■

8.4 Maximum a Posteriori (MAP)

If the distribution of the unknown parameter is known, then we can use the maximum a posteriori (MAP) estimator. The term a posteriori comes from the fact that it is used on maximizing the pdf $p(\theta|z)$, i.e., given the data z. The pdf can be written as

$$p(\theta|z) = \frac{p(z|\theta)p(\theta)}{p(z)}. \tag{8.13}$$

The denominator is not a function of the parameters and can be ignored. The MAP estimator is given by

$$\widehat{\theta}_{\text{MAP}} = \arg\left\{\max_{\theta} p(z|\theta)p(\theta)\right\}.$$

Example 8.6 Use N i.i.d. Gaussian measurements $z(i) \sim N\left(\mu, \sigma^2\right)$ to obtain a map estimate of the mean μ if it is governed by the pdf

$$p(\mu) = \left(2\pi\sigma_\mu^2\right)^{-1/2} \exp\left(-\frac{\mu^2}{2\sigma_\mu^2}\right).$$

Solution

The estimator is obtained using

$$\max_{\mu} p(z|\mu) = \max_{\mu}\left\{-\frac{\mu^2}{2\sigma_\mu^2} - \sum_{i=1}^{N} \frac{[z(i) - \mu]^2}{2\sigma^2}\right\}.$$

The necessary condition for the maximum is

$$\frac{\partial p(\mu)}{\partial \mu} = -\frac{\mu}{\sigma_\mu^2} + \frac{1}{\sigma^2}\sum_{i=1}^{N}[z(i) - \mu] = 0.$$

We obtain the MAP estimator

$$\widehat{\mu}_{\text{MAP}}(N) = \frac{1}{N + \sigma^2/\sigma_\mu^2}\sum_{i=1}^{N} z(i).$$

If there is no information about the parameter, then we can say that $\left(\sigma_\mu^2 \to \infty\right)$ and the MAP estimator becomes

$$\widehat{\mu}_{\text{MAP}}(N) = \overline{z} = \widehat{\mu}_{ML}(N).$$

Thus, if there is no knowledge of the statistics of the parameter, the MAP estimator reduces to MLE. ■

8.5 Numerical Computation of the ML Estimate

To obtain the ML estimator, we equate the score function to zero and solve for the parameter vector. In many problems, the solution for the ML estimator cannot be obtained analytically and numerical methods must be used to obtain the solution. The *score equation* to solve is

$$s(\boldsymbol{\theta}) = \frac{\partial \ln[p(\mathbf{z}|\boldsymbol{\theta})]}{\partial \boldsymbol{\theta}} = \mathbf{0}$$

where $s(\boldsymbol{\theta})$ is the score function. Newton's method uses the first-order approximation

$$s(\boldsymbol{\theta}^* + \Delta\boldsymbol{\theta}) \approx s(\boldsymbol{\theta}^*) + \frac{\partial s(\boldsymbol{\theta})}{\partial \boldsymbol{\theta}}_{\boldsymbol{\theta}^*} \Delta\boldsymbol{\theta} = \mathbf{0}.$$

Starting with an initial estimate of the parameter vector $\boldsymbol{\theta}_k$ at time k, we define the perturbation as

$$\Delta\boldsymbol{\theta} = \boldsymbol{\theta}_{k+1} - \boldsymbol{\theta}_k.$$

We can use the approximation to iteratively update the estimate until it converges to a maximum of the likelihood. The iteration equation is

$$\boldsymbol{\theta}_{k+1} = \boldsymbol{\theta}_k - \left(\left. \frac{\partial s(\boldsymbol{\theta})}{\partial \boldsymbol{\theta}} \right]_{\boldsymbol{\theta}_k} \right)^{-1} s(\boldsymbol{\theta}_k), \quad k = 0, 1, 2, \ldots \tag{8.14}$$

where it is assumed that matrix remains invertible.

The method has two major disadvantages:

(1) The results depend on the initial parameter estimate and if the estimate is poor the method will either not converge or will converge to a local maximum.
(2) The method may not converge if the data used to estimate the parameter has a large variance.
(3) The matrix inversion in each step makes the method computationally costly.

The derivative of the score function is the Hessian of the likelihood function,

$$\frac{\partial s(\boldsymbol{\theta})}{\partial \boldsymbol{\theta}} = \frac{\partial^2 \ln[p(\mathbf{z}|\boldsymbol{\theta})]}{\partial \boldsymbol{\theta} \partial \boldsymbol{\theta}^T}$$

which is the observed information, and the negative of its expectation is the Fisher information matrix

$$I(\boldsymbol{\theta}) = -E\left\{ \frac{\partial s(\boldsymbol{\theta})}{\partial \boldsymbol{\theta}} \right\} = -E\left\{ \frac{\partial^2 \ln[p(\mathbf{z}|)]}{\partial \boldsymbol{\theta} \partial \boldsymbol{\theta}^T} \right\} \left. \frac{\partial s(\boldsymbol{\theta})}{\partial \boldsymbol{\theta}} \right]_{\boldsymbol{\theta}_k} = \left. \frac{\partial^2 \ln[p(\mathbf{z}|)]}{\partial \boldsymbol{\theta} \partial \boldsymbol{\theta}^T} \right]_{\boldsymbol{\theta}_{k-1}}. \tag{8.15}$$

The observed information can be replaced by the Fisher information matrix to give the iteration

$$\boldsymbol{\theta}_{k+1} = \boldsymbol{\theta}_k + I(\boldsymbol{\theta})^{-1} s(\boldsymbol{\theta}_k), \quad k = 0, 1, 2, \ldots. \tag{8.16}$$

This is known as the **scoring method**. The Fisher algorithm has the same disadvantages, and its convergence is slower, but can converge in cases where Newton's method does not. Other numerical algorithms are available in the literature that provide better performance, but they are beyond the scope of this textbook.

Example 8.7 Using the measurements

$$z(k) = A\sin(\omega_0 k) + w(k), k = 0, 1, 2, \ldots, N$$

where A is a known constant and $w(k) \sim \mathbf{N}\left(0, \sigma^2\right)$ is Gaussian white noise; we wish to find the ML estimate of the frequency ω_0.

(a) Write an expression for the score function
(b) Write the update equation for Newton's method to solve for the ML estimate of ω_0
(c) Obtain the fisher information matrix
(d) Generate random data with the numerical values $\omega_0 = 1, A = 5, \sigma = 0.01, N = 5$, then use the data to obtain the ML estimate of ω_0 using both Newton's method and the scoring method. Does either method converge for (i) $\sigma = 1\ 0$, (ii) $N = 30$

Solution

The likelihood function for the data is

$$\begin{aligned} p(z|\omega_0) &= \prod_{k=0}^{N-1} \frac{1}{\sqrt{2\pi\sigma^2}} \exp\left\{\frac{-1}{2\sigma^2}(z(k) - A\sin(\omega_0 k))^2\right\} \\ &= \frac{1}{\left(2\pi\sigma^2\right)^{N/2}} \exp\left\{-\frac{1}{2\sigma^2}\sum_{k=0}^{N-1}(z(k) - A\sin(\omega_0 k))^2\right\}. \end{aligned}$$

The log-likelihood is

$$\ln p(z|\omega_0) = -\frac{N}{2}\ln\left(2\pi\sigma^2\right) - \frac{1}{2\sigma^2}\sum_{k=0}^{N-1}(z(k) - A\sin(\omega_0 k))^2$$

The score function is

$$\frac{\partial \ln p(\mathbf{z}|\omega_0)}{\partial \omega_0} = \frac{1}{\sigma^2}\sum_{k=0}^{N-1} Ak\cos(\omega_0 k)(z(k) - Ak\sin(\omega_0 k))$$
$$= \frac{A}{\sigma^2}\sum_{k=0}^{N-1} k\left(z(k)\cos(\omega_0 k) - \frac{A}{2}\sin(2\omega_0 k)\right).$$

Note that the expression for the score function becomes more elaborate as N increases. This makes it more difficult to solve for the root numerically.

The Hessian is

$$\frac{\partial^2 \ln p(\mathbf{z}|\omega_0)}{\partial \omega_0^2} = -\frac{A}{\sigma^2}\sum_{k=0}^{N-1} k^2(z(k)\sin(\omega_0 k) + A\cos(2\omega_0 k)).$$

We need to solve the highly nonlinear score equation

$$\sum_{k=0}^{N-1} k\left(z(k)\cos(\omega_0 k) - \frac{Ak}{2}\sin(2\omega_0 k)\right) = 0.$$

Starting with an initial guess $\omega_0^{(0)}$ at $k = 0$, we have the equation

$$\omega_0^{(k+1)} = \omega_0^{(k)} + \frac{\sum_{k=0}^{N-1} k\big(z(k)\cos(\omega_0 k) - \frac{A}{2}\sin(2\omega_0 k)\big)}{\sum_{k=0}^{N-1} k^2(z(k)\sin(\omega_0 k) - A\cos(2\omega_0 k))}.$$

The Fisher information is

$$-E\left\{\frac{\partial^2 \ln p(\mathbf{z}|\omega_0)}{\partial \omega_0^2}\right\} = E\left\{\frac{A}{\sigma^2}\sum_{k=0}^{N-1} k^2(z(k)\sin(\omega_0 k) + A\cos(2\omega_0 k))\right\}$$
$$= \frac{A}{\sigma^2}\sum_{k=0}^{N-1} k^2(A\sin(\omega_0 k)\sin(\omega_0 k) + A\cos(2\omega_0 k))$$
$$= \frac{A^2}{\sigma^2}\sum_{k=0}^{N-1} k^2\big(\sin^2(\omega_0 k) + \cos(2\omega_0 k)\big)$$
$$= \frac{A^2}{\sigma^2}\sum_{k=0}^{N-1} k^2\ \cos^2(\omega_0 k).$$

Newton's method does not converge for $\sigma = 0.01, N = 5$, while the scoring approach does. However, neither method converges if the noise standard deviation is increased to 10 or if the number of points is increased to 30.

```
% ExampleNL_MLE: Does not converge
clear
A=5;omega=1;
sigma=0.01; %Noise standard deviation
N=5; % Number of points, k=0:1:N-1 time grid
omega0=10.; % % Initialized with large number, then used to save estimate
omega_hat=1.5; % Initial estimate, updated every step
Tol=1.E-3;
w=5*randn(N,1)*sigma;
% Generate data
for i=1:N
    i1=i-1;
    z(i)=A*sin(omega_hat*i1)+w(i);
end
count=0;
while abs(omega0-omega_hat)>Tol
    count=count+1;
omega0=omega_hat;
sum_n=0;sum_d=0;
for i=1:N
    i1=i-1;
    cosw=cos(omega0*i1);sinw=sin(omega0*i1);
    sum_n=sum_n-i1*cosw*(z(i)-A*sinw);
    sum_d=sum_d-A*i1*i1*(z(i)*sinw+A*cos(2*omega0*i1));
% sum_d=sum_d+A^2*i1*i1*cosw^2;
end
omega_hat=omega0-sum_n/sum_d;
end
'Estimate',omega_hat
```

8.5.1 MATLAB MLE

The MATLAB statistical toolbox can be used to obtain the MLE for a given set of measurements as follows.

> > **x = randn(100,1)**; % Standard normal.

> > **[p,pci] = mle('normal', x)** % P = parameter vector, pci = confidence interval for p, 95% confidence.

p =

0.0912 0.9220

Pci =

– 0.0927 0.8136

0.2751 1.0765

Problems

8.1. Find the maximum-likelihood estimate of the parameter of an exponential distribution given N i.i.d. measurements.

$$p(x|\lambda) = \lambda^{-1} e^{-x/\lambda},\ x \geq 0 \text{ and zero elsewhere}$$

Calculate the value of the estimate for the data then check your answer with the MATLAB command **mle**.

$$\begin{bmatrix} 9.76367\ 4.3166\ 0.4402\ 1.1647\ 0.2068\ 2.1099\ 16.6614\ 0.8178\ 0.3415\ 1.9376 \end{bmatrix}$$

8.2. Find the maximum-likelihood estimate of the parameters of a Laplacian distribution given N i.i.d. measurements.

8.3. Find the maximum-likelihood estimate of the parameters of a Maxwell distribution

$$p(x) = \frac{1}{\sigma^3\sqrt{2\pi}} \exp\left(-\frac{x^2}{2\sigma^2}\right)$$

given N i.i.d. measurements.

8.4. The number of photons k released by a photodetector is governed by the Poisson distribution

$$p_K(k|a) = \frac{a^k e^{-a}}{k!},\ k = 0, 1, 2, \ldots$$

Obtain the maximum-likelihood estimate of the parameter a using N i.i.d. measurements of k. Comment on your result referring to the properties of the Poisson distribution.

8.5. A vehicle randomly departs from its nominal orbit in a plane perpendicular to the direction of motion with zero-mean Gaussian perturbations with variance σ^2 in the directions of the vertical and horizontal axes. Find the maximum-likelihood estimate of the variance σ^2 based on N measurements of the distance.

Hint: The random perturbation is governed by a Rayleigh distribution.

8.6. The mean time between received messages at a receiving station is known to be governed by an exponential distribution

$$p_T(t/a) = \begin{cases} ae^{-at}, & t \geq 0 \\ 0, & t < 0 \end{cases}$$

Obtain a maximum-likelihood estimate of the parameter a using N i.i.d. measurements of t. Explain why this estimate is intuitively appealing using the known properties of the exponential distribution.

8.7. If $x \sim \mathcal{N}(m_x, \sigma_x^2)$ is a Gaussian random variable and $y = e^x$,

(a) Show that the pdf of y is the log-normal distribution

$$f_Y(y) = \begin{cases} \frac{1}{y\sigma_x\sqrt{2\pi}} \exp\left(-\frac{(\ln y - m_x)^2}{2\sigma_x^2}\right), & y > 0 \\ 0, & \text{elsewhere} \end{cases}$$

(b) Find the mean and variance of y in terms of $\left(m_x, \sigma_x^2\right)$.
(c) Find the maximum-likelihood estimate of the parameters of the log-normal distribution given N i.i.d. measurements.

8.8. The lifetime of an electric car battery is governed by the Weibull distribution

$$p_X(x|\alpha, \beta) = \alpha\beta x^{\beta-1} e^{-\alpha x^{\beta}}, x > 0, \alpha, \beta > 0$$

(a) Let the parameters of the distribution be $\alpha = 1, \beta = 2$. Generate 100 data points using the MATLAB statistics toolbox, then plot and save your results. **Hint**: Use the MATLAB Statistics Toolbox command **random**.
(b) Derive the maximum-likelihood necessary and sufficient **conditions** for the parameters of the Weibull distribution. Obtain an expression for the maximum-likelihood estimate for α with one condition and explain how the second condition can be used to obtain the maximum-likelihood estimate of β. Obtain the maximum-likelihood estimates with the MATLAB Statistics Toolbox command **mle**.

8.9. A set of m machines in a workshop is used randomly, with only one machine used at a time, such that the probability of using the ith machine is $p_i, i = 1, \ldots, m$. Design an experiment to determine the probabilities using maximum-likelihood estimation. What data must be collected and what are the maximum-likelihood estimates? **Hint**: Use the multinomial distribution and collect data on the number of times each machine is used.

8.10. The number of positive results in a random experiment is governed by the binomial distribution

$$p_K(k|p) = \frac{n!}{k!(n-k)!} p^k (1-p)^{n-k}, k = 0, 1, 2, \ldots$$

(a) Generate 100 data points using the MATLAB statistics toolbox with $p = 0.3, n = 50$, using the MATLAB Statistics Toolbox command **random**, then find the maximum-likelihood estimate with the MATLAB Statistics Toolbox command **mle**.
(b) Derive an expression for the maximum-likelihood estimate, then obtain the estimate of the parameter p using the 100 i.i.d. measurements of Part (a). Comment on your result referring to the properties of the binomial distribution.

8.11. Given a data set $z(i), i = 1, \ldots, N$, of normally distributed measurements with unknown mean m and known variance σ^2, find the maximum-likelihood estimator for the mean if the mean is known to have a lower bound $\boldsymbol{m_L}$.

8.12. Consider the exponential family of distributions is in the form

$$p(\boldsymbol{x}|\boldsymbol{\theta}) = h(\boldsymbol{x})\exp\left(\boldsymbol{c}(\boldsymbol{\theta})^T\boldsymbol{t}(\boldsymbol{x}) - a(\boldsymbol{\theta})\right)$$

where $\boldsymbol{\theta}$ is a vector of natural parameters, $\boldsymbol{c}(\boldsymbol{\theta})$ is a vector of parameters, and $\boldsymbol{t}(\boldsymbol{x})$ is a vector of sufficient statistics. Problem 5.15 shows that

$$a(\boldsymbol{\theta}) = \ln\left\{\int_{-\infty}^{\infty} h(\boldsymbol{x})\exp\left(\boldsymbol{c}(\boldsymbol{\theta})^T\boldsymbol{t}(\boldsymbol{x})\right)d\boldsymbol{x}\right\}$$

(a) Find the derivative of $a(\boldsymbol{\theta})$
(b) Using the result of part (a), show that the ML estimate for the exponential family is the solution of the equation

$$\boldsymbol{t}(\boldsymbol{x}) = E\{\boldsymbol{t}(\boldsymbol{x}|\boldsymbol{\theta})$$

8.13. Given N i.i.d. Bernoulli distributed measurements with pmf

$$p(z|\theta) = \theta^z(1-\theta)^{1-z}, z \in \{0, 1\}$$

Use (8.12) to obtain the ML estimate of the parameter θ of the distribution.

8.14. The mean θ of the normal distribution governing grades in an engineering class is exponentially distributed with parameter λ. Find the MAP estimate of the mean of the distribution based on N i.i.d. samples of the normal distribution. Show that as $\lambda \to \infty$, the estimate reduces to the maximum-likelihood estimate, i.e., the sample mean.

8.15. Using the measurements

$$z(k) = a_2x^k + a_1x^{k-1} + a_0 + w(k), k = 0, 1, 2, \ldots, N$$

where $a_i, i = 0, 1, 2$, are known constants and $w(k) \sim \mathbf{N}\left(0, \sigma^2\right)$ is Gaussian white noise, find the ML estimate of the constant x.

(a) Write an expression for the score function
(b) Write the update equation for Newton's method to solve for the ML estimate of x
(c) Obtain the Fisher information matrix
(d) Generate random data with $x = 0.5$, then use it to obtain the ML estimate of x using both Newton's method and the scoring method for the following parameter values

$$a_2 = 0.2,\ a_1 = 0.1,\ a_0 = 0.3,\ \sigma = 0.01,\ N = 10$$

Does either method converge for (i) $\sigma = 2$, (ii) $N = 50$.

Bibliography

1. Casella, G., & Berger, R. (1990). *Statistical inference.* Duxbury Press.
2. DeGroot, M. H., & Schervish, M. J. (2012). *Probability and statistics.* Addison-Wesley.
3. Goldberger, A. S. (1964). *Economic theory.* Wiley.
4. Hogg, R. V., McKean, J. W., & Craig, A. T. (2005). *Introduction to mathematical statistics.* Prentice-Hall.
5. Kay, S. M. (1993). *Fundamentals of statistical signal processing: Estimation theory* (Vol. I). Prentice-Hall.
6. Lindsey, J. K. (1996). *Parametric statistical inference.* Clarendon Press.
7. Mendel, J. (1995). *Lessons in estimation theory and signal processing, communications, and control.* Prentice-Hall.

Chapter 9
Minimum Mean-Square Error Estimation

9.1 Minimum Mean-Square Error

Given a zero-mean discrete-time wide-sense stationary random signal $x_i, i = 1, 2, \ldots, n, \quad E\{x_i\} = 0$, that is measurable with noisy wide-sense stationary measurements $z_i, i = 1, 2, \ldots, n$. The measurements are collected to form a measurement vector

$$\boldsymbol{z} = \begin{bmatrix} z_1 \ z_2 \ \ldots \ z_n \end{bmatrix}^T = \boldsymbol{x} + \boldsymbol{v}, \tag{9.1}$$

where the values of the random signal from the vector $\boldsymbol{x} = \begin{bmatrix} x_1 \ x_2 \ \ldots \ x_n \end{bmatrix}^T$. We assume, WLOG, that the measurements are zero-mean since the mean can be subtracted from the measurements if nonzero.

We seek an estimate of the signal using a finite impulse response (FIR) filter with gain matrix $K \in \mathcal{R}^{n \times n}$ whose input is the measurement vector $\boldsymbol{z}$ and whose output is the estimate of the signal

$$\hat{\boldsymbol{x}} = K\boldsymbol{z}. \tag{9.2}$$

The estimation error is

$$\boldsymbol{e} = \boldsymbol{x} - \hat{\boldsymbol{x}} = \boldsymbol{x} - K\boldsymbol{z}. \tag{9.3}$$

As a measure of the total error for the record of the signal, we have the square of the norm of the error vector

$$\|\boldsymbol{e}\|^2 = \|\boldsymbol{x} - \hat{\boldsymbol{x}}\|^2 = (\boldsymbol{x} - K\boldsymbol{z})^T(\boldsymbol{x} - K\boldsymbol{z}) = \boldsymbol{x}^T\boldsymbol{x} - 2\boldsymbol{x}^T K\boldsymbol{z} + \left(\boldsymbol{z}^T K^T\right)(K\boldsymbol{z}).$$

M. S. Fadali, *Introduction to Random Signals, Estimation Theory, and Kalman Filtering*, https://doi.org/10.1007/978-981-99-8063-5_9

Using the trace property of compatible products of matrices A and B, $\mathrm{tr}[AB] = \mathrm{tr}[BA]$, gives

$$\|\boldsymbol{e}\|^2 = \mathrm{tr}\left[\boldsymbol{x}\boldsymbol{x}^T\right] - 2\mathrm{tr}\left[\boldsymbol{z}\boldsymbol{x}^T K\right] + \mathrm{tr}\left[K\boldsymbol{z}\boldsymbol{z}^T K^T\right]. \tag{9.4}$$

The Wiener approach for a discrete-time signal is to select the gain K that minimizes the mean-square error

$$E\left\{\|\boldsymbol{e}\|^2\right\} = E\left\{\mathrm{tr}\left[\boldsymbol{x}\boldsymbol{x}^T\right] - 2\mathrm{tr}\left[\boldsymbol{z}\boldsymbol{x}^T K\right] + \mathrm{tr}\left[K\boldsymbol{z}\boldsymbol{z}^T K^T\right]\right\}.$$

Because expectation and trace are commutative, $E\{\mathrm{tr}[A]\} = \mathrm{tr}[E\{A\}]$, the mean-square error is

$$E\left\{\|\boldsymbol{e}\|^2\right\} = \mathrm{tr}[C_{xx}] - 2\mathrm{tr}\left[C_{zx}K\right] + \mathrm{tr}\left[KC_{zz}K^T\right] \tag{9.5}$$

The minimization of the mean-square error requires the trace formulas

$$\frac{\partial \mathrm{tr}[AB]}{\partial A} = B^T,\ A, B \text{ square} \tag{9.6}$$

$$\frac{\partial \mathrm{tr}\left[ACA^T\right]}{\partial A} = 2AC,\ C \text{ symmetric} \tag{9.7}$$

Using the trace formulas, we obtain the derivative and equate to zero

$$\frac{\partial E\left\{\|\boldsymbol{e}\|^2\right\}}{\partial K} = -2C_{zx}^T + 2KC_{zz} = [0]. \tag{9.8}$$

The necessary condition for an optimal filter is

$$K = C_{zx}^T C_{zz}^{-1} = C_{xz}C_{zz}^{-1} \tag{9.9}$$

The sufficient condition for optimality is a positive definite second derivative using (9.6)

$$\frac{\partial\left\{-2C_{zx}^T + 2KC_{zz}\right\}}{\partial K} = 2C_{zz} > 0,\ i = 1, \ldots, n \tag{9.10}$$

9.1.1 Orthogonality

The minimum mean-square error is the one that fully exploits all measurements to minimize the error. This can only be achieved if the error in the hyperplane of the data is equal to zero. In other words, for an optimal estimate the estimation error must be orthogonal to the hyperplane of the data

$$E\{(x - Kz)z^T\} = C_{xz} - KC_{zz} = [0] \tag{9.11}$$

This **orthogonality condition** gives the optimal controller obtained by direct minimization of the square error

$$K = C_{xz}C_{zz}^{-1}. \tag{9.12}$$

The corresponding mean error vector is

$$\boldsymbol{e} = \boldsymbol{x} - \hat{\boldsymbol{x}} = \boldsymbol{x} - Kz = \boldsymbol{x} - C_{xz}C_{zz}^{-1}z \tag{9.13}$$

In many applications, we are interested in estimating a state vector $\boldsymbol{x}_k \in \mathcal{R}^n$ at time k that is not directly measurable using a sequence of measurements $\{z_i \in \mathcal{R}^m, i = 0, 1, \ldots, l\}$ that contain information about the state vector. This motivates three state estimation problems:

(1) Filtering: estimate the state $\boldsymbol{x}_k$ using measurements $\{z_i \in \mathcal{R}^m, i = 0, 1, \ldots, l\}, l = k$
(2) Prediction: estimate the state $\boldsymbol{x}_k$ using measurements $\{z_i \in \mathcal{R}^m, i = 0, 1, \ldots, l\}, l < k$
(3) Smoothing: estimate the state $\boldsymbol{x}_k$ using measurements $\{z_i \in \mathcal{R}^m, i = 0, 1, \ldots, l\}, l > k$.

To extend the results to a sequence of vectors $\{\boldsymbol{x}_i \in \mathcal{R}^n, i = 0, 1, 2, \ldots\}$, we first prove the following important result.

Theorem 9.1 Fundamental Theorem of Estimation Theory *The minimum mean-square error estimator is the posterior conditional expectation*

$$\hat{\boldsymbol{x}}_k = E\{\boldsymbol{x}_k|z_{0:l}\} \tag{9.14}$$

$z_{0:l} = col\{z_i \in \mathcal{R}^m, i = 0, 1, \ldots, l\}$ *is the vector of all measurements up to and including time* k.

Proof The state estimate depends on all available measurements. The mean square error given all the measurements is

$$\text{MSE} = E\left\{[\boldsymbol{x}_k - \hat{\boldsymbol{x}}_k]^T[\boldsymbol{x}_k - \hat{\boldsymbol{x}}_k]|z_{0:l}\right\} \tag{9.15}$$

Expand and complete squares

$$\text{MSE} = E\left\{\boldsymbol{x}_k^T\boldsymbol{x}_k|z_{0:l}\right\} + [\hat{\boldsymbol{x}}_k - E\{\boldsymbol{x}_k|z_{0:l}\}]^T[\hat{\boldsymbol{x}}(k) - E\{\boldsymbol{x}_k|z_{0:l}\}] - E\left\{\boldsymbol{x}_k^T|z_{0:l}\right\}E\{\boldsymbol{x}_k|z_{0:l}\}.$$

The first and last term do not depend on the estimator $\hat{\boldsymbol{x}}_k$, and the middle has its minimum zero value if $\hat{\boldsymbol{x}}_k = E\{\boldsymbol{x}_k|\boldsymbol{z}_{0:l}\}$.

■

Next, we show how the orthogonality condition applies to a sequence of vectors $\{\boldsymbol{x}_i, i = 0, 1, 2, \ldots\}$. We denote the estimate of the kth vector obtained using the set of measurements $\boldsymbol{z}_{1:l}$ by $\hat{\boldsymbol{x}}_{k|l}$.

Theorem 9.2 Optimal Unbiased Estimator *Let $\boldsymbol{x}_k$ be a random vector with mean $\boldsymbol{m}_x$ and $\boldsymbol{z}_{1:l}$ be the vector of all measurements up to time l with mean $\boldsymbol{m}_z$. If the two vectors are jointly Gaussian, then the minimum-mean-square error estimator of $\boldsymbol{x}_k$ is*

$$\hat{\boldsymbol{x}}_{k|l} = \boldsymbol{m}_x + K\left[\boldsymbol{z}_{0:l} - \boldsymbol{m}_z\right] \tag{9.16}$$

$$K = C_{xz}C_{zz}^{-1} \tag{9.17}$$

with covariance matrix

$$C_{x|z} = C_{xx} - C_{xz}C_{zz}^{-1}C_{zx}. \tag{9.18}$$

The estimator is unbiased, unique, and its estimation error $\tilde{\boldsymbol{x}}_{k|l} = \boldsymbol{x}_k - \hat{\boldsymbol{x}}_{k|l}$ is orthogonal to $\boldsymbol{z}_{0:l}$.

$$E\left\{\tilde{\boldsymbol{x}}_{k|l}\boldsymbol{z}_{0:l}^T\right\} = [0] \tag{9.19}$$

Proof From the properties of the multivariate normal distribution (see Chap. 2), both $\boldsymbol{x}_k$ and $\boldsymbol{z}_{0:l}$ are Gaussian because they are jointly Gaussian and the conditional mean of $\boldsymbol{x}_k$ is

$$\hat{\boldsymbol{x}}_{k|l} = \boldsymbol{m}_{x|z} = \boldsymbol{m}_x + C_{xz}C_{zz}^{-1}(\boldsymbol{z}_{0:l} - \boldsymbol{m}_z)$$

$$= C_{xz}C_{zz}^{-1}\boldsymbol{z}_{0:l} + \left(\boldsymbol{m}_x - C_{xz}C_{zz}^{-1}\boldsymbol{m}_z\right),$$

where C_{xz} is the cross-covariance and C_{zz} is the covariance of $\boldsymbol{z}_{0:l}$. Thus, $\hat{\boldsymbol{x}}_{k|l}$ is in the desired form and we only need to prove orthogonality to establish that it is the optimal estimate. To show orthogonality, we substitute for the optimal estimate, expand the product, then complete the square

$$E\left\{\left(\boldsymbol{x}_k - \hat{\boldsymbol{x}}_{k|l}\right)\boldsymbol{z}_{0:l}^T\right\} = E\left\{\left(\boldsymbol{x}_k - \boldsymbol{m}_x - C_{xz}C_{zz}^{-1}(\boldsymbol{z}_{0:l} - \boldsymbol{m}_z)\right)\boldsymbol{z}_{0:l}^T\right\}$$

$$= E\left\{\boldsymbol{x}_k\boldsymbol{z}_{0:l}^T\right\} - \boldsymbol{m}_x E\left\{\boldsymbol{z}_{0:l}^T\right\} - C_{xz}C_{zz}^{-1}E\left\{(\boldsymbol{z}_{0:l} - \boldsymbol{m}_z)(\boldsymbol{z}_{0:l} - \boldsymbol{m}_z)^T\right\} - C_{xz}C_{zz}^{-1}E\{\boldsymbol{z}_{0:l} - \boldsymbol{m}_z\}\boldsymbol{m}_z^T.$$

Recall two identities. The first is the cross-covariance formula for two random vectors

$$\begin{aligned} C_{xz} &= E\{(\boldsymbol{x}_k - \boldsymbol{m}_x)(\boldsymbol{z}_{0:l} - \boldsymbol{m}_z)^T\} \\ &= E\{\boldsymbol{x}_k \boldsymbol{z}_{0:l}^T\} - \boldsymbol{m}_x E\{\boldsymbol{z}_{0:l}^T\} = E\{\boldsymbol{x}\boldsymbol{z}_{0:l}^T\} - \boldsymbol{m}_x \boldsymbol{m}_z^T. \end{aligned}$$

The second is the definition of the covariance matrix for the set of measurements.

$$C_{zz} = E\{(\boldsymbol{z}_{0:l} - \boldsymbol{m}_z)(\boldsymbol{z}_{0:l} - \boldsymbol{m}_z)^T\}.$$

Hence, we have the orthogonality condition

$$E\{(\boldsymbol{x}_k - \hat{\boldsymbol{x}}_{k|l})\boldsymbol{z}_{0:l}^T\} = C_{xz} - C_{xz}C_{zz}^{-1}C_{zz} - [0] = [0].$$

To show that the estimator is unbiased, we take the expectation of the estimation error

$$E\{\boldsymbol{x}_k - \hat{\boldsymbol{x}}_{k|l}\} = E\{\boldsymbol{x}_k - \boldsymbol{m}_x - C_{xz}C_{zz}^{-1}(\boldsymbol{z}_{0:l} - \boldsymbol{m}_z)\} = 0.$$

The covariance matrix corresponds to the Gaussian conditional density of Chap. 2 and is given by

$$C_{x|z} = C_{xx} - C_{xz}C_{zz}^{-1}C_{zx}.$$

To prove uniqueness, we write optimal estimate in the form

$$\hat{\boldsymbol{x}}_{k|l} = K^o \boldsymbol{z}_{0:l} + \boldsymbol{b}^o$$

$$K^o = C_{xz}C_{zz}^{-1}, \boldsymbol{b}^o = \boldsymbol{m}_x - C_{xz}C_{zz}^{-1}\boldsymbol{m}_z.$$

Then we show that if the estimator with $(K, \boldsymbol{b})$ satisfies the orthogonality condition and is unbiased, then $(K, \boldsymbol{b}) = (K^o, \boldsymbol{b}^o)$. We assume that

$$E\{(\hat{\boldsymbol{x}}_{k|l} - K\boldsymbol{z}_{0:l} - \boldsymbol{b})\boldsymbol{z}_{0:l}^T\} = [0], E\{\hat{\boldsymbol{x}}_{k|l} - K\boldsymbol{z}_{0:l} - \boldsymbol{b}\} = 0.$$

The optimum linear estimate satisfies the same conditions of orthogonality and unbiasedness

$$E\{(\hat{\boldsymbol{x}}_{k|l} - K^o\boldsymbol{z}_{0:l} - \boldsymbol{b}^o)\boldsymbol{z}_{0:l}^T\} = [0], E\{\hat{\boldsymbol{x}}_{k|l} - K^o\boldsymbol{z}_{0:l} - \boldsymbol{b}^o\} = 0.$$

Subtracting one orthogonality condition from the other and substituting from the covariance identity

$$C_{zz} = E\{\boldsymbol{z}\boldsymbol{z}^T\} - \boldsymbol{m}_z\boldsymbol{m}_z^T$$

gives

$$\left(K - K^o\right)E\left\{z_{0:l}z_{0:l}^T\right\} + \left(b - b^o\right)E\left\{z_{0:l}^T\right\} = \left(K - K^o\right)\left(C_{zz} + m_z m_z^T\right) + \left(b - b^o\right)m_z^T = [0].$$

We rewrite the equation as

$$\begin{bmatrix} K - K^o & (K - K^o)m_z + b - b^o \end{bmatrix}\begin{bmatrix} C_{zz} \\ m_z^T \end{bmatrix} = [0].$$

Because equality to zero is true for any vector of measurements $z_{1:k}$, we have the identities

$$K - K^o = [0]$$

$$b - b^o = 0.$$

■

Note the estimator of Theorem 9.2 is the best linear unbiased estimator (BLUE) for Gaussian statistics. For other statistics, the optimal estimator may be nonlinear and may require knowledge of higher-order statistics and not just the mean and covariance.

9.1.2 Bayesian Estimation

Minimizing the minimum mean-square error is one of many options for selecting the optimum estimator. A more general approach that includes the mean-square error as a special case is Bayesian estimation where we minimize the posterior (given the data) expected value known as Bayes risk

$$J = E\left\{C\left(x - \hat{x}\right)\right\} = \int_{-\infty}^{\infty} C(\tilde{x})p_{x|z}(x|z)\mathrm{d}x. \tag{9.20}$$

Bayes risk is the expected value of Bayes loss $C(\tilde{x})$. Clearly, if Bayes loss is the square of the norm of the error vector, then Bayes estimation gives the minimum mean-square error estimator. More generally, we minimize the weighted mean square error with weighting matrix W

$$J = E\left\{\tilde{x}^T W\tilde{x}\right\} = \int_{-\infty}^{\infty} \tilde{x}^T W\tilde{x}\, p_{x|z}(x|z)\mathrm{d}x. \tag{9.21}$$

The following fact shows that we need not go beyond minimizing the minimum means square error in many cases, including the case of multivariate Gaussian statistics.

Fact: For a symmetric convex cost and a symmetric conditional density $p_{x|z}$ that is symmetric about the mean, all Bayesian estimates are the same.

The fact follows from the fact that symmetric costs will all have the same minimizing estimator if the conditional density is symmetric. For a rigorous proof see (Sorenson 1970).

9.2 Batch Versus Recursive Computation

So far, we assumed that all data are available at the time of computation. For online computation, data measured at each sampling point $\{z_1, z_2, z_3, \ldots\}$ and the estimator must be updated as more data becomes available. Batch computation requires repeating all the calculations if a new data point becomes available. Recursive computation updates the calculated estimator without the need to repeat all the previously performed calculations. Hence, recursive computation is more efficient for online computation as demonstrated by the following example.

Example 9.1 Mean Computation Compare batch computation of the mean of a progressive recording of random measurements $\{z_1, z_2, z_3, \ldots\}$ to recursive computation of the mean as more data becomes available.

Solution

Batch processing

- Initialization $i \leftarrow 1, z \leftarrow z_1, \hat{m}_i \leftarrow z$
- While $i < i_{final}$ do
  - Increment and read $i \leftarrow i + 1, z \leftarrow z_i$
  - Calculate estimate of mean at time i

$$\hat{m}_i \leftarrow \frac{1}{i}\sum_{j=1}^{i} z_j$$

- End

Recalculate when new data arrives.

Recursive computation:

- Initialization $i \leftarrow 1, z \leftarrow z_1, \hat{m}_i \leftarrow z$
- While $i < i_{final}$ do
  - Increment and read $i \leftarrow i + 1, z \leftarrow z_i$

– Calculate estimate of mean at time i

$$\hat{m}_i \leftarrow \frac{(i-1)\hat{m}_{i-1} + z_i}{i}$$

– End

The advantages of recursive computation are:

- Memory saving because only the mean is stored at any time point while for batch processing the entire data record must be stored.
- More efficient computation because, unlike batch processing, the mean computation of the earlier operation is not repeated.

9.3 The Discrete Kalman Filter

The Kalman filter is an algorithm for the optimal minimum mean-square recursive estimation of the state of a dynamic system. Because its estimates are also unbiased, the Kalman filter is the best linear unbiased estimator (BLUE) of the state. The algorithm requires:

- An initial state estimate and its error covariance matrix.
- Noisy measurements with known noise statistics.
- A state-space model of the dynamic system.

The state-space model is in the form

$$\boldsymbol{x}_{k+1} = \phi_k \boldsymbol{x}_k + \boldsymbol{w}_k \tag{9.22}$$

$$\boldsymbol{z}_k = H_k \boldsymbol{x}_k + \boldsymbol{v}_k, \tag{9.23}$$

where

$\boldsymbol{x}_k = n \times 1$ state vector at t_k.
$\phi_k = n \times n$ state-transition matrix at t_k.
$\boldsymbol{z}_k = m \times 1$ measurement vector at t_k
$H_k = m \times n$ measurement matrix at t_k.
The noise processes in the model are.
$\boldsymbol{w}_k = n \times 1$ zero-mean white Gaussian process noise vector at t_k

$$E\left\{\boldsymbol{w}_k \boldsymbol{w}_i^T\right\} = \begin{cases} Q_k, & i = k \\ 0, & i \neq k \end{cases} \tag{9.24}$$

$\boldsymbol{v}_k = m \times 1$ zero-mean white Gaussian measurement noise vector at t_k

$$E\{\boldsymbol{v}_k\boldsymbol{v}_i^T\} = \begin{cases} R_k, & i = k \\ 0, & i \neq k \end{cases} \tag{9.25}$$

The process and measurement noise are assumed uncorrelated to simplify the derivation

$$E\{\boldsymbol{w}_k\boldsymbol{v}_i^T\} = 0, \forall i, k \tag{9.26}$$

Theorem 9.2 tells us that for Gaussian noise the optimal estimator is affine in the measurements and provides a batch formula for its calculation. To replace the batch formula for the linear estimator with a recursive estimator, the form of the estimator must change to

$$\hat{\boldsymbol{x}}_{k|k} = K_{x,k}\hat{\boldsymbol{x}}_{k|k-1} + K_k\boldsymbol{z}_k$$

with the a priori estimate depending on the measurements $z_{1:k-1}$. For an unbiased estimator, assuming an initial unbiased state estimate, we must have

$$\hat{\boldsymbol{x}}_{k|k} = (I_n - K_kH_k)\hat{\boldsymbol{x}}_{k|k-1} + K_k\boldsymbol{z}_k = \hat{\boldsymbol{x}}_{k|k-1} + K_k\left[\boldsymbol{z}_k - H_k\hat{\boldsymbol{x}}_{k|k-1}\right] \tag{9.27}$$

This is easily proved by finding the expected value of the estimation error and is left as an exercise.

To derive the formula for the recursive estimator, we need to find the means and covariance matrices of $\boldsymbol{x}_k$ and $\boldsymbol{z}_k$ given the earlier measurements $\boldsymbol{z}_{0:k-1}$, as well as their cross-covariance. Assuming that the initial estimate is unbiased, we know that subsequent estimators of the form (9.27) will also be unbiased. This gives the expectation

$$\boldsymbol{m}_x = E\{\boldsymbol{x}_k|\boldsymbol{z}_{0:k-1}\} = \hat{\boldsymbol{x}}_{k|k-1} \tag{9.28}$$

and its covariance matrix is denoted as $C_{xx} = P_{k|k-1}$. The matrix is known as the error covariance because it is given by the expectation

$$P_{k|k-1} = E\left\{\tilde{\boldsymbol{x}}_{k|k-1}\tilde{\boldsymbol{x}}_{k|k-1}^T\right\} \tag{9.29}$$

where the error $\tilde{\boldsymbol{x}}_{k|k-1} = \boldsymbol{x}_k - \hat{\boldsymbol{x}}_{k|k-1}$ is zero-mean for an unbiased estimator.

The optimal estimate of the measurement vector $\boldsymbol{z}_k$ is the conditional mean

$$\hat{\boldsymbol{z}}_{k|k-1} = E\{\boldsymbol{z}_k|\boldsymbol{z}_{0:k-1}\} = E\{H_k\boldsymbol{x}_k + \boldsymbol{v}_k\} = H_k\hat{\boldsymbol{x}}_{k|k-1} \tag{9.30}$$

The difference between the measurement and its predicted value based on $\hat{\boldsymbol{x}}_{k|k-1}$ is known as the innovations or measurement residual

$$\tilde{\boldsymbol{z}}_k = \boldsymbol{z}_k - \hat{\boldsymbol{z}}_{k|k-1} = \boldsymbol{z}_k - H_k\hat{\boldsymbol{x}}_{k|k-1} = H_k\tilde{x}_{k|k-1} + \boldsymbol{v}_k \tag{9.31}$$

The innovations sequence is zero-mean, and $\hat{z}_{k|k-1}$ is therefore an unbiased estimator of the measurements. It is also a Gaussian white sequence. The proof of the properties of the innovations is left as an exercise.

Substituting for the measurement vector gives

$$\tilde{z}_k = H_k \boldsymbol{x}_k + \boldsymbol{v}_k - H_k \hat{\boldsymbol{x}}_{k|k-1}$$

$$= H_k \tilde{\boldsymbol{x}}_{k|k-1} + \boldsymbol{v}_k.$$

Note the that two terms in the innovations are orthogonal. This is because the measurement noise is white and the estimate $\hat{\boldsymbol{x}}_{k|k-1}$ is orthogonal to the noise vector $\boldsymbol{v}_k$ that appears in z_k, and because the measurement and process noise are orthogonal making $\hat{\boldsymbol{x}}_{k|k-1}$ orthogonal to $\boldsymbol{x}_k$. The covariance matrix of the measurement vector is the sum of two terms

$$\begin{aligned} C_{zz} &= E\left\{\tilde{z}_k \tilde{z}_k^T\right\} = H_k E\left\{\tilde{\boldsymbol{x}}_{k|k-1} \tilde{\boldsymbol{x}}_{k|k-1}^T\right\} H_k^T + E\left\{\boldsymbol{v}_k \boldsymbol{v}_k^T\right\} \\ &= H_k P_{k|k-1} H_k^T + R_k \end{aligned} \tag{9.32}$$

The cross-correlation of $\boldsymbol{x}_k$ and z_k is

$$\begin{aligned} C_{xz} &= E\left\{\tilde{\boldsymbol{x}}_k \tilde{z}_k^T\right\} = E\left\{\tilde{\boldsymbol{x}}_{k|k-1} \tilde{\boldsymbol{x}}_{k|k-1}^T\right\} H_k^T + E\left\{\tilde{\boldsymbol{x}}_k \boldsymbol{v}_k^T\right\} \\ &= P_{k|k-1} H_k^T \end{aligned} \tag{9.33}$$

In summary, the joint density of $\boldsymbol{x}_k$ and z_k given $z_{0:k-1}$ is the Gaussian distribution.

$$\mathcal{N}\left(\begin{bmatrix} \hat{\boldsymbol{x}}_{k|k-1} \\ H_k \hat{\boldsymbol{x}}_{k|k-1} \end{bmatrix}, \begin{bmatrix} P_{k|k-1} & P_{k|k-1} H_k^T \\ H_k P_{k|k-1} & H_k P_{k|k-1} H_k^T + R_k \end{bmatrix}\right) \tag{9.34}$$

Applying Theorem 2.6 for the step from time t_k to time t_{k+1} with the estimate updated with z_k gives the **corrector**

$$\hat{\boldsymbol{x}}_{k|k} = \hat{\boldsymbol{x}}_{k|k-1} + K_k\left[z_k - H_k \hat{\boldsymbol{x}}_{k|k-1}\right] \tag{9.35}$$

$$K_k = C_{xz} C_{zz}^{-1} = P_{k|k-1} H_k^T \left(H_k P_{k|k-1} H_k^T + R_k\right)^{-1} \tag{9.36}$$

The error covariance matrix for the estimate is

$$P_{k|k} = C_{x|z} = C_{xx} - C_{xz} C_{zz}^{-1} C_{zx} = P_{k|k-1} - K_k H_k P_{k|k-1}.$$

This gives the compact form of the error covariance

$$P_{k|k} = [I_n - K_k H_k] P_{k|k-1}. \tag{9.37}$$

The trace of the error covariance matrix is particularly interesting

$$\mathrm{tr}\left[P_{k|k}\right] = \mathrm{tr}\left[E\left\{\tilde{\boldsymbol{x}}_{k|k}\tilde{\boldsymbol{x}}_{k|k}^T\right\}\right] = \mathrm{tr}\left[E\left\{\tilde{\boldsymbol{x}}_{k|k}^T\tilde{\boldsymbol{x}}_{k|k}\right\}\right]$$

where we used the identity $\mathrm{tr}[AB] = \mathrm{tr}[BA]$ for any two compatible matrices A and B. Hence, the trace of the covariance matrix is the sum of the mean square errors

$$\mathrm{tr}\left[P_{k|k}\right] = \sum_{i=1}^{n} E\{\tilde{x}_i^2\} \tag{9.38}$$

The root mean square error is the square root

$$\sqrt{\mathrm{tr}\left[P_{k|k}\right]} = \sqrt{\sum_{i=1}^{n} E\{\tilde{x}_i^2\}} \tag{9.39}$$

The correction using the current measurement z_k, which gives the posterior estimate its name, can only be recursively applied in conjunction with a one-step predictor that provides an a priori estimate at time t_{k+1} based on measurements up to time k. The optimal estimate is

$$\begin{aligned}\hat{\boldsymbol{x}}_{k+1|k} &= E\{\boldsymbol{x}_{k+1}|\boldsymbol{z}_{0:k}\} = E\{\phi_k\boldsymbol{x}_k + \boldsymbol{w}_k|\boldsymbol{z}_{0:k}\} \\ &= \phi_k\hat{\boldsymbol{x}}_{k|k}.\end{aligned}$$

Its covariance matrix is

$$\begin{aligned}P_{k+1|k} &= E\left\{\tilde{\boldsymbol{x}}_{k+1|k}\tilde{\boldsymbol{x}}_{k+1|k}^T\right\} \\ &= \phi_k E\left\{\tilde{\boldsymbol{x}}_{k|k}\tilde{\boldsymbol{x}}_{k|k}^T\right\}\phi_k^T + E\left\{\boldsymbol{w}_k\boldsymbol{w}_k^T\right\},\end{aligned}$$

that is,

$$P_{k+1|k} = \phi_k P_{k|k}\phi_k^T + Q \tag{9.40}$$

The predictor–corrector set of equations requires initial estimates $\hat{\boldsymbol{x}}_0$ and its covariance $\boldsymbol{P}_0$ but then be iterated to recursively estimate the state as shown in Fig. 9.1. Note that the corrector equations are computed without using the optimal gain expression.

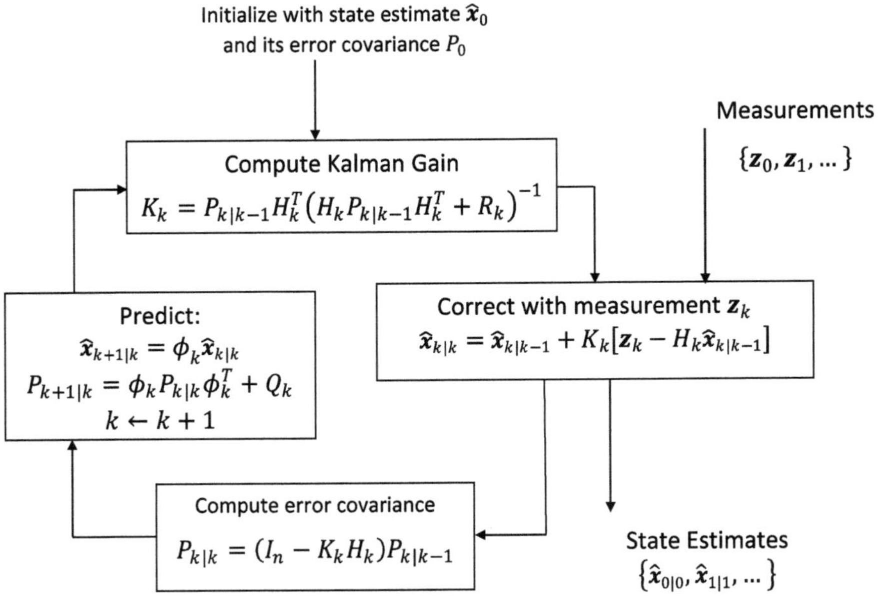

Fig. 9.1 Block diagram for the discrete-time Kalman filter

9.4 Expressions for the Error Covariance

The error covariance for the corrector can be written in different forms which, while algebraically equivalent, have different characteristics in numerical computation because of computational errors. Starting with (9.36), we derive four expressions for the error covariance. Substituting the optimal gain expression gives

$$P_{k|k} = P_{k|k-1} - P_{k|k-1}H_k^T\left(H_kP_{k|k-1}H_k^T + R_k\right)^{-1}H_kP_{k|k-1}. \tag{9.41}$$

We can also obtain an expression from basic principles starting with the estimation error

$$\tilde{\boldsymbol{x}}_{k|k} = \boldsymbol{x}_k - \hat{\boldsymbol{x}}_{k|k} = \boldsymbol{x}_k - (I_n - K_kH_k)\hat{\boldsymbol{x}}_{k|k-1} - K_k\boldsymbol{z}_k$$

$$= (I_n - K_kH_k)\tilde{\boldsymbol{x}}_{k|k-1} + K_k\boldsymbol{v}_k.$$

The two terms are orthogonal because of the assumption that noise is white and process noise and measurement noise are uncorrelated. This gives the Joseph form for the error covariance

$$P_{k|k} = E\left\{\tilde{\boldsymbol{x}}_{k|k}\tilde{\boldsymbol{x}}_{k|k}^T\right\} = (I_n - K_kH_k)P_{k|k-1}(I_n - K_kH_k)^T + K_kR_kK_k^T \tag{9.42}$$

The symmetrical structure of the Joseph form gives it the best numerical computation properties among all forms of the error covariance matrix. In addition, the Joseph form is applicable to any gain K_k because it was derived from the general error expressions without invoking the optimality condition. Expanding the Joseph form gives

$$P_{k|k} = P_{k|k-1} - K_k H_k P_{k|k-1} - P_{k|k-1} H_k^T K_k^T + K_k \big(H_k P_{k|k-1} H_k^T + R_k\big) K_k^T \tag{9.43}$$

Because the optimal gain includes a symmetric matrix

$$K_k = P_{k|k-1} H_k^T \underbrace{\left(H_k P_{k|k-1} H_k^T + R_k\right)^{-1}}_{\text{symmetric}},$$

the last three terms of the Joseph form are equal

$$P_{k|k-1} H_k^T K_k^T = \underbrace{P_{k|k-1} H_k^T \left(H_k P_{k|k-1} H_k^T + R_k\right)^{-1}}_{K_k} H_k P_{k|k-1}$$

$$= K_k \big(H_k P_{k|k-1} H_k^T + R_k\big) \underbrace{\left(H_k P_{k|k-1} H_k^T + R_k\right)^{-1} H_k P_{k|k-1}}_{K_k^T}$$

Cancel two of the three equal terms gives the forms

$$P_{k|k} = P_{k|k-1} - K_k \big(H_k P_{k|k-1} H_k^T + R_k\big) K_k^T$$

$$= (I_n - K_k H_k) P_{k|k-1}.$$

Substituting from (9.40) in Joseph form has

$$P_{k|k} = (I_n - K_k H_k)\big(\phi_k P_{k-1|k-1} \phi_k^T + Q\big)(I_n - K_k H_k)^T + K_k R_k K_k^T$$

$$\begin{aligned} &= (I_n - K_k H_k)\phi_k P_{k-1|k-1} (I_n - K_k H_k)^T \phi_k^T \\ &\quad + (I_n - K_k H_k) Q (I_n - K_k H_k)^T + K_k R_k K_k^T. \end{aligned}$$

The resulting equation

$$P_{k|k} = \overline{\phi}_k P_{k-1|k-1} \overline{\phi}_k^T + \overline{Q}_k \tag{9.44}$$

$$\overline{\phi} = (I_n - K_k H_k)\phi_k$$

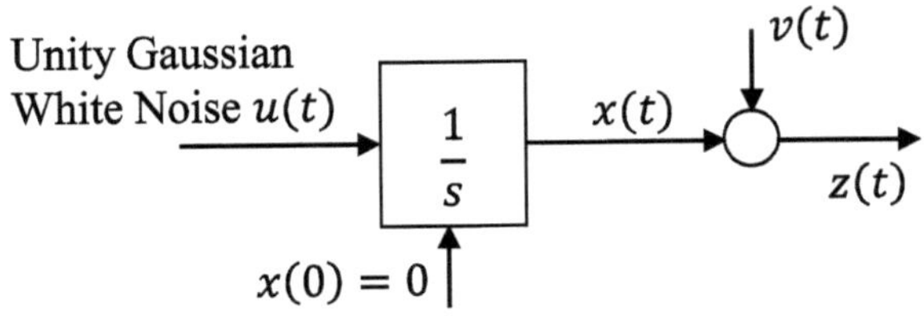

Fig. 9.2 Block diagram of Wiener process

$$\overline{Q}_k = (I_n - K_k H_k)Q(I_n - K_k H_k)^T + K_k R_k K_k^T$$

is known as the Lyapunov equation and is valid for any predictor–corrector and not just for the optimal gain. If the optimal gain is substituted in (9.39), then the result is substituted in (9.40), we get the Riccati equation

$$P_{k+1|k} = \phi_k \Big[P_{k|k-1} - P_{k|k-1} H_k^T \left(H_k P_k^- H_k^T + R_k \right)^{-1} H_k P_{k|k-1} \Big] \phi_k^T + Q_k \quad (9.45)$$

We emphasize that while all expressions are equal, in terms of numerical computation, they behave differently with the Joseph form being the preferred form although it appears more complicated than other forms.

Example 9.2 Example: Wiener Process Design a Kalman filter to estimate the state of a Wiener process of Fig. 9.2 with measurement noise variance of 0.25, and show the calculation of the estimator, Kalman gain and error covariance matrices for the first two iterations. Use a unity sampling period.

Solution

The Wiener process is the integral of unity white noise with zero initial condition. The shaping filter is an integrator with transfer function and impulse response

$$G(s) = \frac{1}{s} \Leftrightarrow g(t) = 1.$$

The system differential equation

$$\dot{x}(t) = u(t)$$

Is discretized to get

$$x_{k+1} = x_k + w_k, \ \phi_k = 1.$$

The measurement equation is

$$z_k = x_k + v_k, \ H_k = 1.$$

The covariance matrix of the process noise is

$$Q_k = E\{w_k^2\} = E\left\{\left(\int_0^{\Delta t} 1.u(\xi)\mathrm{d}\xi\right)\left(\int_0^{\Delta t} 1.u(\eta)\mathrm{d}\eta\right)\right\}$$

$$= \int_0^1\int_0^1 E\{u(\xi)u(\eta)\}\mathrm{d}\xi\,\mathrm{d}\eta$$

$$= \int_0^1\int_0^1 \delta(\xi-\eta)\mathrm{d}\xi\,\mathrm{d}\eta = 1.$$

Because the initial state is known exactly $\hat{\boldsymbol{x}}_0 = 0$, the error covariance is $P_0 = 0$ and the variance of the measurement noise is $R = 0.25$.

With the initial conditions, we calculate the initial Kalman gain

$$K_0 = P_0 H^T \left(H P_0 H^T + R_0\right)^{-1} = 0/(0 + 1/4) = 0.$$

The state estimate and the corresponding error variance are

$$\hat{x}_{0|0} = \hat{x}_0 + K_0\left(z_0 - H\hat{x}_0\right) = 0 + 0(z_0 - 0) = 0$$

$$P_{0|0} = (I - K_0 H)P_0 = (1-0)(0) = 0.$$

The predicted state at time 0 and the corresponding error variance are

$$\hat{x}_{1|0} = \phi\hat{x}_{0|0} = 1.0 = 0$$

$$P_{1|0} = \phi P_{0|0}\phi^T + Q = 1.(0).1 + 1 = 1.$$

The gain at $k = 1$ is

$$K_1 = P_{1|0} H_1^T \left(H_1 P_{1|0} H_1^T + R_1\right)^{-1} = 1/(1 + 1/4)^{-1} = 4/5.$$

We update the estimate and obtain its error variance

$$x_{1|1} = x_{1|0} + K_1\left(z_1 - Hx_{1|0}\right) = 0 + \frac{4}{5}(z_1 - 0) = \frac{4}{5}z_1$$

$$P_{1|1} = (I - K_1 H)P_{1|0} = (1 - 4/5)(1) = 1/5.$$

The predicted estimate at time $k = 2$ and its error covariance are

$$x_{2|1} = \phi x_{1|1} = 1.(4/5)z_1 = (4/5)z_1$$

$$P_{2|1} = \phi P_{1|1}\phi^T + Q = 1.(1/5).1 + 1 = 6/5.$$

■

Example 9.3 With $\Delta t = 0.2$, simulate the Wiener process of Example 9.2 and plot the state versus time together with its estimate and the 2-sigma bounds of the estimate, and plot the mean square estimation error.

Solution

The system can be simulated using MATLAB or Simulink. The error variance of the estimate quickly approaches a steady-state level. The optimal gain in the steady state is $K = 0.8284$ (Figs. 9.3 and 9.4).

Example 9.4 Design a Kalman filer for a general Gauss–Markov process.

Solution

The autocorrelation and PSD of the process are

$$R_{XX}(\tau) = \sigma^2 e^{-\beta|\tau|}, S_{XX}(s) = \frac{2\sigma^2\beta}{-s^2 + \beta^2}.$$

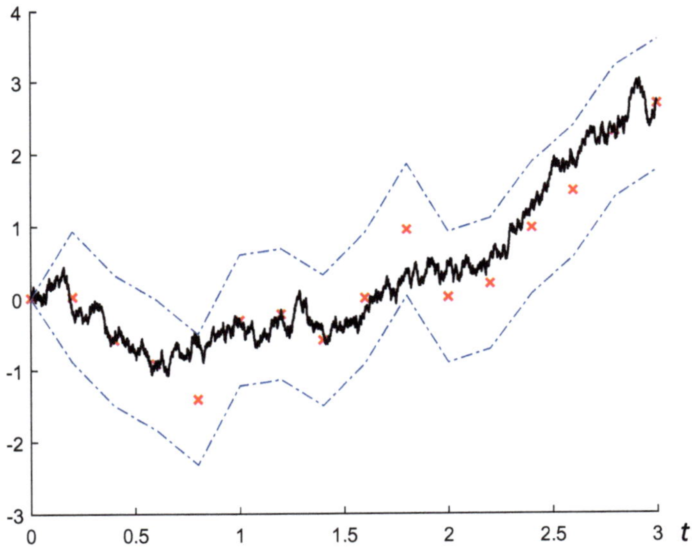

Fig. 9.3 Simulation results for the Wiener process

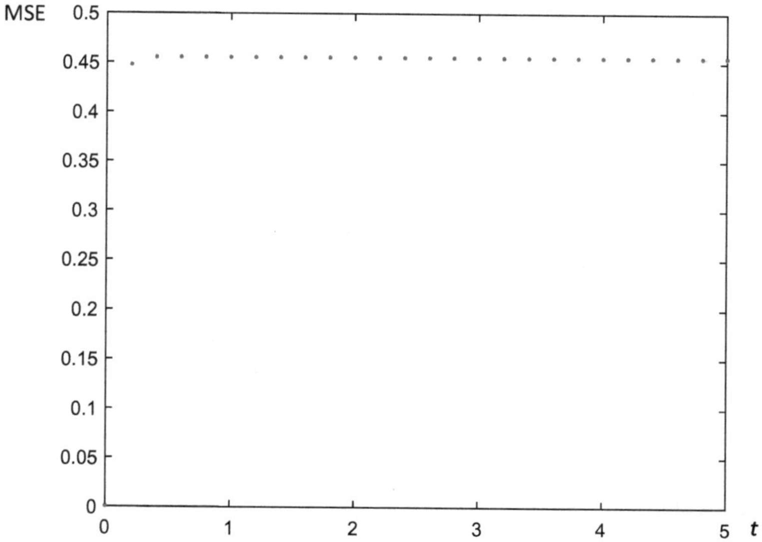

Fig. 9.4 Mean-square estimation error of the Wiener process

Spectral factorization gives the shaping filter and the impulse response

$$G(s) = L(s) = \frac{\sqrt{2\sigma^2\beta}}{s+\beta},\ g(t) = \sqrt{2\sigma^2\beta}e^{-\beta t}.$$

The system can be represented as the state-space model

$$\dot{x} = -\beta x + \sqrt{2\sigma^2\beta}u,\ \ y = x.$$

The covariance matrix for the process noise is

$$Q = \int_0^{\Delta t} e^{A\xi}BB^T e^{A^T\xi}\mathrm{d}\xi = \int_0^{\Delta t}\left(e^{-\beta\Delta t}\sqrt{2\sigma^2\beta}\right)^2\mathrm{d}\xi = \sigma^2\left(1 - e^{-2\beta\Delta t}\right).$$

The discrete state-space model with sampling period Δt is

$$x_{k+1} = e^{-\beta\Delta t}x_k + w_k$$

$$z_k = x_k + v_k.$$

To select initial conditions for the filter, we use the mean and error covariance of the process. The process is zero-mean because

$$\lim_{\tau\to\infty} R_{xx}(\tau) = 0 \Rightarrow E\{x(t)\} = 0.$$

The variance is equal to the mean square of the process

$$R_{xx}(0) = E\{x^2(t)\} = \sigma^2 = var\{x(t)\}.$$

Hence, the filter is initialized with

$$\hat{x}_0 = 0,\ P_0 = \sigma^2.$$

The Kalman gain is given by

$$K_k = P_{k|k-1}H_k^T\left(H_k P_{k|k-1} H_k^T + R_k\right)^{-1} = \frac{P_{k|k-1}}{P_{k|k-1} + R}.$$

The corrected estimate is

$$\hat{\boldsymbol{x}}_{k|k} = \hat{\boldsymbol{x}}_{k|k-1} + K_k\left[z_k - H_k\hat{\boldsymbol{x}}_{k|k-1}\right] = \hat{\boldsymbol{x}}_{k|k-1} + \frac{P_{k|k-1}}{P_{k|k-1} + R_k}\left[z_k - \hat{\boldsymbol{x}}_{k|k-1}\right]$$

and its error variance is

$$P_{k|k} = [1 - K_k H_k]P_{k|k-1} = \left[1 - \frac{P_{k|k-1}}{P_{k|k-1} + R_k}\right]P_{k|k-1} = \frac{P_{k|k-1}R}{P_{k|k-1} + R}.$$

The predictor is

$$\hat{\boldsymbol{x}}_{k+1|k} = \phi_k\hat{\boldsymbol{x}}_{k|k} = e^{-\beta\Delta t}\hat{\boldsymbol{x}}_{k|k},$$

and its error variance is

$$P_{k+1|k} = \phi_k P_{k|k}\phi_k^T + Q = e^{-2\beta\Delta t}P_{k|k} + Q.$$

■

Example 9.5 Simulate the Gauss–Markov process of Example 9.4 and plot the state versus time together with its estimate and the 2-sigma bounds of the estimate, and plot the mean square estimation error. Use $R = 0.25$; variance $\sigma^2 = 0.1$ and $\beta = 0.1\,s^{-1}$ and a sampling period $\Delta t = 0.2$ s.

Solution

The system can be simulated using MATLAB or Simulink. The error variance of the estimate quickly approaches a steady-state level. The optimal gain in the steady state is $K = 0.8274$ (Figs. 9.5 and 9.6).

■

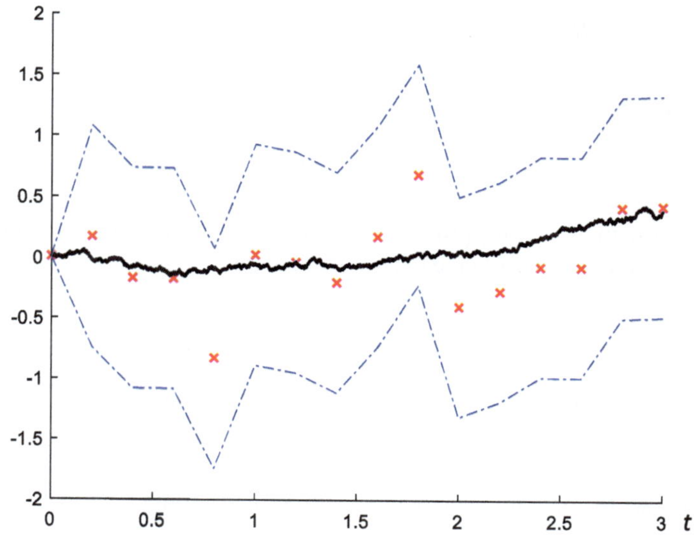

Fig. 9.5 Simulation results for the Gauss–Markov process

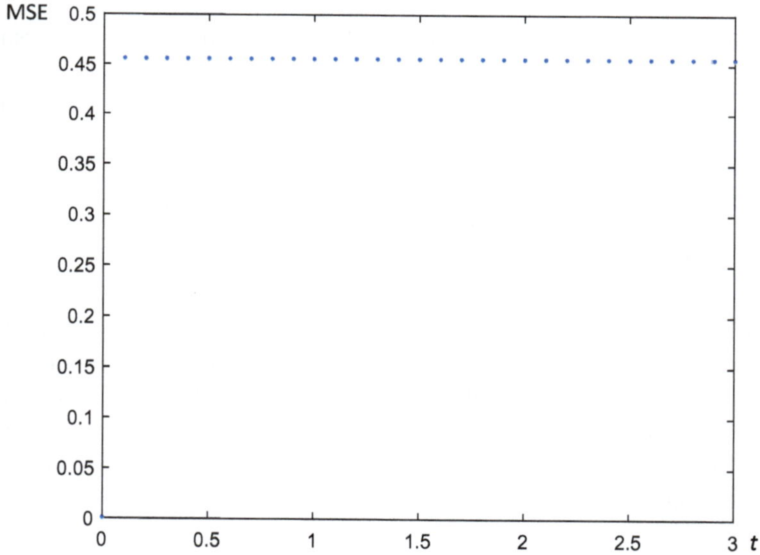

Fig. 9.6 Mean-square estimation error of the Gauss-Markov process

9.4.1 Deterministic Input

If the system has a deterministic input $\boldsymbol{u}_d(t)$, the state equation becomes

$$\dot{\boldsymbol{x}}(t) = A\boldsymbol{x}(t) + B_d\boldsymbol{u}_d(t) + B\boldsymbol{u}(t), \quad \boldsymbol{x}_0 \tag{9.46}$$

For a linear system, we use the principle of superposition to obtain the response. Two options are available to handle the deterministic input

(a) Add the zero-state deterministic response to the predictor equation

$$\hat{\boldsymbol{x}}_{k+1|\boldsymbol{k}} = \phi_k\hat{\boldsymbol{x}}_{k|\boldsymbol{k}} + \int_{t_k}^{t_{k+1}} \phi(t_{k+1}, \tau)B_d(\tau)\boldsymbol{u}_d\mathrm{d}\tau \tag{9.47}$$

(b) (i) Subtract the deterministic output from z_k in the corrector equation

$$\hat{\boldsymbol{x}}_{k|k} = \hat{\boldsymbol{x}}_{k|k-1} + K_k\left[z_k - H_k\hat{\boldsymbol{x}}_{k|k-1} - C\int_{t_k}^{t_{k+1}} \phi(t_{k+1}, \tau)B_d(\tau)\boldsymbol{u}_d\mathrm{d}\tau\right] \tag{9.48}$$

(b) (Compute $\boldsymbol{x}_{k,d}$ separately and add to the KF estimate to obtain the state estimate.

$$\hat{\boldsymbol{x}}_{k,\mathbf{tot}} = \hat{\boldsymbol{x}}_{k|k} + \boldsymbol{x}_{k,d} \tag{9.49}$$

9.4.2 Separation Principle

Kalman filters can be used as state estimators to provide feedback control for a dynamic system. In this state estimator feedback, the estimate is used in place of state measurements to compute the feedback control law. For a linear system with linear state feedback, it can be shown the state estimator and the controller can be designed separately. This is known as the separation principle or the principle of uncertainty equivalence.

9.5 Information Filter

The Kalman filter derived in is known as the covariance filter because it propagates the covariance matrix of the estimator. An alternative formulation propagates the information matrix, which is the inverse of the error covariance matrix. The formulation changes only two equations, the corrector covariance equation and the Kalman gain

$$\left(P_{k|k}\right)^{-1} = \left(P_{k|k-1}\right)^{-1} + H_k^T R_k^{-1} H_k \tag{9.50}$$

$$K_k = P_{k|k} H_k^T R_k^{-1} \tag{9.51}$$

The equations for the information filter are shown in Fig. 9.7. The figure shows that at least two additional matrix inversions are needed to implement the filter when compared to the covariance filter. This makes the filter inefficient, and as a result, the information filter is only used to initialize the covariance filter if the initial covariance matrix is very large, or in theoretical studies.

To derive the equations, we assume that the covariance matrix is positive definite and use the matrix inversion lemma, assuming nonsingular A_1,

$$\left[A_1 + A_2 A_4^{-1} A_3\right]^{-1} = A_1^{-1} - A_1^{-1} A_2 \left[A_4 + A_3 A_1^{-1} A_2\right]^{-1} A_3 A_1^{-1}$$

with the matrices defined as

$$A_1 = \left(P_{k|k-1}\right)^{-1}, A_2 = H^T, A_3 = H, A_4 = R.$$

Substituting in (9.43) gives

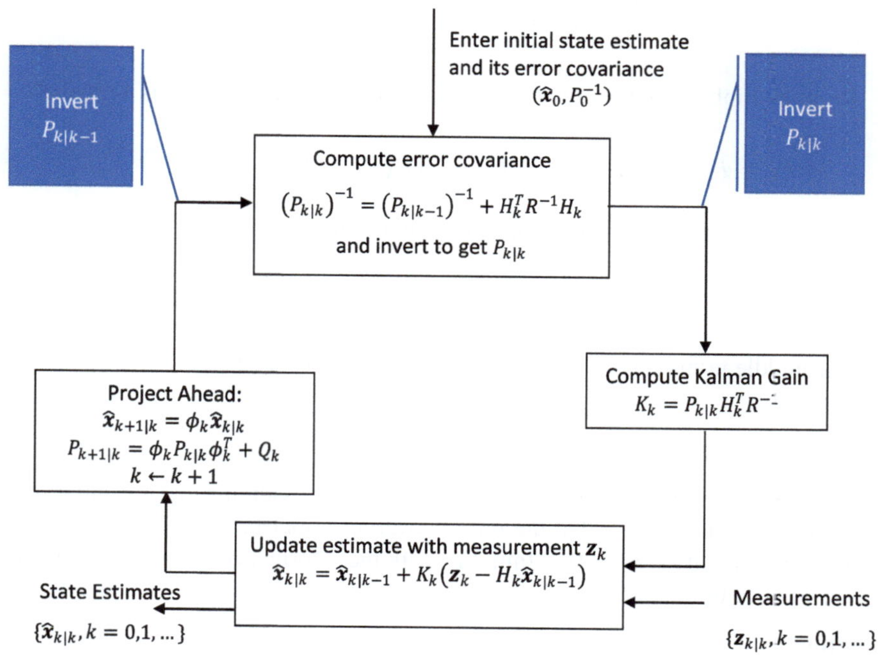

Fig. 9.7 Information filter

$$P_{k|k} = P_{k|k-1} - P_{k|k-1} H_k^T \left(H_k P_{k|k-1} H_k^T + R_k\right)^{-1} H_k P_{k|k-1}$$

$$= \left[\left(P_{k|k-1}\right)^{-1} + H_k^T R_k^{-1} H_k\right]^{-1}.$$

Inverting the matrix gives the covariance update expression

$$\left(P_{k|k}\right)^{-1} = \left(P_{k|k-1}\right)^{-1} + H_k^T R_k^{-1} H_k.$$

To derive the gain expression, we start with the expression

$$K_k = P_{k|k-1} H_k^T \left(H_k P_{k|k-1} H_k^T + R_k\right)^{-1}$$

then substitute for the error covariance matrix

$$P_{k|k} = \left[\left(P_{k|k-1}\right)^{-1} + H_k^T R_k^{-1} H_k\right]^{-1}$$

to obtain

$$K_k = P_{k|k} \underbrace{\left[\left(P_{k|k-1}\right)^{-1} + H^T R^{-1} H\right]}_{\left(P_{k|k}\right)^{-1}} P_{k|k-1} H^T \left(H P_{k|k-1} H^T + R\right)^{-1}.$$

Right multiply with the term $P_{k|k-1} H^T$ then rewrite the last term as a product

$$K_k = P_{k|k} H^T \left[I_m + R^{-1} H P^- H^T\right]\left[R\left(R^{-1} H P_{k|k-1} H^T + I_m\right)\right]^{-1}.$$

Invert the product then cancel and we have the answer

$$K_k = P_{k|k} H^T \left[I_m + R^{-1} H P_{k|k-1} H^T\right]\left[R^{-1} H P_{k|k-1} H^T + I_m\right]^{-1} R^{-1}.$$

The middle terms cancel, and we have the desired result (Fig. 9.7)

$$K_k = P_{k|k} H_k^T R_k^{-1}$$

Example 9.6 Repeat the simulation of the Wiener filter assuming that the initial state is completely unknown so that the initial covariance matrix is $P_0 = 10^{12}$ and the initial state estimate is $x_0 = 0.2$. Initialize with the information filter then switch to the covariance filter after one iteration.

Solution

The simulation results are shown in Fig. 9.8. The Kalman filter. Initialization with the information filter allows the simulation to run successfully in spite of the large initial variance. The Kalman gain quickly reaches a steady-state constant value of $K = 0.8944$ and a steady-state root mean square error of 0.118.

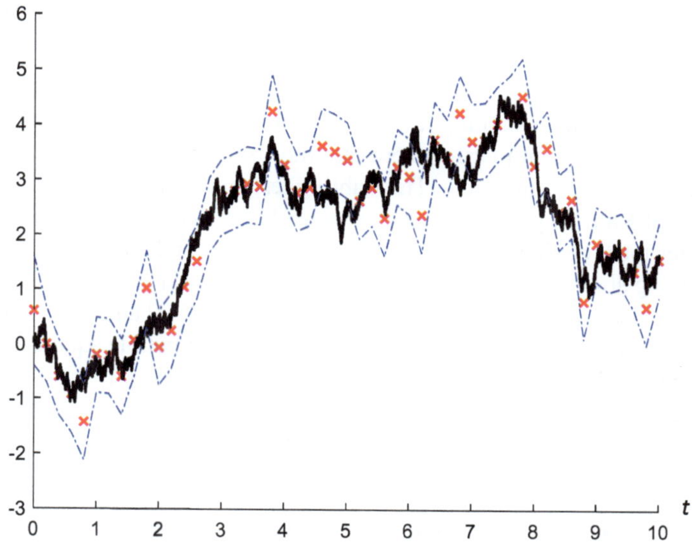

Fig. 9.8 Simulation results for the Wiener process with no knowledge of the initial state

9.6 Steady-State Kalman Filter and Stability

Stability guarantees that the Kalman filter estimates will not diverge from the estimated state. For a linear estimator, stability can be tested by rewriting the predictor

$$\hat{\boldsymbol{x}}_{k+1|k} = \phi_k \hat{\boldsymbol{x}}_{k|k}, \phi_k = \phi(k+1, k)$$

or the corrector

$$\hat{\boldsymbol{x}}_{k|k} = [I_n - K_k H_k]\hat{\boldsymbol{x}}_{k|k-1} + K_k z_k$$

as a recursion. For the predictor, eliminating $\hat{\boldsymbol{x}}_{k|k}$ gives

$$\hat{\boldsymbol{x}}_{k+1|k} = \phi_k [I_n - K_k H_k]\hat{\boldsymbol{x}}_{k|k-1} + \phi_k K_k z_k \tag{9.52}$$

Alternatively, we eliminate $\hat{\boldsymbol{x}}_{k|k-1}$ to obtain

$$\hat{\boldsymbol{x}}_{k+1|k+1} = \left[I_n - K_{k+1} H_{k+1}\right]\phi_k \hat{\boldsymbol{x}}_{k|k} + K_{k+1} z_{k+1} \tag{9.53}$$

The stability of the filter depends on the properties of the matrix

$$\overline{\phi}_k = \phi_k (I_n - K_k H_k), \phi_k = \phi(k+1, k) \tag{9.54}$$

or the matrix

$$\overline{\phi}_k = (I_n - K_{k+1} H_{k+1})\phi_k \tag{9.55}$$

The equation is useful in the case where the matrices K, H and ϕ in (9.55) are constant. This provides a stability test for linear time-invariant systems in the steady state with a constant gain matrix where the filter is stable if the eigenvalues of the matrix $\overline{\phi}$ are inside the unit circle. The stability test applies to any gain and not only to the optimal Kalman gain. However, we must have a known gain to test stability and the filter must be first designed before its stability can be determined. The test tells us nothing about transient error dynamics for a time-varying gain and only tests steady-state behavior. Note that the two matrices $\phi[I_n - KH]$ and $[I_n - KH]\phi$ have the same eigenvalues because, in general for any two compatible square matrices A and B the eigenvalues of AB are the same as the eigenvalues of BA.

To characterize the time evolution of the solution, we need to characterize the behavior of the state-transition matrix using the concept of uniform exponential stability.

Definition 9.1 Uniform Exponential Stability (UES) A linear system is said to be uniformly exponentially stable if there exists a positive constant $\gamma > 0$, and a constant λ, $0 < \lambda < 1$, such that

$$\|\boldsymbol{x}_k\| \le \gamma\lambda^{k-k_0}\left\|\boldsymbol{x}_{k_0}\right\|, \forall k_0, \boldsymbol{x}_{k_0} \tag{9.56}$$

Figure 9.9 depicts the evolution of $\boldsymbol{x}_k$ for a uniformly exponentially stable system. The word uniform signifies that the constants in the definition are independent of time and can be dropped when dealing with LTI systems. It can be shown that uniform exponential stability is equivalent to uniform asymptotic stability for linear time-varying system (see Rugh, p. 106).

Theorem 9.3 *A linear system is UES if and only if*

$$\|\Phi(k, k_0)\|_i \le \gamma\lambda^{k-k_0}, \gamma > 0, 0 < \lambda < 1, \forall k \ge k_0, \forall k_0 \tag{9.57}$$

Proof We need to establish the equivalence

$$\|\Phi(k, k_0)\|_i \le \gamma\lambda^{k-k_0} \Leftrightarrow |\mathbf{x}(k)| \le \gamma\lambda^{k-k_0}\left\|\mathbf{x}_{k_0}\right\|$$

$$\gamma > 0, 0 < \lambda < 1, \forall k \ge k_0, \forall k_0$$

Proof of $\Rightarrow$ (sufficiency)

$$\boldsymbol{x}_k = \Phi(k, k_0)\boldsymbol{x}(k_0), \forall k_0, \boldsymbol{x}_{k_0}$$

Using norm inequalities and the condition of the theorem gives

$$\|\boldsymbol{x}_k\| \le \|\Phi(k, k_0)\|_i \left\|\boldsymbol{x}_{k_0}\right\| \le \gamma\lambda^{k-k_0}\left\|\boldsymbol{x}_{k_0}\right\|.$$

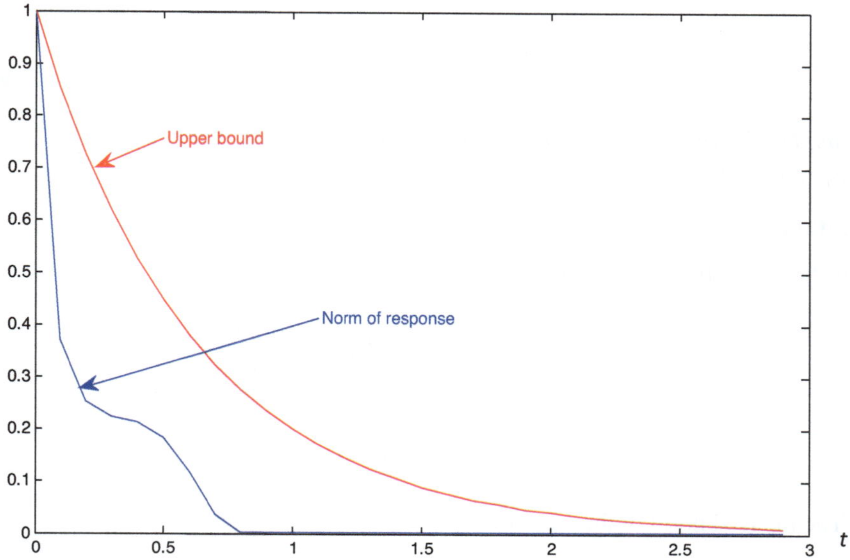

Fig. 9.9 Response bounded above by an exponential decay curve

The bound on the state-transition matrix implies that

$$\mathcal{L}\boldsymbol{im}_{k\to\infty}|\Phi(k,0)|_i = 0$$

Proof of $\Leftarrow$ (necessity)
Assume uniformly exponentially stable:

$$\exists\gamma > 0, \lambda, 0 < \lambda < 1, \text{spt.}\|\mathbf{x}(k)\| \le \gamma\lambda^{k-k_0}\|\mathbf{x}(k_0)\|$$

Consider a state $\boldsymbol{x}_a$ such that

$$\|\boldsymbol{x}_a\| = 1, \|\Phi(k,k_0)\boldsymbol{x}_a\| = \|\Phi(k,k_0)\|_i$$

Setting the initial state $\boldsymbol{x}(k_0) = \boldsymbol{x}_a$

$$\|\boldsymbol{x}_k\| = \|\Phi(k,k_0)\boldsymbol{x}_a\| = \|\Phi(k,k_0)\|_i \times 1$$

By assumption, the state-transition matrix satisfies

$$\|\Phi(k,k_0)\|_i \le \gamma\lambda^{k-k_0}, k \ge k_0$$

Hence, we have the exponential stability condition

$$\boldsymbol{x}_k \leq \gamma \lambda^{k-k_0} \boldsymbol{x}_{k_0}.$$

■

Example 9.7 Determine the stability of the filter for the following matrices $\phi_k = 0.2e^{-0.1k} I_2$, $K = \begin{bmatrix} 0.2 \; 0.4 \end{bmatrix}^T$, $H = \begin{bmatrix} 1 \; 4 \end{bmatrix}$.

Solution

The state-transition matrix for the filter is

$$\overline{\phi}_k = (I_2 - KH)\phi_k = 0.2e^{-0.1k}\left(I_2 - \begin{bmatrix} 0.2 \\ 0.4 \end{bmatrix}\begin{bmatrix} 1 \; 4 \end{bmatrix}\right)$$

$$= 0.2e^{-0.1k}\begin{bmatrix} 0.16 & -0.16 \\ -0.08 & -0.12 \end{bmatrix}.$$

The induced norm of the matrix is

$$\|\Phi(k, k_0)\|_i = 0.2e^{-0.1(k-k_0)}\left\| \begin{matrix} 0.16 & -0.16 \\ -0.08 & -0.12 \end{matrix} \right\|_i$$

$$\leq 0.2291\left(e^{-0.1}\right)^{k-k_0}, 0.2291 > 0, 0 < e^{-0.1} < 1, \forall k \geq k_0, \forall k_0.$$

■

If the filter gain converges to a constant K, or if a filter with constant gain is used, then the state-transition matrix is constant and stability can be tested by determining its eigenvalues. The filter is asymptotically stable if all the eigenvalues are inside the unit circle.

Example 9.8 Determine the stability of the filter for the following matrices

$$\phi = 0.2I_2, K = \begin{bmatrix} 0.2 \; 0.4 \end{bmatrix}^T, H = \begin{bmatrix} 1 \; 4 \end{bmatrix}.$$

Solution

The state-transition matrix for the filter is

$$\overline{\phi} = (I_2 - KH)\phi = 0.2\left(I_2 - \begin{bmatrix} 0.2 \\ 0.4 \end{bmatrix}\begin{bmatrix} 1 \; 4 \end{bmatrix}\right) = \begin{bmatrix} 0.16 & -0.16 \\ -0.08 & -0.12 \end{bmatrix}.$$

The eigenvalues of the matrix are {0.2, 0.16}, and they are inside the unit circle. Therefore, the filter is asymptotically stable.

■

9.6.1 Discrete Lyapunov Equation

The stability analysis for time-varying filters can be based on the error covariance

$$P_{k+1|k} = \phi_k P_{k|k} \phi_k^T + Q_k.$$

Substituting for $P_{k|k}$ from the Joseph form

$$P_{k|k} = (I_n - K_k H_k) P_{k|k-1} (I_n - K_k H_k)^T + K_k R_k K_k^T$$

gives the Lyapunov equation

$$P_{k+1|k} = \overline{\phi}_k P_{k|k-1} \overline{\phi}_k^T + \overline{Q}_k \tag{9.58}$$

with the matrices

$$\overline{\phi}_k = \phi_k (I_n - K_k H_k) \tag{9.59}$$

$$\overline{Q}_k = \phi_k K_k R_k K_k^T \phi_k^T + Q_k \tag{9.60}$$

Substituting for $P_{k|k}$ in the Joseph form gives the Lyapunov equation

$$P_{k|k} = \overline{\phi}_k P_{k-1|k-1} \overline{\phi}_k^T + \overline{Q}_k \tag{9.61}$$

$$\overline{\phi}_{k-1} = (I_n - K_k H_k) \phi_{k-1} \tag{9.62}$$

$$\overline{Q}_{k-1} = \phi_{k-1} K_k R_k K_k^T \phi_{k-1}^T + Q_{k-1} \tag{9.63}$$

This equation applies for any gain K and not just for the optimal gain K. The solution of the Lyapunov equation can be obtained by induction

$$P_{k|k-1} = \Phi(k,0) P_0 \Phi^T(k,0) + \sum_{i=0}^{k-1} \Phi(k,i+1) \overline{Q}_i \Phi^T(k,i+1) \tag{9.64}$$

$$\Phi(k,i) = \overline{\phi}_{k-1} \overline{\phi}_{k-2} \ldots \overline{\phi}_i,\ \Phi(k,k) = I_n \tag{9.65}$$

For constant $\overline{\phi}$, the state-transition matrix is the exponential $\Phi(k,i) = \overline{\phi}^{k-i}$.

$$P_{k+1k} = \overline{\phi}_k P_{kk-1} \overline{\phi}_k^T + \overline{Q}_k,\ \overline{Q}_{ki} < \beta_Q, \forall k \tag{9.66}$$

To analyze the behavior of the solution of the Lyapunov equation, we need the following theorems.

Theorem 9.4 *The linear system*

$$\boldsymbol{x}_k = \Phi(k, k_0)\boldsymbol{x}_{k_0} \tag{9.67}$$

is UES if and only if there exist a finite positive constant β s. t.

$$\sum_{k=k_0}^{\infty} \Phi(k, k_0)_i \leq \beta, k_0 \geq 0 \tag{9.68}$$

Proof **Proof (Necessity)**:

If the system is UES, then by Theorem 9.3 the state-transition matrix satisfies

$$\Phi(k, k_0)_i \leq \gamma\lambda^{k-k_0}, \forall k \geq k_0, \forall k_0$$

$$\gamma > 0, 0 < \lambda < 1.$$

Hence, we have the summation

$$\sum_{k=k_0}^{\infty} \|\Phi(k, k_0)\|_i \leq \sum_{k=k_0}^{\infty} \gamma\lambda^{k-k_0} = \sum_{k=0}^{\infty} \gamma\lambda^k$$
$$= \frac{\gamma}{1-\lambda} = \beta$$

Proof **Proof (Sufficiency)**:

The response of the system satisfies

$$\|\boldsymbol{x}_k\| \leq \|\Phi(k, k_0)\|_i \left\|\boldsymbol{x}_{k_0}\right\|.$$

Adding the response gives the summation

$$\sum_{k=k_0}^{\infty} \|\boldsymbol{x}_k\| = \sum_{k=k_0}^{\infty} \|\Phi(k, k_0)\|_i \left\|\boldsymbol{x}_{k_0}\right\| \leq \beta \left\|\boldsymbol{x}_{k_0}\right\|, \; k_0 \geq 0.$$

Thus, the series converges and consequently

$$\mathcal{L}\boldsymbol{im}_{k\to\infty} \|\boldsymbol{x}_k\| = 0.$$

The system is asymptotically stable.

■

Theorem 9.5 *A linear system is UES if and only if the Lyapunov equation has a steady-state solution.*

Proof The steady-state solution of the Lyapunov equation is

$$P_\infty = \mathcal{L}\boldsymbol{im}_{k\to\infty} P_{k|k-1} = \mathcal{L}\boldsymbol{im}_{k\to\infty}\left\{\Phi(k,0)P_0\Phi^T(k,0) + \sum_{i=0}^{k-1}\Phi(k,i+1)\overline{Q}_i\Phi^T(k,i+1)\right\} \tag{9.69}$$

Proof (Sufficiency) The first term in the solution depends on $\mathcal{L}\boldsymbol{im}_{k\to\infty}\|\Phi(k,0)\|_i$ and goes to zero for a UES system

$$\mathcal{L}\boldsymbol{im}_{k\to\infty}\|\Phi(k,0)\|_i = 0, \sum_{k=k_0}^{\infty}\|\Phi(k,k_0)\|_i \le \beta.$$

Hence, the steady-state solution for a UES system is

$$P_\infty = \mathcal{L}\boldsymbol{im}_{k\to\infty}\sum_{i=0}^{k-1}\Phi(k,i+1)\overline{Q}_i\Phi^T(k,i+1) \tag{9.70}$$

Using norm inequalities gives

$$\begin{aligned}\|P_\infty\|_i &\le \sum_{i=0}^{\infty}\|\Phi(k,i+1)\|_i\left\|\overline{Q}_k\right\|_i\left\|\Phi^T(k,i+1)\right\|_i \\ &< \beta^2\beta_Q < \infty, \left(\left\|\overline{Q}_k\right\|_i < \beta_Q, \forall k\right).\end{aligned}$$

Proof (Necessity) Let the system be UES with an unbounded $\|P_\infty\|_i$, then

$$\sum_{k=k_0}^{\infty}\|\Phi(k,k_0)\|_i \le \beta.$$

However, the inequality ensures that $\|P_\infty\|_i$ remains finite for finite $\left\|\overline{Q}_k\right\|_i$. Hence, $\|P_\infty\|_i$ must remain finite.

For LTI systems, conditions for the stability of the Kalman filter are simpler and are given by the following theorem.

■

Theorem 9.6 *The LTI system is exponentially stable if and only if the solution of the algebraic Lyapunov equation is positive definite symmetric.*

Proof **Sufficiency**

Assume that the steady-state Lyapunov equation is reached, then we have the equation

$$\overline{\phi} P_\infty \overline{\phi}^T - P_\infty = -\overline{Q}$$

$$\overline{\phi} = \phi(I_n - KH), \overline{Q} = \phi K R K^T \phi^T + Q > 0.$$

The matrix $\overline{Q}$ is positive definite since the noise covariance matrices (Q, R) are positive definite. Hence, P_∞ is also positive definite. By Theorem 9.5, the system is exponentially stable if and only if the Lyapunov equation has a steady-state solution.

Necessity

If the system is exponentially stable, then state-transition matrix satisfies $\underset{k\to\infty}{\mathcal{L}\rangle\Updownarrow} \|\Phi(k, 0)\|_i = 0$, and in the steady state, we have the algebraic Lyapunov equation with the symmetric steady-state solution

$$\boldsymbol{P}_\infty = \sum_{k=0}^{\infty} \overline{\phi}^k \, \overline{Q} \, (\overline{\phi}^T)^k.$$

For the LTI case, assuming distinct eigenvalues for simplicity, the state-transition matrix is

$$\overline{\phi}^k = \sum_{i=1}^{n} Z_i \lambda_i^k$$

and the steady-state solution of the Lyapunov equation can be written as

$$\begin{aligned} P_\infty &= \sum_{k=0}^{\infty} \left(\sum_{i=1}^{n} Z_i \lambda_i^k \right) \overline{Q} \left(\sum_{j=1}^{n} Z_j \lambda_j^k \right)^T \\ &= \sum_{i=1}^{n} \sum_{j=1}^{n} Z_i \overline{Q} Z_j^T \sum_{k=0}^{\infty} (\lambda_i \lambda_j)^k. \end{aligned}$$

For an exponentially stable system, $|\lambda_i \lambda_j| < 1$ and

$$P_\infty = \sum_{i=1}^{n} \sum_{j=1}^{n} \frac{1}{1 - \lambda_i \lambda_j} Z_i \overline{Q} Z_j^T.$$

The matric $\overline{Q}$ is positive definite since the noise covariance matrices (Q, R) are positive definite. Hence, P_∞ is also positive definite.

■

Theorem 9.7 *The Lyapunov equation has a symmetric positive definite (semidefinite) solution for any positive semidefinite matrix $\overline{Q}$ such that $\left(\overline{\phi}, \overline{Q}^{1/2}\right)$ is controllable (stabilizable) if and only if $\overline{\phi}$ is asymptotically stable.*

We show that if the matrix P_∞ is positive semidefinite the pair $\left(\overline{\phi}, \overline{Q}^{1/2}\right)$ is stabilizable then the filter is asymptotically stable by contradiction. Assume that $\overline{\phi}$ has an unstable eigenvalue λ with left eigenvector w^T.

$$\overline{\phi} P_\infty \overline{\phi}^T - P_\infty = -\overline{Q}.$$

Proof **Proof (Sufficiency).**
Premultiply and postmultiply by the left eigenvector w^T of the matrix $\overline{\phi}$.

$$w^T \overline{\phi} P_\infty \overline{\phi}^T w - w^T P_\infty w = \left(|\lambda|^2 - 1\right) P_\infty = -w^T \overline{Q}^{\frac{1}{2}} \left(\overline{Q}^{\frac{1}{2}}\right)^T w < 0.$$

For $Q \geq 0$, we have a nonnegative term $\left(|\lambda|^2 - 1\right) P_\infty$ equal to a nonpositive term, which can only be true if both terms are zero. If $w^T \overline{Q}^{\frac{1}{2}} = 0$, then the pair is not stabilizable, and we have a contradiction.

9.6.1.1 Matlab

MATLAB provides a command for the solution of algebraic Lyapuncv equation. The following example shows how the command can be used to test the stability of a filter.

■

Example 9.8 Determine the stability of the filter with

$$\phi = \begin{bmatrix} 0.1 \ 0.2 \\ 0.3 \ 0.4 \end{bmatrix}, \ K = \begin{bmatrix} 5 & -3 \end{bmatrix}^T, \ H = \begin{bmatrix} 1 \ 1 \end{bmatrix}.$$

Solution

```
>> phi=[0.1,0.2;0.3,0.4];H=[1,1] ;% observable
>>k=[5;-3] ; phi_bar= phi*(eye(2)-k*H);
>> P=dlyap(phi_bar,eye(2)) % Pos. Def. P, Q=I
ans =
```

```
1.1402 0.0309
0.0309 1.0101
>> eig(phi*(eye(2)-k*h)) % Fast stable dynamics
ans =
0.2000
0.1000
```

■

9.6.1.2 Discrete Riccati Equation

In this chapter, we derive the recursion for the covariance matrix known as the discrete Riccati equation

$$P_{k+1|k} = \phi_k \Big[P_{k|k-1} - P_{k|k-1} H_k^T \big(H_k P_{k|k-1} H_k^T + R_k \big)^{-1} H_k P_{k|k-1} \Big] \phi_k^T + Q_k.$$

This nonlinear difference equation is obtained by substituting the optimal gain in the Lyapunov equation and it only valid for the optimal Kalman gain.

Consider a time-invariant system with stationary noise. Under certain conditions, the covariance matrix is guaranteed to remain bounded and reach a constant steady state

$$P_\infty = \mathop{\mathcal{L}im}_{k \to \infty} P_{k|k-1}.$$

This gives the algebraic Riccati equation

$$P_\infty = \phi \Big[P_\infty - P_\infty H^T \big(H P_\infty H^T + R \big)^{-1} H P_\infty \Big] \phi^T + Q.$$

The following theorem gives conditions for the existence of P_∞. The condition guarantees that the measurement capture all the unstable dynamics of the system so that any part of the dynamics that the filter cannot follow decays to zero and does not affect steady-state behavior.

Theorem 9.8 (Lewis, p. 100) *If (ϕ, H) is detectable (observable) then for every initial matrix P_0 there is a bounded limiting solution P_∞ to the Riccati equation. The limiting solution is the positive semidefinite (definite) solution of the algebraic Riccati equation.*

■

MATLAB provides a command for the solution of algebraic Riccati equation as in the following example.

Example 9.10 Obtain the solution of the algebraic Riccati equation for the pair

$$\phi = \begin{bmatrix} 0.1 & 0.2 \\ 0.3 & 0.4 \end{bmatrix}, \ H = \begin{bmatrix} 1 & 1 \end{bmatrix}$$

with $Q = I_2$, $R = 0.1$

Solution

The following MATLAB commands give the solution:

```
>> phi=[0.1,0.2;0.3,0.4];H=[1,1]; % Observable
>> Q=eye(2) ; R=0.1;
>> [X,L,G]=dare(phi',H',Q, R)
% P=X, K=G', L=eig(phi*(I-K*H))
X =
1.0072 0.0100
0.0100 1.0167
L =
-0.0212
0.0441
G =
0.1432 0.3339
```

■

Theorem 9.9 (Lewis, p. 101) *Assume that* $Q = Q^{1/2}\left(Q^{1/2}\right)^T \geq 0$, $R > 0$ *and* $\left(\phi, Q^{1/2}\right)$ *is reachable. Then* (H, ϕ) *is detectable* **if and only if**.

i. There is a unique positive definite limiting solution to the algebraic Riccati equation.
ii. The steady-state error system $\overline{A} = \phi(I_n - KH)$

   with steady-state Kalman gain

$$K = P_\infty H^T \left(H P_\infty H^T + R\right)^{-1}$$

is asymptotically stable.

Proof We show that if the pair $\left(\phi, Q^{1/2}\right)$ is stabilizable, then the filter is asymptotically stable.

$$\overline{\phi} P_\infty \overline{\phi}^T - P_\infty = -\overline{Q}.$$

Based on the theory presented in this section, we have the following procedure for testing the stability of the Kalman filter.

Procedure 9.1

1. Obtain the square root matrix $Q^{1/2}$ such that $Q = Q^{1/2}\left(Q^{1/2}\right)^T \geq 0$, $R > 0$

2. Test the reachability of the pair $\left(\phi, Q^{1/2}\right)$.
3. If the pair in step 2 is reachable, test the detectability of the pair (H, ϕ).
4. Use Theorem 9.9 to determine the stability of the Kalman filter.

To test any suboptimal filter, we use the associated Lyapunov equation.

Example 9.12 Test the stability of the Kalman filter for the pair

$$\phi = \begin{bmatrix} 0.1 & 0.2 \\ 0.3 & 0.4 \end{bmatrix}, H = \begin{bmatrix} 1 & 1 \end{bmatrix}$$

with $Q = I_2$, $R = 0.1$

Solution

For an identity matrix, the square root is $Q^{1/2} = I_2$ and the pair $\left(\phi, Q^{1/2}\right)$ is clearly reachable by the controllability rank test because $Q^{1/2}$ is a full-rank square matrix. To test detectability, we can use an observability test because observability is sufficient for detectability. The observability matrix is

$$\mathcal{O} = \begin{bmatrix} H \\ H\phi \end{bmatrix} = \begin{bmatrix} 1 & 1 \\ 0.4 & 0.6 \end{bmatrix}.$$

The matrix is full rank since its determinant is $0.2 \neq 0$. By Theorem 9.9, the filter is stable.

Problems

9.1 Assuming that the initial estimate is unbiased, show that for an unbiased estimator the posterior estimate of the Kalman filter must be in the form

$$\hat{\boldsymbol{x}}_{k|k} = (I_n - K_k H_k)\hat{\boldsymbol{x}}_{k|k-1} + K_k \boldsymbol{z}_k = \hat{\boldsymbol{x}}_{k|k-1} + K_k\left[\boldsymbol{z}_k - H_k\hat{\boldsymbol{x}}_{k|k-1}\right]$$

9.2 Prove that the innovation sequence is zero-mean, Gaussian and white.

9.3 The optimal Kalman gain can be derived using different approaches. One of them is to directly minimizer the means square error using basic multivariable calculus. Obtain the Kalman gain by minimizing the performance measure $E\left\{\tilde{\boldsymbol{x}}_{k|k}^T \tilde{\boldsymbol{x}}_{k|k}\right\} = tr\left[P_{k|k}\right]$.

9.4 Show that the orthogonality conditions hold for the optimal corrector and for the optimal predictor equations of the Gauss–Markov process with process noise variance σ_x^2 and measurement noise variance σ_v^2.

9.5 Show that if the Shaping Filter Response $g_1(t)$ Satisfies

$$\mathcal{L}im_{t\to 0}\, g_1(t) = \mathcal{L}im_{t\to\infty}\, g_1(t)$$

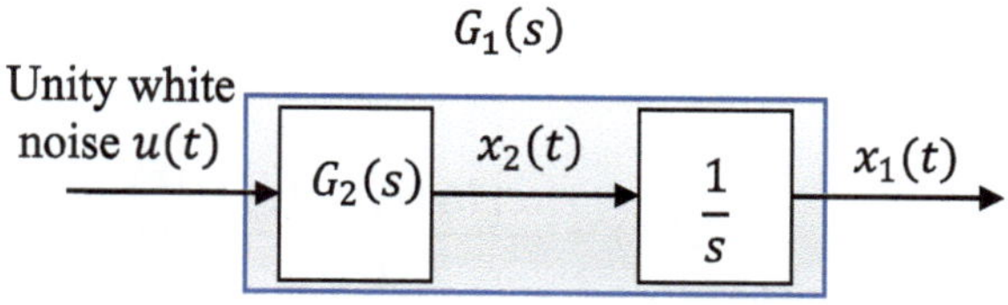

Fig. P9.1 Block diagram of cascade with $G_1(s) = G_2(s)/s$

and $g_2(t) = \mathrm{d}g_1(t)/\mathrm{d}t$, then

$$\underset{t\to\infty}{\mathcal{L}\rangle\Updownarrow} E\{x_1(t)x_2(t)\} = 0$$

Show that for any transfer function of the form (Fig. P9.1)

$$G_2(s) = \frac{c_{n-1}s^{n-1} + \cdots + c_1 s}{s^n + a_{n-1}s^{n-1} + \cdots + a_1 s + a_0} + d$$

$$\underset{t\to\infty}{\mathcal{L}im}\, E\{x_1(t)x_2(t)\} = 0$$

9.6 Show that for the linear state-space model

$$\boldsymbol{x}_{k+1} = \phi_k \boldsymbol{x}_k + \boldsymbol{w}_k$$

with zero-mean white noise $\boldsymbol{w}_k$ with covariance matrix Q the state covariance matrix

$$P_k = E\left\{(\boldsymbol{x}_k - \overline{\boldsymbol{x}}_k)(\boldsymbol{x}_k - \overline{\boldsymbol{x}}_k)^T\right\}, \overline{\boldsymbol{x}}_k = E\{\boldsymbol{x}_k\}$$

is governed by

$$P_{k+1} = \phi_k P_k \phi_k^T + Q$$

9.7 The signal-to-noise ratio is defined as the ratio of the signal power (with zero noise) to the noise power, where the power is the mean square value. Show that for the measurement equation

$$\boldsymbol{z}_k = H_k \boldsymbol{x}_k + \boldsymbol{v}_k$$

with zero-mean white noise $\boldsymbol{v}_k$, the signal-to-noise ratio is

$$SNR(k) = \frac{tr\left[H_k P_k H_k^T\right]}{tr[R]}$$

9.8 Using the orthogonality principle, show that the SNR for the optimal estimator is

$$SNR(k) = \frac{tr\left[H_k\left(P_k - P_{k|k}\right)H_k^T\right]}{tr[R]}$$

Explain how this improves the SNR/
How does this simplify in the case of a scalar measurement $z = \boldsymbol{h}_k^T \boldsymbol{x}_k + v_k$?

9.9 A discrete model of the electricity market is given by

$$\begin{bmatrix} x_1(k+1) \\ x_2(k+1) \\ x_3(k+1) \end{bmatrix} = \begin{bmatrix} 0.998 & 0 & 3.333 \times 10^{-3} \\ 0 & 0.98 & -5 \times 10^{-3} \\ -10^{-5} & 10^{-5} & 1 \end{bmatrix} \begin{bmatrix} x_1(k) \\ x_2(k) \\ x_3(k) \end{bmatrix} + \begin{bmatrix} -6.666 \times 10^{-3} \\ 3.333 \times 10^{-2} \\ 0 \end{bmatrix} + \begin{bmatrix} w_1(k) \\ w_2(k) \\ w_3(k) \end{bmatrix}$$

where $x_1(k)$ is the power output of the supplier, $x_2(k)$ is the power output of the consumer and $x_3(k)$ is the market clearing price (the price when supply matches demand), and $w_i(k)$ are zero-mean Gaussian and white random variations with covariance matrix $Q = 0.5I_3$. The measurement vector is

$$z(k) = \begin{bmatrix} I_2 & 0_{2\times 1} \end{bmatrix} \begin{bmatrix} x_1(k) \\ x_2(k) \\ x_3(k) \end{bmatrix} + \begin{bmatrix} v_1(k) \\ v_2(k) \end{bmatrix}$$

with additive white Gaussian zero-mean Gaussian and white with covariance matrix $R = 0.25I_2$

Design a discrete Kalman filter for the market model.

9.10 For the system of Problem 9.9, if we have no knowledge of the initial state (large initial error covariance)

(a) Check the filter stability before designing it
(b) Simulate the plant and Kalman filter. Plot x, its estimate and the RMS error versus time
(c) Write the equations for the information filter
(d) Start a simulation with the information filter then switch to the covariance filter. Write all the needed equations then simulate the system and obtain plots of the state variables, their estimates, and the RMS error
(e) Repeat for the suboptimal steady-state Kalman gain. Plot the P results together and discuss the difference.

9.11 For the Integrated Gauss–Markov Process with $\beta = 1\,s^{-1}$ and $\sigma^2 = 1$, and the Measurement Noise Variance $R = 0.25$. Using a Sampling Period of 0.1 s.

(a) Check the filter stability before designing it.

(b) Determine suitable initial conditions for the filter.
(c) Simulate the plant and Kalman filter (continuous-time system model and a discrete Kalman filter) and obtain plots of the state variables, their estimates, and the RMS error.
(d) Repeat for the suboptimal steady-state Kalman gain. Plot the RMS error results together and discuss the difference.

9.12 A Manufacturer Determines the Price p_k of a Device Based on the Demand d_k Using the Recursion

$$\boldsymbol{x}_{k+1} = \phi \boldsymbol{x}_k + \boldsymbol{w}_k$$

where

$$\boldsymbol{x}_k = \begin{bmatrix} x_1 \ x_2 \end{bmatrix}^T = \begin{bmatrix} p_k \ d_k \end{bmatrix}^T, \phi = \begin{bmatrix} 1 & \alpha \\ -\beta & \gamma \end{bmatrix}$$

and $\boldsymbol{w}_k = \begin{bmatrix} w_{k1} \ w_{k2} \end{bmatrix}^T$ is a zero-mean, Gaussian, white noise vector with covariance matrix

$$Q = diag\{q_1, q_2\}\delta_k, q_1, q_2 > 0$$

The actual price varies randomly with the vendor and is measured as

$$z_k = \begin{bmatrix} 1 \ 0 \end{bmatrix} \boldsymbol{x}_k + \boldsymbol{v}_k$$

where the noisy measurement z_k includes the additive zero-mean, Gaussian, white noise vector $\boldsymbol{v}_k$ that is independent of the process noise and has the covariance matrix $R = r\delta_k$

Initially, the price and demand are known exactly and are given by (p_0, d_0)

I. Show that a Kalman filter for the system is stable.
II. Write the predictor and corrector equations for the filter at time k.

9.13 For the integrated Gauss–Markov process of Problem 9.11, with $\beta = 1s^{-1}$ and $\sigma^2 = 1$, with no knowledge of the initial state (large initial error covariance, say 10^{12}), and the measurement noise variance $R = 0.25$ and using a sampling period of 0.1 s.

(i) Write the corrector equations for the information filter for the process.
(ii) Simulate the plant and Kalman filter. Plot $\boldsymbol{x}$, its estimate and the trace of P versus time. Start a simulation with the information filter then switch to the covariance filter.

9.14 Consider the discrete Kalman filter of Sect. 9.3, with an added constant bias term

$$\boldsymbol{x}_{k+1} = \phi_k \boldsymbol{x}_k + \boldsymbol{b}_k + \boldsymbol{w}_k$$
$$z_k = H_k \boldsymbol{x}_k + H_k^b \boldsymbol{b}_k + \boldsymbol{v}_k,$$

where $\boldsymbol{b}_k \in \mathcal{R}^n$ is an unknown bias vector at t_k, and $H_k^b \in \mathcal{R}^{m \times n}$ is a bias measurement matrix at t_k.

(i) Obtain an augmented state-space model for the state and bias with the bias equation.

$$\boldsymbol{b}_{k+1} = \boldsymbol{b}_k$$

(ii) Write the discrete Kalman filter for the augmented system, then expand the expressions.

9.15 Design and simulate a Kalman filter for a Wiener process with an added bias of unity with measurement noise variance of 0.1. Use a sampling period of 0.1 s and obtain plots of the state variables, their estimates, and the error covariance matrix.

Bibliography

1. Anderson, B. D. O., & Moore, J. B. (1979). *Optimal filtering*. Prentice Hall.
2. Bar-Shalom, Y., Li, X. R., & Kirubarajan, T. (2001). *Estimation with applications to tracking and navigation: Theory, algorithms and software*. J. Wiley.
3. Brown & Hwang. (2012). *Introduction to random signals and applied Kalman filtering*. Wiley.
4. Jazwinski, A. H. (1970). *Stochastic processes and filtering theory*. Academic Press.
5. Lewis, F. L. (1986). *Optimal estimation: With an introduction to stochastic control theory*. Wiley-Interscience.
6. Mendel, J. M. (1995). *Lessons in estimation theory for signal processing, communications, and control*. Prentice Hall PTR.
7. Rugh, W. J. (1996). *Linear system theory*. Upper Saddle River, NJ: Prentice-Hall.

Chapter 10
Generalizing the Basic Discrete Kalman Filter

10.1 Correlated Noise

The derivation of the Kalman filter of Chap. 9 assumes that process noise and measurement noise are uncorrelated. This assumption is not always true in practice. We consider two approaches that extend the Kalman filter to correlated noise. The first assumes that measurement noise $\boldsymbol{v}_k$ at time k is correlated with the process noise $\boldsymbol{w}_k$. Recall that when discretizing a continuous-time model with noise input $\boldsymbol{u}(t)$ over the interval $\left[t_k, t_{k+1}\right]$ the zero-response is $\boldsymbol{w}_k$. The other argues that it is more appropriate to assume that $\boldsymbol{v}_k$ is correlated with the process noise $\boldsymbol{w}_{k-1}$, that is, with the noise that corresponds to the interval $\left[t_{k-1}, t_k\right]$. Both approaches use the state-space model

$$\boldsymbol{x}_{k+1} = \phi_k \boldsymbol{x}_k + \boldsymbol{w}_k$$

$$\boldsymbol{z}_k = H_k \boldsymbol{x}_k + \boldsymbol{v}_k \tag{10.1}$$

where

$\boldsymbol{x}_k = n \times 1$ state vector at t_k
$\phi_k = n \times n$ state-transition matrix at t_k
$\boldsymbol{z}_k = m \times 1$ measurement vector at t_k
$H_k = m \times n$ measurement matrix at t_k
$\boldsymbol{w}_k = n \times 1$ zero-mean white Gaussian process noise vector at t_k
$\boldsymbol{v}_k = m \times 1$ zero-mean white Gaussian measurement noise vector at t_k.
The noise covariance matrices are

$$E\left\{\boldsymbol{w}_k \boldsymbol{w}_i^T\right\} = \begin{cases} Q_k, & i = k \\ [0], & i \neq k \end{cases}, \quad E\left\{\boldsymbol{v}_k \boldsymbol{v}_i^T\right\} = \begin{cases} R_k, & i = k \\ [0], & i \neq k \end{cases} \tag{10.2}$$

M. S. Fadali, *Introduction to Random Signals, Estimation Theory, and Kalman Filtering*, https://doi.org/10.1007/978-981-99-8063-5_10

10.1.1 Equivalent Model with Uncorrelated Noise

In this section, we assume correlated process and measurement noise

$$E\{\boldsymbol{w}_k\boldsymbol{v}_i^T\} = \begin{cases} S_k \neq [0], & i = k \\ [0], & i \neq k \end{cases} \tag{10.3}$$

To obtain an equivalent model with uncorrelated process and measurement noise, we add a zero term equal to the state equation of the process with a, yet to be chosen, weight matrix D.

$$\boldsymbol{x}_{k+1} = \phi_k\boldsymbol{x}_k + \boldsymbol{w}_k + D_k[\boldsymbol{z}_k - H_k\boldsymbol{x}_k - \boldsymbol{v}_k].$$

We rewrite the state equation as

$$\boldsymbol{x}_{k+1} = \overline{\phi}_k\boldsymbol{x}_k + D_k\boldsymbol{z}_k + \overline{\boldsymbol{w}}_k \tag{10.4}$$

The sate equation has a modified state-transition matrix

$$\overline{\phi}_k = \phi_k - S_k R_k^{-1} H_k \tag{10.5}$$

a modified process noise

$$\overline{\boldsymbol{w}}_k = \boldsymbol{w}_k - D_k\boldsymbol{v}_k \tag{10.6}$$

and a deterministic term $D_k\boldsymbol{z}_k$ given the measurement $\boldsymbol{z}_k$. The noise process can be made uncorrelated to the measurement noise with appropriate choice of the matrix D.

Consider the cross-correlation

$$E\{\overline{\boldsymbol{w}}_k\boldsymbol{v}_i^T\} = \begin{cases} S_k \neq [0], & i = k \\ [0], & i \neq k. \end{cases}$$

For uncorrelated noise, that is for $S_k - D_k R_k = [0]$, let $D_k = S_k R_k^{-1}$. The mean of the corresponding process noise is

$$E\{\overline{\boldsymbol{w}}_k\} = E\{\boldsymbol{w}_k - D_k\boldsymbol{v}_k\} = 0 \tag{10.7}$$

and its covariance is

$$Q_{\overline{w}k} = E\left\{\overline{\boldsymbol{w}}_k\overline{\boldsymbol{w}}_k^T\right\} = E\left\{(\boldsymbol{w}_k - D_k\boldsymbol{v}_k)(\boldsymbol{w}_k - D_k\boldsymbol{v}_k)^T\right\} = Q_k + D_k R_k D_k^T - D_k S_k^T - S_k D_k^T.$$

Substituting $D_k = S_k R_k^{-1}$ gives the covariance

$$Q_{\overline{w}k} = Q_k + S_k R_k^{-1} R_k R_k^{-1} S_k^T - S_k R_k^{-1} S_k^T - S_k R_k^{-1} S_k^T$$

which simplifies to

$$Q_{\overline{w}k} = Q_k - S_k R_k^{-1} S_k^T \tag{10.8}$$

Because we now have a state equation with a noise process that is uncorrelated with the measurement noise, we can use the Kalman filter derived in Chap. 9. The predictor state equation

$$\hat{\boldsymbol{x}}_{k+1|k} = \overline{\phi}_k \hat{\boldsymbol{x}}_{k|k} + S_k R_k^{-1} \boldsymbol{z}_k \tag{10.9}$$

includes a deterministic term and has the error covariance update

$$P_{k+1|k} = \overline{\phi}_k P_{k|k} \overline{\phi}_k^T + Q_{\overline{w}k}.$$

We can rewrite the predictor state equation in a similar form to the corrector equation that uses the original state-transition matrix of the system

$$\hat{\boldsymbol{x}}_{k+1|k} = \phi_k \hat{\boldsymbol{x}}_{k|k} + S_k R_k^{-1} \left[\boldsymbol{z}_k - H_k \hat{\boldsymbol{x}}_{k|k} \right] \tag{10.10}$$

There are no changes in the expressions for the corrector and the Kalman gain because their derivation did not use the assumption of uncorrelated process and measurement noise. The Kalman filter equations are summarized in Fig. 10.1.

Example 10.1 Gauss–Markov Consider a Gauss–Markov process with autocorrelation $R_{xx}(\tau) = \sigma_x^2 e^{-\beta\tau}$. For uniform sampling with sampling period Δt, the corresponding state equation is

$$x_{k+1} = e^{-\beta\Delta t} x_k + w_k$$

with noise covariance $Q = E\{w_k^2\} = \sigma_w^2$. The measurement equation is

$$z_k = x_k + v_k$$

with measurement noise covariance $R = E\{v_k^2\} = \sigma_v^2$. The measurement noise and the process noise are correlated

$$E\{w_k v_k\} = \begin{cases} s \neq 0, & i = k \\ 0, & i \neq k. \end{cases}$$

Write the equations for the discrete Kalman filter initialized with $\hat{x}_0 = 0$, $P_0 = \sigma_x^2$.

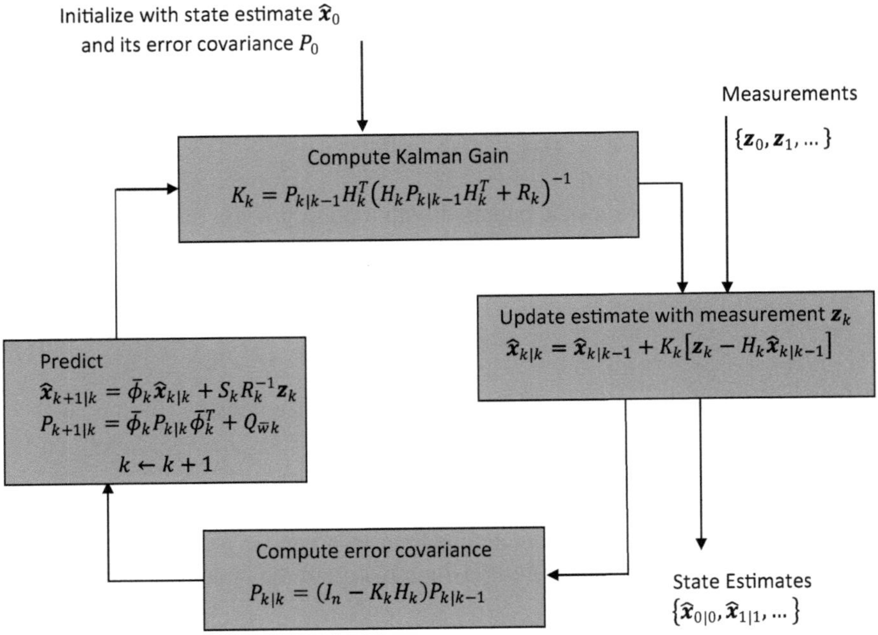

Fig. 10.1 Block diagram for the discrete-time Kalman filter with correlated noise $\boldsymbol{w}_k$ and $\boldsymbol{v}_k$

Solution

We substitute in the filter equations with state-transition matrix $e^{-\beta\Delta t}$ and measurement matrix $H = 1$. The corrector equations are the same as those for the case of uncorrelated measurement and process noise. The optimal gain is

$$K_k = \frac{P_{k|k-1}}{P_{k|k-1} + \sigma_v^2},$$

and the corrected state estimate is

$$\hat{\boldsymbol{x}}_{k|k} = \hat{x}_{k|k-1} + K_k\left(z_k - \hat{x}_{k|k-1}\right).$$

The predictor can be written in the same form as the corrector

$$\hat{x}_{k+1|k} = \phi_k\hat{x}_{k|k} + S_kR_k^{-1}\left[z_k - H_k\hat{x}_{k|k}\right]$$

$$= e^{-\beta\Delta t}\hat{\boldsymbol{x}}_{k|k} + s/\sigma_v^2\left[z_k - \hat{\boldsymbol{x}}_{k|k}\right],$$

and its error covariance is

$$P_{k+1|k} = \overline{\phi}_kP_{k|k}\overline{\phi}_k^T + Q_{\overline{w}k},$$

where

$$\overline{\phi}_k = \phi_k - S_k R_k^{-1} H_k = e^{-\beta\Delta t} - s/\sigma_v^2$$

and the noise covariance is

$$Q_{\overline{w}k} = Q_k - S_k R_k^{-1} S_k^T = \sigma_w^2 - s^2/\sigma_v^2.$$

■

10.1.2 Delayed Process Noise

We write the state equation in the form

$$\boldsymbol{x}_k = \phi_{k-1}\boldsymbol{x}_{k-1} + \boldsymbol{w}_{k-1} \tag{10.11}$$

The effect of process noise on continuous-time system at time k due to $u(t), t \in [t_{k-1}, t_k)$ is represented by $\boldsymbol{w}_{k-1}$

$$E\{\boldsymbol{w}_{k-1}\boldsymbol{v}_i^T\} = \begin{cases} S_k \neq [0], & i = k \\ [0], & i \neq k \end{cases} \tag{10.12}$$

The justification for using $\boldsymbol{w}_{k-1}$ with $\boldsymbol{v}_k$ is that it is the noise that affects the DT system at time k.

The predictor equations are the same as in the case of uncorrelated process and measurement noise with estimator

$$\hat{\boldsymbol{x}}_{k+1|k} = \phi_k \hat{\boldsymbol{x}}_{k|k}$$

and error covariance matrix

$$P_{k+1|k} = \phi_k P_{k|k} \phi_k^T + Q.$$

The corrector equation is also unchanged but the optimal gain and covariance expressions have additional terms. The a posteriori estimation error is obtained by subtracting the estimator

$$\hat{\boldsymbol{x}}_{k|k} = \hat{\boldsymbol{x}}_{k|k-1} + K_k\left[z_k - H_k \hat{\boldsymbol{x}}_{k|k-1}\right]$$

from the process dynamics of (10.11), then substituting for the measurement. The estimation error is

$$\tilde{\boldsymbol{x}}_{k|k} = (I_n - K_k H_k)\tilde{\boldsymbol{x}}_{k|k-1} - K_k v_k \tag{10.13}$$

To show that the two terms in (10.13) are correlated, we first subtract the predicted estimate from the system dynamics of (10.11) to obtain the a priori estimation error

$$\tilde{\boldsymbol{x}}_{k|k-1} = \boldsymbol{x}_k - \hat{\boldsymbol{x}}_{k|k-1} = \phi_{k-1}\tilde{\boldsymbol{x}}_{k-1|k-1} + \boldsymbol{w}_{k-1} \tag{10.14}$$

The cross-correlation of the two terms is

$$\begin{aligned}(I_n - K_k H_k)E\left\{\tilde{\boldsymbol{x}}_{k|k-1}\boldsymbol{v}_k^T\right\}K_k^T &= (I_n - K_k H_k)E\left\{\left[\phi_k\tilde{\boldsymbol{x}}_{k-1|k-1} + \boldsymbol{w}_{k-1}\right]\boldsymbol{v}_k^T\right\}K_k^T \\ &= (I_n - K_k H_k)S_k K_k^T.\end{aligned}$$

The a posteriori error covariance includes the cross-correlation terms and is given by

$$P_{k|k} = E\left\{\tilde{\boldsymbol{x}}_{k|k}\tilde{\boldsymbol{x}}_{k|k}^T\right\}$$

$$\begin{aligned}&= (I_n - K_k H_k)P_{k|k-1}(I_n - K_k H_k)^T + K_k R_k K_k^T \\ &\quad - (I_n - K_k H_k)S_k K_k^T - K_k S_k^T(I_n - K_k H_k)^T\end{aligned} \tag{10.15}$$

Expanding the covariance matrix gives the expression

$$\begin{aligned}P_{k|k} = &(I_n - K_k H_k)P_{k|k-1}(I_n - K_k H_k)^T + K_k R_k K_k^T \\ &- K_k S_k^T - S_k K_k^T - K_k\left[H_k S_k + S_k^T H_k^T\right]K_k^T\end{aligned}$$

which is equal to the covariance matrix for the uncorrelated noise case minus additional terms

$$P_{k|k} = \left[P_{k|k}\right]_{\text{uncorrelated}} - K_k S_k^T - S_k K_k^T - K_k\left[H_k S_k + S_k^T H_k^T\right]K_k^T.$$

The trace of the error covariance matrix is the mean-square error

$$\text{tr}\left[P_{k|k}\right] = \text{tr}\left[E\left\{\tilde{\boldsymbol{x}}_{k|k}\tilde{\boldsymbol{x}}_{k|k}^T\right\}\right] = E\left\{\tilde{\boldsymbol{x}}_{k|k}^T\tilde{\boldsymbol{x}}_{k|k}\right\}$$

and in terms of the error covariance for the uncorrelated case is

$$\text{tr}\left[P_{k|k}\right] = \text{tr}\left[P_{k|k}\right]_{\text{uncorrelated}} - 2\text{tr}\left[K_k S_k^T\right] - \text{tr}\left[K_k\left[H_k S_k + S_k^T H_k^T\right]K_k^T\right].$$

Using the formulas for the derivative of the trace, as in Chap. 9, we differentiate and equate to zero to find the optimal gain. The optimal gain is modified to include the additional term and becomes

$$K_k = \left(P_{k|k-1}H_k^T + S_k\right)\left(H_k P_{k|k-1} H_k^T + R_k + H_k S_k + S_k^T H_k^T\right)^{-1} \tag{10.16}$$

For $S_k = 0$, the optimal gain reduces to the uncorrelated noise optimal gain

$$K_k = P_{k|k-1}H_k^T\left(H_k P_{k|k-1} H_k^T + R_k\right)^{-1}.$$

For the optimal gain K_k, we can show that (Fig. 10.2)

$$P_{k|k} = (I_n - K_k H_k)P_{k|k-1} - K_k S_k^T \tag{10.17}$$

Example 10.2 Gauss–Markov Repeat Example 10.1 if the measurement noise and the process noise are correlated as

$$E\{w_{k-1}v_k\} = \begin{cases} s \neq 0, & i = k \\ 0, & i \neq k. \end{cases}$$

Solution

We substitute in the filter equations with state-transition matrix $e^{-\beta\Delta t}$ and measurement matrix $H = 1$. The predictor equations are the same as those for the case of uncorrelated measurement and process noise.

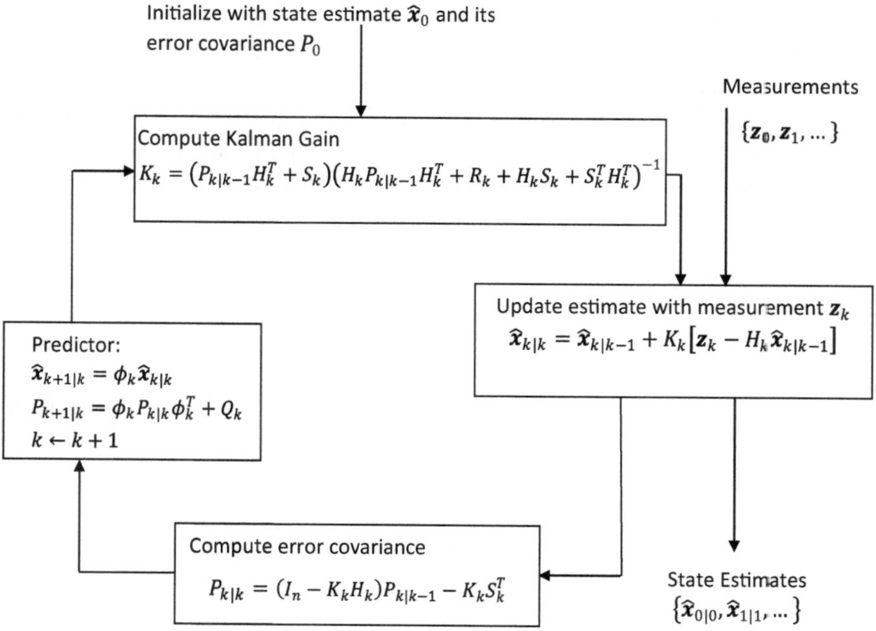

Fig. 10.2 Block diagram for the discrete-time Kalman filter with correlated noise $\boldsymbol{w}_{k-1}$ and $\boldsymbol{v}_k$

$$\hat{x}_{k+1|k} = e^{-\beta\Delta t}\hat{x}_{k|k}$$

$$P_{k+1|k} = e^{-2\beta\Delta t}P_{k|k} + Q_k.$$

The Kalman gain is now

$$K_k = \frac{\left(P_{k|k-1} + s\right)}{P_{k|k-1} + \sigma_v^2 + 2s}.$$

The corrector equations are

$$\hat{\boldsymbol{x}}_{k|k} = \hat{\boldsymbol{x}}_{k|k-1} + K_k\left[z_k - H_k\hat{\boldsymbol{x}}_{k|k-1}\right]$$

$$P_{k|k} = (I_n - K_k H_k)P_{k|k-1} - K_k S_k^T.$$

■

10.2 Colored Noise

If either the process noise, the measurement noise, or both, is colored noise, the Kalman filter of Chap. 9 cannot be used directly. To use the filter, we first find the transfer function of the shaping filter of the colored noise and include it in the model of the system for Kaman filter design. The shaping filter is obtained by spectral factorization as discussed in Chap. 4 and is shown in Fig. 10.3. The colored noise process is the output of the filter $y_n(t)$, and its input is unity Gaussian white noise.

A state-space model is obtained from the transfer function of the shaping filter

$$\dot{\boldsymbol{x}}_n(t) = A_n\boldsymbol{x}_n(t) + B_n\boldsymbol{u}_n(t),$$

$$\boldsymbol{x}_n(t) \in \mathcal{R}^{n_n}, \boldsymbol{u}_n(t) \in \mathcal{R}^{n_u}$$

$$y_n(t) = H_n\boldsymbol{x}_n, \boldsymbol{y}_n(t) \in \mathcal{R}^{n_y} \tag{10.18}$$

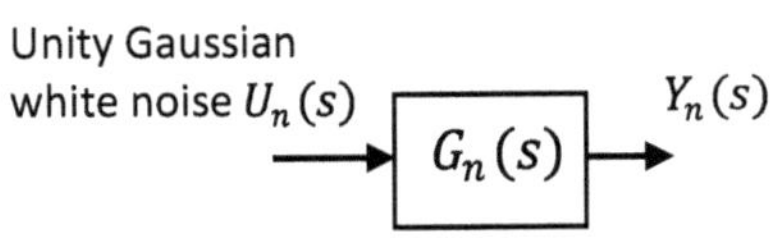

Fig. 10.3 Shaping filter for noise process

Note that with $R = 0$, the gain becomes

$$K_k = P_{k|k-1} H_k^T \left(H_k P_{k|k-1} H_k^T\right)^{-1}$$

and the covariance matrix becomes

$$P_{k|k} = P_{k|k-1} - P_{k|k-1} H_k^T \left(H_k P_{k|k-1} H_k^T\right)^{-1} H_k P_{k|k-1}.$$

If the matrix $H_k P_{k|k-1} H_k^T$ is positive definite then it is invertible but if it is not, a pseudoinverse can be used. However, multiplying the covariance matrix by the measurement matrix gives

$$H_k P_{k|k} H_k^T = H_k P_{k|k-1} H_k^T - H_k P_{k|k-1} H_k^T \left(H_k P_{k|k-1} H_k^T\right)^{-1} H_k P_{k|k-1} H_k^T = 0$$

The next section shows how to cope with this difficulty.

Example 10.3 Consider an armature-controlled DC motor in an angular velocity control system with transfer function

$$G_m(s) = \frac{K_m}{T_m s + 1}$$

and armature voltage as input. The input has added Gauss–Markov noise with PSD

$$S_u = \frac{2\beta\sigma^2}{-s^2 + \beta^2}.$$

The angular velocity is measured with a tachometer and the measured signal includes additive Wiener noise. Obtain a state-space model for an equivalent system with white noise inputs.

Solution

The spectral factorization for the Gauss–Markov process gives

$$S_u = L(s)L(-s) = \left(\frac{\sqrt{2\beta\sigma^2}}{s + \beta}\right)\left(\frac{\sqrt{2\beta\sigma^2}}{-s + \beta}\right).$$

The shaping filter for the Gauss–Markov process

$$G_u(s) = \frac{\sqrt{2\beta\sigma^2}}{s + \beta}$$

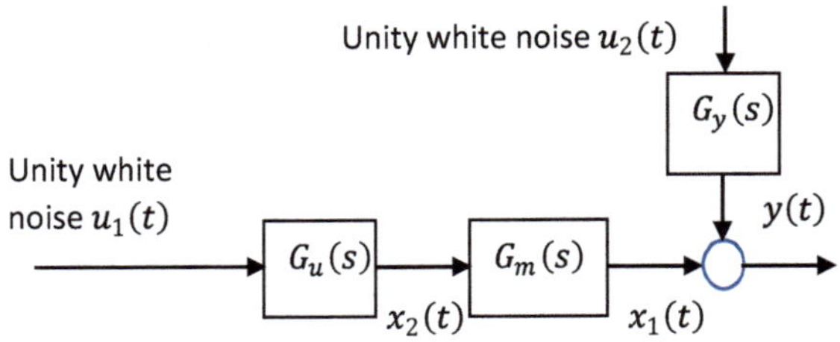

Fig. 10.4 Block diagram of angular velocity control system with colored process noise and colored measurement noise

The shaping filter for the Wiener noise is.

$$G_y(s) = \frac{1}{s}.$$

Using the transfer functions and the block diagram of Fig. 10.4, we write the state equations. The motor state equation is

$$\dot{x}_1 = -\left(\frac{1}{T_m}\right)x_1 + \left(\frac{K_m}{T_m}\right)x_2.$$

The state equation for the Gauss–Markov shaping filter is

$$\dot{x}_2 = -\beta x_2 + \sqrt{2\beta\sigma^2}u_1,$$

and for the Wiener process, it is

$$\dot{x}_3 = u_2.$$

The state-space model for the system is

$$\begin{bmatrix} \dot{x}_1(t) \\ \dot{x}_2(t) \\ \dot{x}_3(t) \end{bmatrix} = \begin{bmatrix} -1/T_m & K_m/T_m & 0 \\ 0 & -\beta & 0 \\ 0 & 0 & 0 \end{bmatrix} \begin{bmatrix} x_1(t) \\ x_2(t) \\ x_3(t) \end{bmatrix} + \begin{bmatrix} 0 & 0 \\ \sqrt{2\beta\sigma^2} & 0 \\ 0 & 1 \end{bmatrix} \begin{bmatrix} u_1(t) \\ u_2(t) \end{bmatrix}$$

$$y(t) = \begin{bmatrix} 1 & 0 & 1 \end{bmatrix} \begin{bmatrix} x_1(t) \\ x_2(t) \\ x_3(t) \end{bmatrix}.$$

Because the measurement equation does not include a noise term, we denote the measurement by $y(t)$ rather than $z(t)$.

■

10.3 Reduced-Order Estimator for Perfect Measurements

In some applications, the measurement noise is negligible, and the measurement can approximately be considered perfect. Perfect measurements also arise when replacing colored measurement noise with a shaping filter with white noise input. The discretized state-space model then has the form

$$\boldsymbol{x}_{k+1} = \phi_k \boldsymbol{x}_k + \boldsymbol{w}_k$$

$$\boldsymbol{y}_k = H_k \boldsymbol{x}_k \tag{10.19}$$

$\boldsymbol{y}_k \in \mathcal{R}^m$. As explained in Sect. 10.2 perfect measurements result in problems with the calculation of the optimal gain of the Kalman filter. To overcome this difficulty, we use the m "perfect measurements" $\boldsymbol{y}_k$ as state variables and estimate the remaining $n - m$ state variables $\boldsymbol{p}_k$

$$\left[\frac{\boldsymbol{y}_k}{\boldsymbol{p}_k}\right] = \left[\frac{H_k}{C_k}\right]\boldsymbol{x}_k \leftrightarrow \boldsymbol{x}_k = L_k\left[\frac{\boldsymbol{y}_k}{\boldsymbol{p}_k}\right] = [L_{1k}|L_{2k}]\left[\frac{\boldsymbol{y}_k}{\boldsymbol{p}_k}\right] \tag{10.20}$$

The matrix $C_k \in \mathcal{R}^{(n-m)\times n}$ is chosen to be linearly independent of H_k so that the transformation matrix $L_k \in \mathcal{R}^{n\times n}$ is invertible but is not unique. The original state variables $\boldsymbol{x}_k$.can be written as

$$\boldsymbol{x}_k = L_{1k}\boldsymbol{y}_k + L_{2k}\boldsymbol{p}_k \tag{10.21}$$

By the fundamental theorem of estimation theory, the minimum-mean-square estimate of $\boldsymbol{x}_k$ is the conditional expectation given the measurement history $\boldsymbol{y}_{1:k} = \{\boldsymbol{y}_k, k = 0, 1, \ldots, k\}$

$$\hat{\boldsymbol{x}}_{k|k} = E\{\boldsymbol{x}_k|\boldsymbol{y}_{0:k}\} = L_{1k}\boldsymbol{y}_k + L_{2k}\hat{\boldsymbol{p}}_{k|k} \tag{10.22}$$

which requires the computation of the unobserved state estimate

$$\hat{\boldsymbol{p}}_{k|k} = E\{\boldsymbol{p}_k|\boldsymbol{y}_{0:k}\} \tag{10.23}$$

Time advancing the output equation

$$\boldsymbol{y}_{k+1} = H_{k+1}\boldsymbol{x}_{k+1}$$

then substituting for $\boldsymbol{x}_{k+1}$ from the state equation, then for $\boldsymbol{x}_k$ from (10.21) gives

$$\boldsymbol{y}_{k+1} = H_{k+1}[\phi_k\boldsymbol{x}_k + \boldsymbol{w}_k] = H_{k+1}\phi_k L_{1k}\boldsymbol{y}_k + H_{k+1}\phi_k L_{2k}\boldsymbol{p}_k + H_{k+1}\boldsymbol{w}_k. \tag{10.24}$$

Consider the weighted difference between $\boldsymbol{y}_{k+1}$ and $\boldsymbol{y}_k$, as a measurement vector

$$y_{u,k+1} = y_{k+1} - H_{k+1}\phi_k L_{1k} y_k$$

$$= H_{k+1}\phi_k L_{2k} p_k + H_{k+1} w_k.$$

Define the unknown part of the measurement as

$$y_{u,k+1} = H_{uk} p_k + H_{k+1} w_k \tag{10.25}$$

where $H_{uk} = H_{k+1}\phi_k L_{2k}$. We now have a new measurement equation in terms of $y_{u,k+1}$. With this measurement differencing, using (10.21), we have the state-space model

$$p_{k+1} = C_{k+1}\phi_k L_{2k} p_k + C_{k+1}\phi_k L_{1k} y_k + C_{k+1} w_k \tag{10.26}$$

We observe that the state update includes a deterministic signal $C_{k+1}\phi_k L_{1k} y_k$ and that the measurement noise is correlated with the process noise with cross-correlation

$$S_k = C_{k+1} E\{w_k w_k^T\} H_{k+1}^T = C_{k+1} Q_k H_{k+1}^T \tag{10.27}$$

For correlated noise, we apply the results of Sect. 10.1.1 with process noise $C_{k+1} w_k$ and measurement noise $H_{k+1} w_k$. The state equation is rewritten as

$$\begin{aligned} p_{k+1} &= C_{k+1}\phi_k L_{2k} p_k + C_{k+1}\phi_k L_{1k} y_k + C_{k+1} w_k + D_k\left[y_{u,k+1} - H_{uk} p_k - H_{k+1} w_k\right] \\ &= \left[C_{k+1}\phi_k L_{2k} - D_k H_{uk}\right] p_k + C_{k+1}\phi_k L_{1k} y_k + D_k y_{u,k+1} + \overline{w}_k \end{aligned}$$

$$\overline{w}_k = \left[C_{k+1} - D_k H_{k+1}\right] w_k.$$

To make the noise process uncorrelated we need to zero the expectation

$$E\{\overline{w}_k w_i^T H_{i+1}^T\} = \begin{cases} \left[C_{k+1} - D_k H_{k+1}\right] Q_k H_{k+1}^T, & i = k \\ [0], & i \neq k. \end{cases}$$

Assuming a full-rank matrix, or otherwise using a pseudoinverse, we have

$$D_k = C_{k+1} Q_k H_{k+1}^T \left(H_{k+1} Q_k H_{k+1}^T\right)^{-1}.$$

The measurement noise covariance is

$$R_k = E\{H_{k+1} w_k w_k^T H_{k+1}^T\} = H_{k+1} Q_k H_{k+1}^T \tag{10.28}$$

which agrees with the expression obtained earlier $D_k = S_k R_k^{-1}$.

This gives an equivalent model with uncorrelated noise and with process noise

$$\overline{w}_{mk} = C_{k+1} w_k - S_k R_k^{-1} H_{k+1} w_k = \left[C_{k+1} - C_{k+1} Q_k H_{k+1}^T R_k^{-1} H_{k+1}\right] w_k = \overline{C}_{k+1} w_k \tag{10.29}$$

$$\overline{C}_{k+1} = C_{k+1}\boldsymbol{w}_k - S_k R_k^{-1} H_{k+1} = C_{k+1}\left[I_n - Q_k H_{k+1}^T R_k^{-1} H_{k+1}\right] \tag{10.30}$$

The process noise covariance is

$$Q_{\overline{w}k} = E\left\{\overline{\boldsymbol{w}}_k \overline{\boldsymbol{w}}_k^T\right\} = \overline{C}_{k+1} Q_k \overline{C}_{k+1}^T \tag{10.31}$$

The state equation with the selected value of D_k is

$$\begin{aligned}\boldsymbol{p}_{k+1} &= C_{k+1}\phi_k L_{2k}\boldsymbol{p}_k + C_{k+1}\phi_k L_{1k}\boldsymbol{y}_k + C_{k+1}\boldsymbol{w}_k \\ &\quad + D_k\left[\boldsymbol{y}_{u,k+1} - H_{uk}\boldsymbol{p}_k - H_{k+1}\boldsymbol{w}_k\right]\end{aligned}$$

$$= \left[C_{k+1}\phi_k L_{2k} - S_k R_k^{-1} H_{uk}\right]\boldsymbol{p}_k + C_{k+1}\phi_k L_{1k}\boldsymbol{y}_k + S_k R_k^{-1}\boldsymbol{y}_{u,k+1} + \overline{\boldsymbol{w}}_k.$$

The predictor estimator for $n - m$ variables $\boldsymbol{p}_k$ is

$$\hat{\boldsymbol{p}}_{k+1|k} = \overline{\phi}_k \hat{\boldsymbol{p}}_{k|k} + C_{k+1}\phi_k L_{1k}\boldsymbol{y}_k + C_{k+1} Q_k H_{k+1}^T R_k^{-1}\boldsymbol{y}_{u,k+1} \tag{10.32}$$

given in terms of the state-transition matrix

$$\begin{aligned}\overline{\phi}_k &= C_{k+1}\phi_k L_{2k} - S_k R_k^{-1} H_{uk} = C_{k+1}\phi_k L_{2k} - C_{k+1} Q_k H_{k+1}^T R_k^{-1} H_{k+1}\phi_k L_{2k} \\ &= C_{k+1}\left[I_n - Q_k H_{k+1}^T R_k^{-1} H_{k+1}\right]\phi_k L_{2k}\end{aligned} \tag{10.33}$$

The predictor error covariance matrix is

$$P_{k+1|k} = \overline{\phi}_k P_{k|k}\overline{\phi}_k^T + Q_{\overline{w}k} = \overline{\phi}_k P_{k|k}\overline{\phi}_k^T + \overline{C}_{k+1} Q_k \overline{C}_{k+1}^T \tag{10.34}$$

The gain matrix is

$$\begin{aligned}K_k &= P_{k|k-1} H_{uk}^T\left[H_{uk} P_{k|k-1} H_{uk}^T + R_k\right]^{-1} \\ &= P_{k|k-1} H_{uk}^T\left[H_{uk} P_{k|k-1} H_{uk}^T + H_{k+1} Q_k H_{k+1}^T\right]^{-1}.\end{aligned} \tag{10.35}$$

The corrector is

$$\hat{\boldsymbol{p}}_{k|k} = \hat{\boldsymbol{p}}_{k|k-1} + K_k\left[\boldsymbol{y}_{u,k+1} - H_{uk}\hat{\boldsymbol{p}}_{k|k-1}\right], \tag{10.36}$$

and its error covariance matrix is

$$P_{k|k} = [I_n - K_k H_{uk}]P_{k|k-1} \tag{10.37}$$

Note that the estimate $\hat{\boldsymbol{p}}_k$ is based on $\boldsymbol{y}_{u,k+1}$, which is not available until time $k+1$. Consequently, the estimate $\hat{\boldsymbol{x}}_k$ is not available until time $k+1$. The equations for the filter are summarized in Fig. 10.5.

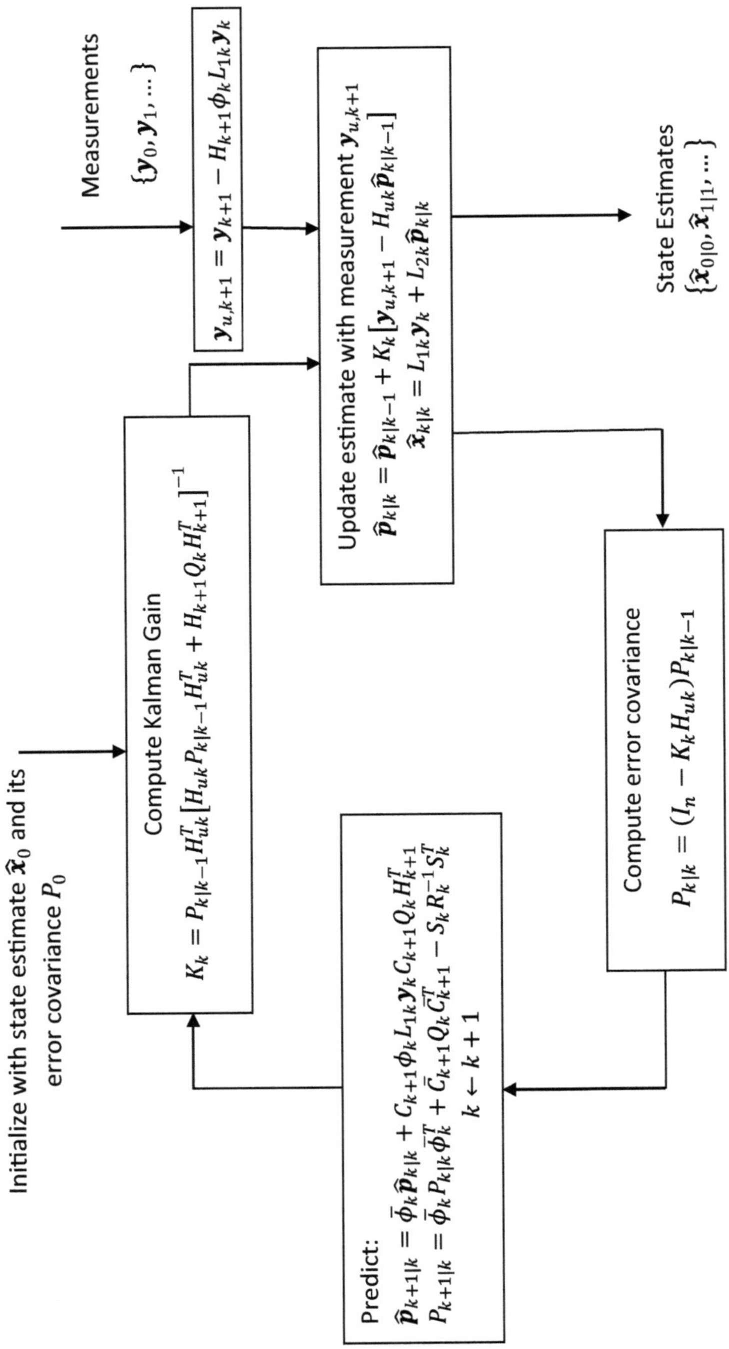

Fig. 10.5 Block diagram for the discrete-time Kalman filter with colored noise $\boldsymbol{w}_k$ and $\boldsymbol{v}_k$

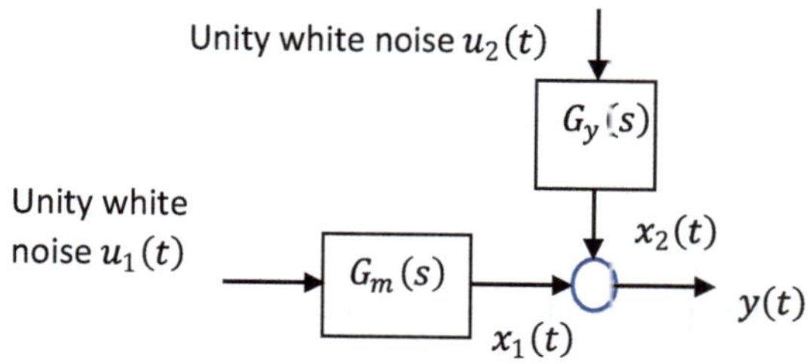

Fig. 10.6 Block diagram of angular velocity control system with colored measurement noise

Example 10.4 Consider the armature-controlled DC motor in an angular velocity control system of Example 10.3 with unity white process noise and Wiener measurement noise. Obtain a state-space model for an equivalent system with white noise inputs and perfect measurements. Use the model to design a reduced-order estimator.

Solution

The system has the block diagram of Fig. 10.6. From Example 10.3, the shaping filter for the Wiener noise is an integrator and the state equations are the motor state equation

$$\dot{x}_1 = -\left(\frac{1}{T_m}\right)x_1 + \left(\frac{K_m}{T_m}\right)u_1$$

and the integrator equation

$$\dot{x}_2 = u_2$$

The state-space model for the system is

$$\begin{bmatrix} \dot{x}_1(t) \\ \dot{x}_2(t) \end{bmatrix} = \begin{bmatrix} -1/T_m & 0 \\ 0 & 0 \end{bmatrix}\begin{bmatrix} x_1(t) \\ x_2(t) \end{bmatrix} + \begin{bmatrix} K_m/T_m & 0 \\ 0 & 1 \end{bmatrix}\begin{bmatrix} u_1(t) \\ u_2(t) \end{bmatrix}$$

$$y(t) = \begin{bmatrix} 1 & 1 \end{bmatrix}\begin{bmatrix} x_1(t) \\ x_2(t) \end{bmatrix}.$$

To obtain the discrete state-space model, we use the Van Loan approach. We first form the matrix

$$M = \left[\begin{array}{c|c} -A & BWB^T \\ 0 & A^T \end{array}\right] = \left[\begin{array}{cc|cc} -1/T_m & 0 & (K_m/T_m)^2 & 0 \\ 0 & 0 & 0 & 1 \\ 0 & 0 & -1/T_m & 0 \\ 0 & 0 & 0 & 0 \end{array}\right]$$

then obtain the matrix exponential

$$e^{M\Delta t} = \left[\begin{array}{cc|cc} e^{-\Delta t/T_m} & 0 & \left(\frac{K_m}{T_m}\right)^2 \Delta t e^{-\Delta t/T_m} & 0 \\ 0 & 1 & 0 & \Delta t \\ 0 & 0 & e^{-\Delta t/T_m} & 0 \\ 0 & 0 & 0 & 1 \end{array}\right].$$

The process noise covariance matrix is

$$Q = \begin{bmatrix} \left(\frac{K_m}{T_m}\right)^2 \Delta t e^{-2\Delta t/T_m} & 0 \\ 0 & \Delta t \end{bmatrix},$$

and the measurement noise variance is

$$\begin{aligned} R = HQH^T &= \begin{bmatrix} 1 & 1 \end{bmatrix} \begin{bmatrix} \left(\frac{K_m}{T_m}\right)^2 \Delta t e^{-2\Delta t/T_m} & 0 \\ 0 & \Delta t \end{bmatrix} \begin{bmatrix} 1 \\ 1 \end{bmatrix} \\ &= \left(1 + \left(\frac{K_m}{T_m}\right)^2 e^{-2\Delta t/T_m}\right) \Delta t. \end{aligned}$$

The cross-correlation

$$S = CQH^T = \begin{bmatrix} 0 & 1 \end{bmatrix} \begin{bmatrix} \left(\frac{K_m}{T_m}\right)^2 \Delta t e^{-2\Delta t/T_m} & 0 \\ 0 & \Delta t \end{bmatrix} \begin{bmatrix} 1 \\ 1 \end{bmatrix} = \Delta t.$$

The discrete state-space model is

$$\begin{bmatrix} x_1(k+1) \\ x_2(k+1) \end{bmatrix} = \begin{bmatrix} e^{-\Delta t/T_m} & 0 \\ 0 & 1 \end{bmatrix} \begin{bmatrix} x_1(k) \\ x_2(k) \end{bmatrix} + \begin{bmatrix} w_1(k) \\ w_2(k) \end{bmatrix}$$

$$y(k) = \begin{bmatrix} 1 & 1 \end{bmatrix} \begin{bmatrix} x_1(k) \\ x_2(k) \end{bmatrix}.$$

We choose the matrix C for a full-rank transformation matrix

$$\begin{bmatrix} y_k \\ \hline p_k \end{bmatrix} = \begin{bmatrix} 1\,1 \\ \hline 0\,1 \end{bmatrix} \boldsymbol{x}_k \leftrightarrow \boldsymbol{x}_k = L \begin{bmatrix} y_k \\ \hline p_k \end{bmatrix} = [L_1|L_2] \begin{bmatrix} y_k \\ \hline p_k \end{bmatrix} = \left[\begin{array}{c|c} 1 & -1 \\ 0 & 1 \end{array}\right] \begin{bmatrix} y_k \\ \hline p_k \end{bmatrix}.$$

We calculate the matrix

$$\overline{C} = C - SR^{-1}H = \begin{bmatrix} 0 & 1 \end{bmatrix} - \frac{\Delta t}{\left[1 + \left(\frac{K_m}{T_m}\right)^2 e^{-2\Delta t/T_m}\right] \Delta t} \begin{bmatrix} 1 & 1 \end{bmatrix}$$

$$= \frac{1}{1+\left(\frac{K_m}{T_m}\right)^2 e^{-2\Delta t/T_m}} \left[-1 \; \left(\frac{K_m}{T_m}\right)^2 e^{-2\Delta t/T_m} \right]$$

and the state-transition matrix

$$\overline{\phi}_k = C\left[I_2 - QH^T R^{-1} H\right]\phi L_2$$

$$= \left[0\;1\right]\left(I_2 - \frac{\begin{bmatrix} \left(\frac{K_m}{T_m}\right)^2 \Delta t e^{-2\Delta t/T_m} & 0 \\ 0 & \Delta t \end{bmatrix}\begin{bmatrix} 1 & 1 \\ 1 & 1 \end{bmatrix}}{\left[1+\left(\frac{K_m}{T_m}\right)^2 e^{-2\Delta t/T_m}\right]\Delta t} \right)\begin{bmatrix} e^{-\Delta t/T_m} & 0 \\ 0 & 1 \end{bmatrix}\begin{bmatrix} -1 \\ 1 \end{bmatrix}$$

$$= \left[0\;1\right]\left(I_2 - \frac{\begin{bmatrix} \left(\frac{K_m}{T_m}\right)^2 e^{-\frac{2\Delta t}{T_m}} & \left(\frac{K_m}{T_m}\right)^2 e^{-\frac{2\Delta t}{T_m}} \\ 1 & 1 \end{bmatrix}}{1+\left(\frac{K_m}{T_m}\right)^2 e^{-\frac{2\Delta t}{T_m}}} \right)\begin{bmatrix} -e^{-\frac{\Delta t}{T_m}} \\ 1 \end{bmatrix}$$

$$= \left(\left[0\;1\right] - \frac{\left[1\;1\right]}{1+\left(\frac{K_m}{T_m}\right)^2 e^{-2\Delta t/T_m}} \right)\begin{bmatrix} -e^{-\Delta t/T_m} \\ 1 \end{bmatrix}$$

$$= \frac{\left[-1 \; \left(\frac{K_m}{T_m}\right)^2 e^{-2\Delta t/T_m} \right]}{1+\left(\frac{K_m}{T_m}\right)^2 e^{-2\Delta t/T_m}}\begin{bmatrix} -e^{-\Delta t/T_m} \\ 1 \end{bmatrix}$$

$$\overline{\phi}_k = e^{-\Delta t/T_m}\frac{1+\left(\frac{K_m}{T_m}\right)^2 e^{-\Delta t/T_m}}{1+\left(\frac{K_m}{T_m}\right)^2 e^{-2\Delta t/T_m}}.$$

The coefficients of the deterministic terms in the state equation are

$$C\phi L_1 = \left[0\;1\right]\begin{bmatrix} e^{-\Delta t/T_m} & 0 \\ 0 & 1 \end{bmatrix}\begin{bmatrix} 1 \\ 0 \end{bmatrix} = 0$$

$$CQH^T R^{-1} = \left[0\;1\right]\frac{\begin{bmatrix} \left(\frac{K_m}{T_m}\right)^2 \Delta t e^{-2\Delta t/T_m} & 0 \\ 0 & \Delta t \end{bmatrix}}{\left[1+\left(\frac{K_m}{T_m}\right)^2 e^{-\frac{2\Delta t}{T_m}}\right]\Delta t}\begin{bmatrix} 1 \\ 1 \end{bmatrix}$$

$$= \frac{1}{1+\left(\frac{K_m}{T_m}\right)^2 e^{-\frac{2\Delta t}{T_m}}} C Q H^T R^{-1}$$

$$= \left[0\ 1\right] \frac{\begin{bmatrix} \left(\frac{K_m}{T_m}\right)^2 \Delta t e^{-2\Delta t/T_m} & 0 \\ 0 & \Delta t \end{bmatrix}}{\left[1+\left(\frac{K_m}{T_m}\right)^2 e^{-\frac{2\Delta t}{T_m}}\right]\Delta t} \begin{bmatrix} 1 \\ 1 \end{bmatrix} = \frac{1}{1+\left(\frac{K_m}{T_m}\right)^2 e^{-\frac{2\Delta t}{T_m}}},$$

and the predictor is

$$\hat{p}_{k+1|k} = \overline{\Phi}_k \hat{p}_{k|k} + C Q H^T R^{-1} y_{u,k+1}$$
$$= e^{-\Delta t/T_m} \frac{1+\left(\frac{K_m}{T_m}\right)^2 e^{-\Delta t/T_m}}{1+\left(\frac{K_m}{T_m}\right)^2 e^{-2\Delta t/T_m}} \hat{p}_{k|k} + \frac{y_{u,k+1}}{1+\left(\frac{K_m}{T_m}\right)^2 e^{-\frac{2\Delta t}{T_m}}}$$

with error covariance matrix

$$P_{k+1|k} = \overline{\phi}^2 P_{k|k} + \overline{C}_{k+1} Q \overline{C}_{k+1}^T = \left(e^{-\frac{\Delta t}{T_m}} \frac{1+\left(\frac{K_m}{T_m}\right)^2 e^{-\frac{\Delta t}{T_m}}}{1+\left(\frac{K_m}{T_m}\right)^2 e^{-\frac{2\Delta t}{T_m}}}\right)^2 P_{k|k}$$
$$+ \frac{1}{\left(1+\left(\frac{K_m}{T_m}\right)^2 e^{-\frac{2\Delta t}{T_m}}\right)^2} \left[-1\ \left(\frac{K_m}{T_m}\right)^2 e^{-2\Delta t/T_m}\right]$$
$$\begin{bmatrix} \left(\frac{K_m}{T_m}\right)^2 \Delta t e^{-2\Delta t/T_m} & 0 \\ 0 & \Delta t \end{bmatrix} \begin{bmatrix} -1 \\ \left(\frac{K_m}{T_m}\right)^2 e^{-2\Delta t/T_m} \end{bmatrix}$$
$$= \left(e^{-\frac{\Delta t}{T_m}} \frac{1+\left(\frac{K_m}{T_m}\right)^2 e^{-\frac{\Delta t}{T_m}}}{1+\left(\frac{K_m}{T_m}\right)^2 e^{-\frac{2\Delta t}{T_m}}}\right)^2 P_{k|k} + \frac{\left(\frac{K_m}{T_m}\right)^2 \Delta t e^{-2\Delta t/T_m}}{1+\left(\frac{K_m}{T_m}\right)^2 e^{-\frac{2\Delta t}{T_m}}}$$

$$H_u = H\phi L_2 = \left[1\ 1\right] \begin{bmatrix} e^{-\Delta t/T_m} & 0 \\ 0 & 1 \end{bmatrix} \begin{bmatrix} -1 \\ 1 \end{bmatrix} = 1 - e^{-\Delta t/T_m}.$$

The Kalman gain is

$$K_k = P_{k|k-1} H_u^T \left[H_u P_{k|k-1} H_u^T + H Q H^T\right]^{-1} = \frac{P_{k|k-1}\left(1 - e^{-\frac{\Delta t}{T_m}}\right)}{P_{k|k-1}\left(1 - e^{-\frac{\Delta t}{T_m}}\right)^2 + \left[1+\left(\frac{K_m}{T_m}\right)^2 e^{-2\Delta t/T_m}\right]\Delta t}.$$

The corrector is

$$\hat{p}_{k|k} = \hat{p}_{k|k-1} + K_k\big[y_{u,k+1} - H_u\hat{p}_{k|k-1}\big]\hat{p}_{k|k-1} + K_k\Big[y_{u,k+1} - \Big(1 - e^{-\Delta t/T_m}\Big)\hat{p}_{k|k-1}\Big],$$

and its error covariance matrix is

$$P_{k|k} = [1 - K_k H_u]P_{k|k-1} = \Big[1 - \Big(1 - e^{-\frac{\Delta t}{T_m}}\Big)K_k\Big]P_{k|k-1}$$

$$= \left[1 - \frac{P_{k|k-1}\Big(1 - e^{-\frac{\Delta t}{T_m}}\Big)^2}{P_{k|k-1}\Big(1 - e^{-\frac{\Delta t}{T_m}}\Big)^2 + \Big[1 + \Big(\frac{K_m}{T_m}\Big)^2 e^{-2\Delta t/T_m}\Big]\Delta t}\right]P_{k|k-1}$$

$$P_{k|k} = \frac{\Big[1 + \Big(\frac{K_m}{T_m}\Big)^2 e^{-2\Delta t/T_m}\Big]\Delta t}{P_{k|k-1}\Big(1 - e^{-\frac{\Delta t}{T_m}}\Big)^2 + \Big[1 + \Big(\frac{K_m}{T_m}\Big)^2 e^{-2\Delta t/T_m}\Big]\Delta t}P_{k|k-1}.$$

■

10.4 Schmidt–Kalman Filter

In some applications, the state matrix and the process noise are both block diagonal. We partition the state into partial state $\boldsymbol{x}_k$ and partial state $\boldsymbol{y}_k$ and write the state equation in the form

$$\begin{bmatrix} \boldsymbol{x}_{k+1} \\ \boldsymbol{y}_{k+1} \end{bmatrix} = \begin{bmatrix} \Phi_k^x & 0 \\ 0 & \Phi_k^y \end{bmatrix}\begin{bmatrix} \boldsymbol{x}_k \\ \boldsymbol{y}_k \end{bmatrix} + \begin{bmatrix} \boldsymbol{w}_{xk} \\ \boldsymbol{w}_{yk} \end{bmatrix} \tag{10.38}$$

with measurement equation

$$z_k = \big[H_k^x | H_k^y\big]\begin{bmatrix} \boldsymbol{x}_k \\ \boldsymbol{y}_k \end{bmatrix} + \boldsymbol{v}_k \tag{10.39}$$

The noise processes are zero-mean Gaussian with process noise covariance matrix

$$Q_k = \begin{bmatrix} Q_k^x & 0 \\ 0 & Q_k^y \end{bmatrix} \tag{10.40}$$

The Schmidt–Kalman filter is a reduced-order Kalman filter that considers but does not estimate the partial state $\boldsymbol{y}_k$ so that it can provide an improvement over completely ignoring part of the dynamics while reducing the computational load. The corrector equation is

$$\hat{\boldsymbol{x}}_{k|k} = \hat{\boldsymbol{x}}_{k|k-1} + K_k^x\left[\boldsymbol{z}_k - H_k^x\hat{\boldsymbol{x}}_{k|k-1}\right] \tag{10.41}$$

We partition the error covariance matrix

$$P_{k|k} = \begin{bmatrix} P_{kkk}^x & P_{k|k}^{xy} \\ P_{kk}^{yx} & P_{k|k}^y \end{bmatrix}, P_{kkk}^{yx} = P_{kk}^{yxT} \tag{10.42}$$

and the gain matrix

$$K_k = \begin{bmatrix} K_k^x \\ K_k^y \end{bmatrix} \tag{10.43}$$

Substituting the partitioned covariance matrix in the gain expression gives

$$K_k = P_{k|k-1}H_k^T\left(H_kP_{k|k-1}H_k^T + R_k\right)^{-1}$$

$$= \begin{bmatrix} P_{k|k-1}^x & P_{k|k-1}^{xy} \\ P_{k|k-1}^{3x} & P_{k|k-1}^y \end{bmatrix}\begin{bmatrix} H_k^{xT} \\ H_k^{yT} \end{bmatrix}\left(\begin{bmatrix} H_k^x | H_k^y \end{bmatrix}\begin{bmatrix} P_{k|k-1}^x & P_{k|k-1}^{xy} \\ P_{k|k-1}^{yx} & P_{k|k-1}^y \end{bmatrix}\begin{bmatrix} H_k^{xT} \\ H_k^{yT} \end{bmatrix} + R_k\right)^{-1}$$

which simplifies to

$$K_k^x = \left(P_{k|k-1}^x H_k^{xT} + P_{k|k-1}^y H_k^{yT}\right)$$
$$\times\left(H_k^x P_{k|k-1}^x H_k^{xT} + P_{k|k-1}^y H_k^{yT} + H_k^x P_{k|k-1}^{xy} H_k^{yT} + H_k^y P_{k|k-1}^{xx} H_k^{xT} + R_k\right)^{-1} \tag{10.44}$$

We substitute in Joseph form of the covariance update

$$\begin{bmatrix} P_{k|k}^x & P_{k|k}^{xy} \\ P_{k|k}^{yx} & P_{k|k}^y \end{bmatrix} = \left(\begin{bmatrix} I_n & 0 \\ 0 & I_n \end{bmatrix} - \begin{bmatrix} K_k^x \\ 0 \end{bmatrix}\left[H_k^x | H_k^y\right]\right)\begin{bmatrix} P_{k|k-1}^x & P_{k|k-1}^{xy} \\ P_{k|k-1}^{1xx} & P_{k|k-1}^y \end{bmatrix}$$
$$\left(\begin{bmatrix} I_n & 0 \\ 0 & I_n \end{bmatrix} - \begin{bmatrix} K_{xk} \\ 0 \end{bmatrix}\left[H_k^x | H_k^y\right]\right)^T + \begin{bmatrix} K_k^x \\ 0 \end{bmatrix} R_k\left[K_k^{xT} | 0\right]$$
$$= \begin{bmatrix} I_n - K_k^x H_k^x & -K_k^x H_y \\ 0 & I_n \end{bmatrix}\begin{bmatrix} P_{hk-1}^x & P_{hk-1}^{xy} \\ P_{hk-1}^{1x} & P_{hk-1}^y \end{bmatrix}\begin{bmatrix} I_n - K_k^x H_k^x & -K_k^x H_k^y \\ 0 & I_n \end{bmatrix}^T$$
$$+ \begin{bmatrix} K_k^x R_k K_k^{xT} & 0 \\ 0 & 0 \end{bmatrix}$$
$$= \begin{bmatrix} \left(I_n - K_k^x H_k^x\right)P_{kk-1}^x - K_k^x H_k^y P_{kk-1}^{\prime x} & \left(I_n - K_k^x H_x\right)P_{kk-1}^{xy} - K_k^x H_k^y P_{kk-1}^y \\ 0 & P_{kk-1}^v \end{bmatrix}$$

$$\begin{bmatrix} \left(I_n - K_k^x H_k^x\right)^T & 0 \\ -H_k^{vi} K_k^{xT} & I_n \end{bmatrix} + \begin{bmatrix} K_k^x R_k K_k^{xT} & 0 \\ 0 & 0 \end{bmatrix}$$

Simplifying gives the covariance update for the corrector

$$\begin{bmatrix} P_{k|k}^x & P_{k|k}^{xy} \\ P_{k|k}^{|x} & P_{k|k}^v \end{bmatrix} = \begin{bmatrix} P_{k|k}^x & (I_n - K_k^x H_{xk}) P_{k|k-1}^{xy} - K_k^x H_k^y P_{k|k-1}^v \\ P_{k|k-1}^{vx} (I_n - K_k^x H_k^x)^T - P_{k|k-1}^v H_k^{yT} K_k^{xT} & P_{k|k-1}^y \end{bmatrix} \tag{10.45}$$

with the submatrices

$$\begin{aligned} P_{k|k}^x &= \left(I_n - K_k^x H_k^x\right) P_{k|k-1}^x \left(I_n - K_k^x H_k^x\right)^T \\ &\quad - K_k^x H_{nk} P_{k|k-1}^{xx} \left(I_n - K_k^x H_k^x\right)^T - \left(I_n - K_k^x H_k^x\right) P_{k|k-1}^{xy} H_k^{yT} K_k^{xT} \\ &\quad + K_k^x \left(H_k^y P_{k|k-1}^y H_k^{yT} + R\right) K_k^{xT} \end{aligned}$$

$$P_{k|k}^{xy} = \left(I_n - K_k^x H_{x\hat{k}}\right) P_{k|k-1}^{xy} - K_k^x H_{\hat{k}}^y P_{k|k-1}^y$$

$$P_{k|k}^y = P_{k|k-1}^y \tag{10.46}$$

We simplify the covariance matrix for the estimate to obtain

$$P_{k|k}^x = \left(I_n - K_k^x H_k^x\right) P_{k|k-1}^x - K_k^x H_k^x P_{k|k-1}^y \tag{10.47}$$

For the predictor, we have the a priori estimate

$$\hat{\boldsymbol{x}}_{k+1|k} = \Phi_k^x \hat{\boldsymbol{x}}_{k|k} \tag{10.48}$$

and the error covariance matrix

$$\begin{aligned} \begin{bmatrix} P_{k+1|k}^x & P_{k+1|k}^{xy} \\ P_{k+1|k}^{yx} & P_{k+1|k}^y \end{bmatrix} &= \begin{bmatrix} \Phi_k^x & 0 \\ 0 & \Phi_k^y \end{bmatrix} \begin{bmatrix} P_{k|k}^x & P_{k|k}^{xy} \\ P_{k|k}^{yx} & P_{k|k}^y \end{bmatrix} \begin{bmatrix} \Phi_k^x & 0 \\ 0 & \Phi_k^y \end{bmatrix}^T + \begin{bmatrix} Q_k^x & 0 \\ 0 & Q_k^y \end{bmatrix} \\ &= \begin{bmatrix} \Phi_k^x P_{k|k}^x \Phi_k^{xT} + Q_k^x & \Phi_k^x P_{k|k}^{yy} \Phi_k^{yT} \\ \Phi_k^y P_{k|k}^{yx} \Phi_k^{xT} & \Phi_k^y P_{k|k}^y \Phi_k^{yT} + Q_k^y \end{bmatrix} \end{aligned}$$

Equating gives the error covariance equations

$$P_{k+1|k}^x = \Phi_k^x P_{k|k}^x \Phi_k^{xT} + Q_k^x$$

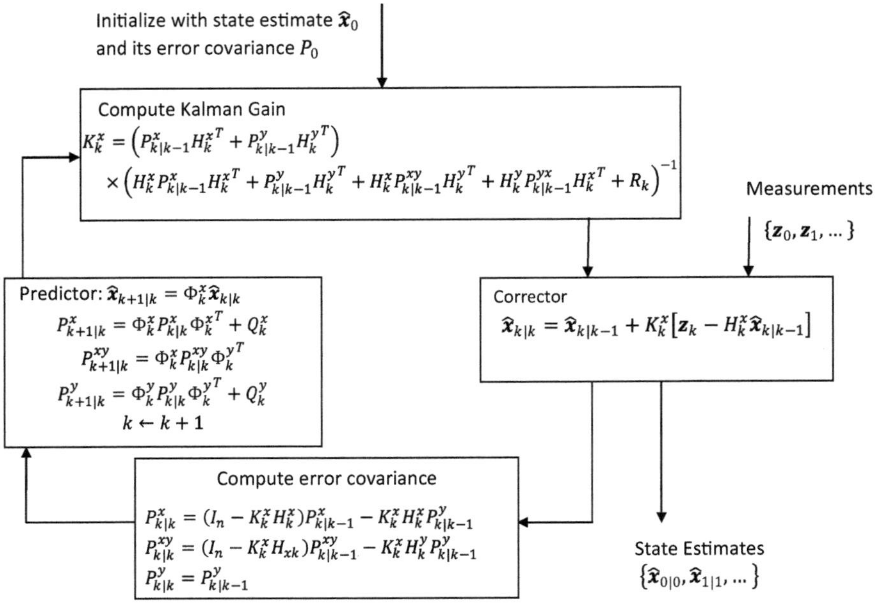

Fig. 10.7 Block diagram for the Schmidt–Kalman filter

$$P_{k+1k}^{\lambda y} = \Phi_k^x P_{k|k}^{xy} \Phi_k^{yT}$$

$$P_{k+1|k}^y = \Phi_k^y P_{k|k}^y \Phi_k^{yT} + Q_k^y \tag{10.49}$$

The Schmidt–Kalman filter loop is depicted in Figs. 10.7, 10.8 and 10.9.

Example 10.5 Consider the process model

$$\begin{bmatrix} x_{k+1} \\ y_{k+1} \end{bmatrix} = \begin{bmatrix} \Phi^x & 0 \\ 0 & \Phi^y \end{bmatrix} \begin{bmatrix} x_k \\ y_k \end{bmatrix} + \begin{bmatrix} w_k^x \\ w_k^y \end{bmatrix}$$

with measurements

$$z_k = \begin{bmatrix} 1 & 1 \end{bmatrix} \begin{bmatrix} x_k \\ y_k \end{bmatrix} + v_k$$

and noise covariances $Q = \text{diag}\{\sigma_x^2, \sigma_y^2\}$, $R = r$. The initial conditions are $\hat{\mathbf{x}}_0 = 0$ and the covariance matrix $P_0 = \text{diag}\{\sigma_x^2, \sigma_y^2\}$. Design a Schmidt–Kalman filter for the system and compare its error to (i) the error if the state variable y_k is completely neglected; (ii) the optimal filter.

Solution

For this system, $H_x = H_y = 1$. The a priori estimate is

$$\hat{\boldsymbol{x}}_{k+1|k} = \Phi_x \hat{\boldsymbol{x}}_{k|k}$$

with error covariance

$$P^x_{k+1k} = \Phi^{x2}_k P^x_{kkk} + \sigma^2_x$$

$$P^{vy}_{k+1k} = \Phi^x_k \Phi^y_k P^n_{kk}$$

$$P^y_{k+1|k} = \Phi^{y2}_k P^y_{k|k} + \sigma^2_y$$

The filter gain is

$$\begin{aligned} K^x_k &= \left(P^x_{k|k-1} H^{sT}_k + P^x_{k|k-1} H^T_k\right) \\ &\quad \times \left(H^x_k P^x_{k|k-1} H^{xT}_k + P^x_{k|k-1} H^x_k + H^x_k P^x_{k|k-1} H^{xT}_k + H^x_k P^{xx}_{kk-1} H^{xT}_k + R_k\right)^{-1} \\ &= \frac{P^x_{k|k-1} + P^y_{k|k-1}}{P^x_{k|k-1} + P^y_{k|k-1} + 2P^{xy}_{k|k-1} + r}. \end{aligned}$$

At any time k, the corrector error covariance is

$$P^x_{k|k} = \left(1 - K^x_k\right) P^x_{k|k-1} - K^x_k P^y_{k|k-1} = \frac{\left(2P^{xy}_{k|k-1} - P^y_{k|k-1} + r\right) P^x_{k|k-1} - \left(P^y_{k|k-1}\right)^2}{P^x_{k|k-1} + P^y_{k|k-1} + 2P^{xy}_{k|k-1} + r}$$

$$P^{xy}_{k|k} = (1 - K_x) P^{xy}_{k|k-1} - K_x P^y_{k|k-1} = \frac{\left(2P^{xy}_{k|k-1} + r\right) P^{xy}_{k|k-1} - \left(P^x_{k|k-1} + P^y_{k|k-1}\right) P^y_{k|k-1}}{P^x_{k|k-1} + P^y_{k|k-1} + 2P^{xy}_{k|k-1} + r}.$$

$$P^y_{k|k} = P^y_{k|k-1}$$

If the state variable y is completely omitted, the term $\left(P^x_{k|k-1} + P^y_{k|k-1}\right) P^y_{k|k-1}$ is removed from the numerator of $P^x_{k|k}$ and the error covariance increases to

$$P^x_{k|k} = \left(1 - K^x_k\right) P^x_{k|k-1} = \frac{\left(2P^{xy}_{k|k-1} + r\right) P^x_{k|k-1}}{P^x_{k|k-1} + P^y_{k|k-1} + 2P^{xy}_{k|k-1} + r}.$$

For the optimal filter, the optimal gain is

$$K_k = P_{k|k-1} H^T_k \left[H_k P_{k|k-1} H^T_k + R_k\right]^{-1}$$

$$= \frac{1}{P^x_{k|k-1} + P^y_{k|k-1} + 2P^{xy}_{k|k-1} + r}\begin{bmatrix} P^x_{k|k-1} + P^{xy}_{k|k-1} \\ P^{xy}_{k|k-1} + P^{yy}_{k|k-1} \end{bmatrix}$$

and K^x_k is

$$K^x_k = \frac{P^x_{k|k-1} + P^{xy}_{k|k-1}}{P^x_{k|k-1} + P^y_{k|k-1} + 2P^{xy}_{k|k-1} + r}.$$

The error covariance is

$$P_{k|k} = (I_2 - K_k H_k) P_{k|k-1}$$
$$= \left(I_2 - \frac{1}{P^x_{k|k-1} + P^y_{k|k-1} + 2P^{xy}_{k|k-1} + r}\begin{bmatrix} P^x_{k|k-1} + P^{xy}_{k|k-1} & P^x_{k|k-1} + P^{xy}_{k|k-1} \\ P^{xy}_{k|k-1} + P^{yy}_{k|k-1} & P^{xy}_{k|k-1} + P^{yy}_{k|k-1} \end{bmatrix}\right)\begin{bmatrix} P^x_{k|k-1} & P^{xy}_{k|k-1} \\ P^{yx}_{k|k-1} & P^y_{k|k-1} \end{bmatrix}.$$

This gives the error covariance

$$P^x_{k|k} = \frac{\left(P^y_{k|k-1} + r\right)P^x_{k|k-1} - {P^{xy}_{k|k-1}}^2}{P^x_{k|k-1} + P^y_{k|k-1} + 2P^{xy}_{k|k-1} + r} = \frac{\det\left(P_{k|k-1}\right) + r P^x_{k|k-1}}{P^x_{k|k-1} + P^y_{k|k-1} + 2P^{xy}_{k|k-1} + r}$$

Compare three filters, the optimal Kalman filter, the Schmidt–Kalman filter, and the reduced-order filter that does not include considered variables. The results show that. While the Schmidt filter is suboptimal, it is significantly better than the filter that ignores the partial state y_k.

10.5 Sequential DKF Computation

For the process and measurement models of (10.1) with zero-mean, white, Gaussian uncorrelated noise, we consider the error covariance update for the information filter (see Chap. 9) in the case of a block diagonal R_k.

$$(P_{kk})^{-1} = (P_{klk-1})^{-1} + H_k^T R_k^{-1} H_k = \left(P_{k\mu-1}\right)^{-1}$$
$$+ \left[H_k^{1\tau} \; H_k^{2\tau} \; \dots \; H_k^{lT} \right] \begin{bmatrix} \left(R_k^1\right)^{-1} & 0 & \dots & 0 \\ 0 & \left(R_k^2\right)^{-1} & \dots & 0 \\ \vdots & \vdots & \ddots & \vdots \\ 0 & 0 & \dots & \left(R_k^l\right)^{-1} \end{bmatrix} \begin{bmatrix} H_k^1 \\ H_k^2 \\ \vdots \\ H_k^l \end{bmatrix}$$

Multiplying the matrices gives an expression including a summation

$$(P_{kk})^{-1} = \left(P_{k|k-1}\right)^{-1} + \sum_{i=1}^{l} H_k^{iT} \left(R_k^i\right)^{-1} H_k^i \tag{10.50}$$

which shows that the computation can be implemented sequentially using the contribution of each measurement separately, or in parallel followed by addition. For a diagonal R_k, the inversion in each term reduces to a simple division and the vector measurement is equivalent to a sequence of scalar measurements. If the matrix is not diagonal, a similarity transformation can reduce it to diagonal form and provide an efficient computational algorithm, particularly if the measurement noise covariance matrix is constant. The transformation leaves the measurement noise terms uncorrelated with each measurement added providing more information about the process. The covariance matrix transformation is achieved by transforming the measurements using the modal decomposition of the covariance matrix

$$R = L\Lambda L^T,\ \Lambda = \text{diag}\{\lambda_1, \ldots, \lambda_m\} \tag{10.51}$$

where L is the modal matrix of eigenvector and $L^T = L^{-1}$. Because the covariance matrix is symmetric positive definite, its eigenvalues are real and positive, i.e., $\lambda_i > 0, i = 1, 2, \ldots, m$. The data transformation is

$$L^T z_k = \overline{z_k} = L^T H \boldsymbol{x}_k + L^T \boldsymbol{v}_k = \overline{H} \boldsymbol{x}_k + \overline{\boldsymbol{v}}_k \tag{10.52}$$

The transformed noise vector is zero-mean with covariance matrix

$$E\left\{\overline{\boldsymbol{v}}_k \overline{\boldsymbol{v}}_k^T\right\} = L^T E\left\{\boldsymbol{v}_k \boldsymbol{v}_k^T\right\} L = L^T R L = \Lambda \tag{10.53}$$

Thus, we have uncorrelated noise $\overline{\boldsymbol{v}}_{ki}$ and diagonal covariance that allow us to process the measurents one at a time.

The summation in the error covariance expression is now in the form

$$\overline{H}_k^T \Lambda_k^{-1} \overline{H}_k = \left[\overline{\mathbf{h}}_k^1\ \overline{\mathbf{h}}_k^2 \ldots \overline{\mathbf{h}}_k^m\right] \begin{bmatrix} \lambda_{k,1}^{-1} & 0 & \cdots & 0 \\ 0 & \lambda_{k,2}^{-1} & \cdots & 0 \\ \vdots & \vdots & \ddots & \vdots \\ 0 & 0 & \cdots & \lambda_{k,m}^{-1} \end{bmatrix} \begin{bmatrix} \mathbf{h}_k^{1T} \\ \mathbf{h}_k^{2T} \\ \vdots \\ \mathbf{h}_k^{mT} \end{bmatrix}$$

where $\mathbf{h}_k^{iT} = i$th. row of $\overline{H}_k$

$$\left(P_{k|k}\right)^{-1} = \left(P_{k|k-1}\right)^{-1} + \sum_{i=1}^{m} \overline{\mathbf{h}}_k^i \overline{\mathbf{h}}_k^{iT} / \lambda_k$$

The formula avoids matrix inversion with sequential covariance computation, and the recursion is equivalent to the covariance form. Thus, we can treat each measurement as a separate corrector for the estimate and add its contribution using a column gain vector. The update of the inverse of the covariance matrix is equivalent to the steps

$$\left(P_k^i\right)^{-1} = \left(P_{k|k-1}\right)^{-1} + \overline{h}_k^i \overline{h}_k^{iT} / \lambda_{k,i}, i = 1, \ldots, m \tag{10.54}$$

with $\left(P_{k|k}\right)^{-1} = \left(P_k^m\right)^{-1}$. Inverting gives the covariance expression

$$P_k^i = \left[\left(P_k^{i-1}\right)^{-1} + \overline{\mathbf{h}}_k^i \overline{\mathbf{h}}_k^{iT} / \lambda_{k,i}\right]^{-1}, i = 1, \ldots, m \tag{10.55}$$

Use the matrix inversion lemma

$$-A_1 + A_2 A_4^{-1} A_3\right]^{-1} = A_1^{-1} - A_1^{-1} A_2 \left[A_4 + A_3 A_1^{-1} A_2\right]^{-1} A_3 A_1^{-1}$$

with

$$A_1 = \left(P_k^{i-1}\right)^{-1}, A_2 = \overline{\mathbf{h}}_k^i, A_3 = \overline{\mathbf{h}}_k^{iT}, A_4 = \lambda_{k,i}$$

to obtain

$$P_k^i = P_k^{i-1} - P_k^{i-1} \overline{\mathbf{h}}_k^i \left(\overline{\mathbf{h}}_k^{iT} P_k^{i-1} \overline{\mathbf{h}}_k^i + \lambda_{k,i}\right)^{-1} \overline{\mathbf{h}}_k^{iT} P_k^{i-1}$$

Defining the gain as

$$\mathbf{k}_k^i = P_k^{i-1} \overline{\mathbf{h}}_k^i / \left[\overline{\mathbf{h}}_k^{iT} P_k^{i-1} \overline{\mathbf{h}}_k^i + \lambda_{k,i}\right] \tag{10.56}$$

gives the date formula

$$P_k^i = \left[I_n - \mathbf{k}_k^i \overline{\mathbf{h}}_k^{iT}\right] P_k^{i-1}, i = 1, \ldots, m \tag{10.57}$$

With the initial condition $P_k^0 = P_{k|k-1}$, we obtain the covariance of the corrector $P_{k|k} = P_k^m$. The corrector is therefore equivalent to m correctors

$$\hat{\mathbf{x}}_k^i = \hat{\mathbf{x}}_k^{i-1} + \mathbf{k}_k^i \left[\overline{\mathbf{z}}_{k,i} - \overline{\mathbf{h}}_k^{iT} \hat{\mathbf{x}}_k^{i-1}\right], \hat{\mathbf{x}}_k^0 = \hat{\mathbf{x}}_{k|k-1} \tag{10.58}$$

Note that the expression for the gain is the same as that of the optimal covariance filter with one measurement. The procedure for sequential filter implementation is summarized as follows.

Procedure 10.1
Initialize with $\left(\hat{\boldsymbol{x}}_k^0 = \hat{\boldsymbol{x}}_{k|k-1}, P_k^0 = P_{k|k-1}\right)$.

For $i = 1, 2, \ldots, m$, calculate

$$P_k^i = \left[I_n - \mathbf{k}_k^i \overline{\mathbf{h}}_k^{iT}\right] P_k^{i-1},\ P_k^0 = P_{k|k-1}$$

$$\mathbf{k}_k^i = P_k^{i-1} \overline{\mathbf{h}}_k^i / \left[\overline{\mathbf{h}}_k^{iT} P_k^{i-1} \overline{\mathbf{h}}_k^i + \lambda_{k,i}\right]$$

$$\hat{\mathbf{x}}_k^i = \hat{\mathbf{x}}_k^{i-1} + \mathbf{k}_k^i \left[\overline{\mathbf{Z}}_{k,i} - \overline{\mathbf{h}}_k^{iT} \hat{\mathbf{X}}_k^{i-1}\right]$$

End

The corrected estimate is $\hat{\boldsymbol{x}}_{k|k} = \hat{\boldsymbol{x}}_k^m$, and its error covariance matrix is $P_{k|k} = P_k^m$. ■

Next, we consider the innovations process for sequential computation. Recall that the innovations process for the Kalman filter is zero-mean Gaussian and white (uncorrelated in time). The innovations for the sequential filter

$$\tilde{\overline{z}}_{k,i} = \overline{z}_{k,i} - \hat{\overline{z}}_{k,i}^{-} = \overline{\mathbf{h}}_k^{iT} \left(\mathbf{x}_k - \hat{\mathbf{x}}_k^{i-1}\right) + \overline{\mathbf{v}}_{k,i} \tag{10.59}$$

have the same properties. The proof is left as an exercise.

Example 10.6 Consider the Gauss–Markov system of Example 10.1

$$x_{k+1} = e^{-\beta t} x_k + w_k$$

with measurement vector

$$\boldsymbol{z}_k = \begin{bmatrix} z_{k,1} & z_{k,2} \end{bmatrix}^T = \begin{bmatrix} 1 \\ 0.2 \end{bmatrix} x_k + \boldsymbol{v}_k.$$

The noise processes are orthogonal, zero-mean, white, Gaussian with statistics

$$w_k \sim \mathcal{N}(0, Q),\ Q = \sigma_w^2.$$

$$\boldsymbol{v}_k \sim \mathcal{N}(0, R),\ R = \sigma_v^2 I_2.$$

Calculate the corrected state estimate $\hat{x}_{0|0}$ and error covariance $p_{0|0}$ using the standard covariance filter equations, then repeat the calculations using sequential computation.

Solution

The initial conditions are

$$\hat{x}_0,\ P_0 = \sigma_x^2.$$

The Kalman gain is

$$K_0 = P_0 H^T \left(H P_0 H^T + R\right)^{-1} = \sigma_x^2 \begin{bmatrix} 1 \\ 0.2 \end{bmatrix} \left(\sigma_x^2 \begin{bmatrix} 1 \\ 0.2 \end{bmatrix} [\, 1 \; 0.2 \,] + \sigma_v^2 I_2 \right)^{-1}$$

$$= \sigma_x^2 [\, 1 \; 0.2 \,] \begin{bmatrix} \sigma_x^2 + \sigma_v^2 & 0.2\sigma_x^2 \\ 0.2\sigma_x^2 & 0.04\sigma_x^2 + \sigma_v^2 \end{bmatrix}^{-1}$$

$$= \sigma_x^2 [\, 1 \; 0.2 \,] \frac{\begin{bmatrix} 0.04\sigma_x^2 + \sigma_v^2 & -0.2\sigma_x^2 \\ -0.2\sigma_x^2 & \sigma_x^2 + \sigma_v^2 \end{bmatrix}}{(\sigma_x^2 + \sigma_v^2)(0.04\sigma_x^2 + \sigma_v^2) - 0.04\sigma_x^4}.$$

This simplifies to

$$K_0 = \frac{\sigma_x^2 [\, \sigma_v^2 \; 0.2\sigma_v^2 \,]}{\sigma_v^2 (1.04\sigma_x^2 + \sigma_v^2)} = \frac{\sigma_x^2 [\, 1 \; 0.2 \,]}{1.04\sigma_x^2 + \sigma_v^2}.$$

The state estimate is

$$\hat{x}_{0|0} = [1 - K_0 H] \hat{x}_0 + K_0 z_0$$

$$= \left[1 - \frac{\sigma_x^2 [\, 1 \; 0.2 \,]}{1.04\sigma_x^2 + \sigma_v^2} \begin{bmatrix} 1 \\ 0.2 \end{bmatrix} \right] \hat{x}_0 + \frac{\sigma_x^2 [\, 1 \; 0.2 \,]}{1.04\sigma_x^2 + \sigma_v^2} z_0$$

$$= \left[1 - \frac{1.04\sigma_x^2}{1.04\sigma_x^2 + \sigma_v^2} \right] \hat{x}_0 + \frac{\sigma_x^2 [\, 1 \; 0.2 \,]}{1.04\sigma_x^2 + \sigma_v^2} z_0.$$

We simplify the expression to obtain the estimate

$$\hat{x}_{0|0} = \left[\frac{\sigma_v^2}{1.04\sigma_x^2 + \sigma_v^2} \right] \hat{x}_0 + \frac{\sigma_x^2 [\, 1 \; 0.2 \,]}{1.04\sigma_x^2 + \sigma_v^2} z_0.$$

The error covariance is

$$P_{0|0} = [1 - K_0 H] P_0 = \frac{\sigma_v^2 \sigma_x^2}{1.04\sigma_x^2 + \sigma_v^2}.$$

For sequential computation, the corrector calculations use one measurement at a time starting with the first measurement. We calculate the first gain

$$k_0^1 = \frac{P_0 h^1}{h^1 P_0 h^1 + \lambda_1} = \frac{\sigma_x^2}{\sigma_x^2 + \sigma_v^2}.$$

The estimate is updated with the first measurement

$$\hat{x}_0^1 = (1 - k_0^1 h^1)\hat{x}_0^0 + k_0^1 z_{0,1}$$

$$= \left[1 - \frac{\sigma_x^2}{\sigma_x^2 + \sigma_v^2}\right]\hat{x}_0 + \frac{\sigma_x^2}{\sigma_x^2 + \sigma_v^2} z_{0,1}$$

$$= \frac{\sigma_v^2}{\sigma_x^2 + \sigma_v^2}\hat{x}_0 + \frac{\sigma_x^2}{\sigma_x^2 + \sigma_v^2} z_{0,1}$$

$$P_0^1 = \left[1 - k_0^1 h^1\right] P_0 = \frac{\sigma_v^2 \sigma_x^2}{\sigma_x^2 + \sigma_v^2}$$

The corrector for the second measurement

$$k_0^2 = \frac{P_0^1 h^2}{h^2 P_0^1 h^2 + \lambda_2} = \frac{0.2\sigma_v^2 \sigma_x^2}{0.04\sigma_v^2 \sigma_x^2 + \sigma_v^2(\sigma_x^2 + \sigma_v^2)}$$

$$= \frac{0.2\sigma_x^2}{1.04\sigma_x^2 + \sigma_v^2}.$$

The estimate is updated with the second measurement

$$\hat{x}_0^2 = (1 - k_0^2 h^2)\hat{x}_0^1 + k_0^2 z_{0,2}$$

$$= \left[1 - \frac{0.04\sigma_x^2}{1.04\sigma_x^2 + \sigma_v^2}\right]\left(\frac{\sigma_v^2}{\sigma_x^2 + \sigma_v^2}\hat{x}_0 + \frac{\sigma_x^2}{\sigma_x^2 + \sigma_v^2} z_{0,1}\right) + \frac{0.2\sigma_x^2}{1.04\sigma_x^2 + \sigma_v^2} z_{0,2}$$

$$= \left[\frac{\sigma_x^2 + \sigma_v^2}{1.04\sigma_x^2 + \sigma_v^2}\right]\left(\frac{\sigma_v^2}{\sigma_x^2 + \sigma_v^2}\hat{x}_0 + \frac{\sigma_x^2}{\sigma_x^2 + \sigma_v^2} z_{0,1}\right) + \frac{0.2\sigma_x^2}{1.04\sigma_x^2 + \sigma_v^2} z_{0,2}.$$

The corrected estimate is

$$\hat{x}_{0|0} = \frac{\sigma_v^2}{1.04\sigma_x^2 + \sigma_v^2}\hat{x}_0 + \frac{\sigma_x^2}{1.04\sigma_x^2 + \sigma_v^2} z_{0,1} + \frac{0.2\sigma_x^2}{1.04\sigma_x^2 + \sigma_v^2} z_{0,2}.$$

The error covariance is

$$P_0^2 = \left[1 - k_0^2 h^2\right] P_0^1 = \left[\frac{\sigma_x^2 + \sigma_v^2}{1.04\sigma_x^2 + \sigma_v^2}\right] \frac{\sigma_v^2 \sigma_x^2}{\sigma_x^2 + \sigma_v^2}$$

$$P_{0|0} = \frac{\sigma_v^2 \sigma_x^2}{1.04\sigma_x^2 + \sigma_v^2}.$$

The equations provide more efficient computation and matrix inversion is eliminated, but the results are identical to the ones obtained using the standard filter equations.

■

10.6 Square Root Filtering

Embedded systems for the implementation of Kalman filters often use shorter word length, which can result in significant computational errors. The computational errors can be reduced by propagating the square roots of covariance matrices rather than the matrices themselves. This is known as square root filtering. The price paid for the improved accuracy is increased computational cost and a more complex algorithm.

Numerical problems arise in Kalman filter computation when a covariance matrix become almost singular. The condition number, that ratio of the maximum to the minimum singular value,

$$\kappa(P) = \|P\| \left\|P^{-1}\right\| = \frac{\sigma_{\max}(P)}{\sigma_{\min}(P)} \tag{10.60}$$

is a measure of how close a matrix is to singularity expressed in terms of the norm of the matrix. The singular values are calculated using the singular value decomposition

$$P = U\Sigma V^{*},\ \Sigma = \mathrm{diag}\{\sigma_1, \ldots, \sigma_n\} \tag{10.61}$$

with orthogonal matrices U and V. The orthogonality of the two matrices implies that

$$PP^{T} = U\Sigma V^{*} V\Sigma U^{*} = U\mathrm{diag}\left\{\sigma_1^2, \ldots, \sigma_n^2\right\}U^{*}$$

$$P^{T}P = V\Sigma U^{*} U\Sigma V^{*} = V\mathrm{diag}\left\{\sigma_1^2, \ldots, \sigma_n^2\right\}V^{*} \tag{10.62}$$

Hence, the eigenvalues of $P^T P$, or PP^T, are

$$\lambda\left(P^{T}P\right) = \lambda\left(PP^{T}\right) = \sigma^2(P).$$

For P symmetric positive semidefinite, we have

$$\sigma^2(P) = \lambda\left(P^2\right) = \lambda^2(P) \geq 0 \tag{10.63}$$

A singular matrix has at least one zero singular value and it condition number is infinite. If the largest singular value is much larger than the smallest and the condition number is large, the matrix behaves as a singular matrix in computation. The condition number has the following properties

$$\kappa(P) \geq 1$$

$$\kappa(\alpha P) = \kappa(P) \tag{10.64}$$

The MATLAB command to calculate the condition number is
≫ **cond(P)**
The error covariance matrix can become ill-conditioned if one state variable has a much higher uncertainty than another. The matrix is then computationally singular.

We show that the square of a matrix has a smaller condition number. The square root of a symmetric positive definite matrix is obtained using the Cholesky decomposition. The MATLAB command to obtain the decomposition
≫ **Ps = chol(P)**
gives an upper triangular matrix corresponding to the decomposition $P = P_s^T P_s$. We use the decomposition $P = P_s P_s^T$ and the MATLAB command
≫ **Ps = chol(P,'lower').**
so that P_s is a lower triangular matrix. The singular values of P_s satisfy

$$\sigma^2(P_s) = \lambda\left(P_s^T P_s\right) = \lambda\left(P_s P_s^T\right) \geq 0 \tag{10.65}$$

The singular values of a symmetric matrix P satisfy.

$$\sigma^2(P) = \lambda\left(P^T P\right) = \lambda\left(P^2\right) = \lambda^2(P) = \lambda\left(P_s P_s^T P_s P_s^T\right) = \sigma^4(P_s) \geq 0.$$

The condition number of P_s.

$$\kappa(P_s) = \frac{\sigma_{\max}(P_s)}{\sigma_{\min}(P_s)} = \sqrt{\frac{\sigma_{\max}(P)}{\sigma_{\min}(P)}} = \sqrt{\kappa(P)}.$$

Because the condition number is lower bounded by unity, we have

$$\sqrt{\kappa(P)} \leq \kappa(P_s) \tag{10.66}$$

with equality only in the uninteresting case of a matrix whose singular values are all equal whose condition number is unity. The following numerical example demonstrates the change in condition number when using the square root matrix.

Example 10.7 Compare the condition number of the covariance matrix $P = \text{diag}\left\{10^7, 10^{-7}\right\}$ to that of its square root.

Solution

The matrix is positive definite with only two singular values. The condition number is

$$\kappa(P) = \frac{\sigma_{\max}(P)}{\sigma_{\min}(P)} = \frac{10^7}{10^{-7}} = 10^{14}.$$

This is likely to cause problems in computation and using the square root matrix reduces the condition number to

$$\kappa(P_s) = \sqrt{\kappa(P)} = 10^7.$$

This is still a large condition number, but it is significantly smaller than the condition number of the covariance matrix. While there is a limit to the improvement in the condition number achievable by using the square root matrix, the improvement can reduce computational errors.

We consider the process

$$\boldsymbol{x}_k = \phi(k, k-1)\boldsymbol{x}_{k-1} + \boldsymbol{w}_{k-1} \tag{10.67}$$

where $\phi(k, k-1)$ is the state-transition matrix, and $\boldsymbol{w}(k)$ is a zero-mean white Gaussian process noise with covariance matrix $Q_k = E\{\boldsymbol{w}_k\boldsymbol{w}_k^T\}$.

The error covariance matrix for the predictor is updated using

$$P_{k+1|k} = \phi(k+1, k)P_{k|k}\phi^T(k+1, k) + Q_k.$$

We rewrite the equation in terms of the square roots of the covariance matrices

$$P^s_{k+1|k}P^{s^T}_{k+1|k} = \phi(k+1, k)P^s_{k|k}P^{s^T}_{k|k}\phi^T(k+1, k) + Q^s_kQ^{sT}_k \tag{10.68}$$

which is equivalent to

$$\begin{bmatrix} P^s_{k+1,k} & 0 \end{bmatrix}\begin{bmatrix} P^s_{k+1,k} \\ 0 \end{bmatrix} = \begin{bmatrix} \phi(k+1, k)P^s_{k|k} & Q^s_k \end{bmatrix}T^TT\begin{bmatrix} P^T_{k,k}\phi^T(k+1, k) \\ Q^{sT}_k \end{bmatrix} \tag{10.69}$$

with T an orthogonal and nonunique matrix.

The factorization shows that

$$T\begin{bmatrix} P^{S^T}_{k|k}\phi^T(k+1, k) \\ Q^{sT}_k \end{bmatrix} = \begin{bmatrix} P^{S^T}_{k+1k} \\ 0 \end{bmatrix}$$

Multiplying by the orthogonal matrix T^T gives

$$\begin{bmatrix} P^s_{kk}\phi^T(k+1, k) \\ Q^{sT}_k \end{bmatrix} = T^T\begin{bmatrix} P^{s^T}_{k+1k} \\ 0 \end{bmatrix} \tag{10.70}$$

This shows that by factorizing the LHS into the product of an orthogonal matrix and a matrix with positive definite upper submatrix and zero lower submatrix gives the square root of the predictor covariance matrix. The desired matrix can be obtained by multiplying by the transpose of the orthogonal matrix and deleting the zero submatrix. The following theorem guarantees the success of the factorization for a full-rank matrix.

Theorem 10.1 *For any full-rank $m \times n$ matrix A, $m \geq n$, there exists a unique $m \times m$ orthogonal matrix Q $\left(\mathrm{Q}^T\mathrm{Q} = I_m\right)$ and a unique $m \times n$ upper triangular matrix R such that*

$$\mathrm{A} = \mathrm{QR} \tag{10.71}$$

QR factorization algorithms perform the desired factorization (see Golub & Van Loan, p.223). Several such algorithms are available, including Householder transformation, Givens transformation, Gram–Schmidt, and modified Gram–Schmidt. The factorization is performed using the MATLAB command

≫ [Q,R] = qr(A) % QR decomposition.

To solve for square root matrix of (10.70), we use QR factorization to find any orthogonal matrix $T^T = \mathrm{Q}$ that corresponds to R in (10.71) such that

$$\begin{bmatrix} P_{k|k}^{s^T}\phi^T(k+1,k) \\ Q_k^s \end{bmatrix} = \mathrm{Q}\begin{bmatrix} \mathrm{R} \\ 0_{n\times n} \end{bmatrix}$$

$$P_{k+1|k}^{s^T} = \mathrm{R} = n \times n \text{ matix} \tag{10.72}$$

Next, we use the equations developed for sequential computation in Sect. 10.1 If the covariance matrix of the measurement noise is not diagonal, we need to transform the data to obtain the rows of the transformed measurement matrix $\overline{\mathbf{h}}_k^{iT}$ and a diagonal covariance matrix R for the measurement noise. The resulting algorithm is Potter's algorithm.

We begin with the expression for the gain

$$\mathbf{k}_k^i = \frac{P_k^{i-1}\overline{\mathbf{h}}_k^i}{\overline{\mathbf{h}}_k^{iT} P_k^{i-1}\overline{\mathbf{h}}_k^i + \lambda_{k,j}} = \frac{P_{ks}^{i-1}P_{ks}^{i-1T}\overline{\mathbf{h}}_k^i}{\overline{\mathbf{h}}_k^{iT} P_{ks}^{i-1}P_{ks}^{i-1T}\overline{\mathbf{h}}_k^i + \lambda_{k,i}} \tag{10.73}$$

To simplify the expression, we define

$$\psi_k^i = P_{ks}^{i-1T}\overline{\mathbf{h}}_k^i, a_k^i = 1/\left[\psi_k^{iT}\psi_k^i + \lambda_{k,i}\right] \tag{10.74}$$

and rewrite the gain as

$$\boldsymbol{k}_k^i = a_k^i P_{ks}^{i-1}\psi_k^i \tag{10.75}$$

and the covariance matrix as

$$P_k^i = \left[I_n - \boldsymbol{k}_k^i\overline{\boldsymbol{h}}_k^{iT}\right]P_k^{i-1} = \left[I_n - a_k^i P_{ks}^{i-1}\psi_k^i\overline{\boldsymbol{h}}_k^{iT}\right]P_{ks}^{i-1}P_{ks}^{i-1T} \tag{10.76}$$

Using the definition of ψ_k^i of (10.74), we write

$$P_k^i = P_{ks}^{i-1}\left[I_n - a_k^i \psi_k^i \overline{\boldsymbol{h}}_k^{iT} P_{ks}^{i-1}\right] P_{ks}^{i-1T} = P_{ks}^{i-1}\left[I_n - a_k^i \psi_k^i \psi_k^{iT}\right] P_{ks}^{i-1T} \tag{10.77}$$

The following identity, whose proof is left as an exercise,

$$\left[I_n - a_k^i \psi_k^i \psi_k^{iT}\right] = \left[I_n - a_k^i \gamma_k^i \psi_k^i \psi_k^{iT}\right]^2 \tag{10.78}$$

allows us to write

$$P_k^i = P_{ks}^i P_{ks}^{iT} = P_{ks}^{i-1}\left[I_n - a_k^i \gamma_k^i \psi_k^i \psi_k^{iT}\right]^2 P_{ks}^{i-1T}$$

We now have an expression for the square root matrix

$$P_{ks}^i = P_{ks}^{i-1}\left[I_n - a_k^i \gamma_k^i \psi_k^i \psi_k^{iT}\right] \tag{10.79}$$

Based on the above equations, we have the following algorithm.

Potter's Algorithm

1. Compute the a priori square root *lower triangular* matrix and estimate and initialize $\hat{\boldsymbol{x}}_k^0 = \hat{\boldsymbol{x}}_{k|k-1},\ P_k^{s0} = P_{k|k-1}^s$
2. Calculate for $i = 1 \ldots, m$

$$\psi_k^i = P_{ks}^{i-1T} \overline{\boldsymbol{h}}_k^i,\ a_k^i = \frac{1}{\psi_k^{iT} \psi_k^i + \lambda_{k,i}}, \quad \boldsymbol{k}_k^i = a_k^i P_{ks}^{i-1} \psi_k^i$$

$$\gamma_k^i = 1 \Big/ \left[1 \pm \sqrt{a_k^i \lambda_{i,k}}\right], \quad P_{ks}^i = P_{ks}^{i-1}\left[I_n - a_k^i \gamma_k^i \psi_k^i \psi_k^{iT}\right]$$

$$\widehat{\boldsymbol{x}}_k^i = \left[I_n - \boldsymbol{k}_k^i \overline{\boldsymbol{h}}_k^{iT}\right] \widehat{\boldsymbol{x}}_k^{i-1} + \boldsymbol{k}_k^i \overline{z}_{k,i}, \quad \widehat{\boldsymbol{x}}_k^0 = \widehat{\boldsymbol{x}}_{k|\,k-1},$$

3. Set $P_{k|k}^s = P_{ks}^m, \hat{\boldsymbol{x}}_{k|k} = \hat{\boldsymbol{x}}_k^m$

Example 10.8 Consider the discretized double integrator system with

$$P_0 = \begin{bmatrix} 0.1 & 0.1 \\ 0.1 & p_2 \end{bmatrix}, \phi = \begin{bmatrix} 1 & 1 \\ 0 & 1 \end{bmatrix}, H = I_2, Q = 0.5 I_2, R = \text{diag}\{1, 10^{12}\}$$

Plot and compare the RMS error of perfect computation, the standard covariance Kalman filter, and the square root filter for (i) $p_2 = 10$, $(i) p_2 = 100$.

Solution

The results of the simulation show that the square root filter has a smaller error than the covariance filter initially (see Figs. 10.8 and 10.9). However, the error for both filter is approximately equal in the steady state.

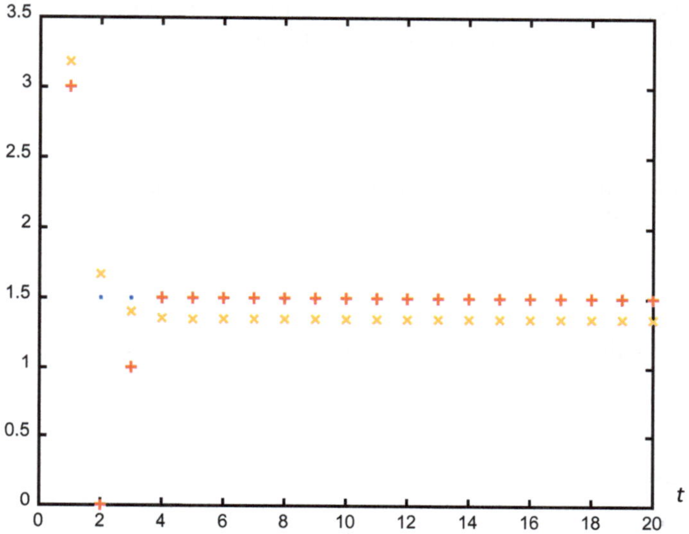

Fig. 10.8 RMS error with the covariance filter (+), the square root filter (.), and symbolic manipulation with $p_2 = 10$

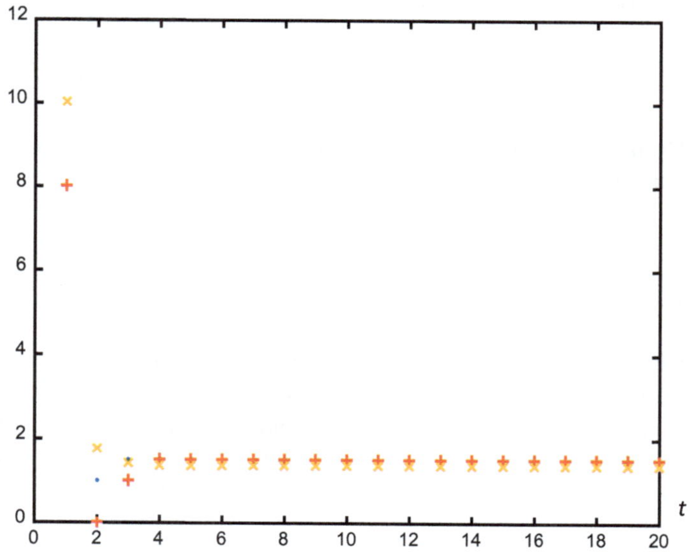

Fig. 10.9 RMS error with the covariance filter (+), the square root filter (.), and symbolic manipulation with $p_2 = 100$

This is not equal to the actual gain value $K_{k+1} = \frac{1}{2+r}\begin{bmatrix} 1 \\ 0 \end{bmatrix}$ but is much better result than the one calculated without square root filtering.

Square root filtering increases the precision of the Kalman filter but is computationally more costly and harder to program. Although this was a more important issue in the early days of Kalman filtering with less powerful computers, it is still useful for practical implementation in embedded systems.

Problems

10.1. Show that the innovations process $\tilde{\bar{z}}_{k,i}$ is zero-mean, Gaussian, and white.

10.2. Show that if the conditions

$$||\Delta P_{k|k}|| \le \varepsilon\kappa\left(P_{k|k}\right)$$

are satisfied $\forall k$, then in the steady state the covariance equation

$$P_{k+1|k} = \phi P_{k|k}\phi^T + Q$$

gives

$$||\Delta P_{k+1|k}|| \le \varepsilon\phi^2$$

10.3. Show that the covariance update equation of sequential computation can be written in the Joseph form

$$P_k^i = \left[I_n - \mathbf{k}_k^i \bar{\mathbf{h}}_k^{iT}\right] P_k^{i-1} \left[I_n - \mathbf{k}_k^i \bar{\mathbf{h}}_k^{iT}\right]^T + \lambda_{k,i}\mathbf{k}_k^i \mathbf{k}_k^{iT}$$

10.4. Design a sequential Kalman filter for the discrete model of the electricity market of Problem 9.9, given by

$$\begin{bmatrix} x_1(k+1) \\ x_2(k+1) \\ x_3(k+1) \end{bmatrix} = \begin{bmatrix} 0.998 & 0 & 3.333\times 10^{-3} \\ 0 & 0.98 & -5\times 10^{-3} \\ -10^{-5} & 10^{-5} & 1 \end{bmatrix} \begin{bmatrix} x_1(k) \\ x_2(k) \\ x_3(k) \end{bmatrix} + \begin{bmatrix} -6.666\times 10^{-3} \\ 3.333\times 10^{-2} \\ 0 \end{bmatrix} + \begin{bmatrix} w_1(k) \\ w_2(k) \\ w_3(k) \end{bmatrix}$$

where $x_1(k)$ is the power output of the supplier, $x_2(k)$ is the power output of the consumer and $x_3(k)$ is the market clearing price (the price when supply matches demand), and $w_i(k)$ are zero-mean Gaussian and white random variations with covariance matrix $Q = 0.5I_3$. The measurement vector is

$$z(k) = \begin{bmatrix} I_2 & 0_{2\times 1} \end{bmatrix} \begin{bmatrix} x_1(k) \\ x_2(k) \\ x_3(k) \end{bmatrix} + \begin{bmatrix} v_1(k) \\ v_2(k) \end{bmatrix}$$

with additive white Gaussian zero-mean Gaussian and white with covariance matrix $R = 0.25 I_2$, with correlated noise

$$E\{\boldsymbol{w}_k \boldsymbol{v}_k^T\} = \begin{bmatrix} 0.1 & 0 \\ 0 & 0.1 \\ 0.1 & 0 \end{bmatrix}.$$

10.5. Design and simulate a Kalman filter for a Gauss–Markov process with Wiener process noise and Wiener measurement noise with two measurements, the process noise and the Gauss–Markov process. Use $\beta = 2s^{-1}, \sigma^2 = 2$, amd a sampling period $\Delta t = 0.1s$.

10.6. Repeat the simulation of Problem 9.11 with sequential computation.

10.7. Prove the following identity that is used in the derivation of Potter's algorithm.

$$\left[I_n - a_k^i \psi_k^i \psi_k^{iT}\right] = \left[I_n - a_k^i \gamma_k^i \psi_k^i \psi_k^{iT}\right]^2$$

10.8. (a) Show that for $\phi = I_n$, $H = I_n Q_k = 0_{n\times n}$, the Riccati equation reduces to

$$P_{k+1|k} = P_{k|k-1} - P_{k|k-1}\left(P_{k|k-1} + R\right)^{-1} P_{k|k-1} = R\left(P_{k|k-1} - R\right)^{-1} P_{k|k-1}$$

(b) For $R = I_n$ and the diagonal elements $p_{k|k-1}^{ii}, i = 1 \ldots, n$, of $P_{k|k-1}$ much larger than unity, show that computing $P_{k+1|k}$ using its two expressions gives different answers.

10.9. Consider the manufacturer of Problem 9.12 with the price p_k of a device based on the demand d_k using the recursion

$$\boldsymbol{x}_{k+1} = \phi \boldsymbol{x}_k + \boldsymbol{w}_k$$

where

$$\boldsymbol{x}_k = \begin{bmatrix} x_1 & x_2 \end{bmatrix}^T = \begin{bmatrix} p_k & d_k \end{bmatrix}^T, \phi = \begin{bmatrix} 1 & \alpha \\ -\beta & \gamma \end{bmatrix}$$

and $\boldsymbol{w}_k = \begin{bmatrix} w_{k1} & w_{k2} \end{bmatrix}^T$ is a zero-mean, Gaussian, white noise vector with covariance matrix

$$Q = \text{diag}\{q_1, q_2\}\delta_k, q_1, q_2 > 0$$

The actual price varies randomly with the vendor and is measured as

$$z_k = \boldsymbol{x}_k + \boldsymbol{v}_k$$

where the noisy measurement z_k includes the additive zero-mean, Gaussian, white noise vector $\boldsymbol{v}_k$ that is independent of the process noise and has the covariance matrix $R = \text{diag}\{r_1, r_2\} = \text{diag}\{1, 10^6\}$. The initial covariance matrix is $P_0 = \begin{bmatrix} 0.1 & 0.1 \\ 0.1 & 100 \end{bmatrix}$. Compare the standard covariance filter to its square root formulation for the calculation of the gain matrix K_0

(i) Calculate the gain matrix K_0 using symbolic manipulation with the covariance filter equations, then substitute the numerical values to obtain K_0.

(ii) Calculate the gain matrix K_0 numerically using the covariance filter equations with only 16 bits and 4 exponent bits. Use the MATLAB command

```
q =quantizer('Mode','float','RoundMode','nearest','OverflowMode', ...
'saturate','Format',[12 4])
```

(iii) Calculate the gain vectors $\boldsymbol{k}_0^i$, $i = 1, 2$, using Potter's algorithm

(iv) Compare the three results obtained in (i-iii)

(v) Obtain the change in the value of K_0 if $P_0 = \text{diag}\{1.01, 100\}$

(vi) Obtain the change in the value of K_0 if $P_0 = \text{diag}\{0.015, 100\}$ if it is calculated using square root filtering.

Bibliography

1. Brown, R. G., & Hwang, P. Y. C. (2012). *Introduction to random signals and applied Kalman filtering* (4th Ed.). J. Wiley.
2. Golub, G. H., & Van Loan, C. F. (1996). *Matrix computation* (3rd Ed.). Johns Hopkins.
3. Kailath, T., Sayed, A. H., & Hassibi, B. (2000). *Linear estimation*. Prentice Hall.
4. Mendel, J. M. (1995). *Lessons in estimation theory for signal processing, communications, and control*. Prentice Hall PTR.
5. Simon, D. (2006). *Optimal state estimation: Kalman, H∞, and nonlinear approaches*. Wiley Interscience.
6. Jazwinski, A. H. (1970). *Stochastic processes and filtering theory*. Academic Press.

7. Lewis, F. L. (1986). *Optimal estimation: With an introduction to stochastic control theory*. Wiley-Interscience.
8. Rugh, W. J. (1996). *Linear system theory*. Prentice-Hall.

Chapter 11
Prediction and Smoothing

11.1 Prediction

The standard formulation of the Kalman filter is a one-step predictor and a corrector. We can generalize the predictor to extend beyond one step. We use the usual linear discrete-time model

$$\boldsymbol{x}_{k+1} = \phi_k \boldsymbol{x}_k + \boldsymbol{w}_k$$

$$z_k = H_k x_k + v_k \tag{11.1}$$

Because we need to write the state-transition matrix for a time advance of more than one step, we use the notation $\phi(k+l, k)$ for l time steps. With this notation $\phi_k = \phi(k+1, k)$, $\phi(k, k) = I_n$, and $\phi(k+l, k+1)\ \phi(k+1, k) = \phi(k+l, k), l \geq 1$. The predictor of the discrete Kalman filter is written as

$$\widehat{x}_{k+1|k} = \phi(k+1, k)\widehat{x}_{k|k} \tag{11.2}$$

$$P_{k+1|k} = \phi(k+1, k) P_{k|k} \phi^T(k+1, k) + Q_k \tag{11.3}$$

Applying the predictor formula recursively N times gives

$$\widehat{x}_{k+N|k} = \phi(k+N, k)\widehat{x}_{k|k} \tag{11.4}$$

$$P_{k+N|k} = \phi(k+N, k) P_{k|k} \phi^T(k+N, k) + Q_{k+N,k} \tag{11.5}$$

where $Q_{k+N,k}$ is the covariance of the **cumulative** effect of white noise inputs from k to $k+N$. To evaluate the cumulative noise term, we consider the estimation error, which is the difference between the state vector and its estimate at time $k+N$. The

M. S. Fadali, *Introduction to Random Signals, Estimation Theory, and Kalman Filtering*, https://doi.org/10.1007/978-981-99-8063-5_11

state vector is the solution of the state equation, which we obtain by induction. We substitute successive values of k

$$\begin{aligned}\boldsymbol{x}_{k+1} &= \phi(k+1,k)\boldsymbol{x}_k + \boldsymbol{w}_k \\ \boldsymbol{x}_{k+2} &= \phi(k+2,k+1)\boldsymbol{x}_{k+1} + \boldsymbol{w}_{k+1} \\ &= \phi(k+2,k)\boldsymbol{x}_k + \sum_{i=k}^{k+2-1} \phi(k+2,i+1)\boldsymbol{w}_i\end{aligned}$$

The solution is in the form

$$\boldsymbol{x}_{k+N} = \phi(k+N,k)\boldsymbol{x}_k + \sum_{i=k}^{k+N-1} \phi(k+N,i+1)\boldsymbol{w}_i \tag{11.6}$$

The N-step ahead prediction error is

$$\tilde{\boldsymbol{x}}_{k+N|k} = \boldsymbol{x}_{k+N} - \hat{\boldsymbol{x}}_{k+N|k} = \phi(k+N,k)\tilde{\boldsymbol{x}}_{k|k} + \sum_{i=k}^{k+N-1} \phi(k+N,i+1)\boldsymbol{w}(i)$$

The two terms are orthogonal because the estimates are orthogonal to the process noise and $\boldsymbol{x}(k+N)$ is orthogonal to earlier process noise terms $\boldsymbol{w}(i), i = k, k+1, \ldots, k+N-1$. The error expression gives the covariance matrix

$$\begin{aligned}P_{k+N|k} &= E\left\{\tilde{\boldsymbol{x}}_{k+N|k}\tilde{\boldsymbol{x}}_{k+N|k}^T\right\} \\ &= \phi(k+N,k)P(k|k)\phi^T(k+N,k) + \sum_{i=k}^{k+N-1} \phi(k+N,i+1)Q_i\phi^T(k+N,i+1)\end{aligned} \tag{11.7}$$

The cumulative noise term is

$$Q_{k+N,k} = \sum_{i=k}^{k+N-1} \phi(k+N,i+1)Q_i\phi^T(k+N,i+1) \tag{11.8}$$

The $Q_{k+N,k}$ expression is difficult to compute and it is easier to iteratively use the one-step covariance recursion to compute the error covariance

$$P_{j|k} = \phi(j,j-1)P_{j-1|k}\phi^T(j,j-1) + Q_{j-1},\ j = k+1,\ldots,k+N \tag{11.9}$$

The state update uses the one-step predictor equation of the Kalman filter. The derivation of the formula is left as an exercise.

Two forms of prediction are possible:

1. k Fixed and N Increases: In this case, the predicted state estimates are $\widehat{\boldsymbol{x}}_{k+i|k}, i = 1, 2, \ldots$ and the prediction continues using a single-step predictor as in the

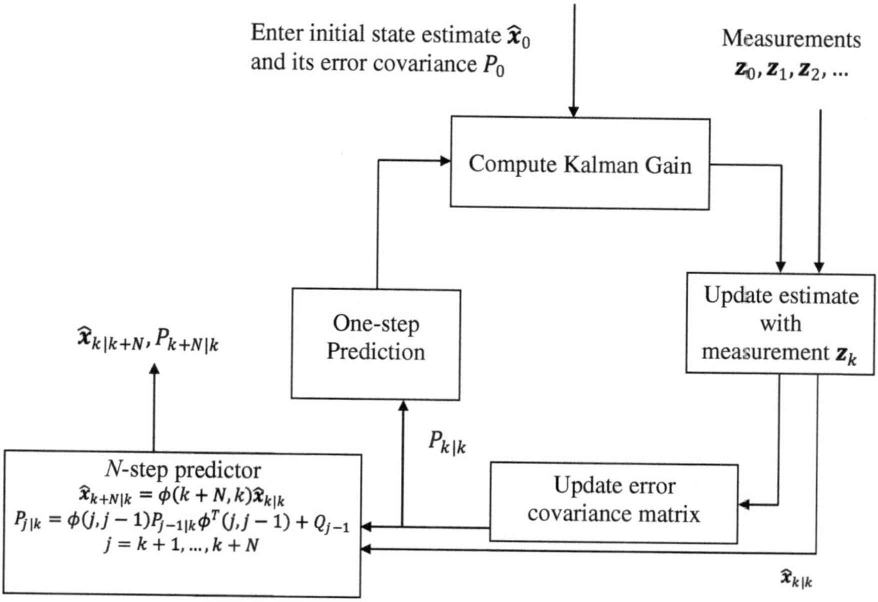

Fig. 11.1 Predictor block diagram with N fixed and increasing k

standard Kalman filter until the quality of the estimates becomes unacceptable. The covariance matrix provides a measure of the predictor quality and allows us to determine the maximum prediction horizon N for which an acceptable prediction is feasible.

2. N Fixed and k Increases: In this case, the predictor is appended to the filter loop as shown in Fig. 11.1. The error covariance matrix $P_{k+N|k}$ provides a measure of prediction quality.

Recall that the trace of the error covariance is

$$\operatorname{tr}\left[P_{k+N|k}\right] = \sum_{i=1}^{n} E\left\{e_{i,k+N|k}^2\right\} \tag{11.10}$$

The error covariance can be used to assess the prediction quality offline before implementing the predictor.

Example 11.1 Design a predictor for the system to predict the state three steps ahead

$$\phi = \begin{bmatrix} 1 & 0 \\ -.02 & -0.3 \end{bmatrix}, H = \begin{bmatrix} 1 \; 0 \end{bmatrix}$$

with noise covariance matrices $Q = I_2$, $R = 1$.

Design another predictor that uses the a posteriori estimate at time k and its covariance to progressively predict future states. Explain when this process should stop if the mean-square error cannot exceed a small positive number ϵ.

Solution

For three steps ahead, we need the state-transition matrix

$$\phi^3 = \begin{bmatrix} 1 & 0 \\ -.02 & -0.3 \end{bmatrix}^3 = \begin{bmatrix} 1 & 0 \\ -0.158 & -0.027 \end{bmatrix}$$

The predictor is

$$\widehat{\boldsymbol{x}}_{k+3|k} = \phi(k+3, k)\widehat{\boldsymbol{x}}_{k|k} = \begin{bmatrix} 1 & 0 \\ -0.158 & -0.027 \end{bmatrix}\widehat{\boldsymbol{x}}_{k|k}$$

The error covariance is updated with the recursion

$$P_{j|k} = \phi P_{j-1|k}\phi^T + Q = \begin{bmatrix} 1 & 0 \\ -.02 & -0.3 \end{bmatrix} P_{j-1|k} \begin{bmatrix} 1 & -.02 \\ 0 & -0.3 \end{bmatrix} + I_2$$

$$j = k+1, \ldots, k+3$$

The Kalman filter equations include the one-step predictor

$$\widehat{\boldsymbol{x}}_{k+1|k} = \phi(k+1, k)\widehat{\boldsymbol{x}}_{k|k} = \begin{bmatrix} 1 & 0 \\ -.02 & -0.3 \end{bmatrix}\widehat{\boldsymbol{x}}_{k|k}$$

with covariance matrix

$$\begin{aligned} P_{k+1|k} &= \begin{bmatrix} p^{11}_{k|k-1} & p^{12}_{k|k-1} \\ p^{12}_{k|k-1} & p^{22}_{k|k-1} \end{bmatrix} = \phi P_{k|k}\phi^T + Q \\ &= \begin{bmatrix} 1 & 0 \\ -.02 & -0.3 \end{bmatrix} P_{k|k} \begin{bmatrix} 1 & -.02 \\ 0 & -0.3 \end{bmatrix} + I_2 \\ &= \begin{bmatrix} p^{11}_{k|k} + 1 & -0.02p^{11}_{k|k} - 0.3p^{12}_{k|k} \\ -0.02p^{11}_{k|k} - 0.3p^{12}_{k|k} & 4 \times 10^{-4}p^{11}_{k|k} + 0.012p^{12}_{k|k} + 0.09p^{22}_{k|k} + 1 \end{bmatrix} \end{aligned}$$

The Kalman gain is

$$K_k = \frac{P_{k|k-1}H^T}{HP_{k|k-1}H^T + R} = \frac{P_{k|k-1}\begin{bmatrix} 1 \\ 0 \end{bmatrix}}{\begin{bmatrix} 1 & 0 \end{bmatrix} P_{k|k-1}\begin{bmatrix} 1 \\ 0 \end{bmatrix} + 1} = \frac{\begin{bmatrix} p^{11}_{k|k-1} \\ p^{12}_{k|k-1} \end{bmatrix}}{p^{11}_{k|k-1} + 1}$$

The corrector is

$$\widehat{\boldsymbol{x}}_{k|k} = \widehat{\boldsymbol{x}}_{k|k-1} + K_k\left[z_k - H\widehat{\boldsymbol{x}}_{k|k-1}\right] = \widehat{\boldsymbol{x}}_{k|k-1} + K_k\left[z_k - \left[\,1\;0\,\right]\widehat{\boldsymbol{x}}_{k|k-1}\right]$$

$$= \widehat{\boldsymbol{x}}_{k|k-1} + \frac{\begin{bmatrix} p^{11}_{k|k-1} \\ p^{12}_{k|k-1} \end{bmatrix}}{p^{11}_{k|k-1} + 1}\left[z_k - \widehat{\boldsymbol{x}}_{1k|k-1}\right]$$

with the covariance matrix

$$P_{k|k} = [I_2 - K_k H]P_{k|k-1} = \frac{\begin{bmatrix} 1 - p^{11}_{k|k-1} & 0 \\ -p^{12}_{k|k-1} & 1 \end{bmatrix}}{p^{11}_{k|k-1} + 1} P_{k|k-1}$$

clearly, this predictor is appended to a Kalman filter that estimates the current state and its covariance matrix given the measurements up to the measurement at time k.

The mean square error is given by the trace of the covariance matrix and the stopping condition for the predictor is

$$tr\left[P_{k|k}\right] = p^{11}_{k|k} + p^{22}_{k|k} \leq \epsilon$$

Prediction is terminated once the above condition is violated.

■

11.2 Smoothing

Smoothers provide estimates for the state over an interval where measurements have been previously recorded. There are three types of smoothers:

1. Fixed-interval smoothing (Fig. 11.2): Given measurements $z_j, j = 0, 1, \ldots, N-1$, where N is a fixed positive integer, estimate $\widehat{\boldsymbol{x}}_{k|N}, k = 0, \ldots, N-1$.

   The smoother uses all the data to estimate the state at each point after an experiment and is useful for offline data processing.
2. Fixed-point smoothing: Estimate $\widehat{\boldsymbol{x}}_{k|j}$ at a fixed positive integer k using measurements $z_j, j = k+1, k+2, \ldots$.

Fig. 11.2 Fixed-interval smoothing

Fig. 11.3 Fixed-point smoothing

Fig. 11.4 Fixed-lag smoothing

The smoother improves the estimate at a fixed-point k as more data becomes available. It can be carried out in real time, but the estimate obtained is delayed by a multiple of the sampling period T. The smoother is useful in applications where an event occurred at time k and continues to be characterized as more data is collected. For example, the fixed-point smoother if a fault occurs at time k but data collection continues after the occurrence of the fault (Fig. 11.3).

3. Fixed-lag smoothing (Fig. 11.4): Estimate $\hat{\boldsymbol{x}}_{k-L|k}, k = 0, 1, \ldots,$ where L is a fixed positive integer delay and k is the current time. The current time changes and the estimate can be obtained in real time but the estimate obtained is delayed by $L \times T$ for a sampling period T. Fixed-lag smoothers provide better estimates than Kalman filters at the desired point, even with a modest lag, at an additional computational cost. This can make them attractive in applications where a better estimate that that provided by filters is needed.

11.3 Fixed-Point Smoothing

Given a measurement set $\{z_i, i = 0, \ldots, k, k+1, k+2, \ldots\}$, the objective is to determine the estimate $\hat{\boldsymbol{x}}_{k|j} = E\{\boldsymbol{x}_k | z_{0:j}\}, j = k+1, k+2, \ldots$ where k is a fixed positive integer and the associated error covariance matrix

$$P_{k|j} = E\left\{\left(\boldsymbol{x}_k - \hat{\boldsymbol{x}}_{k|j}\right)\left(\boldsymbol{x}_k - \hat{\boldsymbol{x}}_{k|j}\right)^T | z_{0:j}\right\}, j \geq k \tag{11.11}$$

The problem can be posed as a Kalman filter design by augmenting the state with the fixed-point state $\boldsymbol{x}_k$. The fixed augmenting state $\boldsymbol{x}_j^a$ has the state equation

$$\boldsymbol{x}_{j+1}^a = \boldsymbol{x}_j^a \tag{11.12}$$

and is initialized with $\boldsymbol{x}_k^a = \boldsymbol{x}_k$ so that $\boldsymbol{x}_{j+1}^a = \boldsymbol{x}_j^a = \boldsymbol{x}_k$ and its estimate is $\widehat{\boldsymbol{x}}_{j+1|j}^a = \widehat{\boldsymbol{x}}_{k|j}$, $\forall j \geq k$. The augmented state vector is $\overline{\boldsymbol{x}}_j = col\left\{\boldsymbol{x}_j, \boldsymbol{x}_j^a\right\}$ evolves with the state-space model

$$\begin{bmatrix} \boldsymbol{x}_{j+1} \\ \boldsymbol{x}_{j+1}^a \end{bmatrix} = \begin{bmatrix} \Phi_j & [0] \\ [0] & I_n \end{bmatrix} \begin{bmatrix} \boldsymbol{x}_j \\ \boldsymbol{x}_j^a \end{bmatrix} + \begin{bmatrix} I_m \\ [0] \end{bmatrix} \boldsymbol{w}_j$$

$$z_j = \begin{bmatrix} H_j & [0] \end{bmatrix} \begin{bmatrix} \boldsymbol{x}_j \\ \boldsymbol{x}_j^a \end{bmatrix} + \boldsymbol{v}_j \tag{11.13}$$

The KF for the augmented system corrects the state $\boldsymbol{x}_j^a$ with every new measurement but does not change it in the KP predictor. We apply the KF to the augmented system to obtain the estimate $\widehat{\boldsymbol{x}}_{j+1|j}^a = \widehat{\boldsymbol{x}}_{k|j}$ and its covariance matrix $P_{j+1|j}^{aa} = P_{k|j}$

$$P_{j+1|j}^{aa} = E\left\{\left(\boldsymbol{x}_{j+1}^a - \widehat{\boldsymbol{x}}_{j+1|j}^a\right)\left(\boldsymbol{x}_{j+1}^a - \widehat{\boldsymbol{x}}_{j+1|j}^a\right)^T |z_{0:j}\right\} \tag{11.14}$$

The predictor equation for the augmented state vector is

$$\begin{bmatrix} \widehat{\boldsymbol{x}}_{j+1|j} \\ \widehat{\boldsymbol{x}}_{j+1|j}^a \end{bmatrix} = \begin{bmatrix} \Phi_j & [0] \\ [0] & I_n \end{bmatrix} \begin{bmatrix} \widehat{\boldsymbol{x}}_{j|j} \\ \widehat{\boldsymbol{x}}_{j|j}^a \end{bmatrix} = \begin{bmatrix} \Phi_j & [0] \\ [0] & I_n \end{bmatrix} \left\{ \begin{bmatrix} \widehat{\boldsymbol{x}}_{j|j-1} \\ \widehat{\boldsymbol{x}}_{j|j-1}^a \end{bmatrix} + \begin{bmatrix} K_j \\ K_j^a \end{bmatrix} \widetilde{z}_j \right\} \tag{11.15}$$

where $\widetilde{z}_j$ is the innovations

$$\widetilde{z}_j = z_j - H_j \widehat{\boldsymbol{x}}_{j|j-1} \tag{11.16}$$

The predictor can be rewritten as

$$\widehat{\boldsymbol{x}}_{j+1|j}^a = \widehat{\boldsymbol{x}}_{j|j-1}^a + K_j^a \widetilde{z}_j$$

$$\widehat{\boldsymbol{x}}_{k|j} = \widehat{\boldsymbol{x}}_{k|j-1} + K_j^a \widetilde{z}_j \tag{11.17}$$

Recall that the predictor covariance matrix for the Kalman filter satisfies

$$P_{k+1|k} = \phi_k P_{k|k} \phi_k^T + Q_k \tag{11.18}$$

The predictor for the augmented system is partitioned as

$$\overline{P}_{j+1|j} = \begin{bmatrix} P_{j+1|j} & {P_{j+1|j}^a}^T \\ P_{j+1|j}^a & P_{j+1|j}^{aa} \end{bmatrix} = \begin{bmatrix} \Phi_j & [0] \\ [0] & I_n \end{bmatrix} \begin{bmatrix} P_{j|j} & {P_{j|j}^a}^T \\ P_{j|j}^a & P_{j|j}^{aa} \end{bmatrix} \begin{bmatrix} \Phi_j^T & [0] \\ [0] & I_n \end{bmatrix} + \begin{bmatrix} Q_j & [0] \\ [0] & [0] \end{bmatrix} \tag{11.19}$$

This yields the covariance equations

$$\begin{bmatrix} P_{j+1|j} & {P^a_{j+1|j}}^T \\ P^a_{j+1|j} & P^{aa}_{j+1|j} \end{bmatrix} = \begin{bmatrix} \Phi_j P_{j|j} \Phi_j^T + Q_j & \Phi_j {P^a_{j|j}}^T \\ P^a_{j|j} \Phi_j^T & P^{aa}_{j|j} \end{bmatrix} \tag{11.20}$$

The corrector equation is

$$\begin{aligned} \begin{bmatrix} \widehat{\boldsymbol{x}}_{j|j} \\ \widehat{\boldsymbol{x}}^a_{j|j} \end{bmatrix} &= \left[I_{2n} - \begin{bmatrix} K_j \\ K_j^a \end{bmatrix} \begin{bmatrix} H_j & [0] \end{bmatrix} \right] \begin{bmatrix} \widehat{\boldsymbol{x}}_{j|j-1} \\ \widehat{\boldsymbol{x}}^a_{j|j-1} \end{bmatrix} + \begin{bmatrix} K_j \\ K_j^a \end{bmatrix} \boldsymbol{z}_j \\ &= \begin{bmatrix} I_n - K_j H_j & [0] \\ -K_j^a H_j & I_n \end{bmatrix} \begin{bmatrix} \widehat{\boldsymbol{x}}_{j|j-1} \\ \widehat{\boldsymbol{x}}^a_{j|j-1} \end{bmatrix} + \begin{bmatrix} K_j \\ K_j^a \end{bmatrix} \boldsymbol{z}_j \end{aligned} \tag{11.21}$$

The corrector error covariance for the Kalman filter satisfies

$$P_{k|k} = (I_n - K_k H_k) P_{k|k-1} \tag{11.22}$$

The corrector error covariance for the augmented system is partitioned as

$$\begin{aligned} \overline{P}_{j|j} &= \begin{bmatrix} P_{j|j} & {P^a_{j|j}}^T \\ P^a_{j|j} & P^{aa}_{j|j} \end{bmatrix} = \begin{bmatrix} I_n - K_j H_j & [0] \\ -K_j^a H_j & I_n \end{bmatrix} \begin{bmatrix} P_{j|j-1} & {P^a_{j|j-1}}^T \\ P^a_{j|j-1} & P^{aa}_{j|j-1} \end{bmatrix} \\ &= \begin{bmatrix} (I_n - K_j H_j) P_{j|j-1} & (I_n - K_j H_j) {P^a_{j|j-1}}^T \\ P^a_{j|j-1} - K_j^a H_j P_{j|j-1} & P^{aa}_{j|j-1} - K_j^a H_j {P^a_{j|j-1}}^T \end{bmatrix} \end{aligned} \tag{11.23}$$

The optimal gain for the Kalman filter is

$$K_k = P_{k|k-1} H_k^T \left(H_k P_{k|k-1} H_k^T + R_k \right)^{-1} \tag{11.24}$$

For the augmented system, we have the gain

$$\begin{bmatrix} K_j \\ K_j^a \end{bmatrix} = \begin{bmatrix} P_{j|j-1} & P^a(j|j-1)^T \\ P^a_{j|j-1} & P^{aa}_{j|j-1} \end{bmatrix} \begin{bmatrix} H_j^T \\ [0] \end{bmatrix} \left(H_j P_{j|j-1} H_j^T + R_j \right)^{-1}$$

which simplifies to

$$\begin{bmatrix} K_j \\ K_j^a \end{bmatrix} = \begin{bmatrix} P_{j|j-1} H_j^T \left(H_j P_{j|j-1} H_j^T + R_j \right)^{-1} \\ P^a_{j|j-1} H_j^T \left(H_j P_{j|j-1} H_j^T + R_j \right)^{-1} \end{bmatrix} \tag{11.25}$$

verify that the corrector error covariance matrix is symmetric, we expand the lower left term and substitute for K_j^a to obtain

$$P^a_{j|j-1} - K^a_j H_j P_{j|j-1} = P^a_{j|j-1} - P^a_{j|j-1} H_j^T \left(H_j P_{j|j-1} H_j^T + R_j\right)^{-1} H_j P_{j|j-1}$$
$$= P^a_{j|j-1} - P^a_{j|j-1} H_j^T K_j^T = P^a_{j|j-1}\left(I_n - H_j^T K_j^T\right) \tag{11.26}$$

The covariance of interest for smoothing is $P^{aa}_{j+1|j} = P_{k|j}$ and is given by

$$P_{k|j} = P^{aa}_{j+1|j} = P^{aa}_{j|j} = P^{aa}_{j|j-1} - K^a_j H_j P^{a\,T}_{j|j-1} \tag{11.27}$$

This can be written as

$$P_{k|j} = P_{k|j-1} - K^a_j H_j P^{a\,T}_{j|j-1} \tag{11.28}$$

The fixed-point smoother equations are:

$$\hat{\boldsymbol{x}}_{k|j} = \hat{\boldsymbol{x}}_{k|j-1} + K^a_j \tilde{\boldsymbol{z}}_j,\ j \ge k,\ IC\hat{\boldsymbol{x}}_{k|k-1}$$

$$P^a_{j+1|j} = P^a_{j|j-1}\left(I_n - K_j H_j\right)^T \Phi_j^T,\ IC\,P_{k|k-1}$$

$$K_j = P_{j|j-1} H_j^T \left(H_j P_{j|j-1} H_j^T + R_j\right)^{-1}$$

$$P_{j+1|j} = \Phi_j\left(I - K_j H_j\right) P_{j|j-1} \Phi_j^T + j \tag{11.29}$$

The following procedure yields the desired fixed-point smoother.

Procedure 11.1 Fixed-point smoother (see Fig. 11.5).

1. Run the standard Kalman filter until time k to obtain $\hat{\boldsymbol{x}}_{k|k-1}$, $P_{k|k-1}$
2. Initialize the filter with

$$\hat{\boldsymbol{x}}^a_{k|k-1} = \hat{\boldsymbol{x}}_{k|k-1},\ P^a_{k|k-1} = P_{k|k-1}$$

$$j = k$$

3. Calculate the gain matrix

$$K^a_j = P^a_{j|j-1} H_j^T \left(H_j P_{j|j-1} H_j^T + R_j\right)^{-1}$$

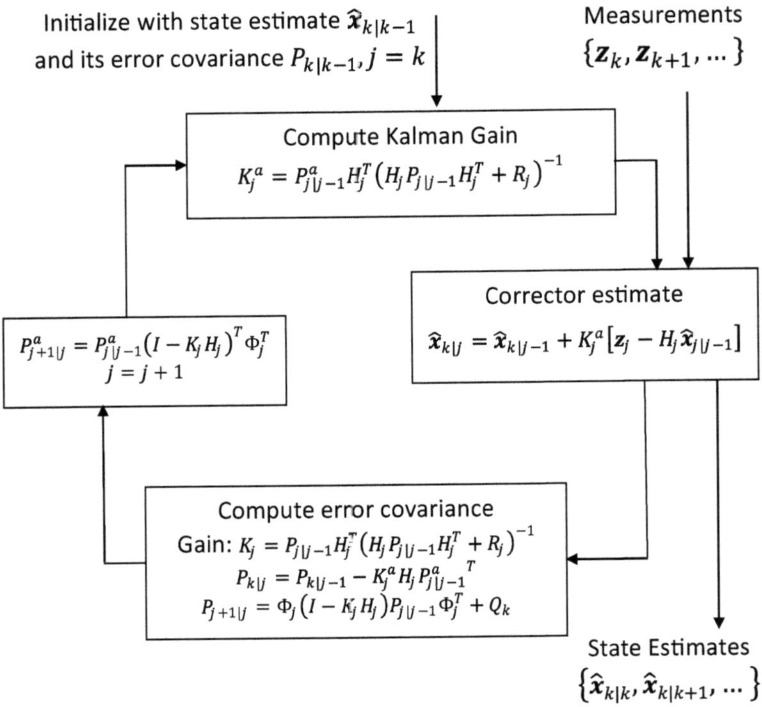

Fig. 11.5 Block diagram for the fixed-point smoother

4. Calculate the estimates and the associated error covariance matrices

$$\hat{\boldsymbol{x}}_{k|j} = \hat{\boldsymbol{x}}_{k|j-1} + K_j^a \left[z_j - H_j \hat{\boldsymbol{x}}_{j|j-1}\right]$$

$$P_{k|j} = P_{k|j-1} - K_j^a H_j P_{j|j-1}^a$$

5. Calculate the covariance matrices then go to Step 3.

$$K_j = P_{j|j-1} H_j^T \left(H_j P_{j|j-1} H_j^T + R_j\right)^{-1}$$

$$P_{j+1|j} = \Phi_j \left(I - K_j H_j\right) P_{j|j-1} \Phi_j^T + Q_j$$

$$P_{j+1|j}^a = P_{j|j-1}^a \left(I - K_j H_j\right)^T \Phi_j^T$$

$j = j + 1$

11.3.1 Properties of Fixed-Point Smoother

- The smoother is a linear discrete-time system of dimension n driven by the innovation $\tilde{z}_k$.
- The fixed-point smoother is time-varying even if the system is LTI because K_j^a is time-varying.
- The improvement due to smoothing is

$$\frac{tr\left[P_{k|j}\right] - tr\left[P_{k|k-1}\right]}{tr\left[P_{k|k-1}\right]} \times 100\% \tag{11.30}$$

The improvement increases monotonically with j (i.e., with more measurements).

Example 11.2 Design a fixed-point smoother for a Wiener process with measurement variance R

Solution

From Example 9.2, we have the discretized state equation

$$x_{k+1} = x_k + w_k$$

The measurement equation is

$$z_k = x_k + v_k$$

The covariance matrix of the process noise is $Q = E\left\{w_k^2\right\} = 1$. Thus, we have $\phi_k = 1$ and $H = 1$. Because the initial state is known exactly $\hat{\boldsymbol{x}}_0 = 0$, the error covariance is $P_0 = 0$. We follow Procedure 11.1:

1. Run the standard Kalman filter until time k to obtain $\hat{\boldsymbol{x}}_{k|k-1}$, $P_{k|k-1}$
2. Initialize the filter with

$$P_{k|k-1}^a = P_{k|k-1},\ P_{k|k-1}^{aa} = P_{k|k-1}$$

$$\hat{\boldsymbol{x}}_{k|k-1}^a = \hat{\boldsymbol{x}}_{k|k-1},\ j = k$$

3. Calculate the gain matrix

$$K_j^a = \frac{P_{j|j-1}^a}{P_{j|j-1} + R}$$

4. Calculate the estimates and the associated error covariance matrix

$$\hat{\boldsymbol{x}}_{k|j} = \hat{\boldsymbol{x}}_{k|j-1} + K_j^a\left[z_j - \hat{\boldsymbol{x}}_{j|j-1}\right]$$

$$P_{k|j} = P_{k|j-1} - K_j^a P_{j|j-1}^a$$

5. Calculate the covariance matrices then go to Step 3.

$$P_{j+1|j} = \Phi_j\big(I - K_j H_j\big) P_{j|j-1} \Phi_j^T + Q_j$$

$$K_j = \frac{P_{j|j-1}}{P_{j|j-1} + R}$$

$$P_{j+1|j}^a = P_{j|j-1}^a\big(1 - K_j\big) = \frac{P_{j|j-1} R}{P_{j|j-1} + R}$$

$j = j + 1$

■

11.4 Fixed-Lag Smoother

Fixed-lag smoothing provides an estimate of $\hat{\boldsymbol{x}}_{k-L|k}$, $k = 0, 1, \ldots$, where L is a fixed positive integer. The state augmentation approach can be used to derive the fixed-lag smoother, but this gives a high dimensional solution. If the lag L is not large, this approach is feasible despite of its computational cost. We augment the state to obtain $L + 1$ state vectors defined as

$$\boldsymbol{x}_{k+1}^{(1)} = \boldsymbol{x}_k, \boldsymbol{x}_{k+1}^{(2)} = \boldsymbol{x}_{k-1}, \ldots, \boldsymbol{x}_{k+1}^{(i+1)} = \boldsymbol{x}_{k+1-i}, i = 1, \ldots, L + 1 \tag{11.31}$$

Although the smoother can yield smoothed estimates for all intermediate states, the desired state estimate is for the last state

$$\boldsymbol{x}_{k+1}^{(L+1)} = \boldsymbol{x}_{k-L} \tag{11.32}$$

We then design a Kalman filter for the augmented system to obtain the estimate and error covariance for $\boldsymbol{x}_{k+1}^{(L+1)}$ given the measurements $\boldsymbol{z}_{0:N}$. The augmented model has the state equation

$$\begin{bmatrix} \boldsymbol{x}_{k+1} \\ \boldsymbol{x}_{k+1}^{(1)} \\ \boldsymbol{x}_{k+1}^{(2)} \\ \vdots \\ \boldsymbol{x}_{k+1}^{(L+1)} \end{bmatrix} = \begin{bmatrix} \phi_k & 0 & \cdots & 0 & 0 \\ I_n & 0 & \cdots & 0 & 0 \\ 0 & I_n & \cdots & 0 & 0 \\ \vdots & \vdots & \ddots & \vdots & \vdots \\ 0 & 0 & \cdots & I_n & 0 \end{bmatrix} \begin{bmatrix} \boldsymbol{x}_k \\ \boldsymbol{x}_k^{(1)} \\ \boldsymbol{x}_k^{(2)} \\ \vdots \\ \boldsymbol{x}_k^{(L+1)} \end{bmatrix} + \begin{bmatrix} I_n \\ 0 \\ 0 \\ \vdots \\ 0 \end{bmatrix} \boldsymbol{w}_k \tag{11.33}$$

and the measurement equation

$$z_k = \begin{bmatrix} H_k \ 0 \cdots 0 \ 0 \end{bmatrix} \begin{bmatrix} \boldsymbol{x}_k \\ \boldsymbol{x}_k^{(1)} \\ \boldsymbol{x}_k^{(2)} \\ \vdots \\ \boldsymbol{x}_k^{(L+1)} \end{bmatrix} + \boldsymbol{v}_k \tag{11.34}$$

Clearly, the above concise form of the model is not useful for computation and equations and the definitions provide a more efficient form to be used in designing a Kalman filter that yields the smoothed estimates. We partition the covariance matrix as

$$P_{k|k} = \begin{bmatrix} P_{k|k} & P_{k|k}^{1^T} & \cdots & P_{k|k}^{L^T} & P_{k|k}^{L+1^T} \\ P_{k|k}^{1} & P_{k|k}^{11} & \cdots & P_{k|k}^{1L^T} & P_{k|k}^{1,L+1^T} \\ P_{k|k}^{2} & P_{k|k}^{12} & \cdots & P_{k|k}^{2L^T} & P_{k|k}^{2,L+1^T} \\ P_{k|k}^{3} & P_{k|k}^{13} & \cdots & 0 & P_{k|k}^{3,L+1^T} \\ \vdots & \vdots & \ddots & \vdots & \vdots \\ P_{k|k}^{L+1} & P_{k|k}^{1,L+1} & \cdots & P_{k|k}^{L,L+1} & P_{k|k}^{L+1,L+1^T} \end{bmatrix} \tag{11.35}$$

We recall the Kalman gain equation

$$K_k = P_{k|k-1} H_k^T \left(H_k P_{k|k-1} H_k^T + R_k \right)^{-1}$$

For the fixed-lag smoother, this becomes

$$K_k = \begin{bmatrix} P_{k|k-1} H_k^T \\ P_{k|k-1}^{1} H_k^T \\ P_{k|k-1}^{2} H_k^T \\ \vdots \\ P_{k|k-1}^{L+1} H_k^T \end{bmatrix} \left(H_k P_{k|k-1} H_k^T + R_k \right)^{-1}$$

The corrector equation for the filter is

$$\hat{\boldsymbol{x}}_{k|k} = \hat{\boldsymbol{x}}_{k|k-1} + K_k \left[z_k - H_k \hat{\boldsymbol{x}}_{k|k-1} \right]$$

For the fixed-lag smoother, we have

$$\begin{bmatrix} \hat{\boldsymbol{x}}_{k|k} \\ \hat{\boldsymbol{x}}^{(1)}_{k|k} \\ \hat{\boldsymbol{x}}^{(2)}_{k|k} \\ \vdots \\ \hat{\boldsymbol{x}}^{(L+1)}_{k|k} \end{bmatrix} = \begin{bmatrix} \hat{\boldsymbol{x}}_{k|k} \\ \hat{\boldsymbol{x}}^{(1)}_{k|k} \\ \hat{\boldsymbol{x}}^{(2)}_{k|k} \\ \vdots \\ \hat{\boldsymbol{x}}^{(L+1)}_{k|k} \end{bmatrix} + \begin{bmatrix} P_{k|k-1} H_k^T \\ P^1_{k|k-1} H_k^T \\ P^2_{k|k-1} H_k^T \\ \vdots \\ P^{L+1}_{k|k-1} H_k^T \end{bmatrix} \left(H_k P_{k|k-1} H_k^T + R_k\right)^{-1} \left[z_k - \hat{\boldsymbol{x}}_{k|k}\right] \tag{11.36}$$

The corrector error covariance matrix is

$$P_{k|k} = (I - K_k H_k) P_{k|k-1}$$

For the fixed-lag smoother, we have

$$P^s_{k|k} = \begin{bmatrix} P_{k|k} & P^{1^T}_{k|k} & \cdots & P^{L^T}_{k|k-1} & P^{L+1^T}_{k|k-1} \\ P^1_{k|k} & P^{11}_{k|k} & \cdots & P^{1L^T}_{k|k} & P^{1,L+1^T}_{k|k} \\ P^2_{k|k} & P^{12}_{k|k} & \cdots & P^{2L^T}_{k|k} & P^{2,L+1^T}_{k|k} \\ P^3_{k|k} & P^{13}_{k|k} & \cdots & 0 & P^{3,L+1^T}_{k|k} \\ \vdots & \vdots & \ddots & \vdots & \vdots \\ P^{L+1}_{k|k} & P^{1,L+1}_{k|k} & \cdots & P^{L,L+1}_{k|k} & P^{L+1,L+1^T}_{k|k} \end{bmatrix}$$

$$= \left(I - \begin{bmatrix} P_{k|k-1} H_k^T \\ P^1_{k|k-1} H_k^T \\ P^2_{k|k-1} H_k^T \\ \vdots \\ P^{L+1}_{k|k-1} H_k^T \end{bmatrix} \left(H_k P_{k|k-1} H_k^T + R_k\right)^{-1} \left[\, H_k \; 0 \cdots 0 \; 0 \right] \right) P^s_{k|k-1}$$

The predictor equations for the filter are

$$\hat{\boldsymbol{x}}^{(1)}_{k+1|k} = \hat{\boldsymbol{x}}_{k|k}, \boldsymbol{x}^{(2)}_{k+1|k} = \boldsymbol{x}^{(1)}_{k|k}, \ldots, \boldsymbol{x}^{(i+1)}_{k+1|k} = \boldsymbol{x}^{(i)}_{k|k}, i = 1, \ldots, L+1 \tag{11.37}$$

The error covariance matrix is

$$P_{k+1|k} = \Phi_k P_{k|k} \Phi_k^T + Q_k \tag{11.38}$$

which for the smoother has the RHS

$$\begin{bmatrix} \phi_k & 0 & \cdots & 0 & 0 \\ I_n & 0 & \cdots & 0 & 0 \\ 0 & I_n & \cdots & 0 & 0 \\ \vdots & \vdots & \ddots & \vdots & \vdots \\ 0 & 0 & \cdots & I_n & 0 \end{bmatrix} \begin{bmatrix} P_{k|k} & P_{k|k}^{1^T} & \cdots & P_{k|k}^{L^T} & P_{k|k}^{L+1^T} \\ P_{k|k}^{1} & P_{k|k}^{11} & \cdots & P_{k|k}^{1L^T} & P_{k|k}^{1,L+1^T} \\ P_{k|k}^{2} & P_{k|k}^{12} & \cdots & P_{k|k}^{2L^T} & P_{k|k}^{2,L+1^T} \\ P_{k|k}^{3} & P_{k|k}^{13} & \cdots & P_{k|k}^{3,L^T} & P_{k|k}^{3,L+1^T} \\ \vdots & \vdots & \ddots & \vdots & \vdots \\ P_{k|k}^{L+1} & P_{k|k}^{1,L+1} & \cdots & P_{k|k}^{L,L+1} & P_{k|k}^{L+1,L+1} \end{bmatrix} \begin{bmatrix} \phi_k^T & I_n & 0 & \cdots & 0 \\ 0 & 0 & I_n & \cdots & 0 \\ \vdots & \vdots & \vdots & \ddots & 0 \\ 0 & 0 & 0 & \cdots & I_n \\ 0 & 0 & 0 & \cdots & 0 \end{bmatrix}$$

$$+ \begin{bmatrix} Q_k & 0 & 0 & \cdots & 0 \\ 0 & 0 & 0 & \cdots & 0 \\ \vdots & \vdots & \vdots & \ddots & 0 \\ 0 & 0 & 0 & 0 & 0 \\ 0 & 0 & \cdots & 0 & 0 \end{bmatrix}$$

We multiply the last two matrices of the first term to obtain

$$\begin{bmatrix} P_{k|k}\phi_k^T & P_{k|k} & \cdots & P_{k|k}^{L-1^T} & P_{k|k}^{L^T} \\ P_{k|k}^{1}\phi_k^T & P_{k|k}^{11} & \cdots & P_{k|k}^{1L-1^T} & P_{k|k}^{1L^T} \\ P_{k|k}^{2}\phi_k^T & P_{k|k}^{12} & \cdots & P_{k|k}^{2,L-1} & P_{k|k}^{2,L} \\ \vdots & \vdots & \ddots & \vdots & \vdots \\ P_{k|k}^{L+1}\phi_k^T & P_{k|k}^{1,L+1} & \cdots & P_{k|k}^{L-1,L+1} & P_{k|k}^{L,L+1} \end{bmatrix}$$

We now have the expression

$$\begin{bmatrix} P_{k+1|k} & P_{k+1|k}^{1^T} & \cdots & P_{k+1|k}^{L^T} & P_{k+1|k}^{L+1^T} \\ P_{k+1|k}^{1} & P_{k+1|k}^{11} & \cdots & P_{k+1|k}^{1L^T} & P_{k+1|k}^{1,L+1^T} \\ P_{k+1|k}^{2} & P_{k+1|k}^{12} & \cdots & P_{k+1|k}^{2L^T} & P_{k+1|k}^{2,L+1^T} \\ P_{k+1|k}^{3} & P_{k+1|k}^{13} & \cdots & P_{k+1|k}^{3,L^T} & P_{k+1|k}^{3,L+1^T} \\ \vdots & \vdots & \ddots & \vdots & \vdots \\ P_{k+1|k}^{L+1} & P_{k+1|k}^{1,L+1} & \cdots & P_{k+1|k}^{L,L+1} & P_{k+1|k}^{L+1,L+1} \end{bmatrix} = \begin{bmatrix} \phi_k P_{k|k}\phi_k^T + Q_k & \phi_k P_{k|k}^{1^T} & \cdots & \phi_k P_{k|k}^{L-1^T} & \phi_k P_{k|k}^{L^T} \\ P_{k|k}\phi_k^T & P_{k|k} & \cdots & P_{k|k}^{1,L-1^T} & P_{k|k}^{L^T} \\ P_{k|k}^{1}\phi_k^T & P_{k|k}^{11} & \cdots & P_{k|k}^{1,L-1} & P_{k|k}^{1,L} \\ \vdots & \vdots & \ddots & \vdots & \vdots \\ P_{k|k}^{L}\phi_k^T & P_{k|k}^{1,L} & \cdots & P_{k|k}^{L-2,L+1} & P_{k|k}^{L-1,L+1} \end{bmatrix} \tag{11.39}$$

Example 11.3 Write the equations for the fixed-lag smoother for the Wiener process with lag $L = 2$

Solution

The parameter values for the system are $\phi = 1,\ H = 1,\ Q = 1$. The Kalman gain is

$$K_k = \frac{\begin{bmatrix} P_{k|k-1} \\ P_{k|k-1}^{11} \\ P_{k|k-1}^{12} \\ P_{k|k-1}^{13} \end{bmatrix}}{P_{k|k-1} + R_k}$$

The corrected estimate is

$$\begin{bmatrix} \widehat{x}_{k|k} \\ \widehat{x}_{k|k}^{(1)} \\ \widehat{x}_{k|k}^{(2)} \\ \widehat{x}_{k|k}^{(3)} \end{bmatrix} = \begin{bmatrix} \widehat{x}_{k|k-1} \\ \widehat{x}_{k|k-1}^{(1)} \\ \widehat{x}_{k|k-1}^{(2)} \\ \widehat{x}_{k|k-1}^{(3)} \end{bmatrix} + \begin{bmatrix} P_{k|k-1} \\ P_{k|k-1}^{11} \\ P_{k|k-1}^{12} \\ P_{k|k-1}^{13} \end{bmatrix} \frac{z_k - \widehat{x}_{k|k}}{P_{k|k-1} + R_k}$$

with covariance matrix

$$P_{k|k}^{s} = \begin{bmatrix} P_{k|k-1}\phi_k^T & P_{k|k-1} & P_{k|k}^{1^T} & P_{k|k}^{2^T} \\ P_{k|k}^{1}\phi_k^T & P_{k|k}^{1} & P_{k|k}^{11} & P_{k|k}^{12^T} \\ P_{k|k}^{2}\phi_k^T & P_{k|k}^{11} & P_{(k|k)}^{12} & P_{(k|k)}^{2,L} \\ & & \ddots & \vdots \\ P_{k|k}^{2}\phi_k^T & P_{k|k}^{12} & & \\ P_{k|k}^{,3}\phi_k^T & P_{(k|k)}^{13} & P_{(k|k)}^{L-1,L+1} & P_{(k|k)}^{L,L+1} \end{bmatrix}$$

$$- \begin{bmatrix} P_{k|k-1} \\ P_{k|k-1}^{11} \\ P_{k|k-1}^{12} \\ P_{k|k-1}^{13} \end{bmatrix} \frac{1}{P_{k|k-1} + R_k} \begin{bmatrix} P_{k|k-1} & P_{k|k-1}^{1^T} & P_{k|k-1}^{2^T} & P_{k|k-1}^{3^T} \end{bmatrix}$$

The predictor equations for the filter are

$$\widehat{\boldsymbol{x}}_{k+1|k}^{(1)} = \widehat{\boldsymbol{x}}_{k|k}, \quad \boldsymbol{x}_{k+1|k}^{(2)} = \boldsymbol{x}_{k|k}^{(1)}, \quad \boldsymbol{x}_{k+1|k}^{(3)} = \boldsymbol{x}_{k|k}^{(2)}$$

with error covariance matrix

$$\begin{bmatrix} P_{k+1|k} & P_{k+1|k}^{1^T} & P_{k+1|k}^{2^T} & P_{k+1|k}^{3^T} \\ P_{k+1|k}^{1} & P_{k+1|k}^{11} & P_{k+1|k}^{12^T} & P_{k+1|k}^{13^T} \\ P_{k+1|k}^{2} & P_{k+1|k}^{12} & P_{k+1|k}^{22} & P_{k+1|k}^{23^T} \\ P_{k+1|k}^{3} & P_{k+1|k}^{13} & P_{k+1|k}^{23} & P_{k+1|k}^{33} \end{bmatrix} = \begin{bmatrix} P_{k|k} & P_{k+1|k}^{1^T} & P_{k+1|k}^{2^T} & P_{k+1|k}^{3^T} \\ P_{k|k} & P_{k+1|k}^{11} & P_{k+1|k}^{12^T} & P_{k+1|k}^{13^T} \\ P_{k|k}^{1} & P_{k|k}^{12} & P_{k|k}^{22} & P_{k|k}^{23^T} \\ P_{k+1|k}^{2} & P_{k+1|k}^{13} & P_{k+1|k}^{23} & P_{k|k}^{33} \end{bmatrix}$$

■

11.4.1 *Properties of Fixed-Lag Smoother*

The following are properties of the fixed-lag smoother:

- The smoother is a linear discrete-time system of dimension nL driven by the innovation $\tilde{z}_k$.
- The smoother gives estimates for all lags, $i = 1, \cdots, L-1$, with no extra effort.
- The fixed-lag smoother is time-invariant for a time-invariant system.
- The improvement increases monotonically with the lag L (i.e., with more measurements).
- The computational cost is acceptable for a short lag L if n is not large.

11.5 Fixed-Interval Smoothing

Fixed-interval smoothing is implemented by first forward Kalman filtering and saving the data then performing a backward sweep as shown in Fig. 11.6. Using the fundamental theorem of estimation theory, the minimum mean square estimate is

$$\hat{\boldsymbol{x}}_{k|N} = E\{\boldsymbol{x}_k|z_{0:N}\}, k = 0, 1, \ldots, N \tag{11.40}$$

We repeat the Kalman filter equations below for convenience.

Fixed-Interval Smoothing: Forward Sweep

The fixed-interval smoother provides the minimum mean square estimate of the state $\boldsymbol{x}_k$ given the measurements $\{z_k, k = 0, 1, \ldots, N$. From the fundamental theorem of estimation theory, this is given by the expectation $E\{\boldsymbol{x}_k|z_{0:N}\}$. To obtain the estimate, we first perform a forward sweep using the covariance filter, then a backward sweep that improves the estimates using additional measurements.

For the forward sweep, we initialize the discrete Kalman filter with $\hat{x}_0$, P_0 and use the usual predictor–corrector with the predictor equations

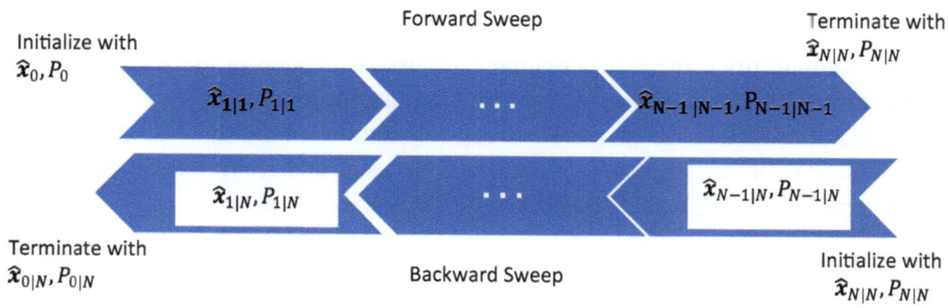

Fig. 11.6 Forward–backward fixed-interval smoothing

$$\hat{\boldsymbol{x}}_{k+1|k} = \phi_k \hat{\boldsymbol{x}}_{k|k}$$

$$P_{k+1|k} = \phi_k P_{k|k} \phi_k^T + Q_k$$

The Kalman gain is

$$K_k = P_{k|k-1} H_k^T \left(H_k P_{k|k-1} H_k^T + R_k\right)^{-1}$$

and the corrector equations are

$$\hat{\boldsymbol{x}}_{k|k} = \hat{\boldsymbol{x}}_{k|k-1} + K_k\left[z_k - H_k \hat{\boldsymbol{x}}_{k|k-1}\right]$$

$$P_{k|k} = (I_n - K_k H_k) P_{k|k-1}$$

In the forward sweep, we save the a priori and the a posteriori estimates with their covariance matrices

$$\hat{\boldsymbol{x}}_{k+1|k}, P_{k+1|k}, k = 0, \ldots, N-1$$

$$\hat{\boldsymbol{x}}_{k|k}, P_{k|k}, k = 0, \ldots, N$$

Fixed-Interval Smoothing: Backward Sweep
We initialize the backward sweep with $\hat{\boldsymbol{x}}_{N|N}$, $P_{N|N}$, then correct the estimate with

$$\hat{\boldsymbol{x}}_{k|N} = \hat{\boldsymbol{x}}_{k|k} + A_k\left\{\hat{\boldsymbol{x}}_{k+1|N} - \hat{\boldsymbol{x}}_{k+1|k}\right\}, k = N-1, N-2, \ldots, 0 \tag{11.41}$$

$$A_k = P_{k|k} \Phi_k^T P_{k+1|k}^{-1} \tag{11.42}$$

The associated error covariance matrix is

$$P_{k|N} = P_{k|k} + A_k\left\{P_{k+1|N} - P_{k+1|k}\right\} A_k^T, k = N-1, N-2, \ldots, 0 \tag{11.43}$$

For the Kalman filter, the error covariance matrix is needed in the forward sweep as part of the Kalman gain calculation. By contrast, the smoothing error covariance matrix is not needed in the backward sweep calculations of the state estimates. Both the calculations of the state estimates and the covariance matrices progress recursively with the pair $\left(\hat{\boldsymbol{x}}_{k|N}, P_{k|N}\right)$ calculated using $\left(\hat{\boldsymbol{x}}_{k+1|N}, P_{k+1|N}\right)$.

Example 11.4 Write the equations of the fixed-interval smoother in the interval [0, 1] for the scalar system

$$x_{k+1} = x_k + w_k$$

$$z_k = x_k + v_k$$

$$E\{w_j w_k\} = \alpha r \delta_{jk}, E\{v_j v_k\} = r \delta_{jk}, \alpha, q, r > 0$$

$$E\{w_j v_k\} = 0, \forall j, k$$

with measurements z_0, z_1. Initialize the filter with $\hat{x}_0 = 0$, P_0. Discuss the effect of the level of measurement noise on the correction in the estimate due to smoothing.

Solution

$k = 0$:

Forward sweep: Initialize with $\hat{x}_0$, P_0 and calculate the gain

$$K_0 = \frac{P_0}{P_0 + r}$$

The corrected estimate at time 0 is

$$\hat{x}_{0|0} = (1 - K_0)\hat{x}_0 + K_0 z_0 = \frac{P_0 z_0}{P_0 + r}$$

with error variance

$$P_{0|0} = (1 - K_0)P_0 = \frac{r P_0}{P_0 + r}$$

$k = 1$:

The predicted estimate at time 1 is

$$\hat{x}_{1|0} = \phi \hat{x}_{0|0} = \frac{P_0 z_0}{P_0 + r}$$

with error variance.

$$P_{1|0} = \phi^2 P_{0|0} + q = \frac{r P_0}{P_0 + r} + \alpha r = r \frac{(1 + \alpha)P_0 + \alpha r}{P_0 + r}$$

The Kalman gain at time one is

$$K_1 = \frac{P_{1|0}}{P_{1|0} + r} = \frac{(1 + \alpha)P_0 + \alpha r}{(2 + \alpha)P_0 + (1 + \alpha)}$$

The corrected estimate at time 1 is

$$\hat{x}_{1|1} = (1 - K_1)\hat{x}_{1|0} + K_1 z_1 = \frac{(P_0 + r)}{(2+\alpha)P_0 + (1+\alpha)} \times \frac{P_0 z_0}{P_0 + r} + K_1 z_1$$

$$= \frac{P_0 z_0 + [(1+\alpha)P_0 + \alpha r]z_1}{(2+\alpha)P_0 + r(1+\alpha)}$$

with error variance

$$P_{1|1} = (1 - K_1)P_{1|0} = \frac{(P_0 + r)}{(2+\alpha)P_0 + (1+\alpha)} \times r\frac{(1+\alpha)P_0 + \alpha r}{P_0 + r}$$

$$= r\frac{(1+\alpha)P_0 + \alpha r}{(2+\alpha)P_0 + (1+\alpha)}$$

Backward Sweep: Initialize with $\hat{x}_{1|1}$, $P_{1|1}$ and calculate the gain

$$A_0 = P_{0|0}\phi P_{1|0}^{-1} = \frac{rP_0}{P_0 + r} \times \frac{1}{r} \times \frac{P_0 + r}{(1+\alpha)P_0 + \alpha r} = \frac{P_0}{(1+\alpha)P_0 + \alpha r}$$

The smoothed estimate at time 0 is

$$\hat{x}_{0|1} = \hat{x}_{0|0} + A_0\{\hat{x}_{1|1} - \hat{x}_{1|0}\}$$

$$= \frac{P_0 z_0}{P_0 + r} + \frac{P_0}{(1+\alpha)P_0 + \alpha r}\left\{\frac{P_0 z_0 + [(1+\alpha)P_0 + \alpha r]z_1}{(2+\alpha)P_0 + r(1+\alpha)} - \frac{P_0 z_0}{P_0 + r}\right\}$$

$$= \frac{P_0\left[z_0(1+\alpha)P_0^2 + z_1\right]}{(2+\alpha)P_0 + r(1+\alpha)}$$

The correction due to smoothing is the difference

$$\hat{x}_{0|1} - \hat{x}_{0|0} = \frac{P_0\left[z_0(1+\alpha)P_0^2 + z_1\right]}{(2+\alpha)P_0 + r(1+\alpha)} - \frac{P_0 z_0}{P_0 + r}$$

$$= \frac{P_0[(z_1 - z_0)P_0 + r z_1]}{[(2+\alpha)P_0 + r(1+\alpha)][P_0 + r]}$$

The corresponding error covariance is

$$P_{0|1} = P_{0|0} + A_0\left(P_{1|1} - P_{1|0}\right) = \frac{P_0 r}{(2+\alpha)P_0 + r(1+\alpha)}$$

To examine the effect of poor measurements, we let $r \to \infty, \alpha \ll 1$, which gives $\hat{x}_{0|1} - \hat{x}_{0|0} \to 0$, with the error covariance

$$P_{0|1} = \frac{P_0}{1+\alpha}$$

Thus, the smoother correction is insignificant if the measurement noise is excessive.

To examine the effect of accurate measurements, we let $r \to 0$ to obtain $P_{0|1}$let$r \to 0$ and

$$\hat{x}_{0|1} - \hat{x}_{0|0} = \frac{z_1 - z_0}{2 + \alpha}$$

which gives a significant correction with a small error variance.

■

Example 11.5 For the Gauss–Markov random position x with autocorrelation $R_x(\tau) = e^{-|\tau|}$ and measurement noise variance $1m^2$, design a smoother with a fixed interval of length 1 s with 51 points, $N = 50$ Use a sampling period of 0.02 s.

Solution

The shaping filter for the process has the transfer function

$$G(s) = \frac{2}{s + 1}$$

and the corresponding differential equation is

$$\dot{x} = -x + 2u$$

The discrete state equation is

$$x_{k+1} = \phi x_k + w_k$$

and the measurement equation is

$$z_k = x_k + v_k$$

with parameters

$$\phi = e^{-0.02},\ H_k = 1,\ R = 1m^2$$

The covariance matrix for the process noise is

$$Q = 2\int_0^{0.02} e^{-2\eta} \mathrm{d}\eta = 1 - e^{-0.04}$$

The filter is initialized with $\hat{x}_0 = 0$, $P_0 = 1$

The forward equations for the filter are

$$K_k = \frac{P_{k|k-1}}{P_{k|k-1} + 1}$$

and the corrector equations are

$$\widehat{x}_{k|k} = \widehat{x}_{k|k-1} + K_k\left[z_k - \widehat{x}_{k|k-1}\right]$$

$$P_{k|k} = (1 - K_k)P_{k|k-1}$$

The predictor equations are

$$\widehat{x}_{k+1|k} = e^{-0.02}\widehat{x}_{k|k}$$

$$P_{k+1|k} = e^{-0.04}P_{k|k} + Q_k$$

In the forward sweep, we save

$$\widehat{x}_{k+1|k}, \quad P_{k+1|k}, k = 0, \ldots, N-1$$

$$\widehat{x}_{k|k}, \quad P_{k|k}, k = 0, \ldots, N$$

The backward sweep equations are

$$\widehat{x}_{k|N} = \widehat{x}_{k|k} + A_k\left\{\widehat{x}_{k+1|N} - \widehat{x}_{k+1|k}\right\}$$

$$A_k = e^{-0.02}\frac{P_{k|k}}{P_{k|k-1}}$$

with variance

$$P_{k|N} = P_{k|k} + e^{-0.04}\left\{P_{k+1|N} - P_{k+1|k}\right\}\left(\frac{P_{k|k}}{P_{k|k-1}}\right)^2$$

$$k = N-1, \ldots, 0$$

We perform Monte Carlo simulation to generate the measurements $z_{0:50} = \{z_i, i = 0, \ldots, 50\}$. The results of the computer simulation are shown in Figs. 11.7 and 11.8. We obtain a smaller mean square error for the smoother. In addition, the smoother "smoothes" the error curve.

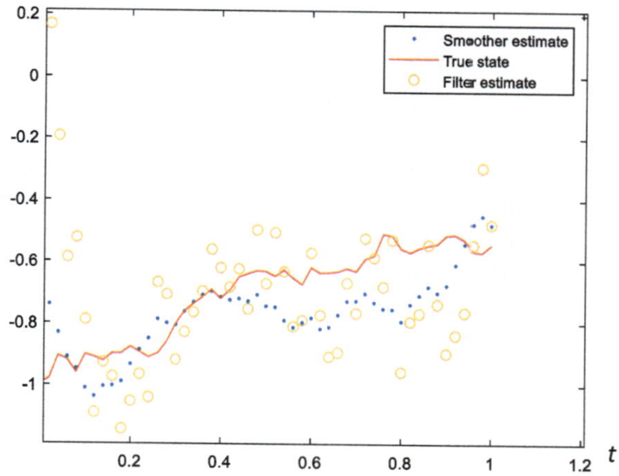

Fig. 11.7 Position and its estimates for Gauss–Markov process

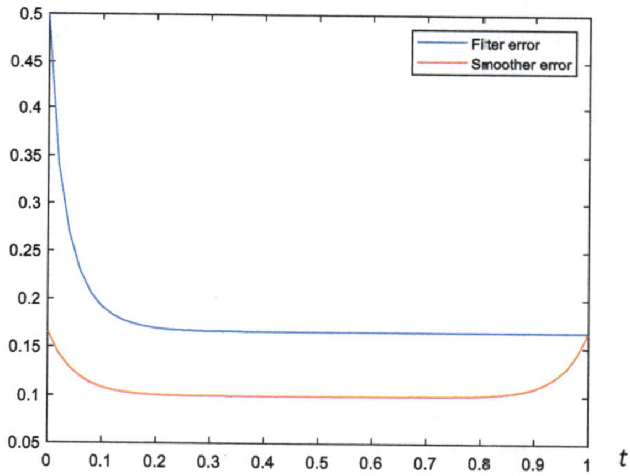

Fig. 11.8 Comparison of filter error and smoother mean square error for Example 11.5

Problems

11.1. Derive Eq. (11.9)

11.2 Show that for an integrator system with zero-mean Gaussian white noise input the variance of the prediction error increases linearly with the length of the prediction period.

11.3 P. H. Leslie developed a model for the female population of a species divided into age classes. Consider a Leslie model with 3 population groups: young, adult, and old. Only the second group can reproduce, and aging moves a percentage of the population from the first to the second group, from the second to the third and from all three groups to death. The resulting state-space model is of the form

$$\begin{bmatrix} x_1(k+1) \\ x_2(k+1) \\ x_3(k+1) \end{bmatrix} = \begin{bmatrix} 0 & 6 & 3.333 \\ 0.6 & 0 & 0 \\ 0 & 0.4 & 0 \end{bmatrix} \begin{bmatrix} x_1(k) \\ x_2(k) \\ x_3(k) \end{bmatrix} + \begin{bmatrix} w_1(k) \\ w_2(k) \\ w_3(k) \end{bmatrix}$$

where $x_1(k), x_2(k), x_3(k)$, are the young, adult, and old populations, respectively, and $w_i(k)$ are random variations that are zero-mean Gaussian and white with covariance matrix $Q = 0.5I_3$

The population count for the total number of females can be obtained as

$$z(k) = \begin{bmatrix} 1 & 1 & 1 \end{bmatrix} \begin{bmatrix} x_1(k) \\ x_2(k) \\ x_3(k) \end{bmatrix} + v(k)$$

with zero-mean white Gaussian measurement noise of variance 0.1 that is orthogonal to the process noise. Design a predictor for the system that will predict the population 2 years from the current time.

11.4 A discrete model of the electricity market is given by

$$\begin{bmatrix} x_1(k+1) \\ x_2(k+1) \\ x_3(k+1) \end{bmatrix} = \begin{bmatrix} 0.998 & 0 & 3.333 \times 10^{-3} \\ 0 & 0.98 & -5 \times 10^{-3} \\ -10^{-5} & 10^{-5} & 1 \end{bmatrix} \begin{bmatrix} x_1(k) \\ x_2(k) \\ x_3(k) \end{bmatrix} + \begin{bmatrix} -6.666 \times 10^{-3} \\ 3.333 \times 10^{-2} \\ 0 \end{bmatrix} + \begin{bmatrix} w_1(k) \\ w_2(k) \\ w_3(k) \end{bmatrix}$$

where $x_1(k)$ is the power output of the supplier, $x_2(k)$ is the power output of the consumer and $x_3(k)$ is the market clearing price (the price when supply matches demand), and $w_i(k)$ are random variations that are zero-mean Gaussian and white with covariance matrix $Q = 0.5I_3$. Write the equations for a two-step predictor for the market clearing price with the measurement vector

$$z(k) = \begin{bmatrix} I_2 & 0_{2\times 1} \end{bmatrix} \begin{bmatrix} x_1(k) \\ x_2(k) \\ x_3(k) \end{bmatrix} + \begin{bmatrix} v_1(k) \\ v_2(k) \end{bmatrix}$$

11.5 Design and simulate a fixed-interval smoother for the integrated Gauss–Markov process with a time constant of 10 s, a variance of 0.1, process noise PSD I_2, and measurement noise covariance matrix 0.2. Simulate the system for a duration of 10 s with a sampling period of 0.2 s.

11.6 Design and simulate a fixed-interval smoother for a unit point mass with zero-mean Gaussian and white random acceleration of standard deviation $0.2m/s^2$, and uncorrelated zero-mean Gaussian and white measurement noise covariance 0.1. Simulate the system for a duration of 10 s with a sampling period of 0.1 s. covariance matrix 0.2. Simulate the system for a duration of 10 s with a sampling period of 0.2 s.

Bibliography

1. Anderson, B. D. O., & Moore, J. B. (1979). *Optimal filtering*. Dover Publications.
2. Baar-Shalom, Y., Li, X. R., & Kirubarajan, T. (2001). *Estimation with applications to tracking and navigation: Theory, algorithms, and software*. Wiley-Interscience.
3. Brown, R. G., & Hwang, P. Y. C. (2012). *Introduction to random signals and applied Kalman filtering* (4th ed.). J. Wiley.

Chapter 12
Nonlinear Filtering

12.1 The Extended and Linearized Kalman Filters

Consider the process and measurement models

$$\begin{aligned} \dot{\boldsymbol{x}}(t) &= \boldsymbol{f}(\boldsymbol{x}(t), \boldsymbol{u}_d(t), t) + \boldsymbol{u}(t) \\ z(t) &= \boldsymbol{h}(\boldsymbol{x}(t), t) + \boldsymbol{v}(t), \end{aligned} \tag{12.1}$$

where

$\boldsymbol{x}(t) = n \times 1$	state vector
$\boldsymbol{u}_d(t) = p \times 1$	reference input vector
$\boldsymbol{u}(t) = p \times 1$	noise vector
$z(t) = z \times 1$	measurement vector
$\boldsymbol{f}, \boldsymbol{h} =$	vectors of known continuous functions.

If the nonlinearity is not severe, then it is possible to linearize the filter dynamics to obtain state estimates. If linearization is about a nominal trajectory, the filter is called the linearized Kalman filter. If linearization is about the estimated trajectory, the filter is called the extended Kalman filter. The linearized Kalman filter performs well if the system dynamics remain close to a known nominal trajectory $\boldsymbol{x}^*(t)$ that corresponds to the deterministic input $\boldsymbol{u}_d(t)$

$$\dot{\boldsymbol{x}}^*(t) = \boldsymbol{f}\big(\boldsymbol{x}^*(t), \boldsymbol{u}_d(t), t\big) \tag{12.2}$$

where linearization provides a good approximation as depicted in Fig. 12.1. If no nominal trajectory is available, then linearization is about the minimum mean-square estimate of the current state and the extended Kalman is used and the equations are identical with $\boldsymbol{x}^*(t)$ replaced by the estimate $\widehat{\boldsymbol{x}}_{k|k}$.

M. S. Fadali, *Introduction to Random Signals, Estimation Theory, and Kalman Filtering*, https://doi.org/10.1007/978-981-99-8063-5_12

Fig. 12.1 Trajectories

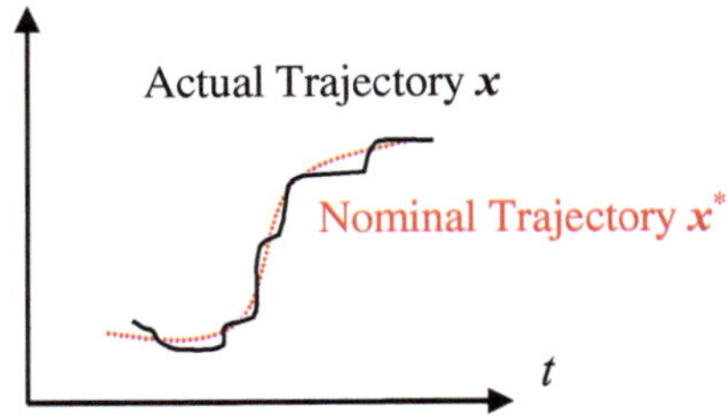

The actual state of the system is

$$x(t) = x^*(t) + \Delta x(t) \tag{12.3}$$

and its derivative is

$$\begin{aligned}\dot{x}^*(t) + \Delta\dot{x}(t) &= f(x^*(t) + \Delta x(t), u_d(t), t) + u(t) \\ &\approx f(x^*, u_d, t) + \left.\frac{\partial f(x, u_d, t)}{\partial x}\right]_{(x^*, u_d)} \Delta x + u(t)\end{aligned}$$

Canceling terms gives the linear approximation

$$\Delta\dot{x}(t) \approx \left.\frac{\partial f(x, u_d, t)}{\partial x}\right]_{(x^*, u_d)} \Delta x + u(t) = A\Delta x + u(t). \tag{12.4}$$

The linear approximation is discretized for use in the Kalan filter

$$\Delta x_{k+1} = \phi_k \Delta x_k + w_k.$$

To linearize the output equation, we use the first-order approximation

$$z(t) = h\big(x^* + \Delta x, t\big) + v(t) \approx h\big(x^*, t\big) + \left.\frac{\partial h(x, t)}{\partial x}\right]_{x^*} \Delta x + v(t).$$

This gives the linear measurement equation

$$\Delta z(t) = z(t) - h\big(x^*, t\big) \approx \left.\frac{\partial h(x, t)}{\partial x}\right]_{x^*} \Delta x + v(t) = H\Delta x + v(t). \tag{12.5}$$

The corrector equations use the linearized model with minor changes to the Kalman filter. The perturbation from the a priori estimate is

$$\Delta\hat{x}_{k|k} = x_k - \hat{x}_{k|k} = \Delta\hat{x}_{k|k-1} + K_k\big[\Delta z_k - H_k \Delta\hat{x}_{k|k-1}\big]. \tag{12.6}$$

The corrector estimate is obtained by adding the perturbation to the predictor estimate

$$\hat{\boldsymbol{x}}_{k|k-1} + \Delta\hat{\boldsymbol{x}}_{k|k} = \hat{\boldsymbol{x}}_{k|k-1} + \Delta\hat{\boldsymbol{x}}_{k|k-1} + K_k\big[z_k - \boldsymbol{h}\big(\hat{\boldsymbol{x}}_{k|k-1}, t_k\big) - H_k\Delta\hat{\boldsymbol{x}}_{k|k-1}\big].$$

The corrector estimate is

$$\begin{aligned}\hat{\boldsymbol{x}}_{k|k} &= \hat{\boldsymbol{x}}_{k|k-1} + K_k\big[z_k - \boldsymbol{h}\big(\hat{\boldsymbol{x}}_{k|k-1}, t_k\big) - H_k\Delta\hat{\boldsymbol{x}}_{k|k-1}\big]\\ &= \hat{\boldsymbol{x}}_{k|k-1} + K_k\big[z_k - \hat{z}_{k|k-1}\big].\end{aligned} \tag{12.7}$$

The corrector estimate is approximately expressed as

$$\begin{aligned}&\hat{\boldsymbol{x}}_{k|k} \approx \hat{\boldsymbol{x}}_{k|k-1} + K_k\big[H_k\Delta\boldsymbol{x}_k + \boldsymbol{v}(k) - H_k\Delta\hat{\boldsymbol{x}}_{k|k-1}\big]\\ &= \boldsymbol{x}_k - \hat{\boldsymbol{x}}_{k|k-1} - K_kH_k\Delta\boldsymbol{x}_k - K_kH_k\Delta\hat{\boldsymbol{x}}_{k|k-1} - K_k\boldsymbol{v}(k)\end{aligned}$$

The estimation error is approximately given by

$$\Delta\hat{\boldsymbol{x}}_{k|k} = (I_n - K_kH_k)\Delta\hat{\boldsymbol{x}}_{k|k-1} - K_k\boldsymbol{v}(k). \tag{12.8}$$

The a posteriori error covariance matrix is

$$P_{k|k} = E\left\{\Delta\hat{\boldsymbol{x}}_{k|k}\Delta\hat{\boldsymbol{x}}_{k|k}^T\right\} = (I_n - K_kH_k)P_{k|k-1}(I_n - K_kH_k)^T + K_kR_kK_k^T. \tag{12.9}$$

The linearized model can be used in the corrector of the standard Kalman filter but the predictor must use the nonlinear dynamics. This requires the discretization of nonlinear systems, which is complicated because we do not have an analytical solution in most cases. The predictor $\widehat{\boldsymbol{x}}_{k+1|k}$ is the numerical solution of the nonlinear state Eq. (12.2) at t_{k+1} with the initial condition $\boldsymbol{x}_k^*$ for the linearized Kalman filter, or $\widehat{\boldsymbol{x}}_{k|k}$ for the extended Kalman filter. Many numerical integration formulas are available for this purpose, and the simplest of those is the Euler forward approximation

$$\dot{\boldsymbol{x}}(t) \approx \frac{\boldsymbol{x}(t_{k+1}) - \boldsymbol{x}(t_k)}{\Delta t} = \boldsymbol{f}(\boldsymbol{x}(t_k), \boldsymbol{u}_d(t_k), t_k) + \boldsymbol{w}(t_k)$$

which gives the update equation

$$\boldsymbol{x}(t_{k+1}) = \boldsymbol{x}(t_k) + \boldsymbol{f}(\boldsymbol{x}(t_k), \boldsymbol{u}_d(t_k), t_k)\Delta t + \boldsymbol{w}(t_k). \tag{12.10}$$

The approximation is simple but can cause numerical instability. Better approximations are available, such the Runge–Kutta formula, but they require more computation. The computation loop of the extended or linearized Kalman filter is shown in Fig. 12.2.

An alternative problem formulation starts with a discrete model, often an approximation of the system dynamics

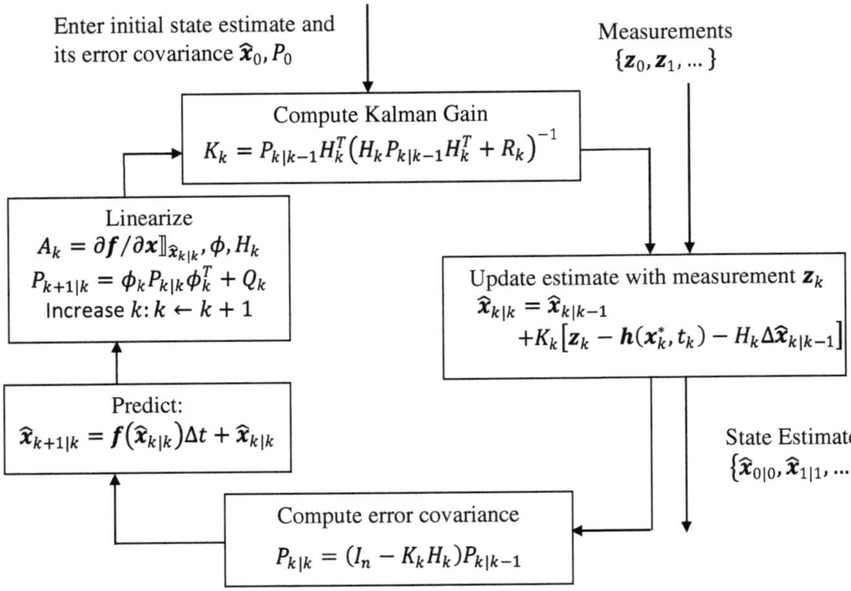

Fig. 12.2 Block diagram of linearized or extended Kalman filter

$$x(t_{k+1}) = f(x(t_k), u_d(t_k), t_k) + w(t_k) \tag{12.11}$$

and its linear approximation

$$x(t_{k+1}) \approx f\left(x^*, u_d, t_k\right) + \left.\frac{\partial f(x, u_d, t_k)}{\partial x}\right]_{x^*(t_k)} \Delta x(t_k) + w(t_k) \tag{12.12}$$

where $\Delta x(t_k) = x(t_k) - x^*(t_k)$. The linearized KF uses the nominal trajectory and the extended Kalman filter uses $x_k^* = \hat{x}_{k|k}$. The predictor is given by

$$\hat{x}_{k+1|k} = E\{x(t_{k+1})|z_{1:k}\} = f\left(\hat{x}_{k|k}, u_d, t_k\right). \tag{12.13}$$

To obtain an expression for the error covariance, we use the linear approximation

$$x(t_{k+1}) \approx f\left(\hat{x}_{k|k}, u_d, t_k\right) + \phi_k \Delta\hat{x}_{k|k} + w(t_k) \tag{12.14}$$

with the perturbation estimate $\Delta\hat{x}_{k|k} = x(t_k) - \hat{x}_{k|k}$. Subtracting the predicted estimate of (12.13) gives the prior error

$$\Delta\hat{x}_{k+1|k} = x(t_{k+1}) - \hat{x}_{k+1|k} = \phi_k \Delta\hat{x}_{k|k} + w(t_k). \tag{12.15}$$

The approximation gives the error covariance matrix. Fig. 12.3 shows the Kalman filter loop of a nonlinear discrete model.

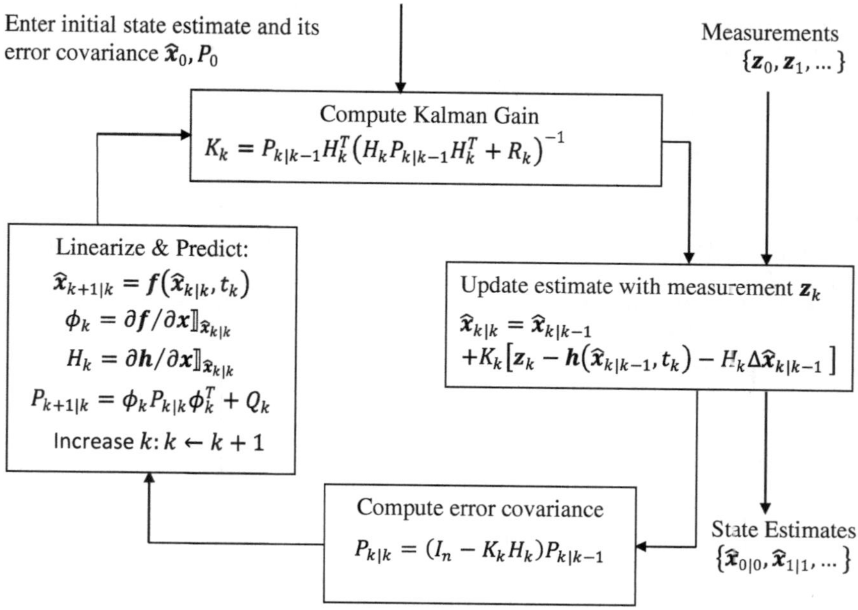

Fig. 12.3 EKF with discrete-time nonlinear model

$$P_{k+1|k} = E\left\{\Delta\hat{x}_{k+1|k}\Delta\hat{x}_{k+1|k}^T\right\} = \phi_k P_{k|k}\phi_k^T + Q_k \tag{12.16}$$

Hybrid Extended Kalman Filter

In most applications, we deal with a continuous-time nonlinear physical system using a discrete Kalman filter using measurement sampling. This requires using an approximation of the nonlinear dynamics and error covariance for the predictor and the filter loop of Fig. 12.4. Because of the continuous nature of the dynamics, one may prefer to use a continuous-time approximation of the covariance matrix derived from the covariance equation of the continuous-time Kalman filter (see Appendix E)

$$\dot{P}(t) = A(t)P(t) + P(t)A^T(t) + B(t)Q(t)B^T(t) - P(t)H^T(t)R^{-1}H(t)P(t). \tag{12.17}$$

The equation includes the covariance of the measurement noise R and the absence of measurement for the predictor makes us let $R \to \infty$ to obtain the equation

$$\dot{P}(t) = A(t)P(t) + P(t)A^T(t) + B(t)Q(t)B^T(t). \tag{12.18}$$

At the sampling points, we use the matrices

$$A_k = A(t_k),\ B_k = B(t_k),\ Q_k = Q(t_k). \tag{12.19}$$

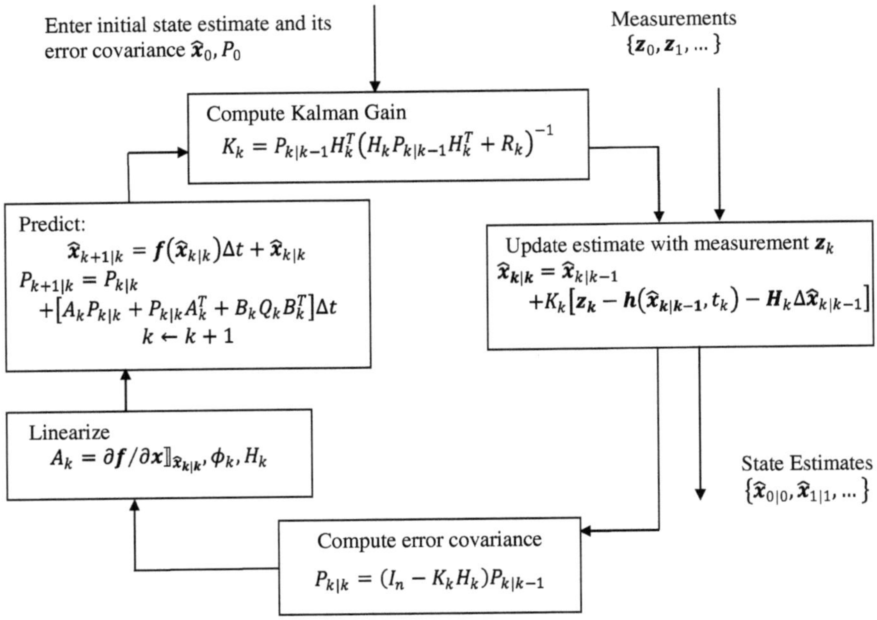

Fig. 12.4 Block diagram of the hybrid Kalman filter

The estimates of the extended Kalman filter can be improved by iteratively applying the corrector equation as shown in Fig. 12.5.

The extended Kalman filter is the most widely used and most popular filter for nonlinear systems. This is because of the relative simplicity of the filter and its ability to provide good estimates if the error propagation is approximately linear. However, the filter is essentially linear and is clearly suboptimal. It attempts to provide the best linear estimate in cases where the optimal filter may be nonlinear. Hence, the filter gives bad results and can even diverge if the linear approximation is bad. In addition, the filter requires the computationally costly calculation of the Jacobian. Although it is possible to use a second order rather than a first-order approximation, the results are not always better than the first-order approximation and require additional computation that may not be justifiable.

Another issue is the question: Which linearization gives better results? The answer depends on the application. In theory, the extended Kalman filter is better on average

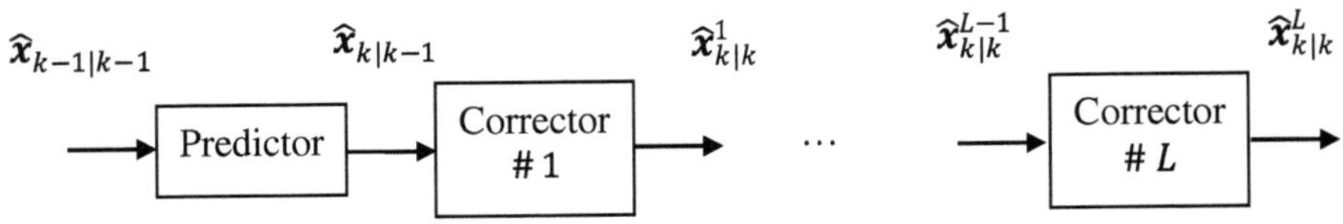

Fig. 12.5 Iteration of the corrector equation for the extended Kalman filter

but the linearized Kalman filter may provide better estimates if the nominal trajectory is a good approximation of the system's trajectory.

Example 12.1 Obtain a linearized model of the nonlinear equations cf motion with polar coordinates (r, θ).

$$\begin{gathered}\ddot{r} - r\dot{\theta}^2 + g\left(\tfrac{R_e}{r}\right)^2 = u_r(t)\\ 2\dot{r}\dot{\theta} + r\ddot{\theta} = u_\theta(t)\end{gathered}$$

where R_e = radius of the earth, (u_r, u_θ)= small white noise random functions in (r, θ) directions.

Solution

The state equations of the system are

$$\begin{gathered}\boldsymbol{x} = \begin{bmatrix} x_1 \; x_2 \; x_3 \; x_4 \end{bmatrix}^T = \begin{bmatrix} r \; \dot{r} \; \theta \; \dot{\theta} \end{bmatrix}^T\\ \dot{x}_1 = x_2\\ \dot{x}_2 = x_1 x_4^2 - \tfrac{K}{x_1^2} + u_r(t)\\ \dot{x}_3 = x_4\\ \dot{x}_4 = -2x_2x_4/x_1 + u_\theta(t)/x_1.\end{gathered}$$

We linearize the equations

$$\Delta\dot{\boldsymbol{x}} = \begin{bmatrix} 0 & 1 & 0 & 0\\ \frac{\partial f_2}{\partial x_1} & 0 & 0 & \frac{\partial f_2}{\partial x_4}\\ 0 & 0 & 0 & 1\\ \frac{\partial f_4}{\partial x_1} & \frac{\partial f_4}{\partial x_2} & 0 & \frac{\partial f_4}{\partial x_4}\end{bmatrix}_{\boldsymbol{x}^*} \Delta\boldsymbol{x} + \begin{bmatrix} 0 & 0\\ 1 & 0\\ 0 & 0\\ 0 & 1/x_1^* \end{bmatrix}\boldsymbol{u}$$

where $\boldsymbol{x}^*$ denotes the nominal state and

$$\begin{aligned}&f_2 = x_1x_4^2 - K/x_1^2 + u_r(t)\\ &f_4 = -2x_2x_4/x_1 + u_\theta(t)/x_1\\ &\partial\{u_\theta/x_1\}/\partial u_\theta = 1/x_1.\end{aligned}$$

Substitute the nominal values

$$x_1^* = r^* = R_0,\, x_2^* = 0,\, x_3^* = \theta^* = \omega_0 t,\, x_4^* = \omega_0$$

to obtain the linear state equations

$$\Delta\dot{\boldsymbol{x}} = \begin{bmatrix} 0 & 1 & 0 & 0\\ \omega_0^2 + \frac{2K}{R_0^3} & 0 & 0 & 2R_0\omega_0\\ 0 & 0 & 0 & 1\\ 0 & -\frac{2\omega_0}{R_0} & 0 & 0\end{bmatrix} \Delta\boldsymbol{x} + \begin{bmatrix} 0 & 0\\ 1 & 0\\ 0 & 0\\ 0 & 1/R_0 \end{bmatrix}\boldsymbol{u}.$$

Both the state and input matrix are constant and are independent of the nominal values. To use the equation in the extended Kalman filter, we must discretize it using the Van Loan approach. This can be done offline before implementing the Kalman filter.

The measurements of the system are

$$\begin{aligned} \boldsymbol{z} &= \begin{bmatrix} z_1 \ z_2 \end{bmatrix}^T = \begin{bmatrix} \gamma \ \alpha \end{bmatrix}^T \\ &= \begin{bmatrix} \sin^{-1}\left(\frac{R_e}{r}\right) \ \alpha_0 - \theta \end{bmatrix}^T = \left[\sin^{-1}\left(\frac{R_e}{x_1}\right) \ \alpha_0 - x_3 \right]^T. \end{aligned}$$

The linearized measurement equation in the vicinity of $x_1^* = r^* = R_0$ is

$$\begin{aligned} \boldsymbol{z}^* = \boldsymbol{h}(\boldsymbol{x}^*) &= \begin{bmatrix} \gamma^* \ \alpha^* \end{bmatrix}^T = \left[\sin^{-1}\left(\frac{R_e}{x_1^*}\right) \ \alpha_0 - x_3^* \right]^T \\ \begin{bmatrix} \Delta z_1 \\ \Delta z_2 \end{bmatrix} &= \begin{bmatrix} -\frac{R_e}{R_0\sqrt{R_0^2 - R_e^2}} & 0 \ 0 \ 0 \\ 0 & 0 \ -1 \ 0 \end{bmatrix} \Delta \boldsymbol{x} + \boldsymbol{v}. \end{aligned}$$

Hence, the H matrix is constant and independent of the nominal state.

■

12.2 Unscented Transformation and the Unscented Kalman Filter

Given a Gaussian random vector $\boldsymbol{x} \in \mathcal{R}^n$ with mean $\boldsymbol{m}_x$ and covariance P_x, then $\boldsymbol{x}$ is completely characterized by its mean and covariance. We seek estimates of the mean $\boldsymbol{m}_y$ and covariance P_y of a random variable $\boldsymbol{y}$ that is defined by a nonlinear transformation of $\boldsymbol{x}$ of the form

$$\boldsymbol{y} = \boldsymbol{h}(\boldsymbol{x}). \tag{12.20}$$

To obtain the desired estimates, we approximate the distribution of $\boldsymbol{x}$ with a set of deterministically chosen sample points $\chi^i, i = 1, 2, \ldots, N$, whose sample mean is $\boldsymbol{m}_x$ and sample covariance is P_x. We nonlinearly transform the sample points using (1) to obtain

$$\mathcal{Y}^i = \boldsymbol{h}(\chi^i), i = 1, 2, \ldots, N. \tag{12.21}$$

It can be shown that the weighted sample mean and sample covariance of the transformed sample points provide estimates of the mean and covariance of $\boldsymbol{y}$ that are accurate up to second-order moments. If the weight associated with the ith sample point is denoted by $W^{(i)}$, then we have the estimate of the mean $\boldsymbol{m}_y$ as

$$\hat{\boldsymbol{m}}_y = \sum_{i=0}^{N} W^{(i)} \mathcal{Y}^i. \tag{12.22}$$

The estimate of the covariance P_y is

$$\hat{P}_y = \sum_{i=0}^{N} W^{(i)} \left(\mathcal{Y}^i - \hat{\boldsymbol{m}}_y\right)\left(\mathcal{Y}^i - \hat{\boldsymbol{m}}_y\right)^T. \tag{12.23}$$

There are many ways of selecting the sample points $\chi^i, i = 1, 2, \ldots, N$. The simplest is known as the basic unscented transformation.

Procedure 12.1 The Basic Unscented Transformation

1. Calculate the Square Root $\sqrt{nP_x}$ of the Matrix nP_x Such that

$$\sqrt{nP_x}^T \sqrt{nP_x} = nP_x. \tag{12.24}$$

2. Select the sample points as

$$\chi^i = \boldsymbol{m}_x + \tilde{P}_x^{(i)}, \; i = 1, \ldots, 2n. \tag{12.25}$$

3. Calculate the Vectors

$$\tilde{P}_x^{(i)} = \left(\sqrt{nP_x}\right)_{(i)}^T, i = 1, \ldots, 2n \tag{12.26}$$

$$\tilde{P}_x^{(n+i)} = -\left(\sqrt{nP_x}\right)_{(i)}^T, i = 1, \ldots, 2n \tag{12.27}$$

where $\left(\sqrt{nP_x}\right)_{(i)}$ is the ith row of $\sqrt{nP_x}$.
4. With the uniform weights,

$$W^{(i)} = \frac{1}{2n}, i = 1, \ldots, 2n \tag{12.28}$$

calculate the estimates of the mean using (12.22) and the covariance using (12.23).

Note that the sample mean for this choice of sample points and weight is equal to the mean $\boldsymbol{m}_x$ and the sample covariance is equal to the covariance P_x. The proof is left as an exercise for the student.

12.2.1 Unscented Kalman Filter

The unscented Kalman filter is a discrete KF that uses an unscented transformation to obtain the mean and covariance updates. The optimal estimate is the posterior expectation

$$\boldsymbol{E}\{\boldsymbol{x}_k | z_{0:k}\} \tag{12.29}$$

where $\boldsymbol{x}_k$ is a $n \times 1$ state vector and $z_{0:k}$ is the data set up to and including time k:

$$z_{0:k} = \{z_k, i = 0, 1, \ldots, k\} \tag{12.30}$$

with z_k a $m \times 1$ measurement vector.

The system is governed by the discrete-time nonlinear model

$$\boldsymbol{x}_{k+1} = \boldsymbol{f}(\boldsymbol{x}_k, \boldsymbol{u}_k) + \boldsymbol{w}_k \tag{12.31}$$

with the nonlinear state-transition vector $\boldsymbol{f}(\boldsymbol{x}_k, \boldsymbol{u}_k)$ a function of the state vector and a deterministic input $\boldsymbol{u}_k$. The model includes a Gaussian white zero-mean process noise vector w_k whose covariance matrix is Q_k. Because physical systems are typically governed by a continuous-time model, an approximate discretization is typically required to obtain a model in the form of (12).

The measurement model is also nonlinear:

$$z_k = \boldsymbol{h}(\boldsymbol{x}_k) + \boldsymbol{v}_k \tag{12.32}$$

with the nonlinear measurement vector $\boldsymbol{h}(\boldsymbol{x}_k)$ and the additive Gaussian white zero-mean measurement noise $\boldsymbol{v}_k$ whose covariance matrix is R_k. We assume uncorrelated measurement and process noise.

The filter is initialized with the Initial state $\hat{\boldsymbol{x}}_0$ and the initial error covariance matrix P_0, then it iteratively alternates between the corrector that uses the incoming measurement and the one-step predictor. The corrector uses $\left(\hat{\boldsymbol{x}}_{k|k-1}, P_{k|k-1}\right)$ as $(\boldsymbol{m}_x, P_x)$ in the calculation of the sample points $\hat{\boldsymbol{x}}_k^{(i)}, i = 0, 1, \ldots, N$. When using the basic unscented transformation, for example, this would require the computations of Eqs. (12.24)–(12.28). These sample points allow us to calculate the estimate of the measurement vector

$$\hat{z}_k^- = \sum_{i=0}^{N} W^{(i)} \boldsymbol{h}\left(\hat{\boldsymbol{x}}_k^{(i)}\right). \tag{12.33}$$

We then calculate an estimate of the measurement error covariance

$$P_{z,k}^{-} = \sum_{i=0}^{N} \left\{ W^{(i)} \left(\boldsymbol{h}\left(\hat{\boldsymbol{x}}_{k}^{(i)}\right) - \hat{\boldsymbol{z}}_{k}^{-} \right) \left(\boldsymbol{h}\left(\hat{\boldsymbol{x}}_{k}^{(i)}\right) - \hat{\boldsymbol{z}}_{k|k-1} \right)^{T} \right\} + R_k \tag{12.34}$$

and the cross-covariance of the state and the measurement

$$P_{xz,k}^{-} = \sum_{i=0}^{N} \left\{ W^{(i)} \left(\hat{\boldsymbol{x}}_{k}^{(i)} - \hat{\boldsymbol{x}}_{k|k-1} \right) \left(\boldsymbol{h}\left(\hat{\boldsymbol{x}}_{k}^{(i)}\right) - \hat{\boldsymbol{z}}_{k|k-1} \right)^{T} \right\}. \tag{12.35}$$

We use the matrices to calculate the gain

$$K_k = P_{xz,k}^{-} \left(P_{z,k}^{-}\right)^{-1}. \tag{12.36}$$

Then we calculate the corrected estimate

$$\hat{\boldsymbol{x}}_{k|k} = \hat{\boldsymbol{x}}_{k|k-1} + K_k \left(\boldsymbol{z}_k - \hat{\boldsymbol{z}}_{k|k-1} \right) \tag{12.37}$$

and the corresponding error covariance matrix

$$P_{k|k} = P_{k|k-1} - K_k P_{z,k}^{-} K_k^T. \tag{12.38}$$

For one-step ahead prediction, we use $\left(\hat{\boldsymbol{x}}_{k|k}, P_{k|k}\right)$ as $(\boldsymbol{m}_x, P_x)$ in the calculation of sample points.

$$\hat{\boldsymbol{x}}_{k+1}^{(i)} = \boldsymbol{f}\left(\hat{\boldsymbol{x}}_{k}^{(i)}, \boldsymbol{u}_k\right), i = 0, 1, \ldots, N. \tag{12.39}$$

When using the basic unscented transformation, for example, this would require the computations of Eqs. (12.24)–(12.28). These sample points allow us to calculate the state estimate

$$\hat{\boldsymbol{x}}_{k+1|k} = \sum_{i=0}^{N} W^{(i)} \hat{\boldsymbol{x}}_{k+1}^{(i)}, W^{(i)}, i = 0, 1, \ldots, N \tag{12.40}$$

and the corresponding error covariance matrix

$$P_{k+1|k} = \sum_{i=1}^{2n} \left\{ W^{(i)} \left(\hat{\boldsymbol{x}}_{k+1}^{(i)} - \hat{\boldsymbol{x}}_{k+1|k} \right) \left(\hat{\boldsymbol{x}}_{k+1}^{(i)} - \hat{\boldsymbol{x}}_{k+1|k} \right)^{T} \right\} + Q_k. \tag{12.41}$$

To initialize the filter, we have one of two options

1. Linearize the system dynamics and use the expected value as $\hat{\boldsymbol{x}}_0$ and the covariance P_0 of the linearized system.

2. Use a zero vector or, if available, a suitable value as $\hat{\boldsymbol{x}}_0$ and a diagonal covariance P_0 with large entries. Large entries indicate large uncertainty in our knowledge of the initial state.

Because the nonlinear filter is suboptimal, the trace of the covariance matrix only gives an approximate estimate of the mean square error. For a more accurate estimate, we can use Monte Carlo simulation.

Example 12.2 Consider a mechanical system with a nonlinear spring governed by the differential equation:

$$\ddot{y} + 0.2\dot{y} + 0.2y + 0.5y^3 = u(t),$$

where y is the displacement, $u(t)$ is a zero-mean white noise process with variance 0.2 and noisy measurements of the displacement are available with additive zero-mean white noise $v(t)$ whose variance is 0.1. Write down the equations of the unscented Kalman filter for the system to obtain estimates of the state variables. Perform one iteration of the corrector and one iteration of the predictor. Assume that the variance matrix for the noise in the discretized model is $0.2I_2$

Solution

We choose the state vector

$$\boldsymbol{x} = \begin{bmatrix} x_1 \ x_2 \end{bmatrix}^T = \begin{bmatrix} y \ \dot{y} \end{bmatrix}^T.$$

The state space-space representation of system is:

$$\begin{aligned} \dot{x}_1 &= x_2 \\ \dot{x}_2 &= -0.2x_2 - 0.2x_1 - 0.05x_1^3 + u(t) \\ z &= x_1 + v. \end{aligned}$$

To obtain a discrete-time model for prediction, we use Euler's forward approximation of the derivative: The nonlinear state equation of the system is discretized as follows:

$$\begin{aligned} x_1(k+1) &= x_2(k)\Delta t + x_1(k) \\ x_2(k+1) &= \left(-0.2x_2(k) - 0.2x_1(k) - 0.05x_1^3(k)\right)\Delta t + x_2(k). \end{aligned}$$

To obtain the initial conditions for the filter, we linearize the nonlinear dynamics in the vicinity of the origin:

$$\begin{aligned} \dot{x}_1 &= x_2 \\ \dot{x}_2 &= -0.2x_2 - 0.2x_1 + u(t). \end{aligned}$$

This gives the transfer functions

$$\frac{X_1}{U} = \frac{1}{s^2 + 0.2s + 0.2}, \frac{X_2}{U} = \frac{s}{s^2 + 0.2s + 0.2}.$$

For zero-mean noise $u(t)$, both state variables are zero-mean. The variances can be calculated using

$$R_{xx}(0) = \frac{1}{2\pi j} \int_{-j\infty}^{j\infty} G(s)G(-s)R_{uu}(s)\mathrm{d}s$$

with $R_{uu}(s) = 0.2$ and the integral is evaluated using the integral table of Appendix I. The two variables can be shown to have zero cross-correlation since $g_2(t)$ is equal to the derivative of $g_1(t)$. The variances obtained are:

$$E\{x_1^2\} = \frac{1}{2(0.2)(0.2)} = 12.5, E\{x_2^2\} = \frac{1}{2(0.2)} = 2.5.$$

These results give the filter initial conditions

$$\hat{\boldsymbol{x}}_0 = \begin{bmatrix} \hat{x}_1(0) \\ \hat{x}_2(0) \end{bmatrix} = \begin{bmatrix} 0 \\ 0 \end{bmatrix}$$
$$P_0 = \begin{bmatrix} 12.5 & 0 \\ 0 & 2.5 \end{bmatrix}.$$

The basic unscented transformation requires the decomposition of the matrix

$$\sqrt{nP_0} = \left(\sqrt{2P_0}\right)^T \sqrt{2P_0} = \begin{bmatrix} 5 & 0 \\ 0 & 2.2361 \end{bmatrix} \begin{bmatrix} 5 & 0 \\ 0 & 2.2361 \end{bmatrix}.$$

We now have the four vectors

$$\tilde{P}_0^{(1)} = \begin{bmatrix} 5 \\ 0 \end{bmatrix}, \quad \tilde{P}_0^{(2)} = \begin{bmatrix} 0 \\ 2.2361 \end{bmatrix}$$
$$\tilde{P}_0^{(3)} = -\tilde{P}_0^{(1)}, \quad \tilde{P}_0^{(4)} = -\tilde{P}_0^{(2)}.$$

We calculate the four vectors

$$\hat{\boldsymbol{x}}_0^{(i)} = \hat{\boldsymbol{x}}_0 + \tilde{P}_0^{(i)}, \ i = 1, 2, 3, 4.$$

Substituting the initial conditions gives

$$\hat{\boldsymbol{x}}_0^{(1)} = \begin{bmatrix} 5 \\ 0 \end{bmatrix}, \hat{\boldsymbol{x}}_0^{(2)} = \begin{bmatrix} 0 \\ 2.2361 \end{bmatrix}$$

$$\hat{\boldsymbol{x}}_0^{(3)} = \begin{bmatrix} -5 \\ 0 \end{bmatrix}, \ \hat{\boldsymbol{x}}_0^{(4)} = \begin{bmatrix} 0 \\ -2.2361 \end{bmatrix}.$$

Since the measurements are the first entry of the state vector, we have

$$\hat{z}_0^- = \frac{1}{4}\sum_{i=1}^{4} \hat{\boldsymbol{x}}_{1,0}^{(i)} = \frac{1}{4}[5 + 0 + (-5) + 0] = 0$$

$$P_{z,0}^- = \frac{1}{4}\sum_{i=1}^{4}\left(\hat{\boldsymbol{x}}_{1,0}^{(i)} - \hat{z}_0^-\right)\left(\hat{\boldsymbol{x}}_{1,0}^{(i)} - \hat{z}_0^-\right)^T + R = \frac{1}{4}\left[5^2 + 0 + (-5)^2 + 0\right] + 0.1 = 12.6$$

$$\begin{aligned} P_{xz0}^- &= \frac{1}{4}\sum_{i=1}^{4}\left(\hat{\boldsymbol{x}}_0^{(1)} - \hat{\boldsymbol{x}}_0\right)\left(\hat{\boldsymbol{x}}_{1,0}^{(i)} - \hat{z}_0^-\right)^T \\ &= \frac{1}{4}\left[\begin{bmatrix} 5 \\ 0 \end{bmatrix} 5 + \begin{bmatrix} 0 \\ 2.2361 \end{bmatrix} 0 + \begin{bmatrix} -5 \\ 0 \end{bmatrix}(-5) + \begin{bmatrix} 0 \\ -2.2361 \end{bmatrix} 0\right] = \begin{bmatrix} 12.5 \\ 0 \end{bmatrix}. \end{aligned}$$

The optimal gain vector is

$$K_0 = P_{xz,0}^-\left(P_{z,0}^-\right)^{-1} = \begin{bmatrix} 12.5 \\ 0 \end{bmatrix}(12.6)^{-1} = \begin{bmatrix} 0.9921 \\ 0 \end{bmatrix}.$$

The corrector estimate at time 0 is

$$\widehat{\boldsymbol{x}}_{0|0} = \widehat{\boldsymbol{x}}_0 + K_0\left(z_0 - \widehat{z}_0^-\right) = 0 + \begin{bmatrix} 0.9921 \\ 0 \end{bmatrix} z_0$$

with covariance

$$P_{0|0} = P_0 - K_0 P_{z,0}^- K_0^T = \begin{bmatrix} 12.5 & 0 \\ 0 & 2.5 \end{bmatrix} - \begin{bmatrix} 0.9921 \\ 0 \end{bmatrix}(12.6)\begin{bmatrix} 0.9921 & 0 \end{bmatrix} = \begin{bmatrix} 0.0922 & 0 \\ 0 & 2.5 \end{bmatrix}.$$

The root mean square error is

$$\text{RMSE} = \sqrt{\text{trace}\left(P_{0|0}\right)} = 1.6122.$$

The basic unscented transformation requires the decomposition of the matrix

$$\sqrt{nP_{0|0}} = \left(\sqrt{2P_{0|0}}\right)^T \sqrt{2P_{0|0}} = \begin{bmatrix} 0.315 & 0 \\ 0 & 1.5811 \end{bmatrix}\begin{bmatrix} 0.315 & 0 \\ 0 & 1.5811 \end{bmatrix}.$$

We calculate the four vectors

$$\hat{\boldsymbol{x}}_0^{(i)} = \hat{\boldsymbol{x}}_{0|0} + \tilde{P}_0^{(i)}, \ i = 1, 2, 3, 4$$

Substituting for $\hat{\boldsymbol{x}}_{0|0}$ and the rows of $\sqrt{2\boldsymbol{P}_{0|0}}$ gives:

$$\hat{\boldsymbol{x}}_0^{(1)} = \begin{bmatrix} 0.9921 \\ 0 \end{bmatrix} z_0 + \begin{bmatrix} 0.315 \\ 0 \end{bmatrix} = \begin{bmatrix} 0.9921z_0 + 0.315 \\ 0 \end{bmatrix}$$

$$\hat{\boldsymbol{x}}_0^{(2)} = \begin{bmatrix} 0.9921 \\ 0 \end{bmatrix} z_0 + \begin{bmatrix} 0 \\ 1.5811 \end{bmatrix} = \begin{bmatrix} 0.9921z_0 \\ 1.5811 \end{bmatrix}$$

$$\hat{\boldsymbol{x}}_0^{(3)} = \begin{bmatrix} 0.9921 \\ 0 \end{bmatrix} z_0 - \begin{bmatrix} 0.315 \\ 0 \end{bmatrix} = \begin{bmatrix} 0.9921z_0 - 0.315 \\ 0 \end{bmatrix},$$

$$\hat{\boldsymbol{x}}_0^{(4)} = \begin{bmatrix} 0.9921 \\ 0 \end{bmatrix} z_0 - \begin{bmatrix} 0 \\ 1.5811 \end{bmatrix} = \begin{bmatrix} 0.9921z_0 \\ -1.5811 \end{bmatrix}.$$

Using Euler's forward approximation of the derivative:

$$\hat{\boldsymbol{x}}_1^{(i)} = \begin{bmatrix} \hat{\boldsymbol{x}}_{1,1}^{(i)} \\ \hat{\boldsymbol{x}}_{1,2}^{(i)} \end{bmatrix} = \begin{bmatrix} \hat{\boldsymbol{x}}_{0,1}^{(i)} + \hat{\boldsymbol{x}}_{0,2}^{(i)} \Delta t \\ \hat{\boldsymbol{x}}_{0,2}^{(i)} + \left(-0.2\hat{\boldsymbol{x}}_{0,2}^{(i)} - 0.2\hat{\boldsymbol{x}}_{0,1}^{(i)} - 0.05(\hat{\boldsymbol{x}}_{0,1}^{(i)})^3\right) \Delta t \end{bmatrix}, \quad i = 1, 2, 3, 4.$$

This gives the vectors

$$\hat{\boldsymbol{x}}_1^{(1)} = \begin{bmatrix} 0.9921z_0 + 0.315 \\ -0.1\left(0.6456 + 2.1319z_0 + 0.4651z_0^2 + 0.4882z_0^3\right)\Delta t \end{bmatrix}$$

$$\hat{\boldsymbol{x}}_1^{(2)} = \begin{bmatrix} 0.9921z_0 + 1.5811\Delta t \\ 1.5811 - 0.1\left(3.1622 + 1.9842z_0 + 0.4882z_0^3\right)\Delta t \end{bmatrix}$$

$$\hat{\boldsymbol{x}}_1^{(3)} = \begin{bmatrix} 0.9921z_0 - 0.315 \\ -0.1\left(0.6456 + 2.1319z_0 - 0.4651z_0^2 + 0.4882z_0^3\right)\Delta t \end{bmatrix}$$

$$\hat{\boldsymbol{x}}_1^{(4)} = \begin{bmatrix} 0.9921z_0 + 1.5811\Delta t \\ -1.5811 + 0.1\left(3.1622 - 1.9842z_0 - 0.4882z_0^3\right)\Delta t \end{bmatrix}.$$

We calculate the predicted state

$$\hat{\boldsymbol{x}}_{1|0} = \frac{1}{4}\sum_{i=0}^{N} \hat{\boldsymbol{x}}_1^{(i)} = \begin{bmatrix} 0.9921z_0 \\ -0.2058z_0 - 0.048824z_0^3\Delta t \end{bmatrix}$$

To calculate the error covariance, we need the formula

$$P_{1|0} = \frac{1}{4}\sum_{i=1}^{4}\left(\hat{\boldsymbol{x}}_1^{(i)} - \hat{\boldsymbol{x}}_{1|0}\right)\left(\hat{\boldsymbol{x}}_1^{(i)} - \hat{\boldsymbol{x}}_{1|0}\right)^T + Q.$$

Neglecting small terms, we have

$$P_{1|0}=\frac{1}{4}\begin{bmatrix}0.315\\ -10^{-2}(6.4563+0.7383z_0+4.6506z_0^2+10^{-9}z_0^3)\Delta t\end{bmatrix}\begin{bmatrix}0.315\\ -0.01(6.4563+0.7383z_0+4.6506z_0^2+10^{-9}z_0^3)\Delta t\end{bmatrix}^T$$
$$+\begin{bmatrix}1.5811\Delta t\\ 1.5811-0.1(3.1622+0.007383z_0+10^{-11}z_0^3)\Delta t\end{bmatrix}\begin{bmatrix}1.5811\Delta t\\ 1.5811-0.1(3.1622+0.007383z_0+10^{-11}z_0^3)\Delta t\end{bmatrix}^T$$
$$+\begin{bmatrix}-0.315\\ 10^{-2}(6.4563+0.7383z_0+4.6506z_0^2+10^{-9}z_0^3)\Delta t\end{bmatrix}\begin{bmatrix}-0.315\\ 10^{-2}(6.4563+0.7383z_0+4.6506z_0^2+10^{-9}z_0^3)\Delta t\end{bmatrix}^T$$
$$+\begin{bmatrix}1.5811\Delta t\\ -1.5811-0.1(3.1622+0.007383z_0+10^{-11}z_0^3)\Delta t\end{bmatrix}\begin{bmatrix}1.5811\Delta t\\ -1.5811-0.1(3.1622+0.007383z_0+10^{-11}z_0^3)\Delta t\end{bmatrix}^T$$
$$+0.2I_2$$

Assuming a small value for Δt then neglecting small terms, we have

$$P_{1|0}\approx\begin{bmatrix}0.2992 & \Delta t\left(7.4793+0.0023z_0-0.0147z_0^2\right)\\ \Delta t\left(7.4793+0.0023z_0-0.01465z_0^2\right) & 7.6996+0.0233z_0\Delta t\end{bmatrix}.$$

The root mean square error is

$$\text{RMSE}=\sqrt{\text{trace}\left(P_{1|0}\right)}\approx\sqrt{8+0.0233z_0\Delta t}.$$

12.3 Ensemble Kalman Filter

The ensemble Kalman filter obtains its estimates by generating an ensemble of random vectors as shown in the filter loop of Fig. 12.6. Estimation accuracy depends on ensemble size. The size of the ensemble is heuristic but 50–100 samples are often enough for thousands of state variables. Hence, it is particularly suited to application with high-order nonlinear dynamics with highly uncertain initial states and with a large number of measurements and is widely used in weather forecasting. Because the filter does not use linearized equations, it does not require the costly computation of the Jacobian. We consider the nonlinear process

$$\boldsymbol{x}_{k+1}=\phi(\boldsymbol{x}_k)+\boldsymbol{w}_k \tag{12.42}$$

with measurement model

$$z_k=\boldsymbol{h}(\boldsymbol{x}_k)+\boldsymbol{v}_k. \tag{12.43}$$

We initialize the filter with $\hat{\boldsymbol{x}}_0$, P_0 then generate N random state estimates

$$\{\mathcal{X}_{k|k}^{(i)}, i=1,2,\ldots,N]\sim\mathcal{N}\left(\hat{\boldsymbol{x}}_{k|k}, P_{k|k}\right). \tag{12.44}$$

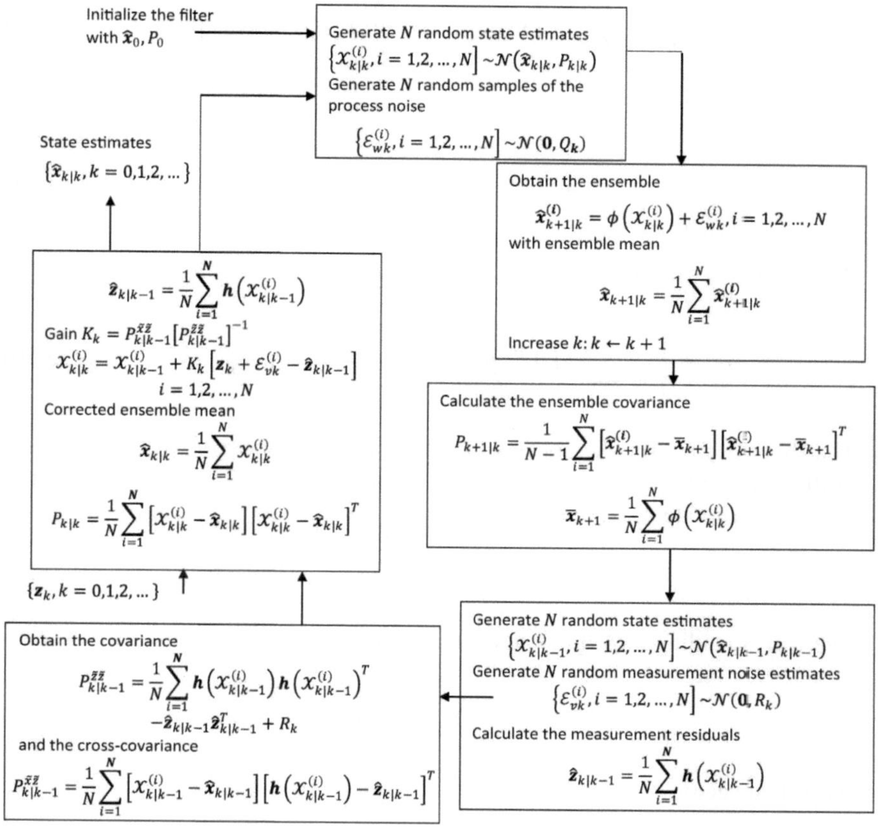

Fig. 12.6 Block diagram of the ensemble Kalman filter

We also generate random samples of the process noise

$$\{\mathcal{E}_{wk}^{(i)}, i = 1, 2, \ldots, N] \sim \mathcal{N}(0, Q_k). \tag{12.45}$$

Adding the noise to the state estimates gives the ensemble

$$\hat{\boldsymbol{x}}_{k+1|k}^{(i)} = \phi\left(\mathcal{X}_{k|k}^{(i)}\right) + \mathcal{E}_{wk}^{(i)}, i = 1, 2, \ldots, N \tag{12.46}$$

with ensemble mean

$$\hat{\boldsymbol{x}}_{k+1|k} = \frac{1}{N}\sum_{i=1}^{N} \hat{\boldsymbol{x}}_{k+1|k}^{(i)}. \tag{12.47}$$

The ensemble covariance is

$$P_{k+1|k} = \frac{1}{N-1}\sum_{i=1}^{N}\left[\widehat{\boldsymbol{x}}_{k+1|k}^{(i)} - \overline{\boldsymbol{x}}_{k+1}\right]\left[\widehat{\boldsymbol{x}}_{k+1|k}^{(i)} - \overline{\boldsymbol{x}}_{k+1}\right]^T \tag{12.48}$$

where

$$\overline{\boldsymbol{x}}_{k+1} = \frac{1}{N}\sum_{i=1}^{N}\phi\left(\mathcal{X}_{k|k}^{(i)}\right). \tag{12.49}$$

An alternative expression for the covariance is

$$P_{k+1|k} = \left\{\frac{1}{N}\sum_{i=1}^{N}\phi\left(\mathcal{X}_{k|k}^{(i)}\right)\phi\left(\mathcal{X}_{k|k}^{(i)}\right)^T\right\} - \hat{\boldsymbol{x}}_{k|k-1}\hat{\boldsymbol{x}}_{k|k-1}^T. \tag{12.50}$$

The above expression is equivalent to (12.48) with division by N instead of $N-1$. The difference for large N is negligible. The proof of the formula is left as an exercise.

Next, we generate N random a priori state estimates

$$\{\mathcal{X}_{k|k-1}^{(i)}, i = 1, 2, \ldots, N]\, \mathcal{N}\left(\hat{\boldsymbol{x}}_{k|k-1}, P_{k|k-1}\right). \tag{12.51}$$

The corresponding measurement estimates are

$$\widehat{\boldsymbol{z}}_{k|k-1} = \frac{1}{N}\sum_{i=1}^{N}\boldsymbol{h}\left(\mathcal{X}_{k|k-1}^{(i)}\right) \tag{12.52}$$

with residual vector

$$\tilde{\boldsymbol{z}}_{k|k-1} = \boldsymbol{z}_k - \hat{\boldsymbol{z}}_{k|k-1} = \boldsymbol{h}(\boldsymbol{x}_k) + \boldsymbol{v}_k - \frac{1}{N}\sum_{i=1}^{N}\boldsymbol{h}\left(\mathcal{X}_{k|k-1}^{(i)}\right). \tag{12.53}$$

We obtain the covariance

$$P_{k|k-1}^{\tilde{z}\tilde{z}} = E\left\{\widetilde{\boldsymbol{z}}_{k|k-1}\widetilde{\boldsymbol{z}}_{k|k-1}^T\right\} = \frac{1}{N}\sum_{i=1}^{N}\boldsymbol{h}\left(\mathcal{X}_{k|k-1}^{(i)}\right)\boldsymbol{h}\left(\mathcal{X}_{k|k-1}^{(i)}\right)^T - \widehat{\boldsymbol{z}}_{k|k-1}\widehat{\boldsymbol{z}}_{k|k-1}^T + R_k \tag{12.54}$$

and the cross-covariance

$$P_{k|k-1}^{\tilde{x}\tilde{z}} = E\left\{\tilde{\boldsymbol{x}}_{k|k-1}\tilde{\boldsymbol{z}}_{k|k-1}^T\right\} = \frac{1}{N}\sum_{i=1}^{N}\left[\mathcal{X}_{k|k-1}^{(i)} - \hat{\boldsymbol{x}}_{k|k-1}\right]\left[\boldsymbol{h}\left(\mathcal{X}_{k|k-1}^{(i)}\right) - \hat{\boldsymbol{z}}_{k|k-1}\right]^T. \tag{12.55}$$

The covariances are used to calculate the filter gain

$$K_k = P_{k|k-1}^{\tilde{x}\tilde{z}} \left[P_{k|k-1}^{\tilde{z}\tilde{z}} \right]^{-1}. \tag{12.56}$$

We generate N random measurement noise estimates

$$\left\{ \mathcal{E}_{vk}^{(i)}, i = 1, 2, \ldots, N \right\} \sim \mathcal{N}(0, R_k) \tag{12.57}$$

and update the N random state estimates

$$\mathcal{X}_{k|k}^{(i)} = \mathcal{X}_{k|k-1}^{(i)} + K_k \left[z_k + \mathcal{E}_{vk}^{(i)} - \hat{z}_{k|k-1} \right], i = 1, 2, \ldots, N. \tag{12.58}$$

The corrected ensemble mean is

$$\hat{\boldsymbol{x}}_{k|k} = \frac{1}{N} \sum_{i=1}^{N} \mathcal{X}_{k|k}^{(i)}, \tag{12.59}$$

and its covariance is

$$P_{k|k} = P_{k|k-1} - K_k P_{k|k-1}^{\tilde{z}\tilde{z}} K_k^T. \tag{12.60}$$

Example 12.3 Write the equations of the ensemble Kalman filter for a Gauss–Markov process with zero-mean Gaussian white measurement noise.

Solution

Consider the Gauss–Markov process with autocorrelation and PSD

$$R_{XX}(\tau) = \sigma^2 e^{-\beta|\tau|}, \; S_{XX}(s) = \frac{2\sigma^2\beta}{-s^2 + \beta^2}.$$

The shaping file transfer function is the causal factor of the PSD

$$G(s) = L(s) = \frac{\sqrt{2\sigma^2\beta}}{s + \beta}, \; g(t) = \sqrt{2\sigma^2\beta} e^{-\beta t}.$$

The corresponding state-space model is

$$\begin{aligned} \dot{x} &= -\beta x + \sqrt{2\sigma^2\beta}\, u \\ y &= x \end{aligned}$$

We discretize the model and obtain

$$
\begin{aligned}
\phi &= e^{-\beta\Delta t},\, H = W = 1,\, B = \sqrt{2\sigma^2\beta} \\
Q(k) &= \int_0^{\Delta t} e^{F\xi} G W G^T e^{F^T\xi} \mathrm{d}\xi \\
&= \int_0^{\Delta t} \left(e^{-\beta\Delta t}\sqrt{2\sigma^2\beta}\right)^2 (1)\mathrm{d}\xi \\
&= \sigma^2\left(1 - e^{-2\beta\Delta t}\right) \\
x_{k+1} &= \phi(x_k) + w_k = e^{-\beta\Delta t} x_k + w_k \\
z_k &= x_k + v_k
\end{aligned}
$$

The predictor steps are as follows:

- Start process with initial conditions ($k = 0$).
- Generate N random state estimates $\{\, \mathcal{X}_{k|k}^{(i)}, i = 1, 2, \ldots, N] \sim \mathcal{N}\left(\hat{\boldsymbol{x}}_{k|k}, P_{k|k}\right)$
- Generate N random noise estimates $\{\, \mathcal{E}_{wk}^{(i)}, i = 1, 2, \ldots, N] \sim \mathcal{N}(0, Q_k)$
- Calculate the ensemble mean

$$
\hat{x}_{k+1|k} = \frac{1}{N}\sum_{i=1}^{N}\left[e^{-\beta\Delta t}\mathcal{X}_k^{(i)} + \mathcal{E}_{wk}^{(i)}\right]
$$

$$
h\left(\mathcal{X}_{k|k-1}^{(i)}\right) = \mathcal{X}_{k|k-1}^{(i)},\, \hat{z}_{k|k-1} = \frac{1}{N}\sum_{i=1}^{N}\mathcal{X}_{k|k-1}^{(i)}
$$

$$
\phi\left(\mathcal{X}_k^{(i)}\right) = \phi\mathcal{X}_{k|k-1}^{(i)} = e^{-\beta\Delta t}\mathcal{X}_{k|k-1}^{(i)}.
$$

The covariance matrices are

$$
\begin{aligned}
P_{k|k-1}^{\tilde{z}\tilde{z}} &= \frac{1}{N}\sum_{i=1}^{N}\boldsymbol{h}\left(\mathcal{X}_{k|k-1}^{(i)}\right)\boldsymbol{h}\left(\mathcal{X}_{k|k-1}^{(i)}\right)^T - \hat{\boldsymbol{z}}_{k|k-1}\hat{\boldsymbol{z}}_{k|k-1}^T + R_k \\
&= \frac{1}{N}\sum_{i=1}^{N}\left(\mathcal{X}_{k|k-1}^{(i)}\right)^2 - \hat{z}_{k|k-1}^2 + R_k \\
P_{k|k-1}^{\tilde{x}\tilde{z}} &= \frac{1}{N}\sum_{i=1}^{N}\left[\mathcal{X}_{k|k-1}^{(i)} - \hat{x}_{k|k-1}\right]\left[\left(\mathcal{X}_{k|k-1}^{(i)}\right) - \hat{z}_{k|k-1}\right]^T
\end{aligned}
$$

The gain matrix

$$
K_k = P_{k|k-1}^{\tilde{x}\tilde{z}}\left[P_{k|k-1}^{\tilde{z}\tilde{z}}\right]^{-1}
$$

is used to generate the points

$$
\mathcal{X}_{k|k}^{(i)} = \mathcal{X}_{k|k-1}^{(i)} + K_k\left[\boldsymbol{z}_k + \mathcal{E}_{vk}^{(i)} - \hat{\boldsymbol{z}}_{k|k-1}\right]
$$

and their average is the a posteriori estimate

$$\hat{x}_{k|k} = \frac{1}{N}\sum_{i=1}^{N} \mathcal{X}_{k|k}^{(i)}.$$

12.4 Bayesian Filter

The Bayesian filter (Fig. 12.7) propagates the pdf of the state given the measurements for the discrete nonlinear process model

$$\boldsymbol{x}_{k+1} = \phi(\boldsymbol{x}_k) + w_k \tag{12.61}$$

with measurement model

$$z_k = \boldsymbol{h}(\boldsymbol{x}_k) + \boldsymbol{v}_k. \tag{12.62}$$

The noise processes are not necessarily Gaussian. For the corrector, we have a recursion for $p(\boldsymbol{x}_k|z_{0:k})$ in terms of $p(\boldsymbol{x}_k|z_{0:k-1})$ and for the predictor a recursion for $p(\boldsymbol{x}_{k+1}|z_{0:k})$ in terms of $p(\boldsymbol{x}_k|z_{0:k})$, where $z_{0:k}$ is the set of measurements up to and including time k.

To derive the pdf update formulas, we first derive some basic probability rules. We derive the rules for probabilities but, as shown in Chap. 2, they can be extended to densities.

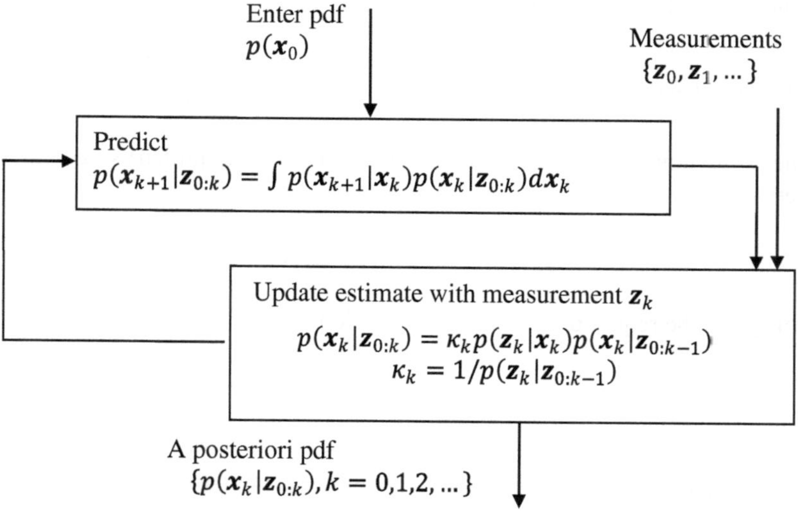

Fig. 12.7 Block diagram of Bayesian filter

Rule 1:

$$(A|B,C)=\frac{P(A,B,C)}{P(B,C)}=\frac{P(B|A,C)P(A,C)/P(C)}{P(B,C)/P(C)}$$
$$=\frac{P(B|A,C)P(A|C)}{P(B|C)}$$

For densities, we have

$$p(a|b,c)=\frac{p(b|a,c)p(a|c)}{p(b|c)}. \tag{12.63}$$

Rule 2:

$$P(A,B|C)=\frac{P(A,B,C)}{P(C)}=\frac{P(B|A,C)P(A,C)}{P(C)}$$
$$=P(B|A,C)P(A|C)$$

For densities, we have

$$p(a,b|c)=p(b|a,c)p(a|c). \tag{12.64}$$

Using the fundamental theorem of estimation theory gives the minimum mean-square estimator as the conditional expectation $\widehat{\boldsymbol{x}}_k = E\{\boldsymbol{x}_k|z_{0:k}\}$ given the history of measurements $z_{0:k}=\{z_i, i=0,1,\ldots,k\}$. For the corrector, we need a recursion for $p(\boldsymbol{x}_k|z_{0:k})$ in terms of $p(\boldsymbol{x}_k|z_{0:k-1})$. We apply the pdf relationship (12.63)

$$p(a|b,c)=\frac{p(b|a,c)p(a|c)}{p(b|c)}$$

to write the conditional pdf as

$$\hat{\boldsymbol{x}}_k=E\{\boldsymbol{x}_k|z_{0:k}\}=p(\boldsymbol{x}_k|z_k,z_{0:k-1})=\frac{p(z_k|\boldsymbol{x}_k,z_{0:k-1})p(\boldsymbol{x}_k|z_{0:k-1})}{p(z_k|z_{0:k-1})}. \tag{12.65}$$

The expression includes a measurement pdf that can be simplified using its Markov property to

$$p(z_k|\boldsymbol{x}_k,z_{0:k-1})=p(z_k|\boldsymbol{x}_k).$$

This gives the required corrector pdf recursion

$$\begin{aligned}p(\boldsymbol{x}_k|z_{0:k})&=\kappa_k p(z_k|\boldsymbol{x}_k)p(\boldsymbol{x}_k|z_{0:k-1})\\ \kappa_k&=1/p(z_k|z_{0:k-1}).\end{aligned} \tag{12.66}$$

For the predictor, we need a recursion for $p(\boldsymbol{x}_{k+1}|z_{0:k})$ in terms of $p(\boldsymbol{x}_k|z_{0:k})$

$$p(\boldsymbol{x}_{k+1}|z_{0:k}) = \int p(\boldsymbol{x}_k, \boldsymbol{x}_{k+1}|z_{0:k})\mathrm{d}\boldsymbol{x}_k$$

Applying the pdf identity to the integrand gives the formula

$$p(\boldsymbol{x}_{k+1}|z_{0:k}) = \int p(\boldsymbol{x}_{k+1}|\boldsymbol{x}_k, z_{0:k})p(\boldsymbol{x}_k|z_{0:k})\mathrm{d}\boldsymbol{x}_k. \tag{12.67}$$

Using the Markov property simplifies the first pdf in the integrand to

$$p(\boldsymbol{x}_{k+1}|\boldsymbol{x}_k, z_{0:k}) = p(\boldsymbol{x}_{k+1}|\boldsymbol{x}_k).$$

We now have the predictor update

$$p(\boldsymbol{x}_{k+1}|z_{0:k}) = \int p(\boldsymbol{x}_{k+1}|\boldsymbol{x}_k)p(\boldsymbol{x}_k|z_{0:k})\mathrm{d}\boldsymbol{x}_k. \tag{12.68}$$

The filter is initialized with the pdf $p(\boldsymbol{x}_0)$

$$p(\boldsymbol{x}_0|z_0) = \kappa_k p(z_0|\boldsymbol{x}_0)p(\boldsymbol{x}_0). \tag{12.69}$$

The filter is not very useful in general because the equations cannot be solved analytically.

12.5 Particle Filters

Most nonlinear filters require the noise processes to be Gaussian. Or approximately Gaussian. However, there are many processes in nature that are non-Gaussian, and Gaussian noise is typically an approximation. Some filters do not function properly if the nonlinearity is severe. Particle filters are able to handle highly nonlinear processes with non-Gaussian noise.

As with other filters, we seek the minimum variance solution to the equation

$$\hat{\boldsymbol{x}}_{k|k} = E\{\boldsymbol{x}_k|z_{0:k}\} = \int \boldsymbol{x}_k p_{x|z}(\boldsymbol{x}_k|z_{0:k})\mathrm{d}\boldsymbol{x}_k. \tag{12.70}$$

To evaluate the integral, we use Monte Carlo integration.

Monte Carlo Integration

Mote Carlo integration is numerical integration using a randomly generated set of points. Unlike other numerical integration approaches, it does not suffer from the "curse of dimensionality"; that is, it does not become intractable as the dimension of the integral increases and the rate of convergence is independent of the dimension of the integral.

Consider the integral of a function $g(x)$ that is continuous on the interval $[a, b]$

$$\int_a^b g(x)\mathrm{d}x. \tag{12.71}$$

To evaluate the integral, we factorize the integrand into the product

$$g(x) = f(x)p(x), x \in [a, b]$$

where $p(x)$ such that $p(x) \geq 0, x \in [a, b]$, and is zero elsewhere and

$$\int_a^b p(x)\mathrm{d}x = 1.$$

Hence, $p(x)$ can be viewed as a pdf and the integral can be written as

$$\int_a^b g(x)\mathrm{d}x = \int_a^b f(x)p(x)\mathrm{d}x = E\{f(x)\}. \tag{12.72}$$

We approximate the continuous pdf with a discrete pdf and use the sample mean as an estimator of $f(x)$ as in the following procedure.

Procedure 12.2: Monte Carlo Integration

- Generate i.i.d. values $X_i \sim p(x), i = 1, \ldots, N$
- Calculate $f(X_i), i = 1, \ldots, N$
- Calculate the estimate of the integral $E\{f(X)\}$.

$$\overline{f(X)} = \frac{1}{N}\sum_{i=1}^{N} f(X_i). \tag{12.73}$$

It is well known (see Chap. 5) that the sample mean $\overline{f(X)}$ is both an unbiased and a consistent estimator of the mean. By the central limit theorem, the estimate is also asymptotically Gaussian. The properties of the Monte Carlo integral are summarized in the following theorem.

Theorem 12.1 *The Monte Carlo integral is an unbiased, consistent, and asymptotically Gaussian estimate of the integral* (12.72) *with standard error (standard deviation of the estimator)*

$$\frac{\sigma_f}{\sqrt{N}}$$

where σ_f is the standard deviation of $f(x)$.

■

The simplest choice of the pdf $p(x)$ is the uniform distribution

$$\int_a^b g(x)\mathrm{d}x = (b-a)\int_a^b \frac{1}{b-a}g(x)\mathrm{d}x = (b-a)E\{g(X)\},\ X \sim U[a,b]. \quad (12.74)$$

Following Procedure 12.2, we generate $X_i \sim U[a,b], i = 1,\ldots,N$, then calculate the estimate of the integral

$$\int_a^b g(x)\mathrm{d}x = \frac{b-a}{N}\sum_{i=1}^{N} g(X_i). \quad (12.75)$$

Example 12.4 Use Procedure 12.2 to estimate π using Monte Carlo integration and the identity

$$\int_0^{\pi} \frac{12}{13+5\cos(x)}\mathrm{d}x = 2\tan^{-1}\left(\frac{2}{3}\tan\left(\frac{\pi}{2}\right)\right) = \pi.$$

Discuss the effect of increasing the number of sample points on the error.

Solution

We first verify that the expression is equal to π using the MATLAB command

```
>> 2*arctan(2*tan((1/2)*x)*(1/3))
```

Next, following Procedure 12.2, we generate samples $X_i \sim U[0,\pi], i = 1,\ldots,N$, and calculate

$$\frac{12}{13+5\cos(X_i)},\ i = 1,\ldots,N$$

for each point then average to obtain an estimate of the mean

$$\pi \approx \frac{1}{N}\sum_{i=1}^{N}\frac{12}{13+5\cos(X_i)}.$$

The following MATLAB script implements the procedure

```
N=100;
sum=0;
for i=1:N
    sum=sum+12/(13+5*cos(rand*pi));
end
pi_hat=sum*pi/N
```

For $N = 100$, the error is 1.5% but the error decreases with a larger number of points. The answer obtained with $N = 10^6$ is

pi_hat =

3.1416

The error is less than 0.002%.

■

Importance Sampling

It is not always possible to generate samples from $p(x)$ but one may be able to generate samples from a similar pdf $q(x)$, known as the proposal or importance density, that has the same support, that is

$$p(x) > 0 \Rightarrow q(x) > 0, \forall x \in \mathcal{R}^n. \tag{12.76}$$

In addition, the choice of $q(x)$ can reduce the variance of the approximation error that we make when we approximate an expected value with Monte Carlo integration. This approach is known as importance sampling.

In importance sampling, we write the integral as

$$\int_a^b g(x)\mathrm{d}x = \int_a^b f(x)p(x)\mathrm{d}x = \int_a^b f(x)\frac{p(x)}{q(x)}q(x)\mathrm{d}x \tag{12.77}$$

where $p(x)/q(x)$ is assumed bounded, then find the expected value of $f(x)p(x)/q(x)$ using Monte Carlo integration. To this end, we generate $X_i \sim q(x), i = 1, \ldots, N$, and form the weighted sum

$$\overline{Y} = \frac{1}{N}\sum_{i=1}^{N} Y_i = \frac{1}{N}\sum_{i=1}^{N} f(X_i)\frac{p(X_i)}{q(X_i)}. \tag{12.78}$$

The name importance sampling is because the summation assigns larger weights to more important ranges of the function $f(.)$. Because the desired density $p(X_i)$ may be unknown, we must normalize the sum as

$$\overline{Y} = \frac{1}{N}\sum\nolimits_{i=1}^{N} f(X_i)W(X_i) \tag{12.79}$$

with normalized weights

$$W(X_i) = \frac{p(X_i)}{q(X_i)} / \sum_{i=1}^{N} \frac{p(X_i)}{q(X_i)}. \tag{12.80}$$

The weights correct for the differences between $p(x)$ and $q(x)$ to obtain a good estimate of the expectation that converges to the true expected value as $N \to \infty$.

Although it may appear that any pdf $q(x)$ that has the same support as $p(x)$ can yield the desired answer, in practice the choice of $q(x)$ is critical and if it is not carefully chosen the weights for most of the point of the sample will be negligible. This implies that a huge sample size will be required to obtain a good estimate.

An example of using importance sampling to calculate the expected value, including a MATLAB script, is provided in [7]. The example is discussed next.

Example 12.5 Use importance sampling to find the expected value of the function

$$E[f(x)] = \int_0^\infty x\, f(x) p(x) \mathrm{d}x$$

with pdf

$$p(x) = \begin{cases} x^{k-1} e^{-x^2/2}, & x \geq 0 \\ 0, & \text{elsewhere} \end{cases}$$

using the proposal distribution

$$q(x) = \begin{cases} \frac{2}{\sqrt{2\pi\sigma^2}} e^{-(x-\mu)^2/2\sigma^2}, & x \geq 0 \\ 0, & \text{elsewhere} \end{cases}$$

and $\mu = 0.8$, $\sigma^2 = 1.5$. Obtain the average of 10 iterations each of samples of sizes $N = 10, 10^2, \ldots, 10^6$, and compare the results. Plot the results together with true value obtained by directly evaluating the expected value using numerical integration.

Solution

We rewrite the expectation as

$$E[f(x)] = \int_0^\infty x f(x) \left[\frac{p(x)}{q(x)}\right] q(x) \mathrm{d}x.$$

We draw samples from the proposal density and calculate an estimate of the expected value using the summation of (12.77) with the normalized weights of (12.78). The samples are generated from a normal distribution with only positive values retained while negative values are discarded. This can be accomplished using the MATLAB statistics toolbox command **normrnd**.

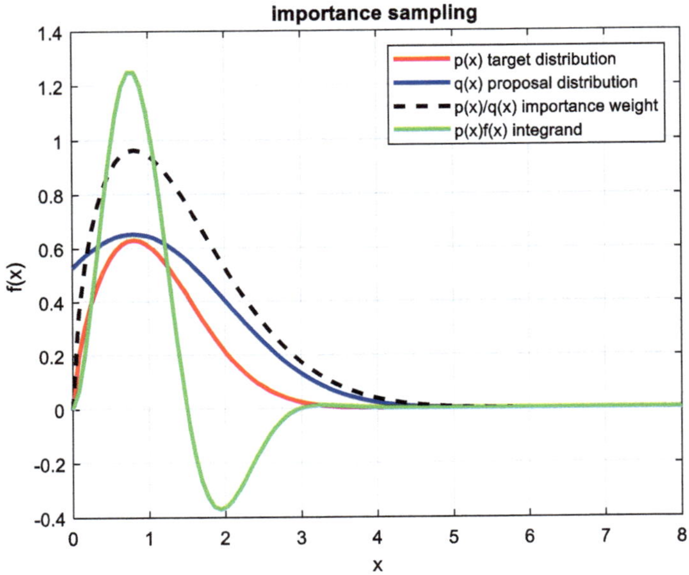

Fig. 12.8 Plot of densities, weights, and integrand

The script provided in [7] gives the results. The plot of Fig. 12.8 shows the target density, the proposal density, the weights, and the integrand in the expectation. We observe the similarity between the proposal and target densities. We also see how the weights increase in the range of larger values of the target density to reshape the proposal density to bring it closer to the integrand. The plot of Fig. 12.9 shows how the estimate becomes progressively closer to the true mean as the sample size increases.

■

Sequential Importance Sampling

Sequential importance sampling (SIS) is an adaptive importance sampling method that generates samples from a sequence of distributions that gradually approach the optimal importance sampling density. This is particularly suited to state-space models where the state evolves based on a known model. In each step, we use importance sampling and the data generated in the preceding step, assuming that there are no big changes in the density in each step. If in the kth step we have the importance density $q_k(x)$, the sample set $\{x_k^{(i)}\}$, and the weights $w_k = p_k(x_k)/q_k(x_k)$, we select the density

$$q_{k+1}(x_{k:k+1}) = q_k(x_k)q_{k+1}(x_{k+1}|x_k) \tag{12.79}$$

so that to obtain the sample set $\left\{x_{k+1}^{(i)}\right\}$ from $q_{k+1}(x_{k:k+1})$, we only need to sample from $q_{k+1}(x_{k+1}|x_k)$. The weights are updated using the recursion

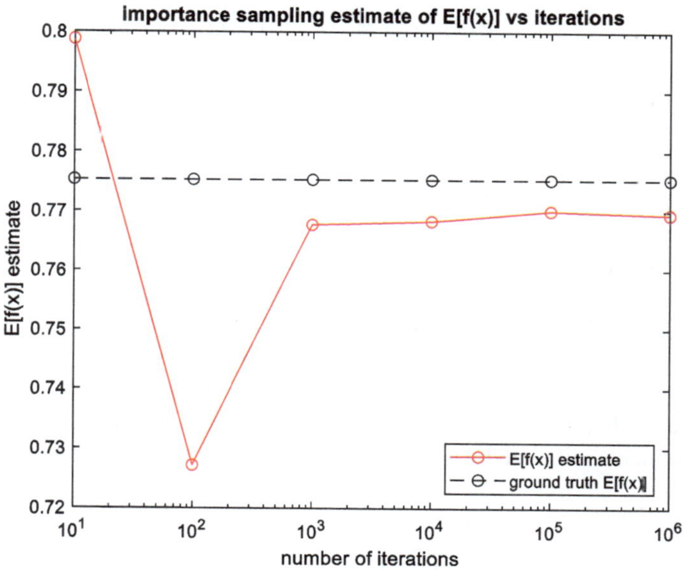

Fig. 12.9 Change of estimate with sample size

$$w_{k+1}\left(x_{1:k+1}^{(i)}\right)=w_k\left(x_{1:k}^{(i)}\right)\frac{p_{k+1}\left(x_{1:k+1}^{(i)}\right)}{p_k\left(x_{1:k}^{(i)}\right)q_{k+1}(x_{k+1}^{(i)}|x_{1:k}^{(i)})}$$

Application to State Estimation

We apply the Monte Carlo integration to minimum mean-square error estimator using the fundamental theorem of estimation theory

$$\widehat{\boldsymbol{x}}_{k|k}=E\{\boldsymbol{x}_k|z_{0:k}\}=\int \boldsymbol{x}_k p_{x|z}(\boldsymbol{x}_k|z_{0:k})\mathrm{d}\boldsymbol{x}_k$$

$z_{0:k} = col\{z_i, i = 0, 1, \ldots, k\}$ = all measurements up to and including time k. To do so, we must (i) find a suitable discrete approximation of the a posteriori density $p_{x|z}(\boldsymbol{x}_k|z_{0:k})$, then (ii) calculate the minimum mean-square error estimator $\widehat{\boldsymbol{x}}(k) = E\{\boldsymbol{x}_k|z_{0:k}\}$ using random samples from the distribution, or *particles*. The distribution, which can be non-Gaussian, is approximated with a discrete distribution

$$p(\boldsymbol{x}_k|z_{0:k})\approx\sum_{i=1}^{n}W_k^{(i)}\delta\left(\boldsymbol{x}_k-\boldsymbol{x}_k^{(i)}\right),\ \sum_{i=1}^{n}W_k^{(i)}=1 \tag{12.81}$$

Different particle filters can be obtained using different discrete approximations of the continuous distribution. The estimate is the conditional mean

$$E(\boldsymbol{x}_k|z_{0:k}) = \sum\nolimits_{i=1}^{n} W_k^{(i)} \boldsymbol{x}_k$$

with the covariance matrix

$$P_k = \sum_{i=1}^{n} W_k^{(i)} \tilde{\boldsymbol{x}}_k^{(i)} \tilde{\boldsymbol{x}}_k^{(i)T}, \tilde{\boldsymbol{x}}_k^{(i)} = \boldsymbol{x}_k^{(i)} - \hat{\boldsymbol{x}}_k \tag{12.82}$$

Unlike ensemble filters, particle filters do not use a simple distribution to generate random samples. They generate points that fit a general pdf using an importance density or proposal density that can easily generate samples.

We recall the equations of the recursive Bayesian filter, which provides a way of recursively updating the posterior pdf, from which the estimate can be obtained. The filter uses a predictor–corrector recursion for the conditional pdf of the state given the measurements initialized with pdf $p(\boldsymbol{x}_0)$. The predictor predicts the conditional pdf

$$p(\boldsymbol{x}_k|z_{0:k-1}) = \int p(\boldsymbol{x}_k|\boldsymbol{x}_{k-1}) p(\boldsymbol{x}_{k-1}|z_{0:k-1}) \mathrm{d}\boldsymbol{x}_{k-1}$$

then the corrector uses the measurements to correct it

$$\begin{aligned} p(\boldsymbol{x}_k|z_{0:k}) &= \kappa_k p(z_k|\boldsymbol{x}_k) p(\boldsymbol{x}_k|z_{0:k-1}) \\ \kappa_k &= 1/p(z_k|z_{0:k-1}) \end{aligned}$$

Substitute to obtain a recursion

$$p(\boldsymbol{x}_k|z_{0:k}) = \kappa_k p(z_k|\boldsymbol{x}_k) \int p(\boldsymbol{x}_k|\boldsymbol{x}_{k-1}) p(\boldsymbol{x}_{k-1}|z_{0:k-1}) \mathrm{d}\boldsymbol{x}_{k-1}. \tag{12.83}$$

We use importance sampling with the importance density $q(\boldsymbol{x}_k|z_{0:k})$ to write

$$p(\boldsymbol{x}_k|z_{0:k}) \propto p(z_k|\boldsymbol{x}_k) \int p(\boldsymbol{x}_k|\boldsymbol{x}_{k-1}) \frac{p(\boldsymbol{x}_{k-1}|z_{0:k-1})}{q(\boldsymbol{x}_k|z_{0:k})} q(\boldsymbol{x}_k|z_{0:k}) \mathrm{d}\boldsymbol{x}_{k-1}. \tag{12.84}$$

Using Monte Carlo integration, we approximate the integral with the summation

$$p(\boldsymbol{x}_k|z_{0:k}) \approx \sum\nolimits_{i=1}^{n} W_k^{(i)} \delta\left(\boldsymbol{x}_k - \boldsymbol{x}_k^{(i)}\right), \sum\nolimits_{i=1}^{n} W_k^{(i)} = 1 \tag{12.85}$$

where the weight is

$$W_k^{(i)} \propto \frac{p\left(\boldsymbol{x}_k^{(i)}|z_{0:k}\right)}{q\left(\boldsymbol{x}_k^{(i)}|z_{0:k}\right)} \tag{12.86}$$

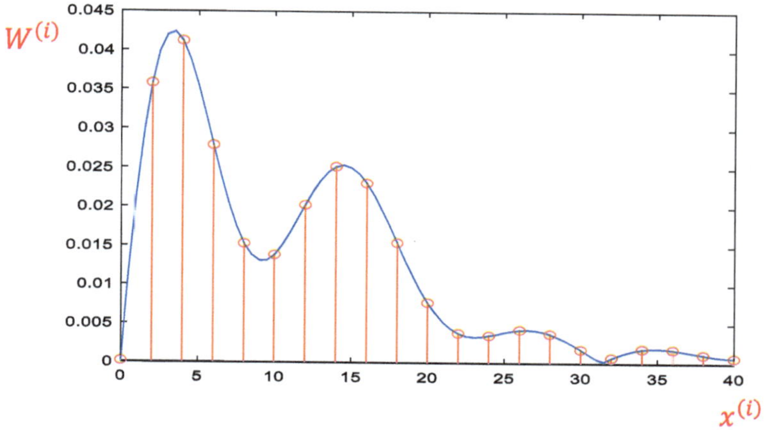

Fig. 12.10 Discrete approximation of continuous pdf

initialized with $p(\boldsymbol{x}_k|\boldsymbol{x}_{k-1}, z_{0:k-1}) = p(\boldsymbol{x}_k|\boldsymbol{x}_{k-1})$. The discrete approximation is depicted in Fig. 12.10.

Sequential Importance Sampling

To reduce the computational load, a recursion is used to update the weights. The algorithm for sequentially updating the weights with importance sampling is known as the sequential importance sampling (SIS) algorithm.

SIS is an adaptive importance sampling algorithm that generates samples from a sequence of distributions that gradually approach the optimal importance sampling density.

To derive the weight recursion, we first choose q to allow the factorization

$$q(\boldsymbol{x}_k|z_{0:k}) = q(\boldsymbol{x}_k|\boldsymbol{x}_{k-1}, z_{0:k}).q(\boldsymbol{x}_{k-1}|z_{0:k-1})$$

We observe that

$$q(\boldsymbol{x}_k|\boldsymbol{x}_{k-1}, z_{0:k}) = q(\boldsymbol{x}_k|\boldsymbol{x}_{k-1}, z_k, z_{0:k-1}) = q(\boldsymbol{x}_k|\boldsymbol{x}_{k-1}, z_k)$$

This gives the update equation

$$q(\boldsymbol{x}_k|z_{0:k}) = q(\boldsymbol{x}_k|\boldsymbol{x}_{k-1}, z_k).q(\boldsymbol{x}_{k-1}|z_{0:k-1}) \tag{12.87}$$

We substitute the update equation in the posterior pdf

$$p(\boldsymbol{x}_k|z_{0:k}) \propto p(z_k|\boldsymbol{x}_k) \int \frac{p(\boldsymbol{x}_k|\boldsymbol{x}_{k-1})}{q(\boldsymbol{x}_k|\boldsymbol{x}_{k-1}, z_k)} \frac{p(\boldsymbol{x}_{k-1}|z_{0:k-1})}{q(\boldsymbol{x}_{k-1}|z_{0:k-1})} q(\boldsymbol{x}_k|z_{0:k}) \mathrm{d}\boldsymbol{x}_{k-1}$$

Thus, the weight is

$$W_k^{(i)} \propto p\left(z_k|\boldsymbol{x}_k^{(i)}\right)\frac{p\left(\boldsymbol{x}_k^{(i)}|\boldsymbol{x}_{k-1}^{(i)}\right)}{q\left(\boldsymbol{x}_k^{(i)}|\boldsymbol{x}_{k-1}^{(i)}, z_k\right)}\frac{p\left(\boldsymbol{x}_{k-1}^{(i)}|z_{0:k-1}\right)}{q\left(\boldsymbol{x}_{k-1}^{(i)}|z_{0:k-1}\right)} \tag{12.88}$$

We now have the weight recursion

$$W_k^{(i)} \propto p\left(z_k|\boldsymbol{x}_k^{(i)}\right)\frac{p\left(\boldsymbol{x}_k^{(i)}|\boldsymbol{x}_{k-1}^{(i)}\right)}{q\left(\boldsymbol{x}_k^{(i)}|\boldsymbol{x}_{k-1}^{(i)}, z_k\right)}W_{k-1}^{(i)} \tag{12.89}$$

where

$p\left(z_k|\boldsymbol{x}_k^{(i)}\right) =$ likelihood function
$p\left(\boldsymbol{x}_k^{(i)}|\boldsymbol{x}_{k-1}^{(i)}\right) =$ transition prior
$q\left(\boldsymbol{x}_k^{(i)}|\boldsymbol{x}_{k-1}^{(i)}, z_k\right) =$ importance density.

The estimator depends on the choice of importance density. A popular choice is the bootstrap filter

$$q\left(\boldsymbol{x}_k|\boldsymbol{x}_{k-1}^{(i)}, z_k\right) = p\left(\boldsymbol{x}_k|\boldsymbol{x}_{k-1}^{(i)}\right). \tag{12.90}$$

This simplifies the weight recursion to

$$W_k^{(i)} \propto W_{k-1}^{(i)}p\left(z_k|\boldsymbol{x}_k^{(i)}\right).$$

The SIS algorithm is as follows.

SIS Algorithm

- Input $\left\{\boldsymbol{x}_k^{(i-1)}, W_k^{(i-1)}, i = 1, \ldots, N\right\}, z_k$

$Sum_w = 0$ % Initialize the sum with zero
for $i = 1 : N$ % Sampling and weight calculation
Draw $\boldsymbol{x}_k^{(i)}$ from $q\left(\boldsymbol{x}_k^{(i)}|\boldsymbol{x}_{k-1}^{(i)}, z_k\right)$ % Sample
$W_k^{(i)} = W_{k-1}^{(i)}p\left(z_k|\boldsymbol{x}_k^{(i)}\right)p\left(\boldsymbol{x}_k^{(i)}|\boldsymbol{x}_{k-1}^{(i)}\right)/q\left(\boldsymbol{x}_k^{(i)}\boldsymbol{x}_{k-1}^{(i)}, z_k\right)$ % Calculate the ith weight
$Sum_w = Sum_w + W_k^{(i)}$ % Update the sum of weights
end
for $i = 1 : N$ %
$W_k^{(i)} = W_k^{(i)}/Sum_w$ Normalize the weights
end

- Output $\left\{\boldsymbol{x}_k^{(i)}, W_k^{(i)}, i = 1, \ldots, N\right\}$

12.6 Degeneracy

A major drawback of particle filters is that updating the weights can lead to degeneracy. This is the phenomenon where normalized weights tend to concentrate into one particle after a number of recursions, which is all other particles degenerate. We define the measure of degeneracy

$$\hat{N}_{eff} = 1/\sum_{i=1}^{N}\left(W_k^{(i)}\right)^2, 1 \leq \hat{N}_{eff} \leq N. \tag{12.91}$$

The measure has the maximum value

$$\hat{N}_{eff} = N, W_k^{(i)} = 1/N, i = 1, 2, \ldots, N.$$

Its worse case value is

$$\hat{N}_{eff} = 1, W_k^{(j)} = 1,$$
$$W_k^{(i)} = 0, i \neq j, i = 1, \ldots, j-1, j+1, \ldots, N.$$

If $\widehat{N}_{eff}$ drops below a threshold value

$$N\text{eff} < N_{\text{thresh}}, \tag{12.92}$$

then resampling is required (Fig. 12.11).

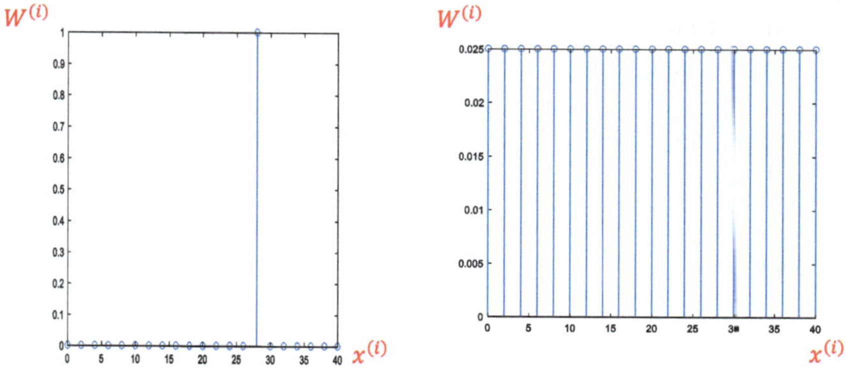

Fig. 12.11 Degeneracy and resampling

Resampling Algorithms

Resampling algorithms are used as a remedy for degeneracy to obtain uniform weighting. One approach is resampling with replacement from the discrete approximation of $p(\boldsymbol{x}_k|z_{0:k})$. Given random samples and weights $\{\boldsymbol{x}^{(i)}, W^{(i)}, i = 1, \ldots, N\}$, resampling accomplishes two objectives:

- Eliminate samples with negligible weights
- Add samples at the same location to obtain a new set of N samples with uniform weights $1/N$.

The following algorithm is commonly used for particle filters.

Systematic Resampling Algorithm

- Generate a random number $u_1 \sim U[0, 1/N]$
- Construct a CDF c from $\{W^{(i)}, i = 1, \ldots, N_T, N_T < N\}$
- Add the weights until their sum exceeds u_1
- Choose the corresponding particle $\boldsymbol{x}_k^i$ as a resampled point $\boldsymbol{x}_k^{j*}$ and save the index i^j of its parent
- Create a discrete uniform pdf where the probability of resampled point $\boldsymbol{x}_k^{j*}$ **is** $\boldsymbol{W}^i$.

It can be shown that the pdf of the resampled set converges to $p(\boldsymbol{x}_k|z_{0:k})$.

Algorithm

- Input $\left\{\boldsymbol{x}_k^{(i)}, W_k^{(i)}, i = 1, \ldots, N_T\right\}$

Draw a starting point $u_1 \sim U\left[0, N^{-1}\right]$ % Generate a random number $u_1 \sim U[0, 1/N]$

% Construct a CDF c from $\{W^{(i)}, i = 1, \ldots, N_T, N_T < N\}$

$c_1 = W_k^{(1)}$ % Initial point

for $i = 2 : N$

$c_i = c_{i-1} + W_k^{(i)}$ % Add weights to construct the CDF

end

for $j = 1 : N$

$u_j = u_1 + N^{-1}(j-1)$ % Add 1/N until the sum exceed c_i

While $u_j > c_i$

$$i = i + 1$$

end

$\boldsymbol{x}_k^{(j)*} = \boldsymbol{x}_k^{(i)}$ % Save the particle

$W_k^{(i)} = N^{-1}$ % Assign uniform weight

$i^j = i$ % Save the index of the particle

End

- Output $\left\{\boldsymbol{x}_k^{(j)*}, W_k^{(j)}, i^j, i = 1, \ldots, N\right\}$

■

It can be shown that the pdf of the resampled set converges to $p(\boldsymbol{x}_k|z_{0:k})$.

Finally, we summarize the steps for implementing a particle filter in the following algorithm.

Particle Filter Algorithm

- Input $\left\{\boldsymbol{x}_{k-1}^{(i)}, , i = 1, \ldots, N\right\}, z_k$

$$Sum_w = 0$$

for $i = 1 : N$ % Importance sampling
  Draw $\boldsymbol{x}_k^{(i)} \sim p\left(\boldsymbol{x}_k|\boldsymbol{x}_{k-1}^{(i)}\right)$
  $W_k^{(i)} = W_{k-1}^{(i)} p\left(z_k|\boldsymbol{x}_k^{(i)}\right)$ % Calculate the importance weights
  $Sum_w = Sum_w + w_k^{(i)}$ % Add the weights to obtain the normalizing constant
  End
  for $j = 1 : N$
  $W_k^{(i)} = W_k^{(i)}/Sum_w$ % Normalize the wights
  end
  Resample to obtain $\boldsymbol{x}_k^{(j)*} = \boldsymbol{x}_k^{(i)}$ using $p\left(\boldsymbol{x}_k|\boldsymbol{x}_{k-1}^{(i)}\right)$

- Output $\left\{\boldsymbol{x}_k^{(i)}, i = 1, \ldots, N\right\}$

MATLAB Implementation of Nonlinear Filters

MATLAB has commands that facilitate the implementation of the nonlinear filters discussed in this chapter. We begin with a discussion of the implementation of particle filters.

There are many types of particle filters, all of which approximate the posterior pdf with a set of discrete points. Figure 12.12, adapted from the MATLAB documentation on particle filters, shows the steps required to implement a particle filter. Particle filters are computationally expensive and should only be used if other filtering approaches are inapplicable.

MATLAB provides extensive commands for particle filter implementation. The command to create a particle filter using the system state-space model is

pf = particleFilter(StateTransitionFcn, MeasurementLikelihoodFcn)

where StateTransitionFcn is a function that calculates the particles (state hypotheses) at the next time step, given the state vector at a time step and is obtained from the state equation. `MeasurementLikelihoodFcn` is a function that calculates the likelihood of each particle based on sensor measurement.

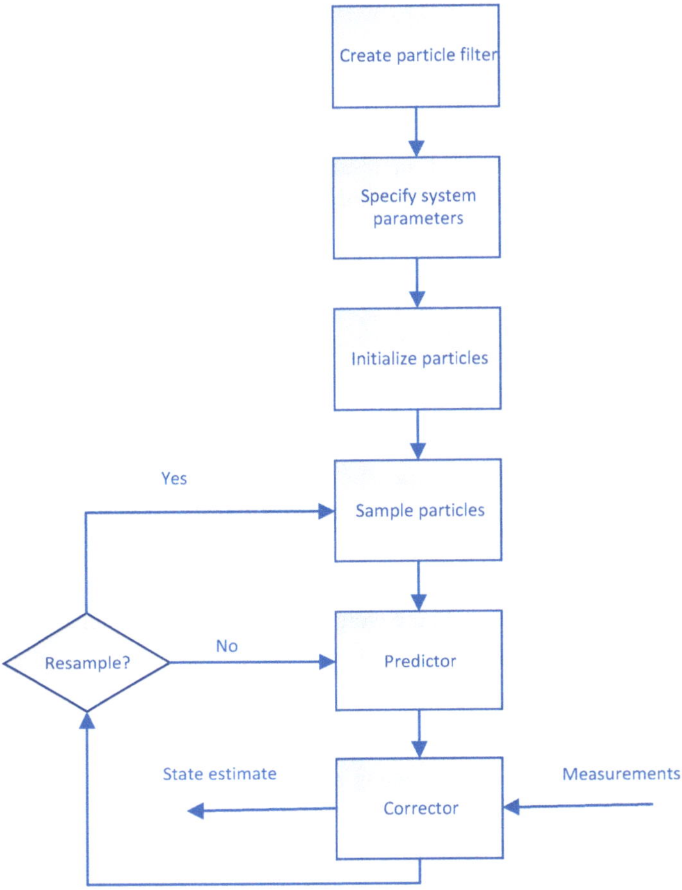

Fig. 12.12 Block diagram of a particle filter

The filter is initialized with the command

initialize(myPF, 1000,[2;0],eye(2));

To obtain the state estimate and the covariance matrix, we use the command

[State, StateCovariance] = getStateEstimate(pf)

The state estimate is calculated using a specified **StateEstimationMethod,** which can either be the default **"mean"** for a weighted mean of the particle values, or **"maxweight,"** for the particle with the maximum weight.

For resampling, MATLAB has four options **"multinomial," "systematic," "stratified," "residual."** The typical choice is "systematic" resampling.

MATLAB uses the same two predictor–corrector commands for the extended, unscented, and particle filter:

[PredictedState,PredictedStateCovariance] = predict(obj)

[CorrectedState,CorrectedStateCovariance] = correct(obj,y(k))

where **obj** specifies the filter used, the state equation, and the measurement equation. For example, it can be set to **pf** as specified above. The commands are used recursively to obtain the state estimate over the period of interest.

The reader is referred to the MATLAB documentation for examples of nonlinear filtering, including particle filters, unscented filters, and extended Kalman filters.

Problems

12.1 The population growth for a single species is described by the model[1]

$$p_{k+1} = p_k + ap_k - bp_k^2 + w_k, \, p_k < (1+a)/b, a < 1$$

where a is the birth rate and b is the death rate. The population is measured with the measurement equation

$$z_k = p_k + v_k.$$

The process and measurement noise are uncorrelated, zero-mean, Gaussian with variances q and r, respectively. Design an extended Kalman filter to estimate the population.

12.2 Design and simulate an extended Kalman filter with sampling period 0.1 s for the nonlinear spring-mass damper system

$$\ddot{x} + x\dot{x} + x + .1x^3 = 1 + u(t)$$
$$z(t) = x(t) + v(t)$$

where $u(t)$ is zero-mean Gaussian white noise with variance 0.4 and $v(t)$ is uncorrelated zero-mean white Gaussian noise of variance 0.2

12.3 For the population model of Problem 12.1, design an unscented Kalman filter. Simulate the filter and plot its RMSE together with that of the steady-state Kalman filter for the linearized model using $q = 1$, $r = 0.2$, and two cases (a) $a = 0.8$, $b = 0.08$, (b) $a = 0.8$, $b = 0.3$. Comment on the results and explain why the error plots are significantly different in the two cases.

12.4 Design an ensemble Kalman filter for the population model of Problem 12.1.

12.5 For the birth rate $a = 0.1$, and the death rate $b = 0.01, 0.05$, simulate the extended Kalman filter and the ensemble Kalman filter and compare the results.

12.6 Obtain the equation for the Bayesian filter if the pdfs are Gaussian and draw a block diagram to summarize its operation. Show that the Kalman filter can be obtained from the Bayesian filter if all pdfs are assumed Gaussian and the system is linear.

[1] M. Shahin, Explorations of Mathematical Models in Biology with MATLAB, J. Wiley, Hoboken, NJ, 2014.

12.7 Use Monte Carol integration to evaluate the series

$$1 - \frac{1}{3^2} + \frac{1}{5^2} - \cdots = \int_0^{\frac{\pi}{2}} \frac{x}{\sin(x)} \mathrm{d}x$$

12.8 Design and simulate an extended Kalman filter for the nonlinear spring-mass-damper system with equation of motion

$$m\ddot{x} + b\dot{x} + k_1 x + k_3 x^3 = f(t) + u(t)$$

with zero-mean, Gaussian unity white random force perturbation $u(t)$. The initial position of the mass is known exactly and is $x = 1$ m. Use noisy displacement measurement with unity zero-mean white Gaussian measurement noise v_k. Use a sampling period of 0.1 s and $f(t) = \sin(t)$

(a) Use the data $m = 1Kg.$, $b = 0.1$ Nt/m/s, $k_1 = 0.1\text{Nt/m}$, $k_3 = 0.8\text{Nt/m}$
(b) Use the data $m = 1Kg.$, $b = 0.1$ Nt/m/s, $k_1 = 0.1\text{Nt/m}$, $k_3 = 1.5\text{Nt/m}$
(c) Compare the results of the two simulations and comment on the performance of the extended Kalman filter in the two cases.

12.9 For the system of Example 12.2, design and simulate an extended Kalman filter, and an unscented Kalman filter and compare the results. Run a large number of simulations (50 or more) and calculate and plot the average RMS error. Compare the results to the approximate RMS error obtained using the covariance matrix.

12.10 Repeat Problem 12.9 using an unscented Kalman filter.

12.11 Prove Eq. (12.50) for the error covariance of the ensemble Kalman filter.
Hint: Use the expressions

$$\overline{\boldsymbol{x}}_{k|k} = X_{k|k}1, 1 = \begin{bmatrix} 1 & 1 & \ldots & 1 \end{bmatrix}^T$$
$$X_{k|k} = \begin{bmatrix} \mathcal{X}_{k|k}^{(1)} & \mathcal{X}_{k|k}^{(2)} & \ldots & \mathcal{X}_{k|k}^{(N)} \end{bmatrix}$$

12.12 If x is a random variable with pdf

$$p(x) = \begin{cases} \frac{1}{2}\cos(x), & x \in [0, \pi/2] \\ 0 & elsewhere \end{cases}$$

Modify the script of [8] to obtain the expected value of the function

$$f(x) = \frac{x^3 + 3x^2 + 2x + 3}{1 + \sin^2(x)}$$

using importance sampling and Markov integration with the importance pdf

$$p(x) = \begin{cases} \frac{1}{\sqrt{(2\pi\sigma^2}} \exp\left\{\frac{(x-m_x)^2}{2\sigma^2}\right\}, & x \in [0, \pi/2] \\ 0 & elsewhere \end{cases}$$

Use $\sigma^2 = 1.5$ and $m_x = 0.9$.

12.13 For the population model of Problem 12.1, design and simulate a particle filter using MATLAB.

Bibliography

1. Brown, R. G., & Hwang, P. Y. C. (2012). *Introduction to random signals and applied Kalman filtering* (4th ed.). J. Wiley.
2. Gillijns, S., Mendoza, O. B., De Moor, B. L. R., Bernstein, D. S., & Ridley, A. A. (2006). What is the ensemble Kalman filter and how well does it work? In *Proceedings of the 2006 American Control Conference*, Minneapolis, MN, June 14–16, 2006.
3. Hogg, R. V., McKean, J. W., Craig, A. T. (2005). *Introduction to mathematical statistics* (6th ed.). Pearson/Prentice-Hall.
4. Khazraj, H., Faria da Silva, F., & Bak, C. L. (2016). A performance comparison between extended Kalman filter and unscented Kalman filter in power system dynamic state estimation. In *2016 51st International Universities Power Engineering Conference (UPEC)*, Coimbra, 2016, pp. 1–6. https://doi.org/10.1109/UPEC.2016.8114125
5. Li, T., Bolic, M., Djuric, P. M. (2015, May) Resampling methods for particle filtering: Classification, implementation, and strategies. *IEEE Signal Processing Magazine, 32*(3), 70–86.
6. Mandel, J. (2007, February). A brief tutorial on the ensemble Kalman filter. In Center for Computational Mathematics Reports, Report No. 242.
7. Ristic, B., Arulampalam, S., & Gordon, N. (2004). *Beyond the Kalman filter: Particle filters for tracking applications.* Artech House.
8. Simon, D. (2006). *Optimal state estimation.* J. Wiley.
9. Smolyakov, V. (2023). Importance sampling. https://www.mathworks.com/matlabcentral/fileexchange/51218-importance-sampling

Chapter 13
The Expectation Maximization Algorithm

13.1 Maximum Likelihood Estimation with Incomplete Data

Maximum-likelihood parameter estimation of a parameter vector $\boldsymbol{\theta}$ is used when all the data $\boldsymbol{x}$ necessary for estimation is available and is governed by a known probability model:

$$p(\boldsymbol{x}|\boldsymbol{\theta}), \boldsymbol{\theta} \in \boldsymbol{\Theta} \subset \mathcal{R}^r, \boldsymbol{x} \in \mathcal{X} \subset \mathcal{R}^n. \tag{13.1}$$

In many practical applications, not all the variables needed for maximum-likelihood (ML) estimation are available but observations $\boldsymbol{y}$ provide data for estimation using a known statistical model. In such cases, the expectation maximization algorithm allows us to use an initial guess of the parameter values to recursively improve our estimate of the parameters of a known distribution in the absence of a complete set of data. The algorithm can also be used in some cases where it is difficult to solve for the ML estimate.

If the complete data is

$$\boldsymbol{x} = \{\boldsymbol{y}, \boldsymbol{z}\} \tag{13.2}$$

$\boldsymbol{x} \in \mathcal{X} \subset \mathcal{R}^n$, which includes the observations $\boldsymbol{y} \in \mathcal{Y} \subset \mathcal{R}^m, m < n$, and the missing data $\boldsymbol{z} \in \mathcal{Z} \subset \mathcal{R}^{n-m}$. Alternatively, we may observe $\boldsymbol{y}$ indirectly through a measurement process:

$$g(\boldsymbol{y}|\theta), \boldsymbol{y} \in \mathcal{Y} \subset \mathcal{R}^m, m < n. \tag{13.3}$$

M. S. Fadali, *Introduction to Random Signals, Estimation Theory, and Kalman Filtering*, https://doi.org/10.1007/978-981-99-8063-5_13

In either case, it is not possible to obtain $\boldsymbol{x}$ because the mapping from $\boldsymbol{x}$ to $\boldsymbol{y}$ is many-to-one and parameter estimation by maximizing the likelihood function as discussed in Chap. 8 is not feasible.

Because the missing data z is random, the log-likelihood of the complete data given the observed data is a random variable, and we can obtain its expectation in terms of the initial guess. Next, we maximize the log-likelihood with respect to the parameters to obtain a new parameter estimate. Repeating the process iteratively yields the desired parameter estimate. The iterative process is known as the expectation maximization algorithms.

The conditional expectation of the log-likelihood given the observations is known as the auxiliary function or Q function and is given by

$$
\begin{aligned}
Q(\boldsymbol{\theta}|\boldsymbol{\theta}^{(k)}) &= E\{\ln[p(\boldsymbol{x}|\boldsymbol{\theta})]|\boldsymbol{y},\boldsymbol{\theta}^{(k)}\} \\
&= \begin{cases} \sum\limits_{z\in\mathcal{Z}} \ln[p(\boldsymbol{x}|\boldsymbol{\theta})]p(\boldsymbol{x}|\boldsymbol{y},\boldsymbol{\theta}^{(k)}), & \text{discrete pdf} \\ \int\limits_{z\in\mathcal{Z}} \ln[p(\boldsymbol{x}|\boldsymbol{\theta})]p(\boldsymbol{x}|\boldsymbol{y},\boldsymbol{\theta}^{(k)})dz, & \text{continuous pdf} \end{cases}.
\end{aligned}
\tag{13.4}
$$

Because the measurements are known and the only random part of the complete data $\boldsymbol{x}$ is z, the auxiliary function can be written in terms of the pdf of the missing data given the observations as

$$
Q(\boldsymbol{\theta}|\boldsymbol{\theta}^{(k)}) = \begin{cases} \sum\limits_{z\in\mathcal{Z}} \ln[p(\boldsymbol{y},z|\boldsymbol{\theta})]p(z|\boldsymbol{y},\boldsymbol{\theta}^{(k)}), & \text{discrete pdf} \\ \int\limits_{z\in\mathcal{Z}} \ln[p(\boldsymbol{y},z|\boldsymbol{\theta})]p(z|\boldsymbol{y},\boldsymbol{\theta}^{(k)})dz, & \text{continuous pdf} \end{cases}.
\tag{13.5}
$$

This is the expectation or E-step of the algorithm.

We then obtain the updated maximum-likelihood parameter estimate using the Q function:

$$
\boldsymbol{\theta}^{(k+1)} = \arg\left\{\max_{\theta} Q(\boldsymbol{\theta}|\boldsymbol{\theta}^{(k)})\right\}, \boldsymbol{\theta} \in \boldsymbol{\Theta} \subset \mathcal{R}^r.
\tag{13.6}
$$

This is the maximization or M-step of the algorithm. The updated estimate $\boldsymbol{\theta}^{(k+1)}$ is fed back to the auxiliary function to replace the old estimate and the process is repeated. The process stops if the change in the estimate drops below a selected tolerance level. The algorithm is illustrated in Fig. 13.1.

For the algorithm to converge to a local maximum, we require the condition

$$
Q(\boldsymbol{\theta}^{(k+1)}|\boldsymbol{\theta}^{(k)}) \geq Q(\boldsymbol{\theta}^{(k)}|\boldsymbol{\theta}^{(k)}), \boldsymbol{\theta} \in \boldsymbol{\Theta} \subset \mathcal{R}^r.
\tag{13.7}
$$

Hence, the maximization step can be written as

$$
\boldsymbol{\theta}^{(k+1)} = \arg\left\{\max_{\theta}[Q(\boldsymbol{\theta}|\boldsymbol{\theta}^{(k)}) - Q(\boldsymbol{\theta}^{(k)}|\boldsymbol{\theta}^{(k)})\right\}, \boldsymbol{\theta} \in \boldsymbol{\Theta} \subset \mathcal{R}^r.
\tag{13.8}
$$

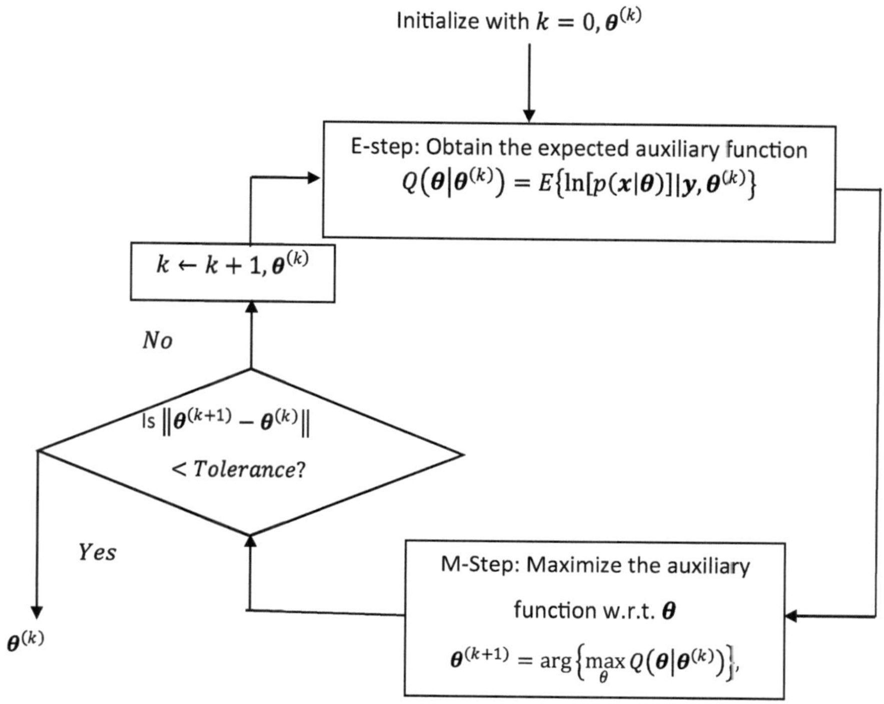

Fig. 13.1 Expectation maximization algorithm

This is equivalent to the maximization of (13.6) and shows that the algorithm converges to a local maximum. For simple problems with a convex search space, this is not a problem but for more general problems convergence to a lccal maximum does not guarantee convergence to a global maximum. Hence, it is recommended to choose several initial estimates for the algorithm and use the results of the run that maximizes the likelihood.

Example 13.1 Fault Characterization A faulty circuit has three different modes of failure but only one of them can be diagnosed. The probability of the first two modes of failure is

$$p_1 = \frac{1-p}{2}, \; p_2 = \frac{1+p}{4}.$$

Use the EM algorithm to determine the parameter p using N records of failure. Write a program to estimate the parameter for the case $p = 0.4$, $N = 100$ failures,with the number of failures in the three modes equal to {28, 34, 38}, respectively. Repeat for the number of failures equal to {34, 31, 35].

Solution

The failure probabilities satisfy

$$p_1 + p_2 + p_3 = 1.$$

Hence, the probability of the third type of failure is

$$p_3 = 1 - p_1 - p_2 = \frac{1+p}{4}.$$

Failure is governed by the trinomial distribution

$$P(f|p) = \frac{N!}{f_1 f_2 f_3} f_1^{p_1} f_2^{p_2} f_3^{p_3}, N = f_1 + f_2 + f_3.$$

Taking the natural log and ignoring terms that do not include the parameter p gives

$$\begin{aligned} f_1 \ln(p_1) + f_2 \ln(p_2) + f_3 \ln(p_3) &= f_1 \ln\left(\frac{1-p}{2}\right) + f_2 \ln\left(\frac{1+p}{4}\right) + f_3 \ln\left(\frac{1+p}{4}\right) \\ &= f_1 \ln(1-p) + f_2 \ln(1+p) + f_3 \ln(1+p) \\ &\quad + p\text{-independent terms.} \end{aligned}$$

Thus, we only need to consider the terms

$$L = f_1 \ln(1-p) + f_2 \ln(1+p) + f_3 \ln(1+p).$$

We can collect the data

$$y_1 = f_1 + f_2, y_2 = f_3.$$

Because f_3 is known, we know the sum

$$f_1 + f_2 = N - f_3.$$

For the EM algorithm, we initialize the process with $k = 0$, $p^{(k)}$.

The expected values of the number of failures of the first two modes given y_1 written in terms of the current parameter estimate can be shown to be

$$\begin{aligned} E\{f_1|y_1\} &= \frac{p_1^{(k)}}{p_1^{(k)} + p_2^{(k)}}(N - f_3) = 2\frac{1-p^{(k)}}{3-p^{(k)}}(N - f_3) \\ E\{f_2|y_1\} &= \frac{p_2^{(k)}}{p_1^{(k)} + p_2^{(k)}}(N - f_3) = \frac{1+p^{(k)}}{3-p^{(k)}}(N - f_3). \end{aligned}$$

The expected values can be substituted in the log-likelihood and the resulting expression can be used in maximum-likelihood estimation. The necessary condition for the maximum is

$$\frac{\partial L}{\partial p} = -\frac{f_1^{(k)}}{1-p} + \frac{f_2^{(k)}}{1+p} + \frac{f_3}{1+p} = 0$$

$$p^{(k+1)} = \frac{f_2^{(k)} + f_3 - f_1^{(k)}}{f_1^{(k)} + f_2^{(k)} + f_3} = \frac{f_2^{(k)} + f_3 - f_1^{(k)}}{N}$$

with

$$f_1^{(k)} = 2\frac{1-p^{(k)}}{3-p^{(k)}}(N - f_3)$$

$$f_2^{(k)} = \frac{1+p^{(k)}}{3-p^{(k)}}(N - f_3).$$

Substituting for $f_1^{(k)}$ and $f_2^{(k)}$ gives

$$p^{(k+1)} = \frac{(3N - 4f_3)p^{(k)} + 4f_3 - N}{N\left(3 - p^{(k)}\right)}.$$

In this simple example, we can solve for the steady-state solution by letting $\hat{p}_{ss} = p^{(k+1)} = p^{(k)}$. This gives

$$\hat{p}_{ss} = \frac{4f_3 - N}{N} or 1.$$

We reject the second answer as unlikely.

The following script implements the EM algorithm. The results can be sensitive to the data.

```
% Example 13_1
clear
k=0;
p=0.4; % True parameter value
pk=0;N=100;
pv=[(1-p)/2,(1+p)/4,(1+p)/4];
psum=0;
% Iterate with different data
data= [28,34,38];
y1=data(1)+data(2);y2=data(3);
error=100;
while error >1.E-6
    f1k=2*(N-y2)*(1-pk)/(3-pk);
    f2k=(N-y2)*(1+pk)/(3-pk);
pplus=(f2k+y2-f1k)/N;
temp=pk;pk=pplus;
k=k+1;
error=norm(temp-pk);
end
pk
```

```
>> Example13_1
pk =
  0.5200
```

The results depend on the second measurement y_2. If the data is changed to [34, 31, 35], the estimate converges to the correct value after 42 iterations. This can be verified by substituting the numerical values in the expression for $\hat{p}_{ss}$. Because the data is generated randomly, the solution obtained may be a local maximum and may not provide a good estimate.

13.2 Exponential Family

The following model can be used to represent many distributions, including the normal distribution:

$$p(\boldsymbol{x}|\boldsymbol{\theta}) = a(\boldsymbol{x})c(\boldsymbol{\theta})e^{\boldsymbol{\pi}^T(\boldsymbol{\theta})\boldsymbol{t}(x)}. \tag{13.9}$$

The family of distribution represented by the model is known as the exponential family. Using this general form simplifies optimization and gives results for the entire exponential family.

Taking the natural log gives

$$\ln[p(\boldsymbol{x}|\boldsymbol{\theta})] = \ln[a(\boldsymbol{x})] + \underbrace{\ln[c(\boldsymbol{\theta})] + \boldsymbol{\pi}^T(\boldsymbol{\theta})\boldsymbol{t}(\boldsymbol{x})}_{\theta-\text{dependent terms}}. \tag{13.10}$$

The first term does not involve the parameter to estimate and can be omitted. Thus, we can use a modified auxiliary function that includes the parameter-dependent terms only

$$Q_m\left(\boldsymbol{\theta}|\boldsymbol{\theta}^{(k)}\right) = E\left\{\ln[c(\boldsymbol{\theta})] + \boldsymbol{\pi}^T(\boldsymbol{\theta})\boldsymbol{t}(\boldsymbol{x})\right\}$$
$$= \begin{cases} \ln[c(\boldsymbol{\theta})] + \boldsymbol{\pi}^T(\boldsymbol{\theta}) \sum\limits_{z\in\mathcal{Z}} \boldsymbol{t}(\boldsymbol{y}, z) p\left(z|\boldsymbol{y}, \boldsymbol{\theta}^{(k)}\right), & \text{discrete} \\ \ln[c(\boldsymbol{\theta})] + \boldsymbol{\pi}^T(\boldsymbol{\theta}) \int\limits_{z\in\mathcal{Z}} \boldsymbol{t}(\boldsymbol{y}, z) p\left(z|\boldsymbol{y}, \boldsymbol{\theta}^{(k)}\right) dz, & \text{continuous} \end{cases}. \tag{13.11}$$

The above expression shows that for the exponential family the expectation step is estimation of $\boldsymbol{t}(\boldsymbol{x})$, which is a sufficient statistic. The EM algorithm reduces to the following two steps:

- E-step: Compute the estimate $\boldsymbol{t}^{(k+1)}(\boldsymbol{x})$

$$\boldsymbol{t}^{(k+1)}(\boldsymbol{x}) = E\left\{\boldsymbol{t}(\boldsymbol{y}, z)|\boldsymbol{y}, \boldsymbol{\theta}^{(k)}\right\} = \sum_{z\in\mathcal{Z}} \boldsymbol{t}(\boldsymbol{y}, z) f\left(z|\boldsymbol{y}, \boldsymbol{\theta}^{(k)}\right). \tag{13.12}$$

- M-step: Obtain the updated maximum-likelihood parameter estimate.

$$\boldsymbol{\theta}^{(k+1)} = \arg\left\{\max_{\theta}\left[\ln[c(\boldsymbol{\theta})] + \boldsymbol{\pi}^T(\boldsymbol{\theta})\boldsymbol{t}^{(k+1)}(\boldsymbol{x})\right]\right\}, \boldsymbol{\theta} \in \boldsymbol{\Theta} \subset \mathcal{R}^r. \tag{13.13}$$

Example 13.2 Normal Distribution Consider the bivariate Gaussian likelihood function for $\boldsymbol{x} = \{y, z\}$, where $E\{\boldsymbol{x}\} = \begin{bmatrix} m_y & 0 \end{bmatrix}^T$ and the covariance of $\boldsymbol{x}$ is $Cov\{\boldsymbol{x}\} = \sigma^2 I_2$. Given i.i.d. measurements $\{y_i, i = 1, \ldots, N\}$, find the ML estimate of the parameter vector $\boldsymbol{\theta}^T = \begin{bmatrix} \theta_1 & \theta_2 \end{bmatrix} = \begin{bmatrix} m_y & \sigma^2 \end{bmatrix}$ using the EM algorithm.

Solution

Since $\boldsymbol{x}$ is Gaussian, that is Gaussian, then they are marginally Gaussian and the likelihood of $\boldsymbol{x}$ is

$$p(\boldsymbol{x}|\boldsymbol{\theta}) = \frac{1}{2\pi\sigma^2} \exp\left\{-\frac{\left(y - m_y\right)^2}{2\sigma^2} - \frac{z^2}{2\sigma^2}\right\}.$$

We expand the likelihood as

$$p(\boldsymbol{x}|\boldsymbol{\theta}) = \frac{1}{2\pi\sigma^2}\exp\left\{-\frac{y^2+z^2-2m_y y+m_y^2}{2\sigma^2}\right\}.$$

Multiplying by 2π, and writing the likelihood in terms of θ_i, $i = 1, 2$, gives

$$p(\boldsymbol{x}|\boldsymbol{\theta}) = \frac{1}{\theta_2}\exp\left\{-\frac{y^2+z^2-2\theta_1 y+\theta_1^2}{2\theta_2}\right\}.$$

The log-likelihood with each measurement y_i is

$$\begin{aligned}\ln[p(y_i,\, z|\boldsymbol{\theta})] &= -\ln[\theta_2] - \frac{\theta_1^2}{2\theta_2} + \begin{bmatrix}\frac{\theta_1}{\theta_2} & -\frac{1}{2\theta_2}\end{bmatrix}\begin{bmatrix} y_i \\ y_i^2+z^2\end{bmatrix}.\\ &= \ln[c(\boldsymbol{\theta})] + \boldsymbol{\pi}^T(\theta)\boldsymbol{t}(\boldsymbol{x})\end{aligned}$$

The sufficient statistic to be estimated is

$$\boldsymbol{t}(\boldsymbol{x}) = \begin{bmatrix} y_i & y_i^2+z^2\end{bmatrix}^T.$$

For N i.i.d. measurements, after dividing by N, we have

$$\frac{1}{N}\sum_{i=1}^{N}\ln[2\pi p(y_i, z|\boldsymbol{\theta})] = \begin{bmatrix}\frac{\theta_1}{\theta_2} & -\frac{1}{2\theta_2}\end{bmatrix}\begin{bmatrix}\overline{y} \\ \overline{y^2}+z^2\end{bmatrix} - \ln[\theta_2] - \frac{\theta_1^2}{2\theta_2}$$

and the sufficient statistic is

$$\boldsymbol{t}(\boldsymbol{x}) = \begin{bmatrix}\overline{y} \\ \overline{y^2}+z^2\end{bmatrix}.$$

Note that scaling with 2π and dividing by N does not affect the results of the algorithm because scaling the function does not alter the location of its maximum.

The EM algorithm for this problem reduces to the following steps:

1. Estimate

$$\begin{aligned}E\{z^2|\boldsymbol{\theta}^{(k)}\} &= \sigma^{2(k)} = \theta_2^{(k)}\\ Q_m\left(\boldsymbol{\theta}|\theta^{(k)}\right) &= \frac{\sum_{i=1}^{N}\ln[2\pi p(y_i, z|\theta)]}{N}\\ &= \begin{bmatrix}\frac{\theta_1}{\theta_2} & -\frac{1}{2\theta_2}\end{bmatrix}\begin{bmatrix}\overline{y} \\ \overline{y^2}+\theta_2^{(k)}\end{bmatrix} - \ln[\theta_2] - \frac{\theta_1^2}{2\theta_2}.\end{aligned}$$

2. Maximization: The necessary condition to maximize the log-likelihood is

$$\frac{\partial Q_m\left(\boldsymbol{\theta}|\theta^{(k)}\right)}{\partial \boldsymbol{\theta}} = \begin{bmatrix} \frac{\overline{y}-\theta_1}{\theta_2} \\ -\frac{1}{\theta_2} + \frac{\theta_1^2+\overline{y^2}+\theta_2^{(k)}-2\overline{y}\theta_1}{2\theta_2^2} \end{bmatrix} = 0.$$

This gives the updated parameter estimate

$$\theta_1^{(k+1)} = \overline{y},\, \theta_2^{(k+1)} = \frac{\overline{y^2} - \overline{y}^2 + \theta_2^{(k)}}{2}.$$

The estimate of the mean is the sample mean and does not change with iteration. The estimate of the variance is the average of the sample variance and the previous estimate of the variance and is therefore positive.

The sufficient condition for a maximum is

$$\left.\frac{\partial^2 Q_m\left(\boldsymbol{\theta}|\theta^{(k)}\right)}{\partial \boldsymbol{\theta}^2}\right|_{\boldsymbol{\theta}^{(k+1)}} = \left.\begin{bmatrix} -\frac{1}{\theta_2} & \frac{\theta_1-\overline{y}}{\theta_2^2} \\ \frac{\theta_1-\overline{y}}{\theta_2^2} & \frac{1}{\theta_2^2} - \frac{\theta_1^2+\overline{y^2}+\theta_2^{(k)}-2\overline{y}\theta_1}{2\theta_2^3} \end{bmatrix}\right|_{\boldsymbol{\theta}^{(k+1)}}$$

$$= \begin{bmatrix} -\frac{1}{\theta_2^{(k+1)}} & 0 \\ 0 & -\frac{4}{\left(\theta_2^{(k+1)}\right)^2} \end{bmatrix} < 0.$$

As $k \to \infty$, the estimate of the mean remains the sample mean $\theta_1^{(\infty)} = \overline{y}$ and the estimate of the variance becomes

$$\theta_2^{(\infty)} = \frac{\overline{y^2} - \overline{y}^2 + \theta_2^{(\infty)}}{2}.$$

Solving for the steady-state estimate gives the sample variance $\theta_2^{(\infty)} = \overline{y^2} - \overline{y}^2$. The example shows that the algorithm converges to the ML estimates, that is the sample mean and the sample variance. ■

Next, we consider the important case of the multivariate Gaussian distribution.

13.3 EM for the Multivariate Normal Distribution

The multivariate normal distribution is the most popular distribution because of its excellent properties and its applicability to many practical applications. We examine the EM algorithm for the distribution through the following example.

Example 13.3 Multivariate Gaussian Distribution Consider a normally distributed vector $\boldsymbol{x} = \{\boldsymbol{y}, \boldsymbol{z}\}$ that includes a measured vector $\boldsymbol{y} \in \mathcal{R}^n$ and unmeasured vector

$z \in \mathcal{R}^m$. The conditional density of the unmeasured vector given the measured vector is

$$p_{z|y}(\boldsymbol{z}|\boldsymbol{y}) = \frac{1}{[2\pi]^{n/2}\sqrt{\det(C_{z|y})}} \exp\left\{-\frac{1}{2}(\boldsymbol{z}-\boldsymbol{m}_{z|y})^T C_{z|y}^{-1}(\boldsymbol{z}-\boldsymbol{m}_{z|y})\right\},$$

where the conditional means is

$$\boldsymbol{m}_{z|y} = E\{z|\boldsymbol{y}\} = \boldsymbol{m}_z + C_{zy}C_y^{-1}(\boldsymbol{y}-\boldsymbol{m}_y)$$

and the conditional covariance matrix is

$$C_{z|y} = C_z - C_{zy}C_y^{-1}C_{yz}$$

The parameters of the distribution are the mean and covariance

$$\boldsymbol{m}_x = col\{\boldsymbol{m}_y, \boldsymbol{m}_z\},\, C_x = \begin{bmatrix} C_y & C_{yz} \\ C_{zy} & C_z \end{bmatrix}.$$

The parameters vector to estimate includes the entries of the mean vector and the covariance matrix

$$\boldsymbol{\theta} = col\{\boldsymbol{m}_y, \boldsymbol{m}_z, \boldsymbol{c}_x\}$$

$$\boldsymbol{c}_x = col\{C_x(i, j), i = 1 : n+m,\, j = i : n+m\}.$$

The number of parameters, recalling that the covariance matrix is symmetric, is

$$(n+m) + (n+m)(n+m+1)/2 = (n+m)(n+m+3)/2.$$

For a set of N measurements $\boldsymbol{y}_i$, $i = 1, \ldots, N$, it is desired to obtain the maximum-likelihood estimate of the parameters. For maximum-likelihood estimation, we need the log-likelihood function:

$$\ln p_x(\boldsymbol{x}|\boldsymbol{m}_x, C_x) = -\frac{Nn}{2}\ln[2\pi] - \frac{N}{2}\ln\det(C_x) - \sum_{i=1}^{N}(\boldsymbol{x}_i - \boldsymbol{m}_x)^T C_x^{-1}(\boldsymbol{x}_i - \boldsymbol{m}_x).$$

It can be shown that the maximum-likelihood parameter estimates for this function are the sample mean and sample covariance, respectively. Because we do not have a complete set of data, we consider the auxiliary function

$$Q\left(\theta|\theta^{(k)}\right) = E_z\left\{\ln[p(\boldsymbol{z}|\boldsymbol{\theta})]|\boldsymbol{y}, \boldsymbol{\theta}^{(k)}\right\}$$

$$= -\ln[2\pi]^{n/2} - \frac{1}{2}\ln\det\left(C_{z|y}\right) - \frac{1}{2}\left(z^{(k)} - m_{z|y}\right)^T C_{z|y}^{-1}\left(z^{(k)} - m_{z|y}\right)$$

where $\theta^{(k)} = col\{m_y^{(k)}, m_z^{(k)}, c_x^{(k)}\}$ is an estimate of the parameters.

The optimal estimate of the unmeasured variables given the measurements in terms of the parameter estimate is the conditional mean

$$z_i^{(k)} = m_z^{(k)} + C_{zy}^{(k)} C_y^{(k)^{-1}} \left(y_i - m_y^{(k)}\right).$$

The estimate $z^{(k)}$ is substituted in the auxiliary function and the results is used for the maximization step. The maximum-likelihood estimate for the mean of the multivariate Gaussian distribution is the sample mean, while the maximum-likelihood estimate of the covariance matrix is the sample covariance.

The sample mean and sample covariance are, respectively

$$\hat{m}_x^{(k)} = \frac{1}{N}\sum_{i=1}^{N} x_i^{(k)}, x_i^{(k)} = col\left\{y_i^{(k)}, z_i^{(k)}\right\}, i = 1, 2, \ldots, N$$

$$C_x^{(k)} = \begin{bmatrix} C_y^{(k)} & C_{yz}^{(k)} \\ C_{zy}^{(k)} & C_z^{(k)} \end{bmatrix} = \frac{1}{N}\sum_{i=1}^{N}\left[x_i^{(k)} - \hat{m}_x^{(k)}\right]\left[x_i^{(k)} - \hat{m}_x^{(k)}\right]^T.$$

Based on the above equations, the following procedure applies the EM Algorithm to the multivariate Gaussian distribution:

1. Start with a guess of $\theta = \theta^{(k)}, k = 0$.
2. Calculate $z_i^{(k)} = m_z^{(k)} + C_{zy}^{(k)} C_y^{(k)^{-1}} (y_i - m_y^{(k)})$
3. Update the parameters using

$$\hat{m}_x^{(k)} = \frac{1}{N}\sum_{i=1}^{N} x_i^{(k)}, x_i^{(k)} = \text{col}\left\{y_i^{(k)}, z_i^{(k)}\right\}, i = 1, 2, \ldots, N$$

$$C_x^{(k)} = \begin{bmatrix} C_y^{(k)} & C_{yz}^{(k)} \\ C_{zy}^{(k)} & C_z^{(k)} \end{bmatrix} = \frac{1}{N}\sum_{i=1}^{N}\left[x_i^{(k)} - \hat{m}_x^{(k)}\right]\left[x_i^{(k)} - \hat{m}_x^{(k)}\right]^T$$

$$k \leftarrow k + 1, \theta = \theta^{(k)}.$$

4. If $\left\|\theta^{(k)} - \theta^{(k-1)}\right\| < \epsilon$ Stop. Otherwise, Go to Step 2.

MATLAB has a command that provides a parameter estimate for a data set with missing data

```
[Param, Covar] = ecmmvnrmle(Data,Design)
```

where:

Data is a $N \times n$ matrix with N samples of a $n \times 1$ random vector. Missing values are represented as NaNs. Only samples that are entirely NaNs are ignored.

Design is either a matrix or a cell array to handle two cases. If $n = 1$, design is a $N \times n_p$ matrix with known values. If $n \geq 1$, design is a cell array of either 1 or N cells each containing a $n \times n_p$ matrix of known values. Design is N cells if the matrices are different for each sample.

The function provides the following:

Param is a $n_p \times 1$ column vector of estimates of the model parameters of the regression model.

Covar is a $n \times n$ matrix of estimates for the covariance of the residuals of the regression.

13.4 Distribution Mixture

Consider a measurement vector $\boldsymbol{y} \in \mathcal{Y} \subset \mathcal{R}^n$ whose pdf is the weighted sum of M Gaussian pdfs

$$p_Y(\boldsymbol{y}|\boldsymbol{\theta}) = \sum_{i=1}^{M} w_i p_i(\boldsymbol{y}|\boldsymbol{\theta}_i). \tag{13.14}$$

Integrating, or adding for the discrete case, shows that

$$\int_{y \in \mathcal{Y}} p_Y(\boldsymbol{y}|\boldsymbol{\theta}) d\boldsymbol{y} = \sum_{i=1}^{M} w_i \int_{y \in \mathcal{Y}} p_i(\boldsymbol{y}|\boldsymbol{\theta}_i) d\boldsymbol{y} = \sum_{i=1}^{M} w_i = 1. \tag{13.15}$$

The constraint on the weights implies that we only need to know the $M-1$ weights $w_i, i = 1, \ldots, M-1$, since

$$w_M = 1 - \sum_{i=1}^{M-1} w_i.$$

The joint pdf of N i.i.d. measurements governed by the mixture is

$$p_Y(\boldsymbol{y}|\boldsymbol{\theta}) = \prod_{i=1}^{N} \sum_{j=1}^{M} w_j p_j(\boldsymbol{y}_i|\boldsymbol{\theta}_j). \tag{13.16}$$

The summation makes it difficult to maximize over the set of parameters and taking log does not simplify the problem.

Assume extra random variables $z_i, i = 1, \ldots, N, z_i \in \{1, \ldots, M\}$ that tell us which pdf corresponds to the ith measurement with probability $P(z_i = i) = w_i$, and define the complete data vector $\boldsymbol{x}_i = col\{\boldsymbol{y}_i, z_i\} \subset \mathcal{R}^{n+1}, i = 1, \ldots, N$. The vector $\boldsymbol{z} = \begin{bmatrix} z_1 \; z_2 \ldots z_N \end{bmatrix}^T$ is a vector of unknown i.i.d. random variables governed by a multinomial pmf with parameters $(1, w_1, \ldots, w_N)$ and w_i is the prior probability of belonging to the ith component of the mixture.

The likelihood function for the complete data $\boldsymbol{x} = \{\boldsymbol{x}_i, i = 1, \ldots N\}$ is

$$p_X(\boldsymbol{x}|\boldsymbol{\theta}) = \prod_{i=1}^{N} p(\boldsymbol{y}_i, z_i|\boldsymbol{\theta}) = \prod_{i=1}^{N} \underbrace{p(\boldsymbol{y}_i|z_i, \boldsymbol{\theta})}_{p_{z_i}(\boldsymbol{y}_i|\boldsymbol{\theta}_{z_i})} \underbrace{p(z_i)}_{w_i}, \sum_{i=1}^{N} p(z_i) = 1, \tag{13.17}$$

where the pdf $p_{z_i}(\boldsymbol{y}_i|\boldsymbol{\theta}_{z_i})$ is fixed once z_i is known. Taking the natural log simplifies the pdf to

$$\ln[p_X(\boldsymbol{x}|\boldsymbol{\theta})] = \sum_{i=1}^{N} \{\ln w_i + \ln[p_{z_i}(\boldsymbol{y}_i|\boldsymbol{\theta}_{z_i})]\}. \tag{13.18}$$

We now have a parameter estimation problem with missing data values $z_i, i = 1, \ldots, N$, and we use the EM algorithm to estimate the parameters

$$\boldsymbol{\theta} = \{w_i, \boldsymbol{\theta}_j, i = 1, \ldots, N, j = 1, \ldots, M\}. \tag{13.19}$$

Given an initial parameter estimate $\boldsymbol{\theta}^{(k)}$, we know the densities

$$p_j\left(\boldsymbol{y}_i|\boldsymbol{\theta}_j^{(k)}\right), i = 1, 2, \ldots, N, j = 1, 2, \ldots, M. \tag{13.20}$$

Using Bayes rule, we have

$$\begin{aligned} p(z_i|\boldsymbol{y}_i, \boldsymbol{\theta}^{(k)}) &= \frac{p(z_i, \boldsymbol{y}_i|\boldsymbol{\theta}^{(k)})}{p(\boldsymbol{y}_i|\boldsymbol{\theta}^{(k)})} = \frac{p(z_i|\boldsymbol{\theta}^{(k)})p(\boldsymbol{y}_i|z_i, \boldsymbol{\theta}^{(k)})}{\sum_{j=1}^{M} p(z_j|\boldsymbol{\theta}^{(k)})p_j\left(\boldsymbol{y}_i|z_j, \boldsymbol{\theta}_j^{(k)}\right)} \\ &= \frac{w_i^{(k)} p_{z_i}(\boldsymbol{y}_i|\boldsymbol{\theta}_{z_i}^{(k)})}{\sum_{j=1}^{M} w_j^{(k)} p_j\left(\boldsymbol{y}_i|\boldsymbol{\theta}_j^{(k)}\right)}, i = 1, 2, \ldots, N. \end{aligned} \tag{13.21}$$

Assuming independently drawn unobserved data, the joint pdf is

$$p(\boldsymbol{z}|\boldsymbol{y}, \boldsymbol{\theta}^{(k)}) = \prod_{i=1}^{N} p(z_i|\boldsymbol{y}_i, \boldsymbol{\theta}^{(k)}) = \prod_{i=1}^{N} \frac{w_i^{(k)} p_{z_i}(\boldsymbol{y}_i|\boldsymbol{\theta}_{z_i}^{(k)})}{\sum_{j=1}^{M} w_j^{(k)} p_j\left(\boldsymbol{y}_i|\boldsymbol{\theta}_j^{(k)}\right)}. \tag{13.22}$$

The auxiliary function is the expected value over the discrete variable $\boldsymbol{z}$ given the measurements and $\boldsymbol{\theta}^{(k)}$

$$Q(\boldsymbol{\theta}|\boldsymbol{\theta}^{(k)}) = E\{\ln[p(\boldsymbol{x}|\theta)]|\boldsymbol{y}, \boldsymbol{\theta}^{(k)}\} = \sum_{z\in\mathcal{Z}} \sum_{i=1}^{N} \ln[p(\boldsymbol{y}, \boldsymbol{z}|\boldsymbol{\theta})] p(\boldsymbol{z}|\boldsymbol{y}, \boldsymbol{\theta}^{(k)}), \quad (13.23)$$

where $\boldsymbol{z} \in \mathcal{Z}$ signifies $z_i \in \{1, 2, \ldots, M\}, i = 1, 2, \ldots, N$. We substitute for $\ln[p(\boldsymbol{y}, \boldsymbol{z}|\boldsymbol{\theta})]$ from (13.17) and for $p(\boldsymbol{z}|\boldsymbol{y}, \boldsymbol{\theta}^{(k)})$ from (13.22) and expand

$$Q(\boldsymbol{\theta}|\boldsymbol{\theta}^{(k)}) = \sum_{z_1=1}^{M} \cdots \sum_{z_N=1}^{M} \sum_{i=1}^{N} \ln[w_{z_i} p_{z_i}(\boldsymbol{y}_i|\theta_{z_i})] \prod_{j=1}^{N} p(z_j|\boldsymbol{y}_j, \boldsymbol{\theta}^{(k)}). \quad (13.24)$$

For a mixture of M distributions where z_i indicates the selected distribution, we write

$$\ln[w_{z_i} p_{z_i}(\boldsymbol{y}_i|\boldsymbol{\theta}_{z_i})] = \sum_{l=1}^{M} \delta_{l,z_i} \ln[w_l p_l(\boldsymbol{y}_i|\boldsymbol{\theta}_l)], \quad (13.25)$$

where δ_{l,z_i} is the Kronecker delta

$$\delta_{l,z_i} = \begin{cases} 1, & z_i = l \\ 0, & z_i \neq l \end{cases}. \quad (13.26)$$

The auxiliary function can now be written as

$$Q(\boldsymbol{\theta}|\boldsymbol{\theta}^{(k)}) = \sum_{l=1}^{M} \sum_{i=1}^{N} \ln[w_l p_l(\boldsymbol{y}_i|\boldsymbol{\theta}_l)] \sum_{z_1=1}^{M} \cdots \sum_{z_N=1}^{M} \delta_{l,z_i} \prod_{j=1}^{N} p(z_j|\boldsymbol{y}_j, \boldsymbol{\theta}^{(k)}). \quad (13.27)$$

Set $z_i = l$ so that $\delta_{l,z_i} = 1$ and the expression reduces to

$$\sum_{z_1=1}^{M} \cdots \sum_{z_N=1}^{M} \delta_{l,z_i} \prod_{j=1}^{N} p(z_j|\boldsymbol{y}_i, \boldsymbol{\theta}^{(k)}) \Bigg|_{z_i=l} = p(l|\boldsymbol{y}_i, \boldsymbol{\theta}^{(k)}) \times \sum_{z_1=1}^{M} \cdots \sum_{z_N=1}^{M} \left[\prod_{\substack{j=1 \\ j\neq i}}^{N} p(z_j|\boldsymbol{y}_j, \boldsymbol{\theta}^{(k)}) \right] \quad (13.28)$$

then use the identity

$$\sum_{z_1=1}^{M} \cdots \sum_{z_N=1}^{M} \prod_{\substack{j=1 \\ j\neq i}}^{N} p(z_j|\boldsymbol{y}_j, \boldsymbol{\theta}^{(k)}) = 1 \quad (13.29)$$

to obtain

$$\sum_{z_1=1}^{M} \cdots \sum_{z_N=1}^{M} \delta_{l,z_i} \prod_{j=1}^{N} p\big(z_j|\boldsymbol{y}_j, \boldsymbol{\theta}^{(k)}\big)\Bigg|_{z_i=l} = p\big(l|\boldsymbol{y}_i, \boldsymbol{\theta}^{(k)}\big). \tag{13.30}$$

The auxiliary function simplifies to

$$Q\big(\boldsymbol{\theta}|\boldsymbol{\theta}^{(k)}\big) = \sum_{l=1}^{M}\sum_{i=1}^{N} \ln\big[w_l p_l\big(\boldsymbol{y}_i|\theta_l\big)\big] p\big(l|\boldsymbol{y}_i, \boldsymbol{\theta}^{(k)}\big). \tag{13.31}$$

We separate terms based on parameters into

$$Q\big(\boldsymbol{\theta}|\boldsymbol{\theta}^{(k)}\big) = \sum_{l=1}^{M} Q_{w_l} + \sum_{l=1}^{M} Q_{\theta_l} \tag{13.32}$$

where

$$Q_{w_l} = \sum_{i=1}^{N} \ln[w_l] p\big(l|\boldsymbol{y}_i, \theta^{(k)}\big) \tag{13.33}$$

$$Q_{\theta_l} = \sum_{i=1}^{N} \ln\big[p_l\big(\boldsymbol{y}_i|\theta_l\big)\big] p\big(l|\boldsymbol{y}_i, \theta^{(k)}\big). \tag{13.34}$$

Including the weight constraint with Lagrange multiplier λ, the necessary condition for a maximum is

$$\begin{aligned}\frac{\partial Q_{w_l}}{\partial w_l} &= \frac{\partial \sum_{i=1}^{N} \ln[w_l] p\big(l|y_i, \boldsymbol{\theta}^{(k)}\big) + \lambda\Big(\sum_{i=1}^{M} w_i - 1\Big)}{\partial w_l} \\ &= \frac{1}{w_l}\sum_{i=1}^{N} p\big(l|y_i, \boldsymbol{\theta}^{(k)}\big) + \lambda = 0.\end{aligned} \tag{13.35}$$

We solve for the weights

$$w_l = -\frac{1}{\lambda}\sum_{i=1}^{N} p\big(l|\boldsymbol{y}_i, \boldsymbol{\theta}^{(k)}\big). \tag{13.36}$$

Adding the weights and using the weight constraint

$$\begin{aligned}\sum_{i=1}^{M} w_i = 1 &= -\frac{1}{\lambda}\sum_{l=1}^{M}\sum_{i=1}^{N} p\big(l|\boldsymbol{y}_i, \boldsymbol{\theta}^{(k)}\big) \\ &= -\frac{1}{\lambda}\sum_{i=1}^{N} 1 = \frac{N}{\lambda}.\end{aligned}$$

Hence, the Lagrange multiplier is

$$\lambda = -N \tag{13.37}$$

and the weights are

$$w_l = \frac{1}{N}\sum_{i=1}^{N} p(l|y_i, \boldsymbol{\theta}^{(k)}), l = 1, 2, \ldots, N. \tag{13.38}$$

Differentiate the auxiliary function to obtain

$$\begin{aligned} \frac{\partial Q_{\theta_l}}{\partial \boldsymbol{\theta}_l} &= \frac{\partial \sum_{i=1}^{N} \ln\big[p_l\big(\boldsymbol{y}_i|\boldsymbol{\theta}_l\big)\big]p\big(l|\boldsymbol{y}_i, \boldsymbol{\theta}^{(k)}\big)}{\partial \boldsymbol{\theta}_l} \\ &= \sum_{i=1}^{N} \frac{p\big(l|\boldsymbol{y}_i, \boldsymbol{\theta}^{(k)}\big)}{p_l\big(\boldsymbol{y}_i|\boldsymbol{\theta}_l\big)} \frac{\partial p_l\big(\boldsymbol{y}_i|\boldsymbol{\theta}_l\big)}{\partial \boldsymbol{\theta}_l} = 0. \end{aligned} \tag{13.39}$$

The expression can be simplified in special cases, including the Gaussian mixture.

13.5 Gaussian Mixture

For the Gaussian case,

$$p_l(\boldsymbol{y}|\boldsymbol{m}_l, \Sigma_l) = \frac{1}{[2\pi]^{n/2}\sqrt{\det(\Sigma_l)}} e^{-\frac{1}{2}(\boldsymbol{y}-\boldsymbol{m}_l)^T \Sigma_l^{-1} (\boldsymbol{y}-\boldsymbol{m}_l)},$$

the parameters of the distribution are

$$\theta_l = \{\boldsymbol{m}_l, \Sigma_l, l = 1, \ldots, M\}.$$

e log-likelihood simplifies the expression to

$$\ln p_l(\boldsymbol{y}|\boldsymbol{m}_l, \Sigma_l) = -\ln(2\pi)^{n/2} - \frac{1}{2}\ln\det(\Sigma_l) - \frac{1}{2}(\boldsymbol{y} - \boldsymbol{m}_l)^T \Sigma_l^{-1} (\boldsymbol{y} - \boldsymbol{m}_l).$$

The necessary conditions for a maximum are obtained by differentiating the auxiliary function

$$\begin{aligned} \frac{\partial p_l\big(\boldsymbol{y}_i|\boldsymbol{m}_l, \Sigma_l\big)}{\partial \boldsymbol{m}_l} &= -\frac{1}{2}(-2)\Sigma_l^{-1}\big(\boldsymbol{y}_i - \boldsymbol{m}_l\big) \\ \frac{\partial Q_{\theta_l}}{\partial \boldsymbol{m}_l} &= \frac{\partial \sum_{i=1}^{N} \ln\big[p_l\big(\boldsymbol{y}_i|\boldsymbol{m}_l\big)\big]p\big(l|\boldsymbol{y}_i, \theta^{(k)}\big)}{\partial \boldsymbol{m}_l} \end{aligned}$$

$$\sum_{i=1}^{N} \Sigma_l^{-1}(\mathbf{y}_i - \mathbf{m}_l) p(l|\mathbf{y}_i, \boldsymbol{\theta}^{(k)}) = 0$$

$$\mathbf{m}_l = \frac{\sum_{i=1}^{N} \mathbf{y}_i p(l|\mathbf{y}_i, \boldsymbol{\theta}^{(k)})}{\sum_{i=1}^{N} p(l|\mathbf{y}_i, \boldsymbol{\theta}^{(k)})}. \tag{13.40}$$

To differentiate w.r.t. the covariance matrix, we need the identities:

(i) The derivative of the log of the determinant of a matrix

$$\frac{\partial \ln \det A}{\partial A} = 2A^{-1} - \mathrm{diag}\{A^{-1}\}. \tag{13.41}$$

(b) Derivative of the trace of a matrix

$$\frac{\partial tr(AB)}{\partial A} = B + B^T - \mathrm{diag}\{B\}. \tag{13.42}$$

(iii) For a symmetric matrix B:

$$\frac{\partial tr(AB)}{\partial A} = 2B - \mathrm{diag}\{B\}. \tag{13.43}$$

Hence, we write the term in the exponent as

$$(\mathbf{y}_i - \mathbf{m}_l)^T \Sigma_l^{-1}(\mathbf{y}_i - \mathbf{m}_l) = tr\{\Sigma_l^{-1} N_{li}\}$$

where $N_{li} = (\mathbf{y}_i - \mathbf{m}_l)(\mathbf{y}_i - \mathbf{m}_l)^T$.

Differentiate and use the identities

$$\begin{aligned}
\ln p_l(\mathbf{y}|\mathbf{m}_l, \Sigma_l) &= -\ln(2\pi)^{\frac{n}{2}} - \frac{1}{2}\ln\det(\Sigma_l) - \frac{1}{2} tr\{\Sigma_l^{-1} N_{li}\} \\
\frac{\partial p_l(\mathbf{y}_i|\mathbf{m}_l, \Sigma_l)}{\partial \Sigma_l} &= \frac{1}{2}\{2\Sigma_l^{-1} - \mathrm{diag}\{\Sigma_l^{-1}\} - (2N_{li} - \mathrm{diag}\{N_{li}\}\} \\
&= \frac{1}{2}\{2M_{li} - \mathrm{diag}\{M_{li}\}\} \\
M_{li} &= \Sigma_l^{-1} - N_{li}.
\end{aligned} \tag{13.44}$$

We now have the necessary condition

$$\begin{aligned}
\frac{\partial Q_{\theta_l}}{\partial \Sigma_l} &= \frac{\partial \sum_{i=1}^{N} \ln[p_l(\mathbf{y}_i|\mathbf{m}_l)] p(l|\mathbf{y}_i, \theta^{(k)})}{\partial \Sigma_l} \\
&= \frac{1}{2}\sum_{i=1}^{N} \{2M_{li} - \mathrm{diag}\{M_{li}\}\} p(l|\mathbf{y}_i, \theta^{(k)}) = 0.
\end{aligned}$$

Define the matrix

$$S = \frac{1}{2}\sum_{i=1}^{N} M_{li}\, p\left(l|\mathbf{y}_i, \theta^{(k)}\right)$$

and write

$$2S - diag\{S\} = [0] \Rightarrow S = [0].$$

Finally, we solve for the estimate

$$\Sigma_l = \frac{\sum_{i=1}^{N}\left(\mathbf{y}_i - \mathbf{m}_l\right)\left(\mathbf{y}_i - \mathbf{m}_l\right)^T p\left(l|\mathbf{y}_i, \theta^{(k)}\right)}{\sum_{i=1}^{N} p\left(l|\mathbf{y}_i, \boldsymbol{\theta}^{(k)}\right)}. \tag{13.45}$$

Summary of EM Steps

1. Initialization

$$k = 0, \boldsymbol{\theta}^{(k)} = \boldsymbol{\theta}^{(0)}$$

2. Iteration

while $\left\|\boldsymbol{\theta}^{(k+1)} - \boldsymbol{\theta}^{(k)}\right\| >$ tolerance.

(a) Update the weights

$$w_l^{(k+1)} = \frac{1}{N}\sum_{i=1}^{N} p\left(l|\mathbf{y}_i, \boldsymbol{\theta}^{(k)}\right), l = 1, 2, \ldots, N$$

(b) Update the mean

$$\mathbf{m}_l^{(k+1)} = \frac{\sum_{i=1}^{N} \mathbf{y}_i\, p\left(l|\mathbf{y}_i, \boldsymbol{\theta}^{(k)}\right)}{\sum_{i=1}^{N} f\left(l|\mathbf{y}_i, \boldsymbol{\theta}^{(k)}\right)}$$

(c) Update the covariance

$$\Sigma_l^{(k+1)} = \frac{\sum_{i=1}^{N}\left(\mathbf{y}_i - \mathbf{m}_l^{(k+1)}\right)\left(\mathbf{y}_i - \mathbf{m}_l^{(k+1)}\right)^T p\left(l|\mathbf{y}_i, \boldsymbol{\theta}^{(k)}\right)}{\sum_{i=1}^{N} p\left(l|\mathbf{y}_i, \boldsymbol{\theta}^{(k)}\right)}$$

End

The following MATLAB commands fit a Gaussian mixture to data is

```
M=3; % Number of Gaussian pdfs of the mixture, depends on the data set
options = statset('MaxIter',100);% Limit iterations to 100
GMModel=fitgmdist((Rerrnew),M,'Options',options);% Fit data Rernew with a
% Gaussian mixture model with M components
```

Example 13.5 The solar generation capacity data shows random variation that can be modeled as a Gaussian mixture of 3 pdfs. Use MATLAB to obtain the Gaussian mixture model for the data provided by NREL at

https://drive.google.com/file/d/1ku5Pf3eO2eFLAcTbNlZ2_PFWFl_8q37T/view?pli=1

Solution

The data is provided in MW and is converted to GW then used to obtain the model. The following script provides the Gaussian mixture mode after 119 iterations:

```
M=3; % 3 terms in sum
options = statset('MaxIter',300);% Maximum number of iterations=100
GMModel=fitgmdist(capacity_mw/10^3,M,'Options',options);%
mu = GMModel.mu;
sigma = GMModel.Sigma;
W=GMModel.ComponentProportion;
cdf_value=0.2;  %
obj = gmdistribution(mu,sigma,W);
pd = makedist('normal');
% qqplot checks the distribution of data
maxvar=max(sigma);
% x = min(capacity_mw):1:max(capacity_mw);
x=-maxvar:.1:5*maxvar;
% creating the distribution functions for each gaussian terms
% hold on
figure(1)
for i=1:M
p1 = pdf('Normal', x, mu(i), sigma(i));
plot(x, p1*W(i))
hold on
end
 hold off
 figure(2)
 histogram(capacity_mw/10^3,34)
figure(3);qqplot(capacity_mw/10^3,pd);
```

The model parameters are:

Means: [4.130290343634414;3.671687226407355;1.783312033451926]

Variances: [0.0072, 0.0094; 1.2475]

Weights: [0.364556233254071, 0.323848988782854, 0.311594777963074]

Figure 13.2 shows the histogram of the data and Fig. 13.3 shows the pdfs of the model. The histogram shows that the data has a bimodal distribution, which justifies the use of a Gaussian mixture model.

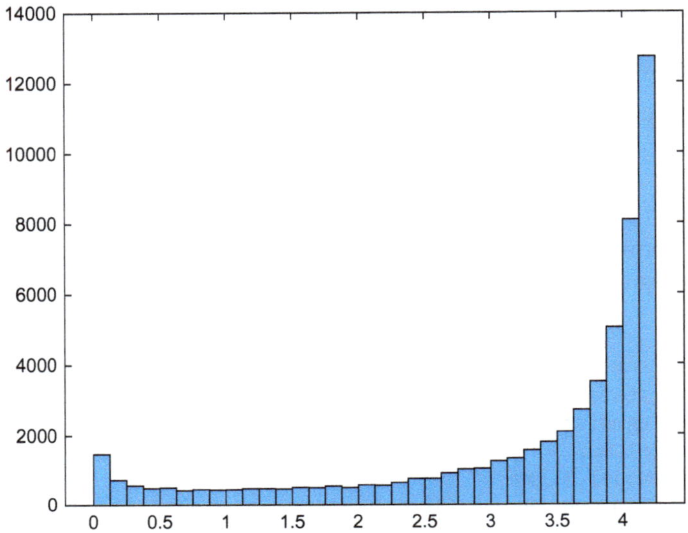

Fig. 13.2 Histogram of the solar capacity data

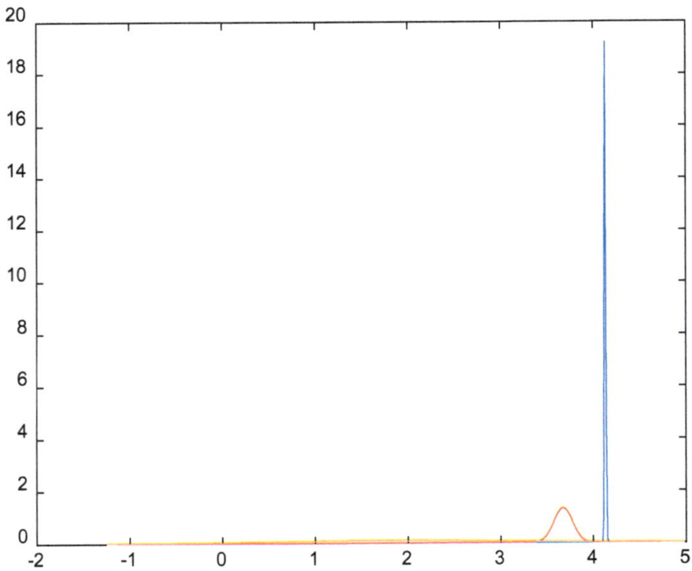

Fig. 13.3 Plot of the pdfs of the Gaussian mixture model

Problems

13.1 Repeat Example 13.1 with the data generated using the MATLAB command mnrnd. Repeat the estimation 10 times and calculate the average estimate.

13.2 Repeat Example 13.2 with the parameters defined as

$$\boldsymbol{\theta}^T = \left[\theta_1\ \theta_2\right] = \left[\frac{m_x}{\sigma_x^2}\ \frac{-1}{2\sigma_x^2}\right], \theta_2 < 0$$

13.3 Write a simple MATLAB script to run the EM algorithm for Example 13.1.

13.4 A chemical process starts with the concentration $m_c \in \mathcal{M}_c$. Due to a fault, the concentration can change to $m_f \in \mathcal{M}_f$. Using a record of the concentration $\{y_0, y_1, \ldots, y_l\}$, the EM algorithm can be used to find the time i_f where the fault occurred. For both the discrete and the continuous case:

(a) Assuming that m_c, m_f, and i_f are mutually independent, simplify the auxiliary function for the concentration with the data set $\boldsymbol{x} = \{\boldsymbol{y}, \boldsymbol{z}\}$, $\boldsymbol{z} = \{m_c, m_f\}$, and $\theta = i_f$.

(b) Eliminate terms that do not influence the result of the maximization step from the auxiliary function.

(c) Write the equations for the equation for the maximization step of the EM algorithm

(d) Consider the EM algorithm for multivariable Gaussian data with $\boldsymbol{x} \in \mathcal{R}^n$, $\boldsymbol{y} \in \mathcal{R}^m$, $m < n$,with

$$\boldsymbol{y} = C\boldsymbol{x} = [C_1|C_2],\ C_1 \in \mathcal{R}^{m\times m}\text{nonsingular}$$

Write the steps of the EM algorithm to estimate the parameters $(\boldsymbol{m}_x, C_x)$.

13.6 The solar irradiation data shows random variation that can be modeled as a Gaussian mixture of 3 pdfs. Use MATLAB to obtain the Gaussian mixture model for the data provided by NREL at

https://drive.google.com/file/d/1ku5Pf3eO2eFLAcTbNlZ2_PFWFl_8q37T/view?pli=1

Bibliography

1. Moon, T. K., & Stirling, W. C. (2000). *Mathematical methods and algorithms for signal processing*. Prentice-Hall.
2. Bilmes, J. A. (1998, April) A gentle tutorial of the EM algorithm and its application to parameter estimation for Gaussian mixture and hidden Markov models. Technical Report TR-97-021, Berkeley, CA.
3. Chen, Y., & Gupta, M. R. (2010, February) EM demystified: An expectation-maximization tutorial. UWEE Technical Report Number UWEETR-2010-0002.

4. Dempster, A. P., Laird, N. M., & Rubin, D. B. (1977). Maximum likelihood estimation from incomplete data via the EM algorithm (with discussion). *Journal of Royal Statistical Society Series B, 39*, 1–38.
5. Koski, T. (2001). *Hidden Markov models for bioinformatics.* Kluwer Academic Publishers.
6. McLachlan, G., & Peel, D. (2000). *Finite mixture models.* Wiley.

Chapter 14
Hidden Markov Models

14.1 Markov Chains

A Markov process is a memoryless process where the future state depends only on the current state and not how it was reached. Thus, for a discrete-time system with a discrete set of state x_k the states have a one-to-one correspondence with the set of integers and the transition probability from state i to state j_1 satisfies

$$P(x_k = i|x_{k-1} = j_1, \ldots x_{k-m} = j_m) = P(x_k = i|x_{k-1} = j_1). \tag{14.1}$$

The transition probability from state j to state i, $P(x_k = i|x_{k-1} = j)$, is denoted by p_{ij}.[1] By the axioms of probability, the sum of all the transition probabilities from any state i must be unity, that is

$$\sum_{i=1}^{n} p_{ij} = 1, \; p_{ij} \in [0, 1], \tag{14.2}$$

where n is the number of states of the Markov chain. The probability of the state equal to i at time k is

$$P(x_k = i) = \sum_{j=1}^{n} P(x_k = i|x_{k-1} = j).P(x_{k-1} = j), i = 1, 2, \ldots, n \tag{14.3}$$

which in equivalent to

[1] In much of the literature, p_{ij} denotes the probability of transition from state i to state j. We use a notation that is more suited to a state equation with state matrix $A = [p_{ij}]$.

M. S. Fadali, *Introduction to Random Signals, Estimation Theory, and Kalman Filtering*, https://doi.org/10.1007/978-981-99-8063-5_14

$$P(x_k = i) =$$

$$\begin{bmatrix} P(x_k = i|x_{k-1} = 1) \ldots P(x_k = i|x_{k-1} = n) \end{bmatrix} \begin{bmatrix} P(x_{k-1} = 1) \\ \vdots \\ P(x_{k-1} = n) \end{bmatrix}. \tag{14.4}$$

By combining the n equation, Markov chain can be described by a state equation that governs probability evolution where the state vector is a vector of probabilities not states. The equation is known as the Chapman–Kolmogorov equation

$$\begin{bmatrix} P(x_k = 1) \\ \vdots \\ P(x_k = n) \end{bmatrix} = \begin{bmatrix} p_{11} & p_{12} & \cdots & p_{1n} \\ \vdots & \vdots & \ddots & \vdots \\ p_{n1} & p_{n2} & \cdots & p_{nn} \end{bmatrix} \begin{bmatrix} P(x_{k-1} = 1) \\ \vdots \\ P(x_{k-1} = n) \end{bmatrix}. \tag{14.5}$$

The matrix

$$A = \begin{bmatrix} a_{11} & a_{12} & \ldots & a_{1n} \\ \vdots & \vdots & \ddots & \vdots \\ a_{n1} & a_{n2} & \ldots & a_{nn} \end{bmatrix} = \begin{bmatrix} p_{11} & p_{12} & \cdots & p_{1n} \\ \vdots & \vdots & \ddots & \vdots \\ p_{n1} & p_{n2} & \cdots & p_{nn} \end{bmatrix} \tag{14.6}$$

is known as the **transition probability matrix, stochastic matrix,** or chain matrix. The Markov chain is **homogeneous** if the matrix is independent of time. It can be shown that matrix has the following properties

(i) The matrix has one eigenvalue equal to unity.
(ii) If A is a stochastic matrix, then A^k is also a stochastic matrix for any positive integer k.

The proof of the properties is left as an exercise.

We define the vector of prior probabilities at time k

$$\boldsymbol{p}_k^x = \begin{bmatrix} p_{k1}^x & p_{k2}^x & \cdots & p_{kn}^x \end{bmatrix}^T = \begin{bmatrix} P(x_k = 1) & P(x_k = 2) & \ldots & P(x_k = n) \end{bmatrix}^T. \tag{14.7}$$

We write the state probability equation in matrix form as

$$\boldsymbol{p}_{k+1}^x = A\boldsymbol{p}_k^x \tag{14.8}$$

with initial condition $\boldsymbol{p}_0^x$.

MATLAB has commands to create a Markov chain, display the number of states, and plot a figure that represents it.

Example 14.1 Generate a plot for the Markov chain of Fig. 14.1 using MATLAB.

Solution

The following script generates the plot of Fig. 14.2 and the number of states.

```
% MATLAB commands
mc = dtmc(A); % Create the Markov chain characterized by the transition matrix P.
numstates = mc.NumStates % Display the number of states in the Markov chain.
Figure;
graphplot(mc); % Plot a directed graph of the Markov chain.
      numstates =
        4
```

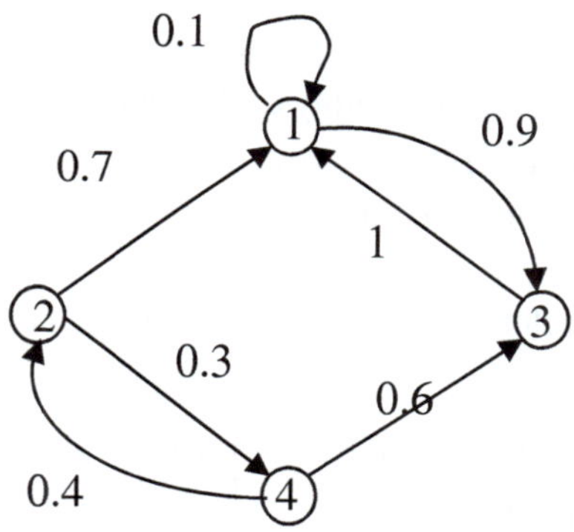

Fig. 14.1 Markov chain example

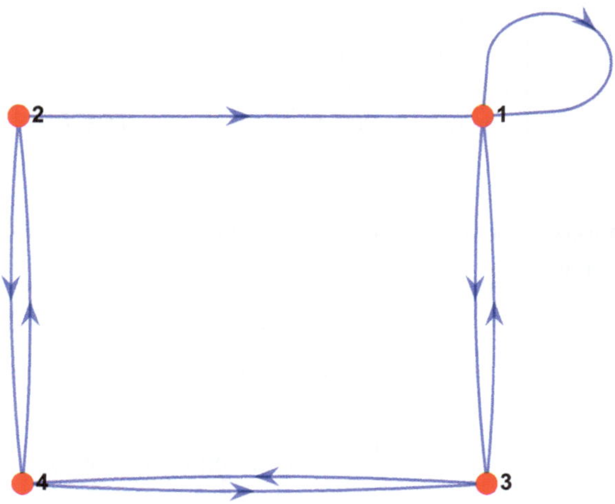

Fig. 14.2 MATLAB plot of Markov chain

The following example shows the evolution of the probability vector using the Chapman–Kolmogorov equation.

Example 14.2 Write the probability state equation for the Markov chain of Fig. 14.1. If all four states are equally likely:

(a) Determine the probability of the sequence of states {2, 1, 3, 1}
(b) Determine the probability of each state at time 2

Solution

$$\begin{bmatrix} P(x_k = 1) \\ P(x_k = 2) \\ P(x_k = 3) \\ P(x_k = 4) \end{bmatrix} = \begin{bmatrix} 0.1 & 0.7 & 1.0 & 0 \\ 0 & 0 & 0 & 0.4 \\ 0.9 & 0 & 0 & 0.6 \\ 0 & 0.3 & 0 & 0 \end{bmatrix} \begin{bmatrix} P(x_{k-1} = 1) \\ P(x_{k-1} = 2) \\ P(x_{k-1} = 3) \\ P(x_{k-1} = 4) \end{bmatrix}$$

(a) The probability of starting at state 2 is $1/4$. We multiply by the transition probabilities to obtain the probability of the sequence.

$$\begin{aligned} P(x_{0:3} = 2, 1, 3, 1) &= P(x_0 = 2)P(x_1 = 1|x_0 = 2)P(x_2 = 3|x_1 = 1)P(x_3 = 1|x_2 = 3) \\ &= (0.25)(0.7)(0.9)(1) = 0.1575. \end{aligned}$$

(b) To determine the vector of probabilities at time 2, we multiply the initial state vector by A^2

$$\begin{bmatrix} P(x_2 = 1) \\ P(x_2 = 2) \\ P(x_2 = 3) \\ P(x_2 = 4) \end{bmatrix} = A^2 \begin{bmatrix} P(x_0 = 1) \\ P(x_0 = 2) \\ P(x_0 = 3) \\ P(x_0 = 4) \end{bmatrix}$$

$$= \begin{bmatrix} 0.1 & 0.7 & 1.0 & 0 \\ 0 & 0 & 0 & 0.4 \\ 0.9 & 0 & 0 & 0.6 \\ 0 & 0.3 & 0 & 0 \end{bmatrix}^2 \begin{bmatrix} 0.25 \\ 0.25 \\ 0.25 \\ 0.25 \end{bmatrix} = \begin{bmatrix} 0.49 \\ 0.03 \\ 0.45 \\ 0.03 \end{bmatrix}$$

■

To train a Markov model, we use experimental or simulation data to determine the transition probabilities

$$a(i, j) = P(\boldsymbol{x} = i|\boldsymbol{x} = j) = \frac{\text{no. of transitions from state } j \text{ to state } i}{\text{no. of transitions from state } j} \tag{14.9}$$

$$c(i, j) = P(\boldsymbol{z} = z_i|\boldsymbol{x} = j) = \frac{\text{no. of outputs } z_i \text{ in state } j}{\text{no. of outputs in state } j}. \tag{14.10}$$

It can be shown that the relative frequencies are the maximum-likelihood estimates of the probabilities. The proof is left as an exercise for the reader.

14.2 Hidden Markov Model

A Markov model where the state is indirectly observable through a stochastic measurement process and is not directly observable is known as a hidden Markov model (HMM). The measurement process z_k is static and is given by

$$P(z_k = i) = \sum_{j=1}^{n} P(z_k = i | x_k = j).P(x_k = j),$$
$$i = 1, 2, \ldots, m, \sum_{j=1}^{n} P(z_k = i | x_k = j) = 1. \tag{14.11}$$

This gives the matrix equation

$$\boldsymbol{p}_k^z = C \boldsymbol{p}_k^x \tag{14.12}$$

$$C = [P(z_k = i | x_k = j)] = \begin{bmatrix} c_{11} & c_{12} & \ldots & c_{1n} \\ \vdots & \vdots & \ddots & \vdots \\ c_{m1} & c_{m2} & \ldots & c_{mn} \end{bmatrix} \tag{14.13}$$

$$\boldsymbol{p}_k^z = \begin{bmatrix} p_{k1}^z & p_{k2}^z & \ldots & p_{km}^z \end{bmatrix}^T = \begin{bmatrix} P(z_k = 1) & P(z_k = 2) & \ldots & P(z_k = m) \end{bmatrix}^T. \tag{14.14}$$

Combining the state probability Eq. (14.8) and the measurement probability, Eq. (14.12) gives a state-space model for the HMM. The HMM is defined by the two matrices (A, C) and the initial condition vector $\boldsymbol{p}_0^x$.

We assume that the measurement z_k is **conditionally independent** of the measurement history $z_{1:k-1}$ given the state x_k, that is (Fig. 14.3)

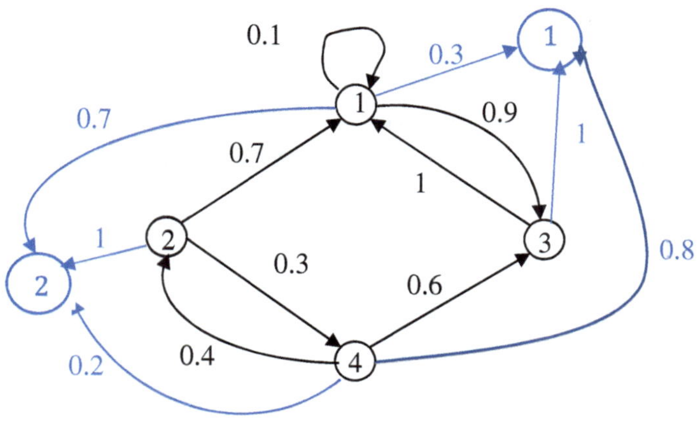

Fig. 14.3 Hidden Markov chain example

$$P(z_k = i|x_k = j, z_{1:k-1}) = A(z_k = i|x_k = j). \tag{14.15}$$

Example 14.3 Write the probability state and observation equations for the HMM of Fig. 14.1. If all four states are equally likely, determine the probability of the sequence of measurements $\{1, 2\}$.

Solution

$$\begin{bmatrix} P(x_k = 1) \\ P(x_k = 2) \\ P(x_k = 3) \\ P(x_k = 4) \end{bmatrix} = \begin{bmatrix} 0.1 & 0.7 & 1.0 & 0 \\ 0 & 0 & 0 & 0.4 \\ 0.9 & 0 & 0 & 0.6 \\ 0 & 0.3 & 0 & 0 \end{bmatrix} \begin{bmatrix} P(x_{k-1} = 1) \\ P(x_{k-1} = 2) \\ P(x_{k-1} = 3) \\ P(x_{k-1} = 4) \end{bmatrix}$$

$$\begin{bmatrix} P(z_k = 1) \\ P(z_k = 2) \end{bmatrix} = \begin{bmatrix} 0.3 & 0 & 1 & 0.8 \\ 0.7 & 1 & 0 & 0.2 \end{bmatrix} \begin{bmatrix} P(x_k = 1) \\ P(x_k = 2) \\ P(x_k = 3) \\ P(x_k = 4) \end{bmatrix}.$$

To determine the probability of the output sequence, we need to find the probabilities of all the state sequences that correspond to it. This excludes some state paths, such as $\{1, 3\}$ and any path starting with $x_0 = 2$ (Table 14.1).

Adding the probability of the mutually exclusive paths gives

$$P(z_{0:1}, x_{0:1}) = 0.0525 + 0.175 + 0.08 = 0.3075.$$

■

Example 14.3 shows that, even for a very simple example, it is difficult to calculate the probability of a sequence of Markov chain outputs. In the next section, we present an algorithm that efficiently calculated the probability.

Table 14.1 Table of probabilities for the output sequence $\{1, 2\}$

State sequence $x_{0:1}$ for $z_{1:2} = \{1, 2\}$	Probability of state sequence $P(x_{0:1})$	Probability of output sequence $P(z_{0:1}\vert x_{0:1})$	Joint probability $P(z_{0:1}, x_{0:1})$
$\{1, 1\}$	$(0.1)(0.25) = 0.025$	$(0.3)(0.7) = 0.21$	$(0.025)(0.21) = 0.0525$
$\{3, 1\}$	$(1)(0.25) = 0.25$	$(1)(0.7) = 0.7$	$(0.25)(0.7) = 0.175$
$\{4, 2\}$	$(0.4)(0.25) = 0.1$	$(0.8)(1) = 0.8$	$(0.1)(0.8) = 0.08$

14.3 The Forward Algorithm

The forward algorithm efficiently calculates several important probabilities including the probability $P(x_k|z_{0:k}, \theta)$, $P(y_{0:k}|\theta)$ given the parameters of an HMM $\theta = (A, C, p_0^x)$. To simplify the notation, we drop the parameters and write the probability as

$$P(y_{0:k}) = \sum\nolimits_{\forall x_k \in [1,n]} P(z_{0:k}, x_k). \tag{14.16}$$

The computation of the summation is intractable because of the large number of possible sequences $x_{1:K}$. The Markov nature of the measurement process allows us to simplify the conditional probability of a measurement sequence. We begin with the expression

$$P(z_{0:k}) = P(z_0)P(z_1, z_2, \ldots z_K|z_0)$$

$$= P(z_0)P(z_1|z_0)P(z_3, \ldots z_K|z_0, z_1).$$

Iteration of the process gives

$$P(z_{0:k}) = P(z_0)\prod\nolimits_{i=1}^{k} P(z_i|z_1, \ldots z_{i-1}).$$

The Markov property reduces the equation to

$$P(z_{0:k}) = P(z_0)\prod\nolimits_{i=1}^{k} P(z_i|z_{i-1}). \tag{14.17}$$

Taking the natural log of the equation, we have

$$\ln P(z_{0:k}) = \ln P(z_0) + \sum\nolimits_{i=1}^{k} \ln P(z_i|z_{i-1}). \tag{14.18}$$

In the following formulation, the forward algorithm follows a predictor–corrector model to update the probability $P(x_{k-1}|z_{0:k-1})$ of the state given the measurements and obtain $P(x_k|z_{0:k})$.

Similarly, it can be shown that

$$P(x_{0:k}) = P(x_0)\prod\nolimits_{i=1}^{k} P(x_i|x_{i-1}). \tag{14.19}$$

Forward Algorithm
Initialization: $k = 0$

$$P(x_0|z_0) = \frac{P(z_0|x_0)P(x_0)}{\sum_{x_0\in[1,n]} P(z_0|x_0)P(x_0)}, \forall x_0 \in [1,n] \tag{14.20}$$

Predictor: Calculate the conditional probability of the state at time k given the measurements up to time $k-1$

$$P(x_k|z_{0:k-1}) = \sum_{x_{k-1}\in[1,n]} P(x_k|x_{k-1})P(x_{k-1}|z_{0:k-1}), \forall x_k \in [1,n]. \tag{14.21}$$

Corrector: Correct the predicted probability using the measurement at time k

$$P(x_k|z_{0:k}) = \frac{P(z_k|x_k)}{\sum_{x_k\in[1,n]} P(z_k|x_k)P(x_k|z_{0:k-1})} P(x_k|z_{0:k-1}), \forall x_k \in [1,n]. \tag{14.22}$$

We can write the equations in vector form as follows.
Predictor:

$$\begin{bmatrix} P(x_k = 1|z_{0:k-1}) \\ P(x_k = 2|z_{0:k-1}) \\ \vdots \\ P(x_k = n|z_{0:k-1}) \end{bmatrix} = A \begin{bmatrix} P(x_{k-1} = 1|z_{0:k-1}) \\ P(x_{k-1} = 2|z_{0:k-1}) \\ \vdots \\ P(x_{k-1} = n|z_{0:k-1}) \end{bmatrix} \tag{14.23}$$

where A is the transition probability matrix

$$A = \begin{bmatrix} P(x_k = 1|x_{k-1} = 1) & P(x_k = 1|x_{k-1} = 2) & \dots & P(x_k = 1|x_{k-1} = n) \\ P(x_k = 2|x_{k-1} = 1) & P(x_k = 2|x_{k-1} = 2) & \dots & P(x_k = 2|x_{k-1} = n) \\ \vdots & \vdots & \ddots & \vdots \\ P(x_k = n|x_{k-1} = 1) & P(x_k = n|x_{k-1} = 2) & \dots & P(x_k = n|x_{k-1} = n) \end{bmatrix}. \tag{14.24}$$

Corrector:

(a) Calculate the unnormalized probability is

$$\boldsymbol{P}_u(k) = \begin{bmatrix} P_u(x_k = 1|z_{0:k}) \\ P_u(x_k = 2|z_{0:k}) \\ \vdots \\ P_u(x_k = n|z_{0:k}) \end{bmatrix} = \begin{bmatrix} P(z_k|x_k = 1) \\ P(z_k|x_k = 2) \\ \vdots \\ P(z_k|x_k = n) \end{bmatrix} * \begin{bmatrix} P(x_k = 1|z_{0:k-1}) \\ P(x_k = 2|z_{0:k-1}) \\ \vdots \\ P(x_k = n|z_{0:k-1}) \end{bmatrix}, \tag{14.25}$$

where $*$ denotes the Hadamard element-by-element vector product.

(b) Calculate the 1-norm for normalization

$$\boldsymbol{P}_u(k)_1 = \sum_{x_k \in [1,n]} P(z_k|x_k)P(x_k|z_{0:k-1}). \tag{14.26}$$

(c) Calculate the corrected probability

$$\begin{bmatrix} P(x_k = 1|z_{0:k}) \\ P(x_k = 2|z_{0:k}) \\ \vdots \\ P(x_k = n|z_{0:k}) \end{bmatrix} = \frac{1}{\boldsymbol{P}_u(k)} \begin{bmatrix} P_u(x_k = 1|z_{0:k}) \\ P_u(x_k = 2|z_{0:k}) \\ \vdots \\ P_u(x_k = n|z_{0:k}) \end{bmatrix} \tag{14.27}$$

The initialization Formula (14.19) is a direct application of Bayes rule. To derive the predictor formula, we write.

$$\begin{aligned} P(x_k|z_{0:k-1}) &= \sum_{x_{k-1} \in [1,n]} P(x_k, x_{k-1}|z_{0:k-1}), \forall x_k \in [1, n] \\ &= \sum_{x_{k-1} \in [1,n]} P(x_k|x_{k-1}, z_{0:k-1})P(x_{k-1}|z_{0:k-1}), \forall x_k. \end{aligned}$$

From the Markov property, $P(x_k|x_{k-1}, y_{0:k-1}) = P(x_k|x_{k-1})$ and (14.20) follows.

To derive the equation for the corrector, we begin with the conditional probability

$$P(x_k|z_{0:k}) = P(x_k|z_{0:k-1}, z_k) = \frac{P(x_k, z_{0:k-1}, z_k)/P(z_{0:k-1})}{P(z_{0:k-1}, z_k)/P(z_{0:k-1})} = \frac{P(x_k, z_k|z_{0:k-1})}{P(z_k|z_{0:k-1})}.$$

We rewrite the numerator as

$$P(x_k, z_k|z_{0:k-1}) = \frac{P(x_k, z_{0:k-1}, y_k)P(x_k, z_{0:k-1})}{P(z_{0:k-1})P(x_k, z_{0:k-1})} = P(z_k|x_k, z_{0:k-1})P(x_k|z_{0:k-1}).$$

Using the conditional independence property of the measurement given the state, this simplifies to

$$P(x_k, z_k|z_{0:k-1}) = P(z_k|x_k, z_{0:k-1})P(x_k|z_{0:k-1}) = P(z_k|x_k)P(x_k|z_{0:k-1}).$$

The denominator of $P(x_k|y_{0:k})$ is obtained by adding over the possible states x_k

$$P(z_k|z_{0:k-1}) = \sum_{x_k \in [1,n]} P(x_k, z_k|z_{0:k-1}) = \sum_{x_k \in [1,n]} P(z_k|x_k)P(x_k|z_{0:k-1}).$$

This completes the proof of the corrector formula.

Example 14.4 Use the forward algorithm to obtain the probability $P(x_1 = 1|z_0 = 1, z_1 = 1)$ for the HMM of Fig. 14.1.

Solution

We initialize the algorithm with

$$P(x_0 = 1|z_0 = 1) = \frac{P(z_0 = 1|x_0 = 1)P(x_0 = 1)}{\sum_{x_0 \in [1,4]} P(z_0 = 1|x_0)P(x_0)}$$
$$= \frac{0.3 \times 0.25}{0.3 \times 0.25 + 1 \times 0.25 + 0.8 \times 0.25} = \frac{1}{7}$$

$$P(x_0 = 3|z_0 = 1) = \frac{P(z_0 = 1|x_0 = 3)P(x_0 = 3)}{\sum_{x_0 \in [1,4]} P(z_0 = 1|x_0)P(x_0)}$$
$$= \frac{1 \times 0.25}{0.3 \times 0.25 + 1 \times 0.25 + 0.8 \times 0.25} = \frac{10}{21}$$

$$P(x_0 = 4|z_0 = 1) = \frac{P(z_0 = 1|x_0 = 4)P(x_0 = 4)}{\sum_{x_0 \in [1,4]} P(z_0 = 1|x_0)P(x_0)}$$
$$= \frac{0.8 \times 0.25}{0.3 \times 0.25 + 1 \times 0.25 + 0.8 \times 0.25} = \frac{8}{21}.$$

Predictor:

$$P(x_1|z_0) = \sum_{x_0 \in [1,4]} P(x_1|x_0)P(x_0|z_0), \forall x_1 \in [1,4]$$

$$P(x_1 = 1|z_0 = 1) = P(x_1 = 1|x_0 = 1)P(x_0 = 1|z_0 = 1) + P(x_1 = 1|x_0 = 3)P(x_0 = 3|z_0 = 1)$$

$$= 0.1 \times \frac{1}{7} + 1 \times \frac{10}{21} = \frac{10.3}{21}$$

$$P(x_1 = 3|z_0 = 1) = P(x_1 = 3|x_0 = 1)P(x_0 = 1|z_0 = 1) + P(x_1 = 3|x_0 = 4)P(x_0 = 4|z_0 = 1)$$

$$= 0.9 \times \frac{1}{7} + 1 \times \frac{8}{21} = \frac{8.27}{21}.$$

Corrector:

$$P(x_1|z_{0:1}) = \frac{P(z_1|x_1)}{\sum_{x_1 \in [1,4]} P(z_1|x_1)P(x_1|z_0)} P(x_1|z_0), \forall x_1 \in [1,4]$$

$$P(x_1 = 1|z_0 = 1, z_1 = 1) = \frac{P(z_1 = 1|x_1 = 1)}{\sum_{x_1 \in [1,4]} P(z_1 = 1|x_1) P(x_1|z_0 = 1)} P(x_1 = 1|z_0 = 1).$$

The denominator is the sum

$$P(z_1 = 1|x_1 = 1)P(x_1 = 1|z_0 = 1) + P(z_1 = 1|x_1 = 3)P(x_1 = 3|z_0 = 1) \\ + (z_1 = 1|x_1 = 4)P(x_1 = 4|z_0 = 1)$$

$$= 0.3 \times \frac{10.3}{21} + 1 \times \frac{8.27}{21} = \frac{11.36}{21}$$

$$P(x_1 = 1|z_0 = 1, z_1 = 1) = \frac{21}{11.36} \times 0.3 = 0.5546.$$

■

14.4 Hidden Markov Modeling

To obtain a model of an HMM, we must determine the matrices (A, C) and the vector of initial conditions $\boldsymbol{p}_0^x$.

- Determine the most likely state sequence.
- Given HMM with parameters θ and measurements $z = z_{1:K}, k = 1, \ldots, K$

$$P(z_{1:K}|, \boldsymbol{x}, \theta) = \prod_{i=1}^{K} P(z_i|\boldsymbol{x}_i)$$

$$P(z_{1:K}, \boldsymbol{x}|\theta) = P(\boldsymbol{y}|, \boldsymbol{x}, \theta) P(\boldsymbol{x}|\theta)$$

$$P(\boldsymbol{x}|\theta) = P(\boldsymbol{x}_1) \prod_{i=2}^{K} P(\boldsymbol{x}_i|\boldsymbol{x}_{i-1})$$

$$P(z_{1:K}, \boldsymbol{x}|\theta) = \prod_{i=1}^{K} P(z_i|\boldsymbol{x}_i) \times P(\boldsymbol{x}_1) \prod_{i=2}^{K} P(\boldsymbol{x}_i|\boldsymbol{x}_{i-1})$$

$$P(\boldsymbol{x}|\theta) = P(\boldsymbol{x}_1) P(\boldsymbol{x}_2, \ldots \boldsymbol{x}_K|\boldsymbol{x}_1)$$

$$= P(\boldsymbol{x}_1) P(\boldsymbol{x}_2|\boldsymbol{x}_1) P(\boldsymbol{x}_3, \ldots \boldsymbol{x}_K|\boldsymbol{x}_1, \boldsymbol{x}_2)$$

$$= P(\boldsymbol{x}_1) \prod_{i=2}^{K} P(\boldsymbol{x}_i|\boldsymbol{x}_1, \ldots \boldsymbol{x}_{i-1}).$$

The Markov property gives

$$P(\boldsymbol{x}|\theta) = P(\boldsymbol{x}_1)\prod\nolimits_{i=2}^{K} P(\boldsymbol{x}_i|\boldsymbol{x}_{i-1}).$$

14.5 The Backward Algorithm

The backward algorithm calculates the probability of any state x_k at any time k in the interval $[0, K]$ given the measurement over the entire interval $z_{0:k}$ and the results of the forward algorithm $P(x_k|z_{0:k})$ for all $k \in [0, K]$. We begin with the ratio

$$P(x_k|z_{0:K})/P(x_k|z_{0:k}). \tag{14.28}$$

Clearly, the ratio is unity at $k = K$, and in general satisfies

$$\begin{aligned}\frac{P(x_k|z_{0:K})}{P(x_k|z_{0:k})} &= \frac{P(x_k, z_{0:K})}{P(z_{0:K})} \times \frac{P(z_{0:k})}{P(x_k, z_{0:k})} \\ &= \frac{P(z_{0:K}|x_k)P(x_k)}{P(z_{0:K})} \times \frac{P(z_{0:k})}{P(z_{0:k}|x_k)P(x_k)}.\end{aligned}$$

By the conditional independence of the observations given the state, we have

$$P(z_{0:K}|x_k) = P(z_{0:k}|x_k)P(z_{k+1:K}|x_k). \tag{14.29}$$

This simplifies the ratio to

$$\frac{P(x_k|z_{0:K})}{P(x_k|z_{0:k})} = P(z_{k+1:K}|x_k) \times \frac{P(z_{0:k})}{P(z_{0:k}, z_{k+1:K})} = \frac{P(z_{k+1:K}|x_k)}{P(z_{k+1:K}|y_{0:k})}$$

with $z_{0:K}$ broken into $(z_{0:k}, z_{k+1:K})$. This gives the conditional probability

$$P(x_k|z_{0:K}) = P(x_k|z_{0:k})\frac{P(z_{k+1:K}|x_k)}{P(z_{k+1:K}|z_{0:k})}. \tag{14.30}$$

Proceeding backwards, we replace k with $k - 1$ to obtain the formula

$$P(x_{k-1}|z_{0:K}) = P(x_{k-1}|z_{0:k-1})\frac{P(z_{k:K}|x_{k-1})}{P(z_{k:K}|y_{0:k-1})}$$

then rewrite it as

$$P(x_{k-1}|z_{0:K}) = P(x_{k-1}|z_{0:k-1})\sum\nolimits_{x_k=1}^{n} \frac{P(x_k, z_{k:K}|x_{k-1})}{P(z_{k:K}|y_{0:k-1})}. \tag{14.31}$$

To simplify the expression, we need to expand the numerator and denominator in the summation of (14.31) using the rule

$$P(AB|C) = \frac{P(ABC)}{P(C)} \times \frac{P(BC)}{P(BC)} = P(A|BC)P(B|C). \tag{14.32}$$

The numerator in the summation can be expanded as

$$P(x_k, z_{k:K}|x_{k-1}) = P(x_k, z_k, z_{k+1:K}|x_{k-1}) = P(z_{k+1:K}|x_{k-1}, x_k, z_k)P(x_k, z_k|x_{k-1})$$

$$= P(z_{k+1:K}|x_{k-1}, x_k, z_k)P(z_k|x_{k-1}, x_k)P(x_k|x_{k-1}).$$

Using the assumptions of conditional independence, we have

$$P(x_k, z_{k:K}|x_{k-1}) = P(z_{k+1:K}|x_k)P(z_k|x_k)P(x_k|x_{k-1})$$

$$= \left[P(x_k|z_{0:k})\frac{P(z_{k+1:K}|x_k)}{P(z_{k+1:K}|z_{0:k})}\right]\frac{P(z_{k+1:K}|z_{0:k})}{P(x_k|z_{0:k})}P(z_k|x_k)P(x_k|x_{k-1}).$$

Using (14.30) gives the expression

$$P(x_k, z_{k:K}|x_{k-1}) = P(x_k|z_{0:K})\frac{P(z_{k+1:K}|z_{0:k})}{P(x_k|z_{0:k})}P(z_k|x_k)P(x_k|x_{k-1}). \tag{14.33}$$

The denominator in the summation of (14.31) is expanded using (14.32) as

$$P(z_{k:K}|z_{0:k-1}) = P(z_k, z_{k+1:K}|z_{0:k-1}) = P(z_{k+1:K}|z_{0:k-1}, z_k)P(z_k|z_{0:k-1})$$

$$= P(z_{k+1:K}|z_{0:k})P(z_k|z_{0:k-1}).$$

Substituting for the numerator and denominator of the summation of (14.31) gives

$$P(x_{k-1}|z_{0:K}) = P(x_{k-1}|z_{0:k-1})\sum_{x_k=1}^{n} P(x_k|z_{0:K})\frac{P(z_{k+1:K}|z_{0:k})}{P(x_k|z_{0:k})}$$

$$\times \frac{P(z_k|x_k)}{P(z_{k+1:K}|z_{0:k})} \times \frac{P(x_k|x_{k-1})}{P(z_k|z_{0:k-1})}.$$

This yields the recursion

$$P(x_{k-1}|z_{0:K}) = P(x_{k-1}|z_{0:k-1})\sum_{x_k=1}^{n} P(x_k|z_{0:K}) \times \frac{P(z_k|x_k)}{P(x_k|z_{0:k})} \times \frac{P(x_k|x_{k-1})}{P(z_k|z_{0:k-1})}. \tag{14.34}$$

We summarize the backward algorithm as follows:

Backward Algorithm

1. Initialize with the results of the forward algorithm with

$$P(z_k|z_{0:k-1}), k = 1, 2, \ldots, K$$

$$P(x_k|z_{0:k}) = 0, 1, 2, \ldots, K.$$

2. Iterate backwards using

$$P(x_{k-1}|z_{0:K}) = P(x_{k-1}|z_{0:k-1}) \sum_{x_k=1}^{n} P(x_k|z_{0:K}) \times \frac{P(z_k|x_k)}{P(x_k|z_{0:k})} \times \frac{P(x_k|x_{k-1})}{P(z_k|z_{0:k-1})}$$

$k = K, K-1, \ldots, 1$.

14.6 The Baum–Welch Algorithm: Application of EM to HMM

The Baum–Welch algorithm uses the forward and backward algorithms together with the EM algorithm to estimate the parameters of an HMM. It begins with an initial estimate of the parameters $\boldsymbol{\theta}^{(0)}$ and recursively improves the estimate. The algorithm terminates when the norm of the change in the estimate drops below a tolerance level $\epsilon \ll 1$.

We assume that the number of possible discrete states is N and that data is collected over a duration K. We define the following probabilities:

$$\alpha_k(j) = P(z_{0:k}, x_k = j|\theta) \tag{14.35}$$

$$\beta_k(j) = P(z_{k+1:K}|x_k = j, \boldsymbol{\theta}) \tag{14.36}$$

$$\gamma_k(j) = P(x_k = j|z_{0:K}, \boldsymbol{\theta}) \tag{14.37}$$

$$\xi_k(i, j) = P(x_k = i, x_{k+1} = j|z_{0:K}, \boldsymbol{\theta}). \tag{14.38}$$

The following facts are used to derive the algorithm.

Facts

(a) Measurement

$$\gamma_k(j) = \frac{\alpha_k(j)\beta_k(j)}{\sum_{j=1}^{N}\alpha_k(j)\beta_k(j)}.$$

(b) Conditional pdf

$$\xi_k(i, j) = \frac{\alpha_k(j) a_{ij} c_j(z_{k+1}) \beta_{k+1}(j)}{\sum_{j=1}^{N} \alpha_k(j)\beta_k(j)}, \forall k, i, \text{ and } j.$$

(c) Transition probability

$$\hat{a}_{ij} = \frac{\sum_{t=1}^{K} \xi_k(i, j)}{\sum_{i=1}^{K} \sum_{k=1}^{N} \xi_k(i, k)}.$$

(d) Measurement probability

$$\hat{c}_j(v_k) = \frac{\sum_{k=1 s..t.z_k = v_k}^{K} \gamma_k(j)}{\sum_{k=1}^{K} \gamma_k(j)}.$$

Proof

(a) Using the definition of conditional density

$$\gamma_k(j) = P(x_k = j | z_{0:K}, \boldsymbol{\theta}) = \frac{P(x_k = j, z_{0:K} | \boldsymbol{\theta})}{P(z_{0:K} | \boldsymbol{\theta})}.$$

The numerator can be written as

$$\begin{aligned} P(x_k = j, z_{0:K} | \boldsymbol{\theta}) &= P(x_k = j, z_{0:k}, z_{k+1:K} | \boldsymbol{\theta}) \\ &= P(x_k = j, z_{0:k} | \boldsymbol{\theta}) P(z_{k+1:K} | x_k = j, z_{0:k}, \boldsymbol{\theta}). \end{aligned}$$

Because the process is Markov, we can write $\gamma_k(j)$ as

$$\gamma_k(j) = \frac{P(z_{0:k}, x_k = j | \theta) P(z_{k+1:K} | x_k = j, z_{0:k}, \boldsymbol{\theta})}{P(z_{0:K} | \boldsymbol{\theta})} = \frac{\alpha_k(j)\beta_k(j)}{P(z_{0:K} | \boldsymbol{\theta})}.$$

The result is obtained by substituting the marginal expression for the denominator

$$P(z_{0:K} | \boldsymbol{\theta}) = \sum_{j=1}^{N} P(x_k = j, z_{0:K} | \boldsymbol{\theta}) = \sum_{j=1}^{N} \alpha_k(j)\beta_k(j).$$

(b) Using the definition of conditional density

$$\xi_k(i, j) = P(x_k = i, x_{k+1} = j | z_{0:K}, \boldsymbol{\theta}) = \frac{P(x_k = i, x_{k+1} = j, z_{0:K} | \boldsymbol{\theta})}{P(z_{0:K} | \boldsymbol{\theta})}.$$

In part (a), we show that the denominator satisfies

$$P(z_{0:K}|\boldsymbol{\theta}) = \sum\nolimits_{j=1}^{N} \alpha_k(j)\beta_k(j).$$

The numerator can be written as

$$P(x_k = i, x_{k+1} = j, z_{0:k+1}, z_{k+2:K}|\boldsymbol{\theta})$$

$$= P(z_{k+2:K}|x_k = i, x_{k+1} = j, z_{0:k+1}, \boldsymbol{\theta})P(x_k = i, x_{k+1} = j, z_{k+1}, z_{0:k}|\boldsymbol{\theta})$$

$$= P(z_{k+2:K}|x_k = i, x_{k+1} = j, z_{0:k+1}, \boldsymbol{\theta})P(z_{k+1}|x_k = i, x_{k+1} = j, z_{0:k}, \boldsymbol{\theta})$$
$$P(x_k = i, x_{k+1} = j, z_{0:k}|\boldsymbol{\theta}).$$

For a Markov process, we have

$$P(z_{k+2:K}|x_k = i, x_{k+1} = j, z_{0:k+1}, \boldsymbol{\theta}) = P(z_{k+2:K}|x_{k+1} = j, \boldsymbol{\theta}) = \beta_{k+1}(j)$$

and measurement independence gives

$$P(z_{k+1}x_k = i, x_{k+1} = j, z_{0:k}, \boldsymbol{\theta}) = P(z_{k+1}x_{k+1} = j, \boldsymbol{\theta}) = b_j(z_{k+1}).$$

Substituting in the joint probability, we obtain

$$\begin{aligned} P(x_k = i, x_{k+1} = j, z_{0:k+1}, z_{k+2:K}|\boldsymbol{\theta}) &= \beta_{k+1}(j)b_j(z_{k+1})P(x_{k+1} = j|x_k = i, z_{0:k}, \boldsymbol{\theta})P(x_k = i, z_{0:k}|\boldsymbol{\theta}) \\ &= \beta_{k+1}(j)b_j(z_{k+1})P(x_{k+1} = j|x_k = i, z_{0:k}, \boldsymbol{\theta})\alpha_k(i). \end{aligned}$$

The Markov property implies that

$$P(x_{k+1} = j|x_k = i, z_{0:k}, \boldsymbol{\theta}) = P(x_{k+1} = j|x_k = i, \boldsymbol{\theta}) = a_{ij.}$$

We now have

$$P(x_k = i, x_{k+1} = j, z_{0:K}|\boldsymbol{\theta}) = \alpha_k(j)a_{ij}b_j(z_{k+1})\beta_{k+1}(j)$$

and the conditional density

$$\xi_k(i, j) = \frac{\alpha_k(j)a_{ij}b_j(z_{k+1})\beta_{k+1}(j)}{\sum_{j=1}^{N}\alpha_k(j)\beta_k(j)}.$$

(c) The transition probability is estimated using

$$\hat{a}_{ij} = \frac{\text{expected number of transitions from state } i \text{ to state } j}{\text{expected number of transitions from state } i}$$

The numerator is the sum for all measurements

$$\sum_{k=0}^{K-1} P(x_k = i, x_{k+1} = j | z_{0:K}, \boldsymbol{\theta}).$$

The following identity is easy to prove

$$P(A|B, C) = P(A, B|C)P(B|C)$$

$$\hat{a}_{ij} = \frac{\sum_{k=1}^{K} \xi_k(i, j)}{\sum_{k=1}^{K} \sum_{j=1}^{N} \xi_k(i, j)}.$$

(d) The measurement probability is estimated using

$$\hat{c}_i(v_k) = \frac{\text{expected number of measurements } v_k \text{ while in state } i}{\text{expected number of times in state } i}.$$

The numerator is estimated as expected number of time where in the sequence of measurements $z_{0:K}$ has the kth measurement $z_k = v_k$ while the system is in state i. The denominator is the expected number of times in state i for the same sequence of measurements. Hence, we write

$$\hat{c}_i(v_k) = \frac{\sum_{k=1, z_k = v_k}^{K} P(x_k = i, z_{0:K} | \boldsymbol{\theta})}{\sum_{k=1}^{K} P(x_k = i, z_{0:K} | \boldsymbol{\theta})}.$$

Dividing the numerator and denominator by $P(z_{0:K})$ gives

$$\begin{aligned}
\hat{c}_i(v_k) &= \frac{\sum_{k=1, z_k = v_k}^{K} P(x_k = i, z_{0:K} | \boldsymbol{\theta}) / P(z_{0:K})}{\sum_{k=1}^{K} P(x_k = i, z_{0:K} | \boldsymbol{\theta}) / P(z_{0:K})} \\
&= \frac{\sum_{k=1, z_k = v_k}^{K} P(x_k = i, z_{0:K} | \boldsymbol{\theta}) / P(x_k = i)}{\sum_{k=1}^{K} P(x_k = i, z_{0:K} | \boldsymbol{\theta}) / P(x_k = i)} \\
&= \frac{\sum_{k=1, z_k = v_k}^{K} P(x_k = i | z_{0:K}, \boldsymbol{\theta})}{\sum_{k=1}^{K} P(x_k = i | z_{0:K}, \boldsymbol{\theta})}.
\end{aligned}$$

The result follows from the definition of $\gamma_k(i)$.

Based on the above facts, we have the following algorithm:

Initialize with $it = 0, \boldsymbol{\theta}^{(it)} = col\{A^{(it)} = [\hat{a}_{ij}], \boldsymbol{c}^{(it)}\}$

1. E-step

$$\gamma_k(j) = \frac{\alpha_k(j)\beta_k(j)}{\sum_{j=1}^{N}\alpha_k(j)\beta_k(j)}, \ \forall k \text{ and } j$$

$$\xi_k(i,j) = \frac{\alpha_k(j)a_{ij}b_j(z_{k+1})\beta_{k+1}(j)}{\sum_{j=1}^{N}\alpha_k(j)\beta_k(j)}, \ \forall k, i, \text{ and } j.$$

2. M-step

$$\hat{a}_{ij}^{(it+1)} = \frac{\sum_{t=1}^{K}\xi_k(i,j)}{\sum_{i=1}^{K}\sum_{k=1}^{N}\xi_k(i,k)}$$

$$\hat{c}_j(v_k) = \frac{\sum_{k=1 s..t.z_k=v_k}^{K}\gamma_k(j)}{\sum_{k=1}^{K}\gamma_k(j)}.$$

3. If $\|\boldsymbol{\theta}^{(it)} - \boldsymbol{\theta}^{(it-1)}\| > \epsilon, it \to it+1$, go to Step 1, else, stop and return

Example 14.5 Given the HMM with matrices

$$A = \begin{bmatrix} 0.95 & 0.1 \\ 0.05 & 0.9 \end{bmatrix}, C = \begin{bmatrix} \frac{1}{6} & \frac{1}{6} & \frac{1}{6} & \frac{1}{6} & \frac{1}{6} & \frac{1}{6} \\ \frac{1}{10} & \frac{1}{10} & \frac{1}{10} & \frac{1}{10} & \frac{1}{10} & \frac{1}{2} \end{bmatrix}^T$$

Use MATLAB to generate data from the HMME, then estimate the parameters of the HMM using the Baum–Welch algorithm.

Solution

MATLAB uses a different notation where the matrices are the transpose of the matrices defined here. The matrix A has entries that are the probability of transition from state i to state j, which is the reverse of how we define the entries, and the row sum is unity. The results are obtained using following MATLAB script.[2]

[2] The example is from the MATLAB manual.

```
% Example BaumWelch: Use hmmtrain(seqs,trans,emis, 'BaumWelch');% default
% or hmmtrain(seqs,trans,emis, 'Viterbi')
trans = [0.95,0.05; % Two state Markov chain: p_ij prob. from i to j
      0.10,0.90]; % A transpose of the notes definition
emis = [1/6, 1/6, 1/6, 1/6, 1/6, 1/6; % Six different observations, state 1
   1/10, 1/10, 1/10, 1/10, 1/10, 1/2];% Six different observations, state 2
% Estimates the transition and emission probabilities for a hidden Markov
% model using the Baum-Welch algorithm
% Take a known Markov model, specified by transition probability matrix
% TRANS and emission probability matrix EMIS, and use it to generate
seq1 = hmmgenerate(200,trans,emis); % First generated sequence
seq2 = hmmgenerate(200,trans,emis);% second generated sequence
seqs = {seq1,seq2}; % Cell array contining 2 sequences seq1 and seq2
% estTR= estimate of transition probabilities P(x(k+1)|x(k) )
% estE=estimate of emission probabilities, P(y|x)
[estTR,estE] = hmmtrain(seqs,trans,emis);
% Given a sequence of states seqs, estimate the transition probs. and the
% emission probabilites.
% seq can be a row vector containing one sequence, a matrix with one row
% per sequence, or a cell array with each cell containing a sequence.
```

The estimates obtained are

$$A = \begin{bmatrix} 0.9425 & 0.0866 \\ 0.0575 & 0.9134 \end{bmatrix}, C = \begin{bmatrix} 0.1682 & 0.1701 & 0.1671 & 0.1603 & 0.1824 & 0.1520 \\ 0.0935 & 0.1211 & 0.1209 & 0.1206 & 0.0914 & 0.4523 \end{bmatrix}^T$$

14.7 Minimum Path Problem

Logs often simplify problem.

- Branch cost is a log-likelihood.

$$P(z, \boldsymbol{x}|\mathrm{M}) = P(z|\boldsymbol{x}, \mathrm{M})P(\boldsymbol{x}|\mathrm{M})$$

$$\boldsymbol{x}^* = \underset{\boldsymbol{x}}{\mathrm{argmax}} P(z, \boldsymbol{x}|\mathrm{M})$$

$$= \underset{\boldsymbol{x}}{\mathrm{argmin}}\{-\ln[P(z, \boldsymbol{x}|\mathrm{M})]\} \tag{14.39}$$

Shortest Path or Maximum Utility
Dynamic programming is an optimization approach that breaks optimization problems into a series of smaller optimization problems, then obtains the overall solution by progressively solving the problems with backward iteration. Several optimization problems can be recast as finding the shortest path through a graph, which can be accomplished using the backward iteration of dynamic programming. If the

optimal solution is obtained by progressing forward, the approach is known as reverse dynamic programming. The Viterbi algorithm is a reverse dynamic programming approach that can be used to find that most likely sequence of states for an HMM using a sequence of observations. To formulate the problem as one of finding the shortest path, the path between nodes of the graph are weighted with the negative of the transition probability. The negative sign is to change the maximization problem to minimization with the transition probability as a distance measure. Alternatively, one could choose the path that maximizes utility directly without the need for the negative sign, with the higher utility associated with a higher probability.

To obtain the optimum solution, we first create a trellis graph. A trellis graph is a directed graph with a starting node, and with nodes forming a vertical slice at each time point. The connections between nodes at each stage are weighted with the transition probabilities between the nodes. Progressing in time, we calculate the maximum utility of reaching each node and save the associated path. The maximum utility from the initial node to node x_k is denoted by

$$\text{Utility}(\boldsymbol{x}_{0:k}) = \max_{x_{0:N}} P(x_{0:N}). \tag{14.40}$$

On reaching the terminal node, we have the optimum path for reaching the node. The following is a summary of the steps of the Viterbi algorithm.

Viterbi Algorithm

- Initialization: $k = 0, \boldsymbol{x}_{0:k} = \boldsymbol{x}_0, \text{Utility}(\boldsymbol{x}_{0:k}) = 1$
- Recursion: Compute the cost and find the survivor and its length then increment time.

$$\text{Utility}(\boldsymbol{x}_{0:k+1}) = \min_{x_k} \{\text{Utility}(\boldsymbol{x}_{0:k}) + \text{Utility}(\boldsymbol{x}_{k:k+1})\}$$

$$\boldsymbol{x}_{0:k+1} = \arg\min_{x_k} \{\text{Utility}(\boldsymbol{x}_{0:k}) + \text{Utility}(\boldsymbol{x}_{k:k+1})\}$$

$k \leftarrow k+1$

- Store: time index (k), survivor terminating in state $x_k(x_{0:k})$, survivor $\text{Utility}(\boldsymbol{x}_{0:k})$, for all states $\boldsymbol{x}_k$, $k = 1, \ldots, M$.

Example 14.6 Over 4 time steps, determine the most likely sequence of states for the Markov chain with transition probability matrix starting and ending in the state 1.

$$A = \begin{bmatrix} 0.1 & 0.7 & 0.6 & 0 \\ 0 & 0 & 0 & 0.4 \\ 0.9 & 0 & 0 & 0.6 \\ 0 & 0.3 & 0.4 & 0 \end{bmatrix}$$

starting and ending with $x_0 = 1$.

Solution

Initialize with $k = 0, x_0 = 1, \text{Utility}(x_{0:k}) = 1$

Step 1:

$$\text{Utility}(x_{0:k+1}) = \max_{x_{k+1}}\{\text{Utility}(\boldsymbol{x}_{0:k}) + \text{Utility}(\boldsymbol{x}_{k:k+1})\}$$

$$\text{Utility}(\{1, 3\}) = \max_{x_1}\{P(x_0 = 1) + P(x_1 = 3|x_0 = 1)\}$$

$$= 1 + 0.9 = 1.9$$

$$\text{Utility}(\{1, 1\}) = \max_{x_1}\{P(x_0 = 1) + P(x_1 = 1|x_0 = 1)\}$$

$$= 1 + 0.1 = 1.1$$

$$k = 1, (x_{0:k}, \text{Utility}(\boldsymbol{x}_{0:k})) = [\{1, 3\}, 1.9], [\{1, 1\}, 1.1]$$

Step 2:

$$\text{Utility}(\boldsymbol{x}_{0:k+1}) = \max_{x_{k+1}}\{\text{Utility}(\boldsymbol{x}_{0:k}) + \text{Utility}(\boldsymbol{x}_{k:k+1})\}$$

$$\max_{x_k}\text{Utility}(\{1, x_k, 1\})$$

$$= \max_{x_{k+1}=1}\{P(x_{0:1} = \{1, 3\}) + P(x_2 = 1|x_1 = 3), P(x_{0:1} = \{1, 1\}) + P(x_2 = 1|x_1 = 1)\}$$

$$= \max(1.9 + 0.6, 1.1 + 0.1\} = \text{Utility}(\{1, 3, 1\}) = 2.5$$

$$\text{Utility}(\{1, 1, 3\}) = \max_{x_1}\{P(x_{0:1} = \{1, 1\}) + P(x_2 = 3|x_1 = 1)\} = 1.1 + 0.6 = 1.7$$

$$\text{Utility}(\{1, 3, 4\}) = \max_{x_1}\{P(x_{0:1} = \{1, 3\}) + P(x_2 = 4|x_1 = 3)\} = 1.9 + 0.4 = 2.3$$

$$k = 2, (x_{0:k}, \text{Utility}(\boldsymbol{x}_{0:k})) = [\{1, 3, 1\}, 2.5], [\{1, 1, 3\}, 1.7], [\{1, 3, 4\}, 2.3]$$

Steps 3 and 4 are left as an exercise. They show that the optimum path is

$$k = 4, (x_{0:k}, \text{Utility}(\boldsymbol{x}_{0:k})) = \{1, 3, 1, 3, 1\}, 4.0$$

The trellis diagram is shown in Fig. 14.4. Note that computing the utilities using the Viterbi algorithm eliminates calculation for suboptimal paths. For example, the

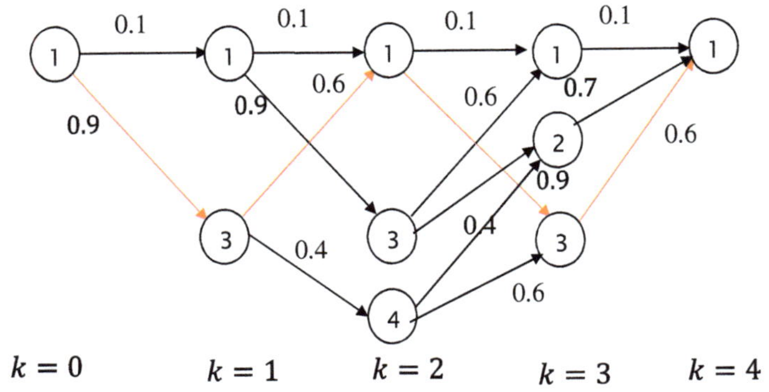

Fig. 14.4 Trellis diagram for the Markov chain

path 1-1-1-1-1 is not considered because getting to state 1 at $k = 2$ is optimal with the path 1-3-1.

■

Because the states are typically only available indirectly through a random measurement process, we need to modify the algorithm to include the measurement statistics. The optimum estimate of the state sequence over an interval $[0, N]$ given a measurement sequence $z_{0:N}$ is

$$\hat{x}_{0:N} = \underset{x_{0:N}}{\operatorname{argmax}} P(x_{0:N}|z_{0:N}). \tag{14.41}$$

Using the definition of conditional probability, we rewrite the estimate as

$$\hat{x}_{0:N} = \underset{x_{0:N}}{\operatorname{argmax}} P(x_{0:N}, z_{0:N})/P(z_{0:N}).$$

However, the estimate is independent of the denominator and term and the estimate is

$$\hat{x}_{0:N} = \underset{x_{0:N}}{\operatorname{argmax}} P(x_{0:N}, z_{0:N}). \tag{14.42}$$

To use the algorithm with an HMM, all that is needed is to redefine the utility as

$$\text{Utility}(\boldsymbol{x}_{0:k}) = \max_{x_{0:k}} P(x_{0:k}, z_{0:k}). \tag{14.43}$$

14.8 MATLAB Commands

MATLAB provides commands to generate data from an HMM, and to estimate the most likely sequence of states. The state-transition matrix and the observation matrix used in MATLAB are the transpose of the ones used here. The following example applies the MATLAB commands to the Markov chain of Example 14.6.

Example 14.7 Generate data using the stochastic matrix of Example 14.6 and the measurement matrix

$$C = \begin{bmatrix} 0.3 & 0.7 & 0.2 & 0.8 \\ 0.7 & 0.3 & 0.8 & 0.2 \end{bmatrix}.$$

Use the Viterbi algorithm command to obtain the most likely sequence of states for the HMM. Generate a diagram for the Markov chain using MATLAB.

Solution

The following MATLAB script gives the solution.

```
% ExampleViterbi
% Hidden Markov model most probable state path
 % hmmviterbi begins with the model in state 1 at step 0.
 % It computes the most likely path beginning in state 1.
 % A is n by n for n states
A=[0.1 0.7 0.6 0;0 0 0 0.4;0.9 0. 0 0.6;0 0.3 0.4 0]; % Transition prob. matrix
C=[0.3 0.7;0.7 0.3;0.2 0.8;0.8 0.2];% Matrix that defines the probability of
% an observation for a given state. C is n by m for n states and m
% observations (transpose of matrix in my notes)
[seq,states] = hmmgenerate(100,A',C); % a_ij is the prob. of transition
% from state i to state j (transpose of my notes)
estimatedStates = hmmviterbi(seq,A',C);
% Alternative with state names (A,B,C,D) instead of numbers
[seq1,states1]=hmmgenerate(10,A',C,'Statenames',{'1';'2';'3';'4'});
estimatesStates1=hmmviterbi(seq1,A',C,'Statenames',{'1';'2';'3';'4'});
```

3 4 3 1 3

■

Problems

14.1 Write the state-space equation for the HMM of Fig. 14.5. If $P(x_0 = 2) = \frac{2}{3}$:

(a) Calculate the probability of each state at time.
(b) Calculate the probability of each output at time 1.

14.2 Show that the stochastic matrix P has a left eigenvector with all unity entries corresponding to a unity eigenvalue where a left eigenvector satisfies
$\mathbf{w}^T P = \lambda \mathbf{w}^T$

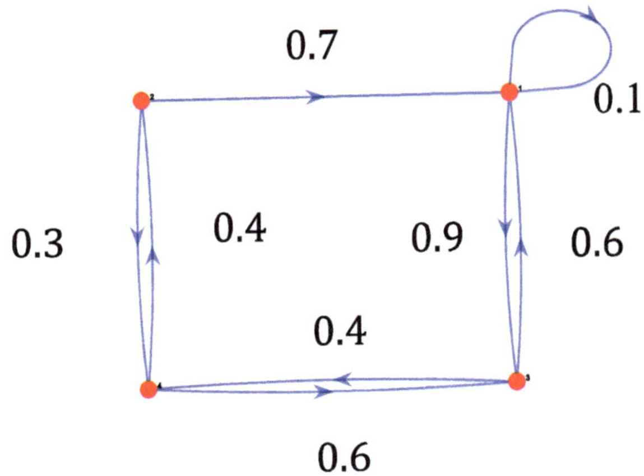

Fig. 14.5 Markov chain diagram for Example 14.7

14.3 Use the forward algorithm to obtain the probability sequence for one time step using the forward algorithm for the HMM of Fig. P14.1 starting at $x_0 = 1$.

14.4 Starting at any state of a Markov chain with N states, show that the maximum-likelihood estimate of the transition probabilities is given by the relative frequency

$$p_{ij} = \frac{n_{ij}}{n_t}, i, j = 1, 2, \ldots, N$$

where n_{ij} = number of transitions from state j to state i is denoted
n_t = total number of transitions.
Hint: The transition is governed by a multinomial distribution.

14.5 Show that conditional independence as defined in (14.15), that is $P(A|BC) = P(A|C)$, is equivalent to $P(AB|C) = P(A|C)P(B|C)$.

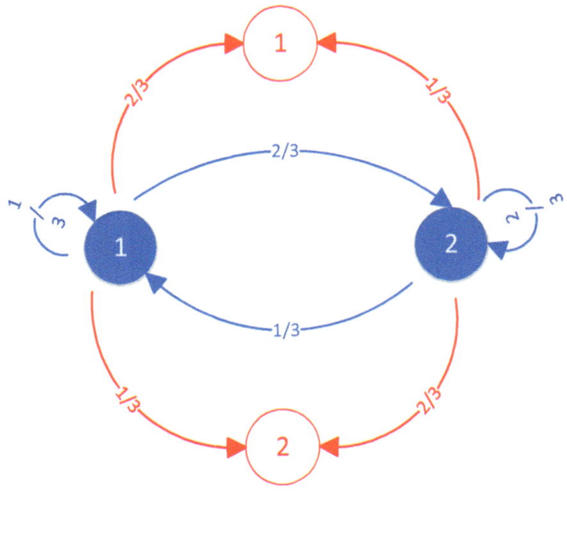

Fig. P14.1 Hidden Markov model

14.6 Show that the forward algorithm yields the equations of the Bayesian filter of Chap. 12 for the case of a continuous state space.

14.7 A two state DNA sequence that transitions randomly between the two unobservable states randomly generates an observable sequence of bases Adenine (A), Cytosine (C), Guanine (G), Thymine (T). One state generates an AT rich sequence and the other generates a CT rich sequence. The transition probability matrix and the emission matrix for the system are given by[3]

$$A = \begin{bmatrix} 0.99 & 0.1 \\ 0.01 & 0.9 \end{bmatrix}, C = \begin{bmatrix} 0.4 & 0.1 & 0.1 & 0.4 \\ 0.05 & 0.4 & 0.5 & 0.05 \end{bmatrix}^T$$

Use MATLAB to generate data from the HMME, then estimate the parameters of the HMM using the Baum–Welch algorithm.

14.8 Complete the calculations of Example 14.6 and verify that the most likely path is $(x_{0:k}, \text{Utility}(\boldsymbol{x}_{0:k})) = \{1, 3, 1, 3, 1\}, 4.0$.

Bibliography

1. DeGroot, M. H., & Schervish, M. J. (2002). *Probability and statistics*. Addison-Wesley.
2. Fraser, A. M. (2008). *Hidden Markov models & dynamical systems*. SIAM.
3. Moon, T. K., & Stirling, W. C. (2000). *Mathematical methods and algorithms for signal processing*. Prentice-Hall.
4. Papoulis, A., & Pillai, S. U. (2002). *Probability, random variables, and stochastic processes*. McGraw-Hill.

[3] S.R. Eddy, "Hidden Markov Models," *Current Opinion in Structural Biology*, Vol 6, pp361-365, 1996.

Appendix A
Table of Integrals

Using Cauchy's residue theorem from complex analysis, Philips[1] evaluated the following integral:

$$I_n = \frac{1}{2\pi j} \int_{-j\infty}^{j\infty} \frac{c(s)}{b(s)b(-s)} \mathrm{d}s,$$

$$c(s) = c_{n-1}s^{2(n-1)} + c_{n-2}s^{2(n-2)} + \cdots + c_1 s^2 + c_0,$$

$$b(s) = b_n s^n + b_{n-1}s^{n-1} + \cdots + b_1 s + b_0.$$

In an Appendix of the same reference, G. R. McClane used the general expression to evaluate integrals up to $n = 7$, for which the highest numerator power is $2(7-1) = 12$. Integrating the power spectral density is a special case of the above integral where

$$c(s) = a(s)a(-s)$$

with $a(s) = a_{n-1}s^{n-1} + a_{n-2}s^{n-2} + \cdots + a_1 s + a_0$.

Expanding the product $a(s)a(-s)$ then equating coefficients allows us to evaluate the integral

$$I_n = \frac{1}{2\pi j} \int_{-j\infty}^{j\infty} \frac{a(s)a(-s)}{b(s)b(-s)} \mathrm{d}s,$$

$$a(s) = a_{n-1}s^{n-1} + a_{n-2}s^{n-2} + \cdots + a_1 s + a_0,$$

---

[1] R. S. Phillips, "RMS-error Criterion in Servomechanism Design," In: Theory of Servomechanisms, H. M. James, N. B. Nicholls, R. S. Phillips, Eds., McGraw-Hill, NY, 1947.

M. S. Fadali, *Introduction to Random Signals, Estimation Theory, and Kalman Filtering*, https://doi.org/10.1007/978-981-99-8063-5

Table A.1 Table of integrals

n	Integral
1	$\dfrac{a_0^2}{2b_0b_1}$
2	$\dfrac{a_0^2b_2+a_1^2b_0}{2b_0b_1b_2}$
3	$\dfrac{a_0^2b_2b_3+b_0b_3(a_1^2-2a_0a_2)+a_2^2b_0b_1}{2b_0b_3(b_1b_2-b_0b_3)}$
4	$\dfrac{a_0^2(b_2b_3b_4-b_1b_4^2)+b_0b_3b_4(a_1^2-2a_0a_2)+b_0b_1b_4(a_2^2-2a_1a_3)+a_3^2(b_0b_1b_2-b_0^2b_3)}{2b_0b_4(b_1b_2b_3-b_0b_3^2-b_1^2b_4)}$

$$b(s)=b_ns^n+b_{n-1}s^{n-1}+\cdots+b_1s+b_0.$$

McLane's results allow us to obtain the desired integral for $n=1,2,\ldots,7$, but the expressions beyond $n=4$ are huge. The results for $n=1,2,3,4$, are given in Table A.1.

Appendix B
Table of Fourier Transforms

f_k	$F(z)$
1. δ_k	1, all z
2. $1, k \geqslant 0$	$(1 - z^{-1})$, $1 < \lvert z \rvert$
3. $k, k \geqslant 0$	$z^{-1}(1 - z^{-1})^{-2}$, $1 < \lvert z \rvert$
4. $k^n, k \geqslant 0$	$\left(-z\frac{d}{dz}\right)^n (1 - z^{-1})^{-1}, \quad 1 < \lvert z \rvert$
5. $\begin{pmatrix} k \\ n \end{pmatrix}, \quad n \leqslant k$	$z^{-n}(1 - z^{-1})^{n+1}, \quad 0 < \lvert z \rvert$
6. $\begin{pmatrix} n \\ k \end{pmatrix}, \quad 0 \leqslant k \leqslant n$	$(1 + z^{-1})^n, \quad 0 < \lvert z \rvert$
7. $\alpha^k, \quad k \geqslant 0$	$(1 - \alpha z^{-1})^{-1}, \quad \lvert\alpha\rvert < \lvert z \rvert$
8. $k^n \alpha^k, \quad k \geqslant 0$	$\left(-z\frac{d}{dz}\right) n (1 - \alpha z^{-1})^{-1}, \quad \lvert\alpha\rvert < \lvert z \rvert$
9. $\alpha^k, \quad k < 0$	$-(1 - \alpha z^{-1})^{-1}, \quad \lvert z \rvert < \lvert\alpha\rvert$
10. $k^n \alpha^k, \quad k < 0$	$-\left(-z\frac{d}{dz}\right)^n (1 - \alpha z^{-1})^{-1}, \quad \lvert z \rvert < \lvert\alpha\rvert$
11. $\alpha^{\lvert k \rvert}, \quad \text{all } k$	$(1 - \alpha^2)\left[(1 - \alpha z)(1 - \alpha z^{-1})\right]^{-1}, \lvert\alpha\rvert < \lvert z \rvert < \left\lvert\frac{1}{\alpha}\right\rvert$
12. $\frac{1}{k'} \quad k > 0$	$-\ln(1 - z^{-1}), \quad 1 < \lvert z \rvert$
13. $\cos \alpha k, \quad k \geqslant 0$	$(1 - z^{-1} \cos \alpha)(1 - 2z^{-1} \cos \alpha + z^{-2})^{-1}, 1 < \lvert z \rvert$
14. $\sin \alpha k, \quad k \geqslant 0$	$z^{-1} \sin \alpha (1 - 2z^{-1} \cos \alpha + z^{-2})^{-1}, \quad 1 < \lvert z \rvert$
15. $\alpha \cos \alpha k + b \sin \alpha k, \quad k \geqslant 0$	$\left[a + z^{-1}(b \sin \alpha - a \cos \alpha)\right]$ $\left(1 - 2z^{-1} \cos a + z^{-2}\right)^{-1}, \quad 1 < \lvert z \rvert$

(continued)

M. S. Fadali, *Introduction to Random Signals, Estimation Theory, and Kalman Filtering*, https://doi.org/10.1007/978-981-99-8063-5

(continued)

f_k	$F(z)$
16. $\alpha\cos\alpha k + \left(\dfrac{d + c\cos\alpha}{\sin\alpha}\right)\sin\alpha k$ $k \geqslant 0$	$(c + dz^{-1})(1 - 2z^{-1}\cos\alpha + z^{-2})^{-1}, \quad 1 < \lvert z\rvert$
17. $\cosh\alpha k, \quad k \geqslant 0$	$\left(1 - z^{-1}\cosh\alpha\right)\left(1 - 2z^{-1}\cosh\alpha + z^{-2}\right)^{-1},$ $\max\left\{\lvert\alpha\rvert, \left\lvert\frac{1}{\alpha}\right\rvert\right\} < \lvert z\rvert$
18. $\sinh\alpha k, \quad k \geqslant 0$	$\left(z^{-1}\sinh\alpha\right)\left(1 - 2z^{-1}\cosh\alpha + z^{-2}\right)^{-1}$ $\max\left\{\lvert\alpha\rvert, \left\lvert\frac{1}{\alpha}\right\rvert\right\} < \lvert z\rvert$

Source R. A. Gabel and R. A. Roberts, *Signals and Linear Systems*, Third Edition John Wiley \& Sons, New York, 1980, p. 186

Appendix C
Table of Two-Sided Laplace and Z-Transforms[2]

Laplace Transform: $F(s) = \int_{-\infty}^{\infty} f(6)e^{-st}\mathrm{d}t$ **z-Transform**: $F(z) = \sum_{-\infty}^{\infty} f_k z^{-k}$.

[2] *The discrete time functions are similar to but not always a sampled form of the continuous-time functions.

M. S. Fadali, *Introduction to Random Signals, Estimation Theory, and Kalman Filtering*, https://doi.org/10.1007/978-981-99-8063-5

No.	Continuous time	Laplace transform	Discrete time	Z-transform
1	$\delta(t)$	1	$\delta(k)$	1
2	$1(t)$	$\frac{1}{s}$, $Re\{s\} > 0$	$1(k)$	$\frac{z}{z-1}$, $\|z\| > 1$
3	$t1(t)$	$\frac{1}{s^2}$, $Re\{s\} > 0$	$k1(k)^3$	$\frac{z}{(z-1)^2}$, $\|z\| > 1$
4	$\frac{t^n}{n!}1(t)$	$\frac{1}{s^{n+1}}$, $Re\{s\} > 0$	$k^n\, 1(k)$	$\left(-z\frac{d}{dz}\right)^n\left(\frac{z}{z-1}\right)$, $\|z\| > 1$
5	$e^{\alpha t}1(t)$	$\frac{1}{s-\alpha}$, $Re\{s\} > \alpha$	$a^k 1(k)^4$	$\frac{z}{z-a}$, $\|z\| > \|a\|$
6	$e^{-\alpha\|t\|}$	$\frac{2\alpha}{-s^2+\alpha}$, $-\alpha < Re\{s\} < \alpha$	$\alpha^{\|k\|}$	$\frac{(1-\alpha^2)z}{(1-\alpha z)((z-\alpha)}$, $\frac{1}{\|a\|} > \|z\| > \|a\|$
7	$\cos(\omega_n t)1(t)$	$\frac{s}{s^2+\omega_n^2}$	$\cos(\omega_n k)1(k)$	$\frac{z[z-\cos(\omega_n)]}{z^2-2z\cos(\omega_n)+1}$, $\|z\| > 1$
8	$\sin(\omega_n t)1(t)$	$\frac{\omega_n}{s^2+\omega_n^2}$	$\sin(\omega_n k)1(k)$	$\frac{z\sin(\omega_n)}{z^2-2z\cos(\omega_n)+1}$, $\|z\| > 1$
9	$te^{at}1(t-1)$	$\frac{1}{(s-a)^2}$, $Re\{s\} > a$	$ka^k 1(k-1)$	$\frac{az}{(z-a)^2}$, $\|z\| > \|a\|$
10	$e^{-\zeta\omega_n t}\sin(\omega_d t)1(t)$	$\frac{\omega_d}{(s+\zeta\omega_n)^2+\omega_d^2}$ $Re\{s\} > \zeta\omega_n$	$e^{-\zeta\omega_n k}\sin(\omega_d k)1(k)$	$\frac{e^{-\zeta\omega_n} z\sin(\omega_d)}{z^2-2ze^{-\zeta\omega_n}\cos(\omega_d)+e^{-2\zeta\omega_n}}$, $\|z\| > e^{-\zeta\omega_n}$
11	$e^{-\zeta\omega_n t}\cos(\omega_d t)1(t)$	$\frac{s+\zeta\omega_n}{(s+\zeta\omega_n)^2+\omega_d^2}$ $Re\{s\} > \zeta\omega_n$	$e^{-\zeta\omega_n k}\cos(\omega_d k)1(k)$	$\frac{e^{-\zeta\omega_n} z\sin(\omega_d)}{z^2-2ze^{-\zeta\omega_n}\cos(\omega_d)+e^{-2\zeta\omega_n}}$, $\|z\| > e^{-\zeta\omega_n}$

3 **Sampling t gives kT whose transform is obtained by multiplying the transform of k by T.

4 ***The function $e^{-\alpha kT}$ is obtained by setting $a = e^{-\alpha T}$.

Appendix D
Computation and Computational Errors

- Floating-point numbers and arithmetic.
- Roundoff errors and their effects.
- Modeling errors
- Optimal error covariance analysis.
- Examples.

State and parameter estimation require numerical computation using a digital computer. Numerical computation always involves errors. The errors include the following types:

1. Rounding errors due to representation using a finite word length.
2. Truncation errors due to arithmetic operations.
3. Errors in the input data.
4. Errors introduced by the operator handling the data.

Digital computers represent the data in estimation problems as **floating-point** numbers F using a word of finite length. Floating-point numbers are characterized by four integers: the base β, the precision t, and two additional numbers to define the range of the exponent $e \in [L, U]$. Computers use binary numbers and the base $\beta = 2$, and typical values are $t = 56, L = -64, U = 64$. Floating-point numbers include 0 and the numbers.

$$f = \pm .d_1 d_2 \ldots d_t \beta^e, \tag{D.1}$$

$$0 < d_i < \beta, d_1 \neq 0, L \leq e \leq U. \tag{D.2}$$

The floating-point numbers are in the range $m \leq |f| \leq M$, where

M. S. Fadali, *Introduction to Random Signals, Estimation Theory, and Kalman Filtering*, https://doi.org/10.1007/978-981-99-8063-5

$$m = 0.1\beta^L, \tag{D.3}$$

$$M = 0.11\ldots 1\beta^U = \beta^U\left(1 - \beta^{-t}\right). \tag{D.4}$$

For the typical values $(\beta, t, L, U) = (2, 56, -64, 64)$, we have

$$f \in \left[0.1 \times 2^{-64}, 2^{64} \times \left(1 - 2^{-56}\right)\right] = \left[2^{-65}, 2^{64} - 2^8\right].$$

Thus, any number outside this range cannot be stored in the computer memory.

Calculations using these numbers are performed using floating-point arithmetic. The operations $\{+, -, \times, \div\}$ must yield an answer that can be represented as a floating-point number.

$$G = \{x \in \mathcal{R} : m \le |x| \le M\} \cup \{0\}. \tag{D.5}$$

Thus, an operation Op that yields a number

$$|a \text{ Op } b| \notin G, \tag{D.6}$$

that is a number outside the range $[m, M]$ or 0, results in an error. If the operation yields a number less than m,

$$|a \text{ Op } b| < m, \tag{D.7}$$

the error is called an underflow error, and if it is greater than M,

$$|a \text{ Op } b| > M. \tag{D.8}$$

If the result of the computation is in the range $[m, M]$ or is zero, then it is mapped to a floating-point number with the mapping $fl : G \to F$, where the mapping depend on the quanitzer used, whether it uses rounding or truncation. The floating-point operator use the rule

$$fl(x) = \begin{cases} \text{rounded arithmetic:nearest } c \in F \text{ to } x \\ \text{If tie, round away from zero.} \\ \text{chopped arithmetic:nearest } c \in F \text{ to } x,\ |c| \le |x| \end{cases}. \tag{D.9}$$

Clearly, rounded arithmetic gives a smaller error on average and is preferred.

Because identical algebraic expressions can require a different set of arithmetic operations, the associated computational errors can be different and algorithms that

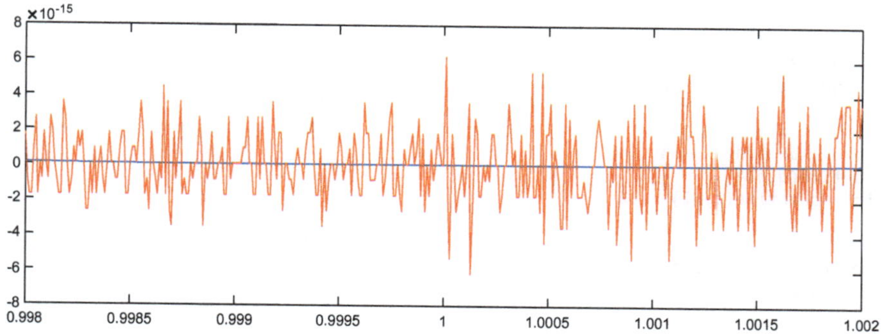

Fig. IV.1 Error plots for two equal expressions.

are equivalent mathematically often behave very differently numerically. The difference in error may be negligible if the operation is performed a few times, but a large number of repetitions can make one expression far more preferable to another.

Example D.1 (Van Loan, 1997) Calculate and plot the error associated with the two forms of the following expression $(x-1)^6 = x^6 - 6x^5 + 15x^4 - 20x^3 + 15x^2 - 6x + 1$, in the range $x \in [0.998, 1.002]$.

Solution

Calculating and plotting the errors with MATLAB show that the two forms behave very differently with larger errors for the expansion. Although the larger error is still very small, the accumulation of the error in a computational loop repeated many times can lead to an unacceptably large error (Fig. IV.1).

MATLAB Commands

MATLAB allows the user to define a quantizer with a specified number of bits and number of bits for the exponent with the command **quantizer**. A number is then quantized with the command **quantize.**

The following script shows how the two commands work:

```
% TestPrec
% Test finite word length commands
p2=100.1234;
q = quantizer('Mode','float','RoundMode','nearest', ...
              'OverflowMode','saturate','Format',[12 4])
% Floating point numbers, rounded to the nearest level, 12 bits, 4 exponent length
% Overflow mode is saturation
quantize(q,p2)
err = quantize(q,p2) - p2
```

The script gives the following output

```
>> TestPrec
q =
     DataMode = float
    RoundMode = nearest
       Format = [12  4]
ans =
  100


err =
  -0.1234
```

Bibliography

1. Linfield, G., & Penny, J. (1999). *Numerical methods using MATLAB* (2nd ed.). Prentice Hall.
2. Van Loan, C. F. (1997). *Introduction to scientific computing*.

Appendix E
The Continuous-Time Kalman Filter

- The continuous-time Kalman filter.
- The Riccati equation.
- Examples.

The continuous-time Kalman filter is known as the Kalman-Bucy filter. It is far less important than its discrete-time counterpart because implementation is digital in almost all applications. However, there are application that require the use of the continuous-time Kalman filter.

E.1 The Optimal Gain

The continuous-time state-space model is

$$\dot{\boldsymbol{x}}(t) = A(t)\boldsymbol{x}(t) + B(t)\boldsymbol{u}(t), \tag{E.1}$$

$$z(t) = H(t)\boldsymbol{x}(t) + \boldsymbol{v}(t), \tag{E.2}$$

where the Gaussian white noise processes satisfy:

$$E\{\boldsymbol{u}(t)\} = \mathbf{0},\ \ E\left\{\boldsymbol{u}(t)\boldsymbol{u}^T(t+\tau)\right\} = Q\delta(\tau), \forall t, \tau, \tag{E.3}$$

$$E\{\boldsymbol{v}(t)\} = \mathbf{0}, E\left\{\boldsymbol{v}(t)\boldsymbol{v}^T(t+\tau)\right\} = R\delta(\tau), \forall t, \tau. \tag{E.4}$$

We assume orthogonal process and measurement noise

M. S. Fadali, *Introduction to Random Signals, Estimation Theory, and Kalman Filtering*, https://doi.org/10.1007/978-981-99-8063-5

$$E\{\boldsymbol{u}(t)\boldsymbol{v}^T(\tau)\} = \mathbf{0}, \forall t, \tau. \tag{E.5}$$

Consequently, the state vector and the measurement noise are also orthogonal

$$E\{\boldsymbol{x}(t)\boldsymbol{v}^T(\tau)\} = \mathbf{0}, \forall t, \tau. \tag{E.6}$$

The measurement data $z(\tau), \tau \in [0, t]$ is used to obtain a state estimate using a filter whose dynamics are based on the system dynamics with a corrective term proportional to the residual

$$\dot{\hat{\boldsymbol{x}}}(t) = A(t)\hat{\boldsymbol{x}}(t) + K(t)[z(t) - H(t)\hat{\boldsymbol{x}}(t)]. \tag{E.7}$$

The filter dynamics can be rewritten in terms of the observer matrix

$$A_o(t) = A(t) - K(t)H(t) \tag{E.8}$$

as

$$\dot{\hat{\boldsymbol{x}}}(t) = A_o(t)\hat{\boldsymbol{x}}(t) + K(t)z(t). \tag{E.9}$$

Substituting from the measurement equation, we write the filter differential equation as

$$\dot{\hat{\boldsymbol{x}}}(t) = A_o(t)\hat{\boldsymbol{x}}(t) + K(t)H(t)\boldsymbol{x}(t) + K(t)\boldsymbol{v}(t).$$

The estimation error is $\tilde{\boldsymbol{x}}(t) = \boldsymbol{x}(t) - \hat{\boldsymbol{x}}(t)$, and the error dynamics is obtained by subtracting the filter state equation from the system state equation

$$\dot{\tilde{\boldsymbol{x}}}(t) = A_o(t)\tilde{\boldsymbol{x}}(t) + B(t)\boldsymbol{u}(t) - K(t)\boldsymbol{v}(t). \tag{E.10}$$

The orthogonality principle states that for minimum mean-square estimation the estimation error is orthogonal to the estimate

$$E\{\tilde{\boldsymbol{x}}(t)\boldsymbol{x}^T(t)\} = \mathbf{0}. \tag{E.11}$$

We also have the equality

$$E\{\tilde{\boldsymbol{x}}(t)\boldsymbol{x}^T(\tau)\} = E\left\{\tilde{\boldsymbol{x}}(t)[\tilde{\boldsymbol{x}}(\tau) + \hat{\boldsymbol{x}}(\tau)]^T\right\},$$

which for $\tau = t$ reduces to

$$E\{\tilde{\boldsymbol{x}}(t)\boldsymbol{x}^T(t)\} = P(t). \tag{E.12}$$

Orthogonality also implies that the estimation error is orthogonal to the measurements

$$E\{\tilde{\boldsymbol{x}}(t)\boldsymbol{z}^T(\tau)\} = \boldsymbol{0}, \forall \tau \le t. \tag{E.13}$$

Substituting from the measurement equation (E.2) in (E.13) gives

$$E\{\tilde{\boldsymbol{x}}(t)\boldsymbol{x}^T(\tau)\}H^T(\tau) + E\{\boldsymbol{x}(t)\boldsymbol{v}^T(\tau)\} - E\{\hat{\boldsymbol{x}}(t)\boldsymbol{v}^T(\tau)\} = \boldsymbol{0}, \forall \tau \le t.$$

Because the state is orthogonal to the measurement noise, we have the equality

$$E\{\tilde{\boldsymbol{x}}(t)\boldsymbol{x}^T(\tau)\}H^T(\tau) = E\{\hat{\boldsymbol{x}}(t)\boldsymbol{v}^T(\tau)\}, \forall \tau \le t. \tag{E.14}$$

The estimate $\hat{\boldsymbol{x}}(t)$ is the solution of the estimator state Equation (E.7)

$$\hat{\boldsymbol{x}}(\tau) = \phi_o(\tau, 0)\hat{\boldsymbol{x}}(0) + \int_0^\tau \phi_o(\tau, \xi)K(\xi)\boldsymbol{z}(\xi)\mathrm{d}\xi. \tag{E.15}$$

We can now rewrite

$$\begin{aligned}
&E\left\{\tilde{\boldsymbol{x}}(t)\left[\tilde{\boldsymbol{x}}(\tau) + \hat{\boldsymbol{x}}(\tau)\right]^T\right\}H^T(\tau) \\
&= E\left\{\left[\phi_o(\tau, 0)\hat{\boldsymbol{x}}(0) + \int_0^\tau \phi_o(\tau, \xi)K(\xi)\boldsymbol{z}(\xi)\mathrm{d}\xi\right]\boldsymbol{v}^T(\tau)\right\}, \forall \tau \le t \\
&= \phi_o(\tau, 0)E\{\hat{\boldsymbol{x}}(0)\boldsymbol{v}^T(\tau)\} + \int_0^\tau \phi_o(\tau, \xi)K(\xi)H(\xi)[E\{\boldsymbol{x}(\xi)\boldsymbol{v}^T(\tau)\} \\
&\quad + E\{\boldsymbol{v}(\xi)\boldsymbol{v}^T(\tau)\}\mathrm{d}\xi, \forall \tau \le t.
\end{aligned}$$

The initial state estimate is known and is not random and the measurement noise is zero-mean white and is orthogonal to the state. In addition, the estimate is orthogonal to the error. This gives

$$\begin{aligned}
E\{\tilde{\boldsymbol{x}}(t)\tilde{\boldsymbol{x}}(\tau)^T\}H^T(\tau) &= \int_0^\tau \phi_o(\tau, \xi)K(\xi)H(\xi)E\{\boldsymbol{v}(\xi)\boldsymbol{v}^T(\tau)\}d\xi, \forall \tau \le t \\
&= \phi_o(t, \tau)K(\tau)R(\tau).
\end{aligned}$$

Letting $\tau \to t$, the state-transition matrix becomes $\phi_o(t, \tau) \to I_n$, and using (E.12) gives

$$P(t)H^T(t) = K(t)R.$$

We solve for the optimal gain

$$K(t) = P(t)H^T(t)R^{-1}. \tag{E.16}$$

E.2 Autocorrelation of the State Vector

The solution of the state equation is

$$\boldsymbol{x}(t) = \phi(t, 0)\boldsymbol{x}(0) + \int_0^t \phi(t, \xi)B(\xi)\boldsymbol{u}(\xi)\mathrm{d}\xi. \tag{E.17}$$

Assume that the initial condition and the input noise are orthogonal, we have the autocorrelation

$$R_{xx}(t) = E\left\{\boldsymbol{x}(t)\boldsymbol{x}^T(t)\right\} = \phi(t, 0)R_{xx}(0)\phi^T(t, 0) + \int_0^t\int_0^t \phi(t, \xi)B(\xi)E\left\{\boldsymbol{u}(\xi)\boldsymbol{u}^T(\eta)\right\}B^T(\eta)\phi^T(t, \eta)\mathrm{d}\xi\mathrm{d}\eta.$$

Using the sifting property of the delta function reduces the double integral to a single integral

$$R_{xx}(t) = \phi(t, 0)R_{xx}(0)\phi^T(t, 0) + \int_0^t \phi(t, \xi)B(\xi)QB^T(\xi)\phi^T(t, \xi)\mathrm{d}\xi. \tag{E.18}$$

The state-transition matrix satisfies the identity

$$\frac{\partial\phi(t, \tau)}{\partial t} = A(t)\phi(t, \tau). \tag{E.19}$$

Differentiating the autocorrelation and using (E.19), we have

$$\dot{R}_{xx}(t) = A(t)\left[\phi(t, 0)R_{xx}(0)\phi^T(t, 0) + \int_0^t \phi(t, \xi)B(\xi)QB^T(\xi)\phi^T(t, \xi)\mathrm{d}\xi\right]$$

$$+\left[\phi(t,0)R_{xx}(0)\phi^T(t,0)+\int\limits_0^t \phi(t,\xi)B(\xi)QB^T(\xi)\phi^T(t,\xi)d\text{dx}i\right]A^T(t)+B(t)QB^T(t),$$

which, using (E.18), can be written as

$$\dot{R}_{xx}(t) = A(t)R_{xx}(t) + R_{xx}(t)A^T(t) + B(t)QB^T(t). \tag{E.20}$$

E.3 The Lyapunov and Riccati Equations

Combining noise vectors, we write the error dynamics (E.10) as

$$\dot{\tilde{\boldsymbol{x}}}(t) = A_o(t)\tilde{\boldsymbol{x}}(t)(t) + \overline{B}(t)\overline{\boldsymbol{u}}(t), \tag{E.21}$$

where

$$\overline{\boldsymbol{u}}(t) = \begin{bmatrix} \boldsymbol{u}(t) \\ \boldsymbol{v}(t) \end{bmatrix}, \overline{B}(t) = \begin{bmatrix} B(t) & -K(t) \end{bmatrix}. \tag{E.22}$$

The covariance matrix for the combined noise vector is

$$\begin{aligned} E\{\overline{B}(t)\overline{\boldsymbol{u}}(t)\overline{\boldsymbol{u}}^T(t)\overline{B}(t)^T\} &= \begin{bmatrix} B(t) & -K(t) \end{bmatrix} \begin{bmatrix} Q & \mathbf{0} \\ \mathbf{0} & R \end{bmatrix} \begin{bmatrix} B(t)^T \\ -K(t)^T \end{bmatrix} \\ &= \overline{B}(t)\overline{Q}(t)\overline{B}(t)^T = B(t)Q(t)B(t)^T + K(t)R(t)K(t)^T \end{aligned} \tag{E.23}$$

with $\overline{Q}(t) = diag\{Q(t), R(t)\}$.

E.3.a The Lyapunov Equation

Applying the derivation of the state autocorrelation to the estimation error equation

$$\dot{\tilde{\boldsymbol{x}}}(t) = A_o(t)\tilde{\boldsymbol{x}}(t) + \overline{B}(t)\overline{\boldsymbol{u}}(t) \tag{E.24}$$

yields the Lyapunov matrix differential equation

$$\dot{P}(t) = A_o(t)P(t) + P(t)A_o(t)^T + \overline{B}(t)\overline{Q}(t)\overline{B}(t)^T. \tag{E.25}$$

Substituting for $A_o(t)$ from (E.8) and for $\overline{B}(t)\overline{Q}(t)\overline{B}(t)^T$ from (E.23) gives the expanded form

$$\dot{P}(t) = [A(t) - K(t)H(t)]P(t) + P(t)[A(t) - K(t)H(t)]^T + B(t)Q(t)B(t)^T + K(t)R(t)K(t)^T.$$

E.3.b The Riccati Equation

Substitute for the Kalman gain of (E.14) in the Lyapunov equation and use the identities

$$K(t)RK^T(t) = P(t)H^T(t)R^{-1}RR^{-1}H(t)P(t),$$

$$K(t)H(t)P(t) = P(t)H^T(t)R^{-1}H(t)P(t).$$

The resulting equation is known as the Riccati matrix differential equation

$$\dot{P}(t) = A(t)P(t) + P(t)A^T(t) + B(t)Q(t)B^T(t) - P(t)H^T(t)R^{-1}H(t)P(t). \tag{E.26}$$

The equations for the filter are summarized in Table E.1.

E.4 Steady-State Filter

Under certain conditions, the steady-state solution P of the ARE exists and is unique and positive definite. As in the discrete-time case, we can use the suboptimal steady-state filter for simpler implementation. The filter uses the equations

$$\dot{\hat{x}}(t) = A\hat{x}(t) + K\big[z(t) - H\hat{x}(t)\big], \tag{E.27}$$

Table E.1 Summary of filter equations

Title	Equation
Kalman gain	$K(t) = P(t)H^T(t)R^{-1}$
Filter state equation	$\dot{\hat{x}}(t) = A(t)\hat{x}(t) + K(t)\big[z(t) - H(t)\hat{x}(t)\big]$
Error covariance equation	$\dot{P}(t) =$ $A(t)P(t)+P(t)A^T(t)+B(t)Q(t)B^T(t)-P(t)H^T(t)R^{-1}H(t)P(t)$

$$K = PH^T R^{-1}, \tag{E.28}$$

where P is the solution of the algebraic Riccati equation

$$[\mathbf{0}] = AP + PA^T + BQB^T - PH^T R^{-1} HP. \tag{E.29}$$

Example E.1 For the equation of motion

$$\ddot{y}(t) = u(t)$$

with noisy position measurements

$$z(t) = x(t) + v(t)$$

solve the algebraic matrix Riccati equation and the steady-state Kalman gain.

Solution

The state-space model of the system is

$$\begin{bmatrix} \dot{x}_1 \\ \dot{x}_2 \end{bmatrix} = \begin{bmatrix} 0 & 1 \\ 0 & 0 \end{bmatrix}\begin{bmatrix} x_1 \\ x_2 \end{bmatrix} + \begin{bmatrix} 0 \\ 1 \end{bmatrix} u(t),$$

$$z(t) = \begin{bmatrix} 1 & 0 \end{bmatrix}\begin{bmatrix} x_1 \\ x_2 \end{bmatrix} + v(t).$$

The algebraic Riccati equation for the problem is

$$\begin{aligned} [0] &= AP + PA^T + BQB^T - PH^T R^{-1} HP \\ &= \begin{bmatrix} 0 & 1 \\ 0 & 0 \end{bmatrix}\begin{bmatrix} p_1 & p_2 \\ p_2 & p_3 \end{bmatrix} + \begin{bmatrix} p_1 & p_2 \\ p_2 & p_3 \end{bmatrix}\begin{bmatrix} 0 & 0 \\ 1 & 0 \end{bmatrix} + \begin{bmatrix} 0 \\ 1 \end{bmatrix} q \begin{bmatrix} 0 & 1 \end{bmatrix} \\ &\quad - \begin{bmatrix} p_1 & p_2 \\ p_2 & p_3 \end{bmatrix}\begin{bmatrix} 1 \\ 0 \end{bmatrix}\left(\frac{1}{r}\right)\begin{bmatrix} 1 & 0 \end{bmatrix}\begin{bmatrix} p_1 & p_2 \\ p_2 & p_3 \end{bmatrix}. \end{aligned}$$

Solve the algebraic Riccati equation

$$0 = 2p_2 - \frac{p_1^2}{r}, 0 = p_3 - \frac{p_1 p_2}{r}, 0 = q - \frac{p_2^2}{r},$$

$$p_2 = \sqrt{qr}, p1 = \left(4qr^3\right)^{1/4}.$$

The Kalman gain is

$$\begin{aligned}K &= PH^T R^{-1}\\ &= \begin{bmatrix} p_1 & p_2 \\ p_2 & p_3 \end{bmatrix}\begin{bmatrix} 1 \\ 0 \end{bmatrix}\left(\frac{1}{r}\right)\\ &= \begin{bmatrix} p_1 \\ p_2 \end{bmatrix}\left(\frac{1}{r}\right) = \begin{bmatrix} (4q/r)^{1/4} \\ (q/r)^{1/2} \end{bmatrix}.\end{aligned}$$

The term p_3 is needed to calculate the steady-state mean square error (trace of P).

Bibliography

1. Burl, J. B. (1999). *Linear optimal control: $\mathcal{H}_2$ and $\mathcal{H}_\infty$ methods.* Addison Wesley Longman.
2. Kalman, R. E., & Bucy, R. S. (1961). New results in linear filtering and prediction theory. *Journal of Basic Engineering, Transactions of the ASME, Series, 83*, 95–108.
3. Kwakernaak, H., & Sivan, R. (1972). *Linear optimal control systems,* Wiley Interscience.

Appendix F
Modes of Convergence

Topics

- Stochastic versus deterministic convergence.
- Modes of stochastic convergence.
- Relations between convergence modes.

F.1 Deterministic Convergence

When a sequence of numbers asymptotically approaches a limit, it is said to converge to the limit. More formally, we define convergence as follows.

Definition F.1 Convergence of a Sequence. A sequence $\{\boldsymbol{a}_i\}$ of complex numbers converges to a point $\boldsymbol{a}$ in C if $\forall \epsilon > \mathbf{0}$, $\exists$ an integer N such that

$$|a_n - a| < \epsilon \text{ whenever } n \geq N.$$

The sequence converges if we can get as close as desired to the limit by going farther down the sequence.

An infinite series is the sum of a sequence

$$\boldsymbol{s} = \sum_{i=1}^{\infty} \boldsymbol{a}_i. \tag{F.1}$$

The nth partial sum of a series is

$$s_n = \sum_{i=1}^{n} a_i. \tag{F.2}$$

M. S. Fadali, *Introduction to Random Signals, Estimation Theory, and Kalman Filtering*, https://doi.org/10.1007/978-981-99-8063-5

Definition F.2 Convergence of a Series A series is convergent if the sequence of partial sums $\{s_n\}$ converges to a limit s. The limit s is the sum of the series.

Deterministic convergence:

Example F.1 Divergent and Convergent Series The following are divergent series:

(a) The sum of integers.

For the sequence of integers $\{1, 2, \ldots, n\}$, the general term is $a_n = n, n = 1, 2, \ldots$ and the sum is

$$s_n = \frac{n(n+1)}{2}.$$

(b) The harmonic series

For the sequence of reciprocals of integers $\{1, 1/2, \ldots, 1/n\}$, the general term is $a_n = \frac{1}{n}, n = 1, 2, \ldots$ and the sum is

$$s_n = \ln(n) + constant.$$

The following are convergent series.

(a) The geometric series.

For the sequence $\{1, a, \ldots, a^n\}$, the general term is $a_n = a^n, n = 0, 1, 2, \ldots,$ and for $|a| < 1$ the sum is

$$s = \frac{1}{1-a}.$$

(b) The exponential series

For the sequence $\left\{1, a, \frac{a^2}{2!} \ldots, \frac{a^n}{n!}\right\}$, the general term is $a_n = a^n, n = 0, 1, 2, \ldots$ and the sum for $|a| < \infty$ is $s = e^a$.

■

F.2 Stochastic Convergence

Many applications of probability and statistics involve a sequence of random variables or a random series. The following is a familiar example.

Example F.2 Sample Mean. A random sample $X_i, i = 1, \ldots, n$, is collected from the same population. Obtain an estimate of the mean.

Solution

We consider $X_i, i = 1, \ldots, n$, as a sequence of n i.i.d. random variables and use the sample mean as estimate of mean

$$Y_n = \frac{1}{n}\sum_{i=1}^{n} X_i. \tag{F.3}$$

■

An important question to consider is the following: How does the estimate change when we add more sample points (increase n)? To assess the behavior of the sample mean for a large sample, we are interested in the limit

$$Y_n = \mathcal{L}im_{n\to\infty} \frac{1}{n}\sum_{i=1}^{n} X_i. \tag{F.4}$$

This type of limit plays an important role in evaluating the performance of estimators as the sample size increases.

Because of their random nature, the usual definitions of convergence of deterministic sequences and deterministic series cannot be applied to random sequence and random series. A stochastic notion of convergence is required. Several definitions of stochastic convergence can be used. Several standard definitions of convergence are available.

Consider a sequence of random vectors $\boldsymbol{X}_1, \boldsymbol{X}_2, \boldsymbol{X}_3, \ldots, \boldsymbol{X}_i = [X_1, X_2, \ldots, X_m]^T, i = 1, 2, 3, \ldots$, with joint cumulative distribution function

$$P_X(\boldsymbol{x}) = P(\boldsymbol{X} \le \boldsymbol{x}) = P(X_i \le x_i, i = 1, \ldots, n). \tag{F.5}$$

The following definitions of convergence characterize convergence of the sequence to a limit $\boldsymbol{X}$.

F.2.i Convergence in Law

$\boldsymbol{X}_n$ converges in law to $\boldsymbol{X}$, if

$$\mathcal{L}im_{n\to\infty} P_{X_n}(\boldsymbol{x}) = P_X(\boldsymbol{x}), \tag{F.6}$$

$\forall \boldsymbol{x}$ where $P_X(\boldsymbol{x})$ **is continuous.**

Converges in law is also called convergence in distribution or weak convergence and is denoted by

$$\boldsymbol{X}_n \xrightarrow{\mathcal{L}} \boldsymbol{X}. \tag{F.7}$$

F.2.ii Convergence in Probability

$\boldsymbol{X}_n$ converges in probability to $\boldsymbol{X}$, if

$$\forall \varepsilon > 0, \underset{n\to\infty}{\mathcal{L}im}\, P(\|\boldsymbol{X}_n - \boldsymbol{X}\| > \varepsilon) = 0. \tag{F.8}$$

It is denoted by

$$\boldsymbol{X}_n \xrightarrow{P} \boldsymbol{X}, \tag{F.9}$$

or

$$\text{plim}\boldsymbol{X}_n = \boldsymbol{X}. \tag{F.10}$$

The following theorem is useful when determining convergence in probability.

Theorem F.1 *Slutsky's Theorem.*
If $\boldsymbol{X}_n \xrightarrow{\mathcal{L}} \boldsymbol{X} \in \mathcal{R}^n$, $\text{plim}\boldsymbol{B}_n = \boldsymbol{B} \in \mathcal{R}^{l\times m}$ *and* $\text{plim}\boldsymbol{a}_n = \boldsymbol{a} \in \mathcal{R}^l$, *then*

(1) $B_n\boldsymbol{X}_n \xrightarrow{\mathcal{L}} B\boldsymbol{X}$.

(2) $\boldsymbol{a}_n + B_n\boldsymbol{X}_n \xrightarrow{\mathcal{L}} a + B\boldsymbol{X}$.

F.2.iii Convergence in r^{th} Mean

$\boldsymbol{X}_n$ converges in the rth mean to $\boldsymbol{X}$, if

$$\underset{n\to\infty}{\mathcal{L}im}\, E\{\|\boldsymbol{X}_n - \boldsymbol{X}\|^r\} = 0, \tag{F.11}$$

where $\|.\|$ denotes the norm of the vector. In other words, the sequence converges to the limit vector $\boldsymbol{X}$ if the distance between the two vectors goes to zero asymptotically. For $r = 2$, which is the most common definition, we have **convergence in quadratic mean**.

Converges in the rth mean is denoted by

$$\boldsymbol{X}_n \xrightarrow{\text{r}} \boldsymbol{X}. \tag{F.12}$$

Convergence in quadratic mean is denoted by

$$\text{l.i.m.}\boldsymbol{X}_n = \boldsymbol{X}. \tag{F.13}$$

F.2.iv Almost Sure Convergence

$\boldsymbol{X}_n$ converges almost surely or with probability 1 to $\boldsymbol{X}$, if

$$P\left\{\underset{n\to\infty}{\mathcal{L}im}\, \boldsymbol{X}_n = \boldsymbol{X}\right\} = 1.$$

This is also called strong convergence and is denoted by

$$\boldsymbol{X}_n \xrightarrow{a.s.} \boldsymbol{X}. \tag{F.14}$$

Example F.3 Variance of Sample Mean Consider the mean of the i.i.d. sequence $X_i, i = 1, 2, \ldots$

$$Y_n = \frac{1}{n}\sum_{i=1}^{n} X_i.$$

The sequence is uncorrelated because it is independent and the variance of its sum is

$$var\left\{\sum_{i=1}^{n} X_i\right\} = \sum_{i=1}^{n} var\{X_i\} = n\sigma_x^2.$$

The variance of the sample mean is

$$\text{var}\{Y_n\} = \frac{1}{n^2} var\left\{\sum_{i=1}^{n} X_i\right\} = \frac{\sigma_x^2}{n}.$$

In the limit as the sample size approaches infinity, we have

$$\underset{n\to\infty}{\mathcal{L}im}\, \text{var}\{Y_n\} = \underset{n\to\infty}{\mathcal{L}im}\, \frac{\sigma_x^2}{n} = 0.$$

For an i.i.d. sequence, each term has the same mean and the asymptotic mean is simply the mean. Because the sample mean is unbiased (its expected value is the mean),

$$E\{Y_n\} = E\{X_i\}.$$

Thus, the sample mean converges in mean square to the population mean. For unbiased, asymptotic mean square error is the asymptotic variance and we have convergence in mean square.

$$\underset{n\to\infty}{\text{l.i.m.}}\, Y_n = E\{X_i\}.$$

The following relations govern the different definitions of convergence:

- Convergence in r^{th} mean $\Rightarrow$ convergence in s^{th} mean, $r > s \geq 1$.
- Convergence in r^{th} mean $\Rightarrow$ convergence in probability.
- Almost sure convergence $\Rightarrow$ convergence in probability.
- Convergence in probability $\Rightarrow$ convergence in distribution.

These relations explain why convergence in law is called weak convergence and almost sure convergence is called strong convergence. Because proving one form of convergence is straightforward in some cases, the above relations can simplify the proof of other forms of convergence. For example, it is often convenient to prove convergence in probability by first proving convergence in mean square. Estimators that converge in probability are called **consistent estimators** and one can prove estimator consistency using convergence in mean square. Thus, we can prove that the sample mean converges in probability to the population mean, which is known as the weak law of large numbers. It also converges almost surely to the population mean, which is known as the strong law of large numbers.

Note that almost sure convergence neither implies nor is implied by means square convergence.

Bibliography

1. Apostol, T. M. (1974). *Mathematical analysis.* Addison Wesley.
2. Casella, G., & Berger, R. (1990). *Statistical inference.* Duxbury Press.
3. Ferguson, T. S. (1996). *A course in large sample theory.* Chapman & Hall.
4. Lindsey, J. K. (1996). *Parametric statistical inference.* Claredon Press.

Appendix G
State-Space Models and Their Properties

This appendix reviews some basic properties of state-space models of linear systems, with emphasis on linear time-invariant (LTI) systems. Proofs are omitted and they can be found in texts on linear systems.

G.1 *Continuous-Time State-Space Models*

Continuous-time linear systems are characterized by the state-space model

$$\dot{\boldsymbol{x}}(t) = A(t)\boldsymbol{x}(t) + B(t)\boldsymbol{u}(t), \tag{G.1}$$

$$\boldsymbol{y}(t) = C(t)\boldsymbol{x}(t) + D(t)\boldsymbol{u}(t), \tag{G.2}$$

where $\boldsymbol{x}(t) \in \mathcal{R}^n, \boldsymbol{u}(t) \in \mathcal{R}^m, \boldsymbol{y}(t) \in \mathcal{R}^l$, and (A, B, C, D) are matrices of compatible dimensions. The matrices are known as

$A =$ state or system matrix.

$B =$ input matrix.

$C =$ Output or measurement matrix.

$D =$ direct transmission or feedforward matrix.

The solution of the state equation comprises two parts. The response due to the initial conditions is the zero-input response $\boldsymbol{x}_{ZI}(t)$. The response due to the input is the zero-response $\boldsymbol{x}_{ZS}(t)$. The total response of the system is the sum

$$\boldsymbol{x}(t) = \boldsymbol{x}_{ZI}(t) + \boldsymbol{x}_{ZS}(t) = \underbrace{\phi(t, t_0)\boldsymbol{x}(t_0)}_{\substack{\text{zero} - \text{input} \\ \text{response}}} + \underbrace{\int_{t_0}^{t} \phi(t, \tau)B\boldsymbol{u}(\tau)\mathrm{d}\tau}_{\text{zero}-\text{state response}} \tag{G.3}$$

M. S. Fadali, *Introduction to Random Signals, Estimation Theory, and Kalman Filtering*, https://doi.org/10.1007/978-981-99-8063-5

The response is in terms of the state-transition matrix $\phi(t, t_0)$, which satisfies the equation

$$\frac{\partial \phi(t, t_0)}{\partial t} = A(t)\phi(t, t_0). \tag{G.4}$$

For the LTI case, the state-transition matrix is the matrix exponential

$$e^{At} = \sum_{i=0}^{\infty} \frac{(At)^i}{i!} = \mathcal{L}^{-1}\{[sI_n - A]^{-1}\}, \tag{G.5}$$

and the solution of the state equation is

$$\boldsymbol{x}(t) = \boldsymbol{x}_{ZI}(t) + \boldsymbol{x}_{ZS}(t) = \underbrace{e^{A(t-t_0)}\boldsymbol{x}(t_0)}_{\substack{\text{zero} - \text{input} \\ \text{response}}} + \underbrace{\int_{t_0}^{t} e^{A(t-\tau)} B\boldsymbol{u}(\tau)\mathrm{d}\tau}_{\text{zero}-\text{state response}}. \tag{G.6}$$

If a matrix has as many eigenvectors as eigenvalues, that is, the geometric multiplicity is equal to the algebraic multiplicity, then the matrix exponential can be written as

$$e^{At} = \sum_{i=1}^{n} Z_i e^{\lambda_i t}, \tag{G.7}$$

where $\lambda_i, i = 1, \ldots, n$, are the eigenvalues of the state matrix and $Z_i, i = 1, \ldots, n$, are its constituent matrices. The expression is valid for all matrices with distinct eigenvalues but for repeated eigenvalues with multiplicity n_i we have

$$e^{At} = \sum_{i=1}^{n} \sum_{j=1}^{n_i - 1} Z_{ij} e^{\lambda_i t} t^j. \tag{G.8}$$

The constituent matrices are the product of the right and left eigenvectors of the state matrix

$$Z_i = \boldsymbol{v}_i \boldsymbol{w}_i^T, i = 1, \ldots, n \tag{G.9}$$

provided that the eigenvectors are appropriately scaled such that

$$WV = \left[\boldsymbol{w}_i^T \boldsymbol{v}_j\right] = I_n. \tag{G.10}$$

The scaling guarantees that

$$\boldsymbol{w}_i^T \boldsymbol{v}_j = \begin{cases} 1, \ i = j \\ 0, \ i \neq j \end{cases}. \tag{G.11}$$

The matrix maps its right eigenvectors to themselves, scaled by the eigenvalues

$$A\boldsymbol{v}_i = \lambda_i \boldsymbol{v}_i, i = 1, 2, \ldots, n. \tag{G.12}$$

Left eigenvectors are the row vectors

$$\boldsymbol{w}_i^T A = \lambda_i \boldsymbol{w}_i^T, i = 1, 2, \ldots, n. \tag{G.13}$$

Constituent matrices have the following properties.

1. Constituent matrices have rank 1.
2. The product of two constituent matrices is

$$Z_i Z_j = \begin{cases} Z_i, i = j \\ \boldsymbol{0}, i \neq j \end{cases}. \tag{G.14}$$

Hence, raising Z_i to any power gives the matrix Z_i, i.e., Z_i is *idempotent.*

3. The sum of the n constituent matrices of a $n \times n$ matrix is equal to the identity matrix

$$\sum_{i=1}^{n} Z_i = I_n. \tag{G.15}$$

4. If a real matrix has complex conjugate eigenvalues $\lambda_{1,2} = \sigma \pm j\omega_d$, the corresponding eigenvectors and constituent matrices are also complex conjugates and the matrix exponential can be written as

$$\begin{aligned} e^{At} &= 2Re\{Z_1 e^{\lambda_1 t}\} \\ &= 2Re\{Z\}e^{\sigma t}\cos(\omega_d t) - 2Im\{Z\}e^{\sigma t}\sin(\omega_d t). \end{aligned} \tag{G.16}$$

The transfer function matrix is written in terms of the state-space quadruple (A, B, C, D) as

$$G(s) = C[sI_n - A]^{-1}B + D. \tag{G.17}$$

G.2 Discrete-Time Systems

Discrete-time systems have the state-space model

$$\boldsymbol{x}(k+1) = \phi(k+1, k)\boldsymbol{x}(k) + B_d\boldsymbol{u}(k), k = 0, 1, \ldots \tag{G.18}$$

$$\boldsymbol{y}(k) = C\boldsymbol{x}(k) + D\boldsymbol{u}(k) \tag{G.19}$$

with the index k indicating that the variable is at time t_k. If the model is obtained by discretizing a continuous-time system, then the state matrix is

$$\phi(k+1, k) = \phi(t_{k+1}, t_k). \tag{G.20}$$

The output and direct transmission matrices are the same for both systems. The input matrix is only defined if the input is constant over the interval

$$B_d = \int_{t_k}^{t_{k+1}} \phi(t, \tau)B\mathrm{d}\tau. \tag{G.21}$$

For a LTI system with uniform sampling and with sampling period Δt, we have

$$\phi(k+1, k) = \phi = e^{A\Delta t}. \tag{G.22}$$

If the input is constant over each sampling period, then the input matrix is

$$B_d = \int_0^T e^{A\tau}B\mathrm{d}\tau. \tag{G.23}$$

For distinct eigenvalues, we have

$$B_d = \begin{cases} \sum\limits_{i=1}^{n} Z_i B\left[\frac{1-e^{\lambda_i T}}{-\lambda_i}\right], & \lambda_i \neq 0 \\ Z_1 BT + \sum\limits_{i=2}^{n} Z_i B\left[\frac{1-e^{\lambda_i T}}{-\lambda_i}\right], & \lambda_1 = 0 \end{cases}. \tag{G.24}$$

The solution of the discrete-time state equation is

$$\boldsymbol{x}(k) = \underbrace{\phi(k, k_0)\boldsymbol{x}(k_0)}_{\substack{\text{zero} - \text{input} \\ \text{response}}} + \underbrace{\sum_{i=k_0}^{k-1} \phi(k, i)B\boldsymbol{u}(i)}_{\text{zero}-\text{state response}}. \tag{G.25}$$

For the LTI case, the solution is

$$\boldsymbol{x}(k) = \phi^{k-k_0}\boldsymbol{x}(k_0) + \sum_{i=k_0}^{k-1} \phi^{k-i-1} B_d \boldsymbol{u}(i). \tag{G.26}$$

G.3 State-Space Realizations

A state-space realization is a state-space model obtained from a differential equation or, for linear time-invariant systems from a transfer function. Canonical state-space realizations can be obtained from transfer functions, and the procedure is almost identical for continuous-time and discrete-time systems. The following are the most common realizations for single–input–single–output discrete-time systems. For continuous-time systems, the complex variable s replaces by z in the transfer function and the state equation is a matrix differential equation, instead of a difference equation.

For a system with input $u(t)$ and output $y(t)$, the realization is based on a transfer function of the form

$$G(z) = \frac{Y(z)}{U(z)} = c_n + G_d(z), \tag{G.27}$$

where

$$G_d(z) = \frac{c_{n-1}z^{n-1} + c_{n-2}z^{n-2} + \cdots + c_1 z + c_0}{z^n + a_{n-1}z^{n-1} + \cdots + a_1 z + a_0}. \tag{G.28}$$

Controllable Canonical Form

The state equation is in the form

$$\boldsymbol{x}(k+1) = \begin{bmatrix} \mathbf{0}_{n-1\times 1} & I_{n-1} \\ -a_0 & -a_1 \; \ldots \; -a_{n-1} \end{bmatrix} \boldsymbol{x}(k) + \begin{bmatrix} \mathbf{0}_{n-1\times 1} \\ 1 \end{bmatrix} u(k) \tag{G.29}$$

with output equation

$$y(k) = \begin{bmatrix} c_0 \; c_1 \; \ldots \; c_{n-1} \end{bmatrix} \begin{bmatrix} x_1(k) \\ x_2(k) \\ \vdots \\ x_n(k) \end{bmatrix} + du(k)$$

$$d = c_n. \tag{G.30}$$

Observable Canonical Form

The form can be obtained from the controllable form as follows:

1. Transpose the state matrix A.
2. Transpose and interchange the input matrix B and the output matrix C.

Clearly, the observable form can be written directly using the coefficients of the transfer function

$$\boldsymbol{x}(k+1) = \begin{bmatrix} \mathbf{0}^T_{1\times n-1} & | -\boldsymbol{a} \\ I_{n-1} & \end{bmatrix} \boldsymbol{x}(k) + \boldsymbol{b}u(k), \tag{G.31}$$

$$y(k) = \left[\mathbf{0}^T_{1\times n-1} | 1\right] \boldsymbol{x}(k) + du(k), \tag{G.32}$$

$$\boldsymbol{a} = \left[a_0\ a_1\ \ldots\ a_{n-1} \right]^T,$$

$$\boldsymbol{b} = \left[b_0\ b_1\ \ldots\ b_{n-1} \right]^T.$$

Diagonal Form

The diagonal form, or parallel realization, is obtained from the partial fractions of the transfer function

$$G(z) = \frac{Y(z)}{U(z)} = d + G_d(z) = d + \sum_{i=1}^{n} \frac{K_i}{z + p_i}. \tag{G.33}$$

The name is due to the diagonal form of the state matrix

$$\begin{bmatrix} x_1(k+1) \\ x_2(k+1) \\ \vdots \\ x_{n-1}(k+1) \\ x_n(k+1) \end{bmatrix} = \begin{bmatrix} -p_1 & 0 & \ldots & 0 & 0 \\ 0 & -p_2 & \ldots & 0 & 0 \\ \vdots & \vdots & \ldots & \vdots & \vdots \\ 0 & 0 & \ldots & -p_{n-1} & 0 \\ 0 & 0 & \ldots & 0 & -p_n \end{bmatrix} \begin{bmatrix} x_1(k) \\ x_2(k) \\ \vdots \\ x_{n-1}(k) \\ x_n(k) \end{bmatrix} + \begin{bmatrix} 1 \\ 1 \\ \vdots \\ 1 \\ 1 \end{bmatrix} u(k), \tag{G.34}$$

$$y(k) = \left[K_1\ K_2\ \ldots\ K_{n-1}\ K_n \right] \begin{bmatrix} x_1(k) \\ x_2(k) \\ \vdots \\ x_{n-1}(k) \\ x_n(k) \end{bmatrix} + du(k). \tag{G.35}$$

G.4 Stability

Definition G.1 An equilibrium point or equilibrium state is an initial state from which the system never departs unless perturbed.

For a nonlinear continuous-time system $\dot{\boldsymbol{x}}(t) = \boldsymbol{f}(\boldsymbol{x}(t))$, an equilibrium state satisfies

$$\boldsymbol{f}(\boldsymbol{x}_e) = \boldsymbol{0}. \tag{G.36}$$

For a nonlinear discrete-time system $\boldsymbol{x}(k+1) = \boldsymbol{f}[\boldsymbol{x}(k)]$, an equilibrium state satisfies

$$\boldsymbol{f}[\boldsymbol{x}_e] = \boldsymbol{x}_e. \tag{G.37}$$

A nonlinear system may have multiple equilibrium points.

For a linear continuous-time system $\dot{\boldsymbol{x}}(t) = A\boldsymbol{x}(t)$, an equilibrium state satisfies

$$A\boldsymbol{x}_e = \boldsymbol{0}. \tag{G.38}$$

For a linear discrete-time $\boldsymbol{x}(k+1) = A\boldsymbol{x}(k)$, an equilibrium state satisfies

$$A\boldsymbol{x}_e = \boldsymbol{x}_e \Leftrightarrow [A - I_n]\boldsymbol{x}_e = \boldsymbol{0}. \tag{G.39}$$

For a continuous-time system, the equilibrium condition is uniquely satisfied at the origin $\boldsymbol{x}_e = 0$ if A is nonsingular that is if the state matrix has no zero eigenvalues. For a discrete-time system, the equilibrium condition is uniquely satisfied at the origin $\boldsymbol{x}_e = 0$ if the matrix $[A - I_n]$ has no unity eigenvalues and is nonsingular.

Definition G.2 A continuous-time LTI system is asymptotically (Hurwitz) stable if and only if all the eigenvalues of its state matrix are in the left half plane.

$$Re\{\lambda_i(\phi)\} < 0, i = 1, 2, \ldots, n. \tag{G.40}$$

A discrete-time LTI system is asymptotically (Schur) stable if and only if all the eigenvalues of its state matrix are inside the unit circle.

$$|\lambda_i(\phi)| < 1, i = 1, 2, \ldots, n. \tag{G.41}$$

G.5 Controllability and Observability

Controllability and observability are important properties that characterize the efficacy of control and measurements, respectively. The definition and theorems give below are for discrete-time systems but the definitions are similar and the tests are identical for continuous-time systems.

Definition G.3 An LTI system is controllable if for any initial state $\boldsymbol{x}(k_0)$ there exists a control sequence $\boldsymbol{u}(k)$, $k_0 = 1, 2, \ldots, k_f - 1$, such that an arbitrary final state $\boldsymbol{x}(k_f)$ can be reached in finite time.

The following three theorems are tests for controllability.

Theorem G.1 *An LTI system is controllable if and only if the products satisfy*

$$\boldsymbol{w}_i^T B_d \neq 0^T, i = 1, 2, \ldots, n, \tag{G.42}$$

$\boldsymbol{w}_i^T = i^{th}$ *left eigenvector,* $B_d =$ *input matrix.*
If $\boldsymbol{w}_i^T B_d = 0^T$, then the ith mode is uncontrollable.

Theorem G.2 *A LTI system is completely controllable if and only if the* $n \times m.n$ *controllability matrix*

$$\mathcal{C} = [B|AB|\ldots|A^{n-2}B|A_d^{n-1}B_d] \tag{G.43}$$

has rank n. If the matrix is not full rank, then its rank deficit is equal to the number of uncontrollable modes and the left eigenvector $\boldsymbol{w}_i^T$ *such that* $\boldsymbol{w}_i^T \mathcal{C} = 0^T$ *corresponds to the uncontrollable mode* $e^{\lambda_i t}$.

MATLAB has a command that computes the controllability matrix and a command to determine its rank.

```
>> C=ctrb(A,B) % Calculate controllability matrix
>> rank(C) % Find the rank of the matrix.
```

Theorem G.3 *A system in normal form is controllable if and only if its input matrix has no zero rows. Furthermore, if the input matrix has a zero row, then the corresponding mode in uncontrollable.*

$$\begin{bmatrix} x_1(k+1) \\ \vdots \\ x_n(k+1) \end{bmatrix} = \begin{bmatrix} \lambda_1 & \cdots & 0 \\ \vdots & \ddots & \vdots \\ 0 & \cdots & \lambda_n \end{bmatrix} \begin{bmatrix} x_1(k) \\ \vdots \\ x_n(k) \end{bmatrix} + \begin{bmatrix} \boldsymbol{b}_1^T \\ \vdots \\ \boldsymbol{b}_n^T \end{bmatrix} \boldsymbol{u}(k). \tag{G.44}$$

An important consideration is whether the uncontrollable modes are stable or unstable, with the latter as undesirable.

Definition G.4 A system is stabilizable if all its uncontrollable modes decay to zero asymptotically.

An alternative statement of the definition is that for a stabilizable system all unstable modes are controllable. Clearly, stability and controllability are independent properties.

Physical systems are often stabilizable but not controllable. This poses no problem if the uncontrollable dynamics decay to zero sufficiently fast.

Observability.

Definition G.5 A system is said to be observable if any initial state $\boldsymbol{x}(k_0)$ can be estimated from the control sequence $\boldsymbol{u}(k), k = k_0, k_0 + 1, \ldots, k_f - 1$, and the measurements $\boldsymbol{y}(k), k = k_0, k_0 + 1, \ldots, k_f$.

Theorem G.4 *A LTI system is observable if and only if $C\boldsymbol{v}_i$ is nonzero for $i = 1, 2, \ldots, n$, where $\boldsymbol{v}_i$ is the ith eigenvector of the state matrix. Furthermore, if the product $C\boldsymbol{v}_i$ is zero then the ith mode is unobservable.*

Theorem G.5 *A LTI system is observable if and only if the $l.n \times n$ observability matrix*

$$\mathcal{O} = \begin{bmatrix} C \\ CA_d \\ \vdots \\ CA_d^{n-1} \end{bmatrix} \tag{G.45}$$

has rank n. If the matrix is not full rank, then there is a right eigenvector $\boldsymbol{v}_i$ such that $\mathcal{O}\boldsymbol{v}_i = 0$ and the rank deficit of the matrix is equal to the number of unobservable modes. Furthermore, if the product $C\boldsymbol{v}_i$ is zero then the ith mode is unobservable.

To calculate the observability matrix and find its rank, MATLAB has the commands.

```
>> o = obsv(A, C) % Calculate observability matrix
>> rank(o) % Find the rank of the matrix.
```

Theorem G.6 *A system in normal form is observable if and only if its output matrix has no zero columns. Furthermore, if the output matrix has a zero column the corresponding mode in unobservable.*

Definition G.6 A system is detectable if all its unobservable modes decay to zero asymptotically.

It follows that if a system is detectable then all unstable modes are observable. Observability and stability are independent properties. Physical systems are typically detectable but not observable. This does not pose a problem if the unobservable modes decay to zero sufficiently fast.

To transform a system to diagonal form, we use the MATLAB commands.

```
>> p=ss(A,B,C,D)
>> [v,d]=eig(A) % v is the matrix of eigenvectors
>> pdiag=ss2ss(p,inv(v)).
```

G.6 Similarity Transformation

Given a state vector $\boldsymbol{x}(k)$, define a new state vector $z(k)$ using the invertible transformation matrix T

$$\boldsymbol{x}(k) = Tz(k) \Leftrightarrow z(k) = T^{-1}\boldsymbol{x}(k). \tag{G.46}$$

Substitute for $\boldsymbol{x}(k)$ in the state equation

$$\boldsymbol{x}(k+1) = A\boldsymbol{x}(k) + \boldsymbol{Bu}(k)$$

to obtain

$$Tz(k+1) = ATz(k) + \boldsymbol{Bu}(k).$$

Premultiplying the state equation by T^{-1} gives the transformed state equation

$$z(k+1) = T^{-1}ATz(k) + T^{-1}B\boldsymbol{u}(k).$$

Substituting in the output equation gives

$$y(k) = CTz(k) + D\boldsymbol{u}(k).$$

Thus, the state-space quadruple for the state vector $z(k)$ is

$$(\mathrm{A}, \mathrm{B}, \mathrm{C}, \mathrm{D}) = \left(T^{-1}AT, T^{-1}B, CT, D\right). \tag{G.47}$$

Because the matrix T defines an invertible transformation, the quadruple for the state vector $\boldsymbol{x}(k)$ can be obtained from the quadruple of $z(k)$ using the inverse transformation matrix $S = T^{-1}$. The transformation from $\boldsymbol{x}(k)$ to $z(k)$ can also be written in terms of the matrix $S = T^{-1}$

$$(\mathrm{A}, \mathrm{B}, \mathrm{C}, \mathrm{D}) = \left(SAS^{-1}, SB, CS^{-1}, D\right), \tag{G.48}$$

which is the form used in MATLAB

» **Pt = ss2ss(P, S).**

Similarity transformation is identical for continuous-time and discrete-time systems and is defined by

$$\boldsymbol{x}(t) = Tz(t) \Leftrightarrow z(t) = T^{-1}\boldsymbol{x}(t). \tag{G.49}$$

Transformation to diagonal form is achieved by using the modal matrix of eigenvector V. This gives the matrices

$$A_z = V^{-1}AV = V^{-1}\left(V\Lambda V^{-1}\right)V = \Lambda$$

$$B_z = V^{-1}B, C_z = CV, D, \tag{G.50}$$

where $\Lambda = \text{diag}\{\lambda_1, \lambda_2, \ldots, \lambda_n\}$, and λ_i are the eigenvalues of A.

For a system that is not completely observable with r unobservable modes, it is possible to transform the system into an observable subsystem and an unobservable subsystem using the observability matrix using the following steps.

1. Obtain the basis vectors of the null space of the observability matrix $\boldsymbol{t}_i, i = 1, \ldots, r < n$.
2. Form an invertible matrix using the basis vectors from Step 1 and $n - r$ additional linearly independent vectors $\boldsymbol{t}_i, i = 1, \ldots, r$.
3. Obtain the transformation matrix

$$T_o = [\boldsymbol{t}_1|\ldots|\boldsymbol{t}_r|\boldsymbol{t}_{r+1}|\ldots|\boldsymbol{t}_n].$$

4. Obtain the realization

$$(\mathrm{A_o}, \mathrm{B_o}, \mathrm{C_o}, \mathrm{D_o}) = \left(T_o^{-1}AT_o, T_o^{-1}B, CT_o, D\right).$$

The steps are accomplished with the following MATLAB script

```
Ob=obsv(A,C);
nn=null(Ob)
T=[nn/nn(1),[eye(2);0 0]];
pn=ss2ss(p,inv(T))
```

Appendix H
Review of Linear Algebra

The appendix presents some important formulas from linear algebra that are used throughout the text.

H.1 Determinant of a Matrix

The determinant of a square matrix is a scalar computed using its entries. For a 1×1 matrix, the determinant is simply the matrix itself. For a 2×2 matrix $A = \left[a_{ij}\right]$, the determinant is

$$\det(A) = a_{11}a_{22} - a_{12}a_{21}.$$

For higher-order matrices, the following definitions are needed to define the determinant.

Minor:

The ij^{th} minor of an $n \times n$ matrix is the determinant of the $n-1 \times n-1$ matrix obtained by removing the ith row and the jth column and is denoted M_{ij}.

Cofactor of a matrix: The ij^{th} cofactor of an $n \times n$ matrix is a signed minor given by

$$C_{ij} = (-1)^{i+j} M_{ij}.$$

The sign of the ij^{th} cofactor is the ij^{th} entry of the matrix

$$\begin{bmatrix} + - + \dots \\ - + - \dots \\ + - + \dots \\ - + - \dots \\ \vdots \vdots \vdots \ddots \end{bmatrix}.$$

M. S. Fadali, *Introduction to Random Signals, Estimation Theory, and Kalman Filtering*, https://doi.org/10.1007/978-981-99-8063-5

The determinant of a square $n \times n$ matrix can be obtained by expansion along any row or column and is given by

$$det(A) = \sum_{i=1}^{n} a_{is} C_{is} \sum_{j=1}^{n} a_{sj} C_{sj}.$$

A **singular matrix** is a matrix whose determinant is zero $\det(A) = 0$ and the matrix is **nonsingular** if its determinant is nonzero $\det(A) \neq 0$.

Properties of determinants

For $n \times n$ matrices A and B, the determinants have the following properties:

$$\det(A) = \det\left(A^T\right),$$

$$\det(\alpha A) = \alpha^n \det(A),$$

$$\det(AB) = \det(A)\det(AB).$$

The following MATLAB command gives the determinant of a matrix.

>> det(A).

Adjoint matrix

The adjoint matrix is the transpose of the matrix of cofactors

$$adj(A) = \left[C_{ij}\right]^T.$$

H.2 Inverse of a Matrix

The inverse of a square matrix is a matrix satisfying

$$A\,A^{-1} = A^{-1}A = I_n.$$

The inverse of the matrix is given by

$$A^{-1} = \frac{adj(A)}{\det(A)}.$$

For MATLAB computation of the inverse, use the command

>> **inv**(**A**)

For multiplication by an inverse, there is no need to invert the matrix. It is preferable to use the command

$$>> \mathbf{A} \backslash \mathbf{B}$$

to calculate $A^{-1}B$ and the command

$$>> \mathbf{A}/\mathbf{B}$$

to calculate AB^{-1}.

H.3 Combinations of Operations

For transposition and inversion, the following rules apply:

Transpose of a product: $(ABC)^T = C^T B^T A^T$.

Inverse of a product: $(ABC)^{-1} = C^{-1}B^{-1}A^{-1}$.

Commutative operations: $\left(A^T\right)^{-1} = \left(A^{-1}\right)^T = A^{-T}$.

Orthogonal matrix:

A matrix whose inverse is equal to its transpose $\mathbf{A}^{-1} = \mathbf{A}^T$, and multiplying the matrix by its transpose gives the identity matrix

$$A^T A = AA^T = I_n.$$

Using the properties of a determinant of a square matrix,

$$\det(I_n) = \det(A)\det\left(A^T\right) = \det(A)^2 = 1.$$

Hence, we have $\det(A) = \pm 1$ for an orthogonal matrix.

Unitary matrix:

A matrix whose inverse is equal to its complex conjugate transpose

$$A^{-1} = A^*.$$

H.4 Trace of a Matrix

The trace of square matrix is the sum of its diagonal entries

$$tr(A) = \sum_{i=1}^{n} a_{ii}.$$

The trace satisfies the following properties:

$$tr\left(A^T\right) = tr(A),$$

$$tr(AB) = tr(BA),$$

$$tr(A + B) = tr(A) + tr(B).$$

H.5 Linearly Independent Vectors

A set of vectors $\{\boldsymbol{x}_i, i = 1, 2, \ldots n\}$ is linearly independent if their weighted sum is only zero if and only if all the weights are set equal to zero

$$\sum_{i=1}^{n} \alpha_i \boldsymbol{x}_i = \boldsymbol{0} \Leftrightarrow \alpha_i = 0, i = 1, 2, \ldots n.$$

Otherwise, the set is said to be **linearly dependent**.

Rank of a Matrix

The row rank of a matrix is the number of linearly independent rows. The column rank is the number of linearly independent columns. Because the two are equal, we only refer to the rank of the matrix, which is the number of linearly independent rows, or the number of linearly independent columns.

Because the column rank and the row rank are equal, the rank of a rectangular $m \times n$ matrix A is $rank(A) \leq \min\{m, n\}$.

A rectangular matrix is said to be **full rank** if $rank(A) = \min\{m, n\}$.

A full-rank square matrix is nonsingular.

H.6 Eigenvalues and Eigenvectors

The eigenvector of a matrix A are vectors that are mapped to themselves when multiplied by the matrix A:

$$A\boldsymbol{v} = \lambda \boldsymbol{v}.$$

In other words

$$[\lambda I_n - A]\boldsymbol{v} = \boldsymbol{0}.$$

The scale factor on the right-hand side of the equation is known as the eigenvalue. For a nonzero solution $\boldsymbol{v}$ to the preceding equation to exist, the premultiplying matrix must be rank deficient—that is, λ must be an eigenvalue of the matrix A.

The eigenvector is defined by a direction or by a specific relationship between its entries. Multiplication by a scalar changes the length but not the direction of the vector.

The eigenvalues of an $\boldsymbol{n} \times \boldsymbol{n}$ matrix are the $\boldsymbol{n}$ roots of the characteristic equation.

Distinct eigenvalues:

$$\lambda_i \neq \lambda_j, i \neq j, i, j \in \{1, 2, \ldots, n\}.$$

Repeated eigenvalues:
$\lambda_i = \lambda_j$, for some $i \neq j$.

Multiplicity of the eigenvalue:
The number of repetitions of the repeated eigenvalue (also known as the algebraic multiplicity).

Spectrum of matrix A: The set of eigenvalues $\{\lambda_i, i = 1, 2, \ldots, n\}$.

Spectral radius of a matrix: Maximum absolute value over all the eigenvalues of the matrix.

Trace in terms of eigenvalues:

$$tr(A) = \sum_{i=1}^{n} \lambda_i.$$

Upper triangular matrix

$$A = \begin{bmatrix} a_{11} & a_{12} & \ldots & a_{1n} \\ 0 & a_{22} & \ldots & a_{2n} \\ \vdots & \vdots & \ddots & \vdots \\ 0 & 0 & \cdots & a_{nn} \end{bmatrix}.$$

Lower triangular matrix

$$A = \begin{bmatrix} a_{11} & 0 & \ldots & 0 \\ a_{21} & a_{22} & \ldots & 0 \\ \vdots & \vdots & \ddots & \vdots \\ a_{n1} & a_{n2} & \cdots & a_{nn} \end{bmatrix}.$$

For lower triangular, upper triangular, and diagonal matrices,

$$\{\boldsymbol{\lambda_i = a_{ii}, i = 1, 2, \ldots, n}\}.$$

H.7 Eigenvalues and Eigenvectors

The eigenvector of a matrix are mapped by the matrix to a scaled version of the eigenvector

$$A\boldsymbol{v} = \lambda \boldsymbol{v}.$$

The scale factor is the eigenvalue of the matrix.
Rewriting (.) as

$$[\lambda I_n - A]\boldsymbol{v} = \mathbf{0}$$

shows that the eigenvalues make the matrix $[\lambda I_n - A]$ rank deficient. Thus, the eigenvalues are the root of the equation

$$\det[\lambda I_n - A] = 0.$$

Normal matrix:
A normal matrix multiplication by the conjugate transpose is commutative

$$A^* A = AA^*.$$

This includes Hermitian (symmetric) matrices as a special case.

The matrix of eigenvectors of a normal matrix can be selected as an orthogonal (unitary) matrix, and the matrix can be written as

$$A = V \Lambda V^*.$$

H.8 Partitioned Matrix

A matrix partitioned into smaller submatrices

$$\begin{bmatrix} A_{11} & A_{12} & \dots & A_{1m} & \dots & A_{1n} \\ A_{21} & A_{22} & \dots & A_{2m} & \dots & A_{2n} \\ \vdots & \vdots & \ddots & \vdots & \dots & \vdots \\ A_{m1} & A_{m2} & \cdots & A_{mm} & \dots & A_{mn} \end{bmatrix}.$$

To transpose of a partitioned matrix, change the rows into columns, then transpose all the submatrices:

$$\begin{bmatrix} A_{11}^T & A_{21}^T & \dots & A_{m1}^T \\ A_{12}^T & A_{22}^T & \dots & A_{m2}^T \\ \vdots & \vdots & \ddots & \vdots \\ A_{1m}^T & A_{2m}^T & \dots & A_{mm}^T \\ \vdots & \vdots & \ddots & \vdots \\ A_{1n}^T & A_{2n}^T & \cdots & A_{mn}^T \end{bmatrix}.$$

Sum/difference of partitioned matrices: add the corresponding submatrix

$$C = A \pm B \Leftrightarrow C_{ij} = A_{ij} + B_{ij}, i = 1, 2, \dots, m, j = 1, 2, \dots, n.$$

Product of partitioned matrices:

Apply the rules of matrix multiplication with the products of matrix entries replaced by the noncommutative products of submatrices.

$$C = AB \Leftrightarrow C_{ij} = \sum_{k=1}^{n} A_{ik} B_{kj}, i = 1, 2, \dots, r, j = 1, 2, \dots, s.$$

Determinant of a partitioned matrix:

$$det\begin{bmatrix} A_1 & A_2 \\ A_3 & A_4 \end{bmatrix} = \begin{cases} det(A_1)det\left(A_4 - A_3 A_1^{-1} A_2\right), \ A_1^{-1} \\ det(A_4)det\left(A_1 - A_2 A_4^{-1} A_3\right), \ A_4^{-1} \end{cases}.$$

Inverse of a partitioned matrix:

$$\begin{bmatrix} A_1 & A_2 \\ A_3 & A_4 \end{bmatrix}^{-1} = \begin{bmatrix} \left(A_1 - A_2 A_4^{-1} A_3\right)^{-1} & -A_1^{-1} A_2 \left(A_4 - A_3 A_1^{-1} A_2\right)^{-1} \\ -A_4^{-1} A_3 \left(A_1 - A_2 A_4^{-1} A_3\right)^{-1} & \left(A_4 - A_3 A_1^{-1} A_2\right)^{-1} \end{bmatrix}.$$

Matrix Inversion Lemma

The following identity can be used in either direction to simplify matrix expressions:

$$\left[A_1 + A_2 A_4^{-1} A_3\right]^{-1} = A_1^{-1} - A_1^{-1} A_2 \left[A_4 + A_3 A_1^{-1} A_2\right] A_3 A_1^{-1}.$$

H.9 Norms

Norm of a Vector

The norm is a measure of size or length of a vector. It satisfies the following axioms, which apply to the familiar concept of length in the plane.

Norm axioms

1. $\|\boldsymbol{x}\| = 0$ if and only if $\boldsymbol{x} = \mathbf{0}$.
2. $\|\boldsymbol{x}\| > 0$ for $\boldsymbol{x} \neq \mathbf{0}$
3. $\|\alpha\boldsymbol{x}\| = |\alpha|\|\boldsymbol{x}\|$
4. $\|\boldsymbol{x}+\boldsymbol{y}\| \le \|\boldsymbol{x}\| + \|\boldsymbol{y}\|$ (triangle inequality).

The most popular vector norms are the $\boldsymbol{l}_p$ norms.

l_∞ **norm**: $\|\boldsymbol{x}\|_\infty = \max_i |x_i|$

l_2 norm: $\|\boldsymbol{x}\|_2^2 = \sum_{i=1}^{n} |x_i|^2$.

l_1 norm: $\|\boldsymbol{x}\|_1 = \sum_{i=1}^{n} |x_i|$.

Equivalent norms

Norms that satisfy the inequality

$$k_1\|\boldsymbol{x}\|_i \le \|\boldsymbol{x}\|_j \le k_2\|\boldsymbol{x}\|_i$$

with finite constants k_1 and k_2. All norms for $n \times 1$ real vectors are equivalent. All equivalent norms are infinite if and only if any one of them is infinite.

Norms are obtained using the following MATLAB commands:

>> norm(a, 2)	% 2-norm (square root of sum of squares)
>> norm(a, 1)	% 1-norm (sum of absolute values)
>> norm(a, inf)	% Infinity-norm (max element)

Matrix Norms

Matrix norms satisfy the norm axioms and can be induced from vector norms.

Induced matrix norms

Norms that are induced from vector norms using the definition

$$\|A\|_i = \max_x \frac{\|A\boldsymbol{x}\|}{\|\boldsymbol{x}\|} = \max_{\|x\|=1} \|A\boldsymbol{x}\|,$$

where ||•|| is any vector norm. The induced matrix norm have the submultiplicative properties

$$\|A\boldsymbol{x}\| \le \|A\|\|\boldsymbol{x}\|,$$

$$\|AB\| \le \|A\|\|B\|.$$

All induced norms are submultiplicative, but not all noninduced norms are submultiplicative.

The following are the induced matrix norms for different vector norms:

l_1 Norm	$\|A\|_1 = \max_j \sum_{i=1}^{m} \|a_{ij}\|$	(Maximum absolute column sum)
l_∞ Norm	$\|A\|_\infty = \max_i \sum_{j=1}^{m} \|a_{ij}\|$	(Maximum absolute row sum)
l_2 Norm	$\|A\|_2 = \max_i \lambda_i^{1/2}(A^T A)$	(Maximum singular value = square root of maximum eigenvalue of $A^T A$)

Other matrix norms include

$$\|A\| = \max_{i,j} |a_{ij}|$$

and the Frobenius norm

$$\|A\|_F = \sqrt{\sum_{i=1}^{m} \sum_{j=1}^{n} |a_{ij}|^2} = \sqrt{tr(A^T A)}.$$

The Frobenius norm is not an induced norm.

H.10 Quadratic Forms

A quadratic form is a function of the form

$$V(\boldsymbol{x}) = \boldsymbol{x}^T P \boldsymbol{x} = \sum_{i=1}^{n} \sum_{j=1}^{n} p_{ij} x_i x_j,$$

where $\boldsymbol{x}$ is an $n \times 1$ real vector and P is an $n \times n$ matrix. The matrix P can be assumed to be symmetric without loss of generality. To show this, assume that P is not symmetric and rewrite the quadratic form in terms of the symmetric component and the skew-symmetric component of P as follows:

$$V(\boldsymbol{x}) = \boldsymbol{x}^T \left(\frac{P + P^T}{2} \right) \boldsymbol{x} + \boldsymbol{x}^T \left(\frac{P - P^T}{2} \right) \boldsymbol{x} = \boldsymbol{x}^T \left(\frac{P + P^T}{2} \right) \boldsymbol{x} + \frac{1}{2} \left(\boldsymbol{x}^T P \boldsymbol{x} - (P\boldsymbol{x})^T \boldsymbol{x} \right).$$

Interchanging the row and column in the last term gives

$$\begin{aligned} V(\boldsymbol{x}) &= \boldsymbol{x}^T \left(\frac{P + P^T}{2} \right) \boldsymbol{x} + \frac{1}{2} \left(\boldsymbol{x}^T P \boldsymbol{x} - \boldsymbol{x}^T P \boldsymbol{x} \right) \\ &= \boldsymbol{x}^T \left(\frac{P + P^T}{2} \right) \boldsymbol{x}. \end{aligned}$$

Thus, if P is not symmetric, we can replace it with its symmetric component without changing the quadratic form.

The sign of a quadratic form for nonzero vectors $\boldsymbol{x}$ can be invariant depending on the matrix P. In particular, the eigenvalues of the matrix P determine the sign of the quadratic form. To see this, we examine the eigenvalues-eigenvector decomposition of the matrix P in the quadratic form. We assume, without loss of generality, that P is symmetric. Hence, its eigenvalues are real and positive and its modal matrix of eigenvectors is orthogonal. The matrix can be written as

$$P = V_p \Lambda V_p^T, \Lambda = diag\{\lambda_1, \lambda_2, \ldots, \lambda_n\}.$$

Using the eigenvalues decomposition of the matrix, we have

$$V(\boldsymbol{x}) = \boldsymbol{x}^T \boldsymbol{V}_p \Lambda \boldsymbol{V}_p^T \boldsymbol{x} = \boldsymbol{y}^T \Lambda \boldsymbol{y} = \sum_{i=1}^{n} \lambda_i y_i^2 > 0,$$

$$\boldsymbol{y} = \begin{bmatrix} y_1 \; y_2 \ldots y_n \end{bmatrix}^T.$$

Because the modal matrix V_p is invertible, there is a unique $\boldsymbol{y}$ vector associated with each $\boldsymbol{x}$ vector. The expression for the quadratic form in terms of the eigenvalues allows us to characterize it and the associated matrix as follows.

Positive definite:

A quadratic form is positive definite if

$$V(\boldsymbol{x}) = \boldsymbol{x}^T P \boldsymbol{x} > \mathbf{0}, \boldsymbol{x} \neq \mathbf{0}.$$

This is true if the eigenvalues of P are all positive, in which case we say that P is a positive definite matrix, and we denote this by $P > 0$.

Negative definite:

A quadratic form is negative definite if

$$V(\boldsymbol{x}) = \boldsymbol{x}^T P \boldsymbol{x} < \mathbf{0}, \boldsymbol{x} \neq \mathbf{0}.$$

This is true if the eigenvalues of P are all negative, in which case we say that P is a negative definite matrix, and we denote this by $P < 0$.

Positive semidefinite:

A quadratic form is positive semidefinite if $V(\boldsymbol{x}) = \boldsymbol{x}^T P \boldsymbol{x} \geq \mathbf{0}, \boldsymbol{x} \neq \mathbf{0}$.

This is true if the eigenvalues of P are all positive or zero, in which case we say that P is a positive semidefinite matrix, and we denote this by $P \geq 0$. Note that in this case, if an eigenvalue λ_i is zero, then the nonzero vector $\mathbf{y}$ with its ith entry equal to 1 and all other entries zero gives a zero value for V. Thus, there is a nonzero vector $\boldsymbol{x} = V_p \boldsymbol{y}$ for which V is zero.

Negative semidefinite:

A quadratic form is negative semidefinite if $\mathrm{V}(\boldsymbol{x}) = \boldsymbol{x}^{\mathrm{T}} \mathrm{P} \boldsymbol{x} > \mathbf{0}, \boldsymbol{x} \neq \mathbf{0}$.

This is true if the eigenvalues of P are all negative or zero, in which case we say that P is a negative semidefinite matrix, and we denote this by $P \leq 0$. In this case, if an eigenvalue λ_i is zero, then V is zero for the nonzero $\boldsymbol{x} = V_p \boldsymbol{y}$, where $\boldsymbol{y}$ is a vector with its ith entry equal to 1 and all other entries are zero.

Indefinite:

If the matrix P has some positive and some negative eigenvalues, then the sign of the corresponding quadratic form depends on the vector $\boldsymbol{x}$, and the matrix is called indefinite.

H.11 Singular Value Decomposition and Pseudoinverses

Any $n \times m$ real matrix A can be decomposed into the product

$$A = U \Sigma V^T,$$

where U is $n \times n$, V is $m \times m$, and Σ is $n \times m$. For a matrix of rank r the matrices in the decomposition satisfy

$$U^{-1} = U^*,$$

$$V^{-1} = V^*,$$

$$\boldsymbol{\Sigma} = \begin{bmatrix} \boldsymbol{\Sigma}_r & \boldsymbol{0}_{r \times (m-r)} \\ \boldsymbol{0}_{(n-r) \times r} & \boldsymbol{0}_{(n-r) \times (m-r)} \end{bmatrix},$$

$$\Sigma_{\mathrm{r}} = diag\{\sigma_1, \sigma_2, \ldots, \sigma_r\}.$$

The terms on the diagonal are real and positive and are known as the singular values of the matrix. The columns of U are called the left singular vectors and the columns of V are called the right singular vectors. We assume without loss of generality that the diagonal terms are arranged such that

$$\sigma_{max} = \sigma_1 \geq \sigma_2 \geq \cdots \geq \sigma_r = \sigma_{min}.$$

If the matrix A is full rank, then the matrix Σ has one of the following forms:

$$\Sigma = \begin{cases} \begin{bmatrix} \Sigma_n & 0_{n \times (m-n)} \end{bmatrix}, & n < m \\ \Sigma_m, & n = m \\ \begin{bmatrix} \Sigma_m \\ 0_{(n-m) \times m} \end{bmatrix}, & n > m \end{cases}.$$

The singular value decomposition is obtained using the MATLAB command

$$>> [\text{U, Sigma, V}] = \text{svd(A)\%A} = \text{U Sigma } V'$$

Eigenvalues and Singular Values

For any $n \times m$ real matrix A we have

$$AA^T = U\Sigma\Sigma^T U^T = Udiag\{\sigma_1^2, \sigma_2^2, \ldots, \sigma_r^2, \mathbf{0}_{1\times(n-r)}\}U^T,$$

$$A^T A = V\Sigma^T \Sigma V^T = Vdiag\{\sigma_1^2, \sigma_2^2, \ldots, \sigma_r^2, \mathbf{0}_{1\times(m-r)}\}V^T.$$

Thus, the nonzero eigenvalues of either of the above products are equal to the singular values of the matrix A.

The determinant of an $n \times n$ matrix A is

$$|det(A)| = \left|\prod_{i=1}^{n} \lambda_i(A)\right| = \prod_{i=1}^{n} \sigma_i(A).$$

The eigenvalues of a normal $n \times n$ matrix A and its singular values are related by

$$|\lambda_i(A)| = \sigma_i, i = 1, 2, \ldots, n.$$

Singular Value Inequalities

For any invertible square matrix, the singular values satisfy

$$\sigma_{max}(A) = 1/\sigma_{min}(A^{-1}),$$

$$\sigma_{min}(A) = 1/\sigma_{max}(A^{-1}).$$

For any two compatible matrices A and B

$$\sigma_{max}(A + B) \le \sigma_{max}(A) + \sigma_{max}(B),$$

$$\sigma_{max}(AB) \le \sigma_{max}(A)\sigma_{max}(B).$$

Pseudoinverse

Using the singular value decomposition, we can define a pseudoinverse for any $n \times m$ real matrix A as

$$A^{\#} = V\Sigma^{\#}U^{*}$$

with the pseudoinverse of the matrix of singular values given by

$$\Sigma^{\#} = \begin{bmatrix} \Sigma_r^{-1} & 0_{r\times(n-r)} \\ 0_{(m-r)\times r} & 0_{(m-r)\times(n-r)} \end{bmatrix},$$

$$\Sigma_r^{-1} = diag\{1/\sigma_1, 1/\sigma_2, \ldots, 1/\sigma_r\}.$$

Note that the inversion is similar to the inversion of the product of three matrices, with the inverse of the middle matrix being a pseudoinverse. For a full-rank matrix, we have

$$\Sigma^{\#} = \begin{cases} \begin{bmatrix} \boldsymbol{\Sigma}_{\boldsymbol{n}}^{-1} \\ 0_{(n-m)\times \boldsymbol{n}} \end{bmatrix}, & n < m \\ \boldsymbol{\Sigma}_{\boldsymbol{n}}^{-1}, & n = m \\ \begin{bmatrix} \boldsymbol{\Sigma}_{\boldsymbol{m}}^{-1} & 0_{\boldsymbol{m}\times(n-\boldsymbol{m})} \end{bmatrix}, & n > m \end{cases}.$$

Clearly, the pseudoinverse of a nonsingular square matrix is simply its inverse. The following MATLAB command computes the pseudoinverse:
>> pinv(A).

Pseudoinverse of a Full-Rank Matrix

For a full-rank matrix, the pseudoinverse of an m by n matrix reduces to the following:

$$A^{\#} = \begin{cases} \left(A^T A\right)^{-1} A^T, & n < m \\ A^{-1}, & n = m \\ A^T\left(AA^T\right)^{-1}, & n > m \end{cases}.$$

The first is a left inverse, the second is the usual matrix inverse, and the third is a right inverse of the matrix. The terms right inverse and left inverse are due to the products

$$\left(A^T A\right)^{-1} A^T A = I_n,$$

$$A^T A\left(A^T A\right)^{-1} = I_m,$$

$$A^{\#} = V\Sigma^{\#}U^T.$$

H.12 Matrix Differentiation/integration

The derivative (integral) of a matrix $A(t) = \left[a_{ij}(t)\right]$ is a matrix whose entries are the derivatives (integrals) of the entries of the matrix.

$$\frac{dA(t)}{dt} = \left[\frac{da_{ij}(t)}{dt}\right],$$

$$\int A(t)\mathrm{d}t = \left[\int a_{ij}(t)\mathrm{d}t\right].$$

Matrix differentiation and integration are subject to the following rules:

Derivative of a product	$\frac{dAB}{dt} = \frac{dA}{dt}B + A\frac{dB}{dt}$
Derivative of the inverse matrix	$\frac{dA^{-1}}{dt} = -A^{-1}\frac{dA}{dt}A^{-1}$

Gradient vector:

The derivative of a scalar function $f(x)$ with respect to the vector x is known as the gradient vector and is given by the $n \times 1$ vector.

$$\frac{\partial f(x)}{\partial \boldsymbol{x}} = \left[\frac{\partial f(x)}{\partial x_i}\right].$$

Some authors define the gradient as a row vector.

Jacobian matrix:

The derivative of an $\boldsymbol{n} \times \mathbf{1}$ vector function $\boldsymbol{f}(\boldsymbol{x})$ with respect to the vector $\boldsymbol{x}$ is known as the Jacobian matrix and is given by the $\boldsymbol{n} \times \boldsymbol{n}$ matrix.

$$\frac{\partial \boldsymbol{f}(x)}{\partial \boldsymbol{x}} = \left[\frac{\partial f_i(x)}{\partial x_j}\right].$$

Gradient of inner product:

$$\frac{\partial \boldsymbol{a}^T\boldsymbol{x}}{\partial \boldsymbol{x}} = \frac{\partial \boldsymbol{x}^T\boldsymbol{a}}{\partial \boldsymbol{x}} = \left[\frac{\partial a_i x_i}{\partial x_i}\right] = [a_i] = \boldsymbol{a}.$$

Gradient matrix of a quadratic form:

$$\frac{\partial \boldsymbol{x}^T P\boldsymbol{x}}{\partial \boldsymbol{x}} = \boldsymbol{x}^T\frac{\partial P\boldsymbol{x}}{\partial \boldsymbol{x}} + \frac{\partial\left(P^T\boldsymbol{x}\right)^T}{\partial \boldsymbol{x}}\boldsymbol{x} = \left(P + P^T\right)\boldsymbol{x}.$$

Because P can be assumed to be symmetric with no loss of generality, we write

$$\frac{\partial \boldsymbol{x}^T P \boldsymbol{x}}{\partial \boldsymbol{x}} = 2P\boldsymbol{x}.$$

Hessian matrix of a quadratic form:
The Hessian or second derivative matrix is given by

$$\frac{\partial^2 \boldsymbol{x}^T P \boldsymbol{x}}{\partial \boldsymbol{x}^2} = \frac{\partial 2P\boldsymbol{x}}{\partial \boldsymbol{x}} = 2\left[\frac{\partial \boldsymbol{p}_i^T \boldsymbol{x}}{\partial x_j}\right] = 2\left[p_{ij}\right] = 2P,$$

where the ith entry of the vector $P\boldsymbol{x}$ is $p_i^T x$

$$P = \left[p_{ij}\right] = \begin{bmatrix} \boldsymbol{p}_1^T \\ \boldsymbol{p}_2^T \\ \vdots \\ \boldsymbol{p}_n^T \end{bmatrix}.$$

H.13 Kronecker Product

The Kronecker product of two matrices A of order $m \times n$ and B of order $p \times q$ is denoted by $\otimes$ and is defined as

$$A \otimes B = \begin{bmatrix} a_{11}B & a_{12}B & \cdots & a_{1n}B \\ a_{21}B & a_{22}B & \cdots & a_{2n}B \\ \vdots & \vdots & \ddots & \vdots \\ a_{m1}B & a_{m2}B & \cdots & a_{mn}B \end{bmatrix}.$$

The resulting matrix is of order $\boldsymbol{mp} \times \boldsymbol{nq}$. The MATLAB command for the Kronecker product is

```
>> kron(A,B).
```

Bibliography

1. Barnett, S. (1984). *Matrices in control theory*. R.E. Krieger.
2. Brogan, W. L. (1985). *Modern control theory*. Prentice Hall.
3. Fadeeva, V. N. (1959). *Computational methods of linear algebra*. Dover.
4. Gantmacher, F. R. (1959). *The theory of matrices*. Chelsea.
5. Noble, B., & Daniel, J. W. (1988). *Linear algebra*. Prentice Hall.

Index

M. S. Fadali, *Introduction to Random Signals, Estimation Theory, and Kalman Filtering*, https://doi.org/10.1007/978-981-99-8063-5

L

M

N

O

P

R

S

MIX
Papier aus verantwortungsvollen Quellen
Paper from responsible sources
FSC® C105338

If you have any concerns about our products,
you can contact us on
ProductSafety@springernature.com

In case Publisher is established outside the EU,
the EU authorized representative is:
Springer Nature Customer Service Center GmbH
Europaplatz 3, 69115 Heidelberg, Germany

Printed by Libri Plureos GmbH
in Hamburg, Germany